QH 441.5 .O74 2012

Organelle genetics

Organelle Genetics

Charles E. Bullerwell

Editor

Organelle Genetics

Evolution of Organelle Genomes
and Gene Expression

 Springer

Editor
Charles E. Bullerwell
Mount Allison University
Departments of Biology and
Biochemistry
York Street 63B
E4L 1G7 Sackville New Brunswick
Canada
charles_bullerwell@yahoo.com

ISBN 978-3-642-22379-2   e-ISBN 978-3-642-22380-8
DOI 10.1007/ 978-3-642-22380-8
Springer Heidelberg Dordrecht London New York

Library of Congress Control Number: 2011937549

Printed on acid-free paper

Springer is part of Springer Science+Business Media (www.springer.com)

Preface

Mitochondria and chloroplasts are eukaryotic organelles that maintain their own genomes. The products of these genomes work in concert with those of the nuclear genome to ensure proper organelle metabolism and biogenesis. The inspiration for this book was to explore the forces that have shaped organelle genomes and the expression of their genes since their divergence from bacterial ancestors in the distant evolutionary past.

In the opening part, the evolutionary origins of these organelles and their diversification throughout Eukarya are explored. In Part II, we take a closer look at organelle genomes and gene contents and explore a critical process in organelle evolution: loss of organelle genes and the loss of functional plastids and mitochondria. Part III explores what drives this gene loss and genome reduction, specifically the role of mutational processes and transfer of organelle-encoded information into the nucleus.

Once genetic information has switched organelles, how does it get back to the compartment where it performs its function? Part IV looks both at the mechanisms for getting nucleus-encoded organelle proteins back to where they do their jobs, as well as getting a feel for what proteins ultimately are located in an organelle – looking specifically at plastids.

In the final three parts we look at transcription and its regulation (Part V), RNA processing (Part VI), and translation and the genetic code (Part VII). Organelles are microcosms of genome evolution, and some bizarre and unexpected means of gene expression were first identified in organelle genomes. Two such features are RNA editing and modifications to the universal genetic code. Both of these topics are highlighted here.

Two overarching themes that we looked to highlight in this book are current techniques used to study organelle genetics and an evolutionary perspective on how and why organelle genomes evolve as they do. As organelle dysfunction plays essential roles in a variety of cellular processes and is an important factor in many diseases, we hope that this book, rather than simply serving as a review of a small portion of the vast topic of organelle genetics, might also inspire researchers to consider evolutionary approaches to understanding cellular function, to appreciate the complexity of organelle gene expression, and hopefully to further explore the ideas presented here.

I would like to thank the many talented authors who have contributed their time and expertise to this volume. I would also like to thank external reviewers who gave comments on manuscript drafts and the editors at Springer Publishing for suggesting the idea for this volume and for asking me to participate.

Welcome to the world of organelle genetics!

Sackville, Canada   Charles E. Bullerwell

Contents

Contributors

Juan D. Alfonzo Department of Microbiology and OSU Center for RNA Biology, The Ohio State University, Columbus, OH 43210, USA, Alfonzo.1@osu.edu

John F. Allen Centre for Eukaryotic Evolutionary Microbiology, Biosciences, College of Life and Environmental Sciences, University of Exeter, Devon EX4 4QJ, UK, j.f.allen@qmul.ac.uk

John M. Archibald Program in Integrated Microbial Biodiversity, Canadian Institute for Advanced Research, Department of Biochemistry and Molecular Biology, Sir Charles Tupper Medical Building, Dalhousie University, Halifax, NS, Canada B3H 4R2, jmarchib@dal.ca

Thomas Börner Institut für Biologie/Genetik, Humboldt-Universität zu Berlin, Chausseestr. 117, 10115 Berlin, Germany, thomas.boerner@rz.hu-berlin.de

Christian Barth Department of Microbiology, La Trobe University, Melbourne, Australia, c.barth@latrobe.edu.au

Natacha Beck Centre Robert Cedergren, Département de Biochimie, Université de Montréal, Boulevard Edouard Montpetit, Montréal, QC, Canada

Gertraud Burger Department of Biochemistry, Robert-Cedergren Centre for Bioinformatics and Genomics, Université de Montréal, 2900 Edouard-Montpetit, Montreal, QC, H3T 1J4, Canada, gertraud.burger@umontreal.ca

Palmiro Cantatore Dipartimento di Biochimica e Biologia Molecolare "Ernesto Quagliariello", Università degli Studi di Bari, Via Orabona, 4, 70125 Bari, Italia; Istituto di Biomembrane e Bioenergetica, CNR, Via Orabona, 4, 70125 Bari, Italia, p.cantatore@biologia.uniba.it

Paul R. Fisher Department of Microbiology, La Trobe University, Melbourne, Australia

Toni Gabaldón Bioinformatics and Genomics Programme, Centre for Genomic Regulation (CRG), and Universitat Pompeu Fabra, Doctor Aiguader, 88, 08003 Barcelona, Spain, tgabaldon@crg.es

Mark van der Giezen School of Biological and Chemical Sciences, Queen Mary, University of London, Mile End Road, London E1 4NS, UK, m.vandergiezen@exeter.ac.uk

Georg Hausner Department of Microbiology, University of Manitoba, Winnipeg, Canada R3T 2N2, hausnerg@cc.umanitoba.ca

Chris J. Jackson School of Botany, University of Melbourne, Melbourne, VIC 3010, Australia, cjackson1245@gmail.com

Kirsten Kehrein Research Group Membrane Biogenesis, University of Kaiserslautern, 67663 Kaiserslautern, Germany; Center for Biomembrane Research, Department for Biochemistry and Biophysics, University of Stockholm, Stockholm, Sweden

Luke A. Kennedy Department of Microbiology, La Trobe University, Melbourne, Australia

Kirsten Krause Department for Arctic and Marine Biology, University of Tromsø, Dramsvegen 201, N-9037 Tromsø, Norway, kirsten.krause@uit.no

Marcel Kuntz Laboratoire de Physiologie Cellulaire & Végétale, CNRS/Université Joseph Fourier/INRA/CEA Grenoble, 17 rue des Martyrs, F-38000 Grenoble, France

B. Franz Lang Centre Robert Cedergren, Département de Biochimie, Université de Montréal, Boulevard Edouard Montpetit, Montréal, QC, Canada, Franz.Lang@Umontreal.ca

Dennis Lavrov Department of Ecology, Evolution and Organismal Biology, Iowa State University, Ames, IA, USA

Karsten Liere Institut für Biologie/Genetik, Humboldt-Universität zu Berlin, Chausseestr. 117, 10115 Berlin, Germany, karsten.liere@rz.hu-berlin.de

Andrew H. Lloyd School of Molecular and Biomedical Science, The University of Adelaide, South-Australia 5005, Australia, andrew.lloyd@adelaide.edu.au

Oliver Mirus Department of Biosciences, Molecular Cell Biology of Plants, Institute for Molecular Biosciences, Goethe University, Max-von-Laue Str. 9, 60438 Frankfurt, Germany

Martin Ott Research Group Membrane Biogenesis, University of Kaiserslautern, 67663 Kaiserslautern, Germany; Center for Biomembrane Research, Department for Biochemistry and Biophysics, University of Stockholm, Stockholm, Sweden, martin.ott@biologie.uni-kl.de, martin.ott@dbb.su.se

Wilson de Paula Centre for Eukaryotic Evolutionary Microbiology, Biosciences, College of Life and Environmental Sciences, University of Exeter, Devon EX4 4QJ, UK, w.depaula@qmul.ac.uk

Paola Loguercio Polosa Dipartimento di Biochimica e Biologia Molecolare "Ernesto Quagliariello", Università degli Studi di Bari, Via Orabona, 4, 70125 Bari, Italia

Marina Roberti Dipartimento di Biochimica e Biologia Molecolare "Ernesto Quagliariello", Università degli Studi di Bari, Via Orabona, 4, 70125 Bari, Italia

Norbert Rolland Laboratoire de Physiologie Cellulaire & Végétale, CNRS/ Université Joseph Fourier/INRA/CEA Grenoble, 17 rue des Martyrs, F-38000 Grenoble, France, norbert.rolland@cea.fr

Mathieu Rousseau-Gueutin School of Molecular and Biomedical Science, The University of Adelaide, South-Australia 5005, Australia, mathieu. rousseau@adelaide.edu.au

Mary Anne T. Rubio Department of Microbiology and OSU Center for RNA Biology, The Ohio State University, Columbus, OH 43210, USA

Enrico Schleiff Department of Biosciences, Cluster of Excellence Macromolecular Complexes, Molecular Cell Biology of Plants, Institute for Molecular Biosciences and Centre of Membrane Proteomics, Goethe University, Max-von-Laue Str. 9, 60438 Frankfurt, Germany, schleiff@bio.uni-frankfurt.de

Anna E. Sheppard School of Molecular and Biomedical Science, The University of Adelaide, South-Australia 5005, Australia; Department of Evolutionary Ecology and Genetics, Zoological Institute Christian-Albrechts University of Kiel, Am Botanischen Garten 1-9, Kiel 24118, Germany, asheppard@zoologie.uni-kiel.de

Toshiharu Shikanai Faculty of Science, Department of Botany, Kyoto University, Kyoto 606-8502 Japan, shikanai@pmg.bot.kyoto-u.ac.jp

Daniel B. Sloan Department of Biology, University of Virginia, Charlottesville, VA 22904, USA, dbs4a@virginia.edu

Sergey V. Steinberg Centre Robert Cedergren, Département de Biochimie, Université de Montréal, Boulevard Edouard Montpetit, Montréal, QC, Canada

Douglas R. Taylor Department of Biology, University of Virginia, Charlottesville, VA 22904, USA, drt3b@virginia.edu

Jeremy N. Timmis School of Molecular and Biomedical Science, The University of Adelaide, South-Australia 5005, Australia, jeremy.timmis@adelaide.edu.au

Ross F. Waller School of Botany, University of Melbourne, Melbourne, VIC 3010, Australia, r.waller@unimelb.edu.au

Andreas Weihe Institut für Biologie/Genetik, Humboldt-Universität zu Berlin, Chausseestr. 117, 10115 Berlin, Germany, andreas.weihe@rz.hu-berlin.de

Part I
Origins of Organelle Genomes

Chapter 1
Mitochondrial Origins

Toni Gabaldón

1.1 Introduction

Mitochondria are cellular organelles surrounded by a double membrane. Mitochondria, or evolutionary-related organelles such as hydrogenomes or mitosomes (see below), have been identified in every eukaryotic organism that has been carefully examined to date. This indicates that the origin of these organelles preceded the diversification of all known groups of eukaryotes, estimated to date back 1.5–2 billion years (Brocks et al. 1999). Regarding this origin, there is now a widespread consensus in that mitochondria originated from an alpha-proteobacterial ancestor (the so-called proto-mitochondrion) by means of an endosymbiotic process. Extant representatives of alpha-proteobacteria constitute a large and highly diversified group, in which a large variety of metabolic capacities and lifestyles can be observed. Most probably, none of the currently existing alpha-proteobacterial species can be regarded as an accurate model for the original proto-mitochondrion, since they thrive in environments that are likely very different from the one that governed the establishment of the initial endosymbiosis. Even modern alpha-proteobacteria that have intra-cellular lifestyles such as the insect endosymbiont *Wolbachia*, or the intracellular pathogens of the genus *Ricketsia*, represent parallel adaptations to intra-cellular life and should be considered different scenarios since, contrary to the proto-mitochondrion, they inhabit full-fledged eukaryotic host cells that already possess mitochondria. Moreover, phylogenetic analyses of mitochondrial genes cannot identify a particular group within the alpha-proteobacteria as the ancestor of mitochondria (Esser et al. 2004). Comparative genomics and phylogenomics have served to circumvent the problem of a lack of an extant model by identifying eukaryotic genes with a clear alpha-proteobacterial ancestry (Gabaldón and Huynen

T. Gabaldón (✉)
Bioinformatics and Genomics Programme, Centre for Genomic Regulation (CRG),
and Universitat Pompeu Fabra, Doctor Aiguader, 88, 08003 Barcelona, Spain
e-mail: tgabaldon@crg.es

C.E. Bullerwell (ed.), *Organelle Genetics*,
DOI 10.1007/978-3-642-22380-8_1, © Springer-Verlag Berlin Heidelberg 2012

2003, 2007b). This has enabled a partial reconstruction of the proto-mitochondrial metabolism, which has shed light on the possible metabolic scenarios that favored the initial symbiosis. In contrast, there is much debate on what was the probable nature of the host cell. The classical view considers a rather developed eukaryotic host, presenting a cell nucleus and a cytoskeleton, and with the ability to perform phagocytosis, which would have enabled the engulfing of the alpha-proteobacterial endosymbiont (de Duve 2007). Alternatively, other authors envision a prokaryotic host, with the mitochondrial endosymbiosis event itself giving rise to the formation of the eukaryotic cell (Martin and Müller 1998), by means of selective pressures that favored the creation of the cell nucleus and other eukaryotic features (Martin and Koonin 2006). Current research is directed into carefully examining upcoming biochemical, cellular, and genomic data on extant organisms, in order to assess the plausibility of the various evolutionary hypotheses that have been put forward. Recent technical developments, especially in genomic sequencing, have enabled access to data from an unprecedented range of diverse organisms, which, in turn, has stimulated great advances in our understanding of the origin and evolution of mitochondria. Here, I will provide an overview on the current knowledge on the evolutionary origins of mitochondria. I will first introduce our current knowledge on the diversity of extant mitochondria and related organelles, to subsequently focus on how this information has served to reconstruct the early phases of mitochondrial evolution.

1.2   Mitochondrial Proteome and Genome Diversity in Extant Eukaryotes

1.2.1   Mitochondrial Proteomes are Highly Diverse

Recent developments in experimental techniques, in particular the use of subcellular proteomics (Au et al. 2007) and also other high-throughput localization analyses, provide the means to catalog the repertoire of proteins that function within the mitochondrion, the so-called mitochondrial proteome. Currently, there is a wealth of data on mitochondrial proteomes for several eukaryotic species, including a broad diversity of eukaryotic species (Mootha et al. 2003; Heazlewood et al. 2004; Prokisch et al. 2006; Pagliarini et al. 2008; Atteia et al. 2009; Li et al. 2009). However, available data on mitochondrial proteomes from microbial eukaryotes, which constitute the bulk of eukaryotic diversity (Keeling et al. 2005), remain poor. Comparative analyses of proteomic species have revealed a large diversity in protein content across species and also across tissues. For instance, mitochondria of human and yeasts share less than 50% of their proteomes (Gabaldón and Huynen 2007b), and a typical pair of distinct mammalian tissues shares ~75% of their protein repertoires (Pagliarini et al. 2008). Reflecting such proteomic diversity, the functional properties of mitochondria can also vary widely.

Biological processes found to be performed, at least in part, within mitochondria include, among many others, energy metabolism, lipid and amino acid metabolism, Fe–S cluster biosynthesis, and secondary metabolism. In addition, mitochondria can often function as centers for cellular homeostasis and apoptotic pathways (Dimmer and Rapaport 2008). Finally, except for the cases in which mitochondria have lost their genome (see below), a significant fraction of the proteome is devoted to processes associated to replication, transcription, and translation of mitochondrially encoded genes. Below, I will focus on the metabolic and informational processes that are important for the discussion of mitochondrial origins.

1.2.2 Energy Metabolism

Early research on mitochondrial metabolism has been much focused on its role within the energetic metabolism of the cell. In mammals, and many other organisms, mitochondria are the main place for the synthesis of adenosine triphosphate (ATP), the so-called energetic currency of the cell. Indeed, in many human tissues, the contribution to ATP production of alternative sources is almost negligible, with more than 95% of the ATP used in the cell originating from mitochondrial pathways. Considering this, mitochondria have been dubbed the power houses of the cell (Vo and Palsson 2007), and this function is the one that is considered most widely in textbooks. ATP is mainly formed in the mitochondrion through the process of oxidative phosphorylation (OXPHOS), enabled by a complex machinery involving dozens of proteins associated to the inner mitochondrial membrane. A first part of the process consists of a transfer of electrons from reduced donors (NADH or succinate) to an electron acceptor (O_2), through an intricate system of redox centers that are carried by four distinct membrane-associated multi-protein complexes (named complexes I–IV). Three of these complexes (I, III, and IV) are able to couple the transfer of electrons to the translocation of protons across the membrane, thereby generating a proton gradient. The dissipation of such gradient is what provides the necessary energy required by a fifth complex, the ATP-synthase, to phosphorylate ADP into ATP. The OXPHOS pathway is present, in a very similar organization, in mitochondria from very distant eukaryotes, including plants, fungi, and metazoans. Moreover, the origin of the core protein components of the OXPHOS complex can be reliably traced back to the alphaproteobacterial ancestor of mitochondria, indicating its presence in the proto-mitochondrion (Kurland and Andersson 2000; Gabaldón and Huynen 2003). Despite a widespread presence of the OXPHOS pathway in mitochondria from diverse eukaryotes, the system is far from being ubiquitous or homogeneous. Indeed, there is an important diversity in terms of presence/absence of specific components of this pathway. For instance, complex I, or NADH–ubiquinone oxidoreductase, has been lost in several lineages, including one apicomplexan and three fungal lineages (Gabaldón et al. 2005; Marcet-Houben et al. 2009). This has been accompanied by the complete loss of the whole electron transport chain in several eukaryotic lineages. Such an extreme adaptation took place independently in at least six lineages, namely Apicomplexa (e.g., *Cryptosporidium*), Microsporidia (e.g., *Enzephalitozoon*), Chytrids (e.g., *Neocallismastix*), Amoebozoa (e.g., *Entamoeba*), Heterolobosae (e.g., *Psalteriomonas*), and Metamonada (e.g., *Giardia*). Interestingly, the complete loss of the mitochondrial OXPHOS pathway is always associated with the disappearance of the mitochondrial genome (Hjort et al. 2010), suggesting that the presence of some mitochondrial OXPHOS components is the sole force driving the retention of the mitochondrial DNA, and its associated machinery required for replication, repair, transcription, and translation of mitochondrially encoded genes (see the following section and chapters of Parts II and III). Even in the groups where the core OXPHOS pathway is present, the specific subunit composition of its electron transport complexes can vary greatly, as a result of an evolutionary expansion of multi-protein complexity (Gabaldón and Huynen 2004; Gabaldón et al. 2005). The functional properties of this OXPHOS pathway can also vary, with the use of alternative NADH dehydrogenases (Marcet-Houben et al. 2009) or alternative electron acceptors (Tielens et al. 2002). However, losses of the entire OXPHOS pathway or of some of its components can always be traced back to mitochondriate ancestors that lost these systems through secondary adaptations to anaerobic environments. This, together with the fact that key components for the electron transport chain are encoded in the mitochondrial genome, indicates that the last common ancestor of all mitochondria possessed this system.

1.2.3   Replication and Translation Machineries

Most mitochondria retain a reduced bacterial-like genome, in which several proteins for the respiratory chain and some RNAs are encoded. For the production of the encoded proteins, mitochondria rely on their own translation machinery, comprising a ribosome, tRNAs, and elongation factors. It is clear that, as other bacteria, the proto-mitochondrion possessed a ribosome and the other components of the translation machinery. The alpha-proteobacterial ancestry of core mitoribosomal subunits is supported by phylogenetic analyses (Kurland and Andersson 2000; Gabaldón and Huynen 2003) and the fact that many are still encoded in mitochondrial genomes. In addition to the core, proto-mitochondrial-derived subunits of the mitoribosomes, these have recruited additional subunits from other evolutionary sources (Smits et al. 2007). This evolutionary expansion, which almost doubled the number of components with respect to the typical alpha-proteobacterial ribosome, has resulted in extensive variations across eukaryotic lineages. Remarkably, although many mitochondria encode bacterial-like tRNAs, the set of proteins required for tRNA modification has been found to be enriched with proteins of non-alphaproteobacterial proteins, at least in mammals (Szklarczyk and Huynen 2010). This indicates that the original alpha-proteobacterial set for tRNA modifications has been gradually replaced. Similarly, the proteins responsible for transcription and replication of the mitochondrial genome seem to have been replaced by proteins of viral origin (Shutt and Gray 2006).

1.2.4 Hydrogenosomes

A particular adaptation of mitochondrial metabolism is that represented by the hydrogenosomes, which possess the ability of producing hydrogen (van der Giezen 2009). These organelles were first thought to constitute distinct organelles of independent origin. Although structural similarities with mitochondria and phylogenetic relationships of mitochondrial and hydrogenosomal proteins had been noted (Bui et al. 1996), the lack of an organellar genome prevented establishing their origins with confidence. The unexpected finding of a chromosome-bearing hydrogenosome from the ciliate *Nyctoterus ovalis*, and the analysis of its sequence, provided definitive molecular evidence for their evolutionary relatedness with mitochondria (Boxma et al. 2005). Interestingly, the presence of the genome in these hydrogenosomes is related to the presence of some membrane-embedded subunits of complex I. Nowadays, it is widely accepted that hydrogenosomes represent diverse mitochondrial adaptations to anaerobic lifestyles that have recursively appeared throughout eukaryotic evolution (Hackstein et al. 2006; van der Giezen 2009). This adaptation has been observed within at least six eukaryotic lineages (Fig. 1.1): stramenopiles (e.g., *Blastocystis*), ciliates (e.g., *N. ovalis*), parabasalids (e.g., *Trichomonas vaginalis*), amoebozoans (e.g., *Mastigamoeba balamuthi*), chytrids (e.g., *Neocalixmatis frontalis*), and, finally, a recently reported metazoan (*Spinoloricus* sp.) that lives in anoxic conditions and that harbors a hydrogenosome-like organelle (Danovaro et al. 2010).

1.2.5 Mitosomes

Not all forms of anaerobic mitochondria are able to produce hydrogen. These other types of extreme mitochondria were initially named differently in the various lineages where they have been described, for instance, the term *crypton* was used to describe apparently cryptic mitochondria in *Entamoeba histolytica* (Mai et al. 1999). Currently, they are all referred to as mitosomes (Tovar et al. 1999), although it is clear that these forms of highly diversified mitochondria are polyphyletic. Indeed, mitosomes have evolved from typical aerobic mitochondria in at least four different lineages (Fig. 1.1), namely microsporidians (e.g., *Encephalitozoon cuniculi*), amoebozoans (e.g., *Entamoeba histolytica*), diplomonads (e.g., *Giardia intestinalis*), and apicomplexans (e.g., *Cryptosporidium parvum*). The functions of these organelles remain to be fully established, but a common denominator is their participation in the synthesis of iron–sulfur (Fe–S) clusters. Iron–sulfur clusters are iron and sulfide ensembles that can be found as prosthetic groups in a variety of

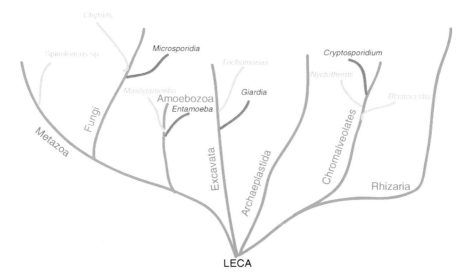

Fig. 1.1 Schematic view of the phylogenetic position of lineages bearing mitosomes (*orange*) or hydrogenosomes (*yellow*), in the context of the main eukaryotic groups (*green*). The eukaryotic tree of life is drawn according to Keeling et al. (2005), showing dubious groupings as multifurcations. Branch lengths and branching positions are only approximate. *LECA* last common eukaryotic ancestor

organellar and cytosolic metalloproteins participating in oxidation–reduction reactions, such as the ferredoxins, NADH dehydrogenase, or hydrogenases. Fe–S cluster assembly is so far the only function attributed to *Giardia* mitosomes (Tovar et al. 2003), and the only functional common denominator of all forms of mitochondria-related organelles. Thus, this function, rather than oxidative phosphorylation, seems to play the key role performed by mitochondria in the broad sense, and seems to be the sole force to prevent the complete loss of this organelle in extreme adaptations.

1.3 Tracing the Origin of Mitochondria

1.3.1 Mitochondria derived from an Alpha-Proteobacterial Ancestor

An endosymbiotic origin of mitochondria has been proposed almost since its discovery. Since the nineteenth century, the German physiologist Robert Altman referred to mitochondria as independent units able to live outside the cell and its presence within the cell being the result of a biological interaction between two different organisms (Altmann 1890). This theory was later revived by Lyn Margulis in the framework of her serial endosymbiosis theory (Margulis 1970) that proposed an endosymbiotic origin for mitochondria and many other eukaryotic structures. Nowadays, the endosymbiotic origin of mitochondria can be considered as being firmly established. There are many lines of evidence that support the origin of mitochondria from bacterial ancestors. Besides similarities with bacteria in terms of shape and metabolism, the presence of an organellar genome within mitochondria constitutes the definitive proof that mitochondria have a bacterial ancestry. With rare exceptions, such as the phage-derived RNA polymerase, the proteins encoded in the mitochondrial genome have their closest relatives among bacteria. From the phylogenetic analysis of those genes (Gray et al. 1999), two main conclusions can be drawn: first, all mitochondria are monophyletic, i.e., they derive from a single endosymbiotic event and, second, the ancestor of the mitochondrion, the so-called proto-mitochondrion, was an alpha-proteobacterium. Notably, this phylogenetic analysis confirmed earlier suggestions based solely on biochemical evidence (John and Whatley 1975). Besides these clearly established facts, there are many open questions regarding the origin of mitochondria, mainly the phylogenetic position of the proto-mitochondrial ancestor within the alpha-proteobacteria (Esser et al. 2004) and the metabolic scenario that fixed the initial symbiosis.

1.3.2 The Phylogenetic Position of Mitochondria Within the Alpha-Proteobacterial Tree is Elusive

Phylogenetic efforts to pinpoint a specific ancestor of mitochondria within the alpha-proteobacteria have been, so far, inconclusive. Initial evolutionary analyses of mitochondrially encoded genes suggested an origin from within the Rickettsiales (Gray et al. 1999; Kurland and Andersson 2000; Emelyanov 2001). However, similar analyses including more genes and genomes failed to provide conclusive results, with different genes providing alternative topologies and phylogenetic affiliations for mitochondria (Esser et al. 2004). There are numerous possible explanations to this apparent incongruence. First, the endosymbiotic event dates back more than 1,500 million years ago and pre-dates the diversification of current eukaryotic groups. Thus, the phylogenetic signal is likely to have been largely erased through time. This is especially true for mitochondrially encoded proteins, since they evolve at faster rates due to a lack of repair mechanisms. Similarly, the branching order of the main eukaryotic groups – representing more recent events – is largely unresolved (Keeling et al. 2005; Hampl et al. 2009). Second, owing to rampant horizontal gene transfer in bacteria, it has been suggested that the mitochondrial ancestor's genome was itself a chimera from different alpha-proteobacterial groups (Esser et al. 2007). Finally, I would like to add a third factor that has not been generally considered: the difficulty to assess affiliation with a specific group of alpha-proteobacterial may also reflect that mitochondria share a common ancestor with several alpha-proteobacterial groups that had not been diversified yet at the time of the mitochondrial endosymbiosis, from which many may have gone extinct. This would place the branching point of the mitochondrial lineage deep within the phylogeny of alpha-proteobacteria rather than at swallow branches and would pose more difficulties to phylogenetic reconstruction. Probably, a combination of all three factors is what sets the problem of specifying the concrete phylogenetic position of alpha-proteobacteria, one of the most challenging problems of early eukaryotic evolution today.

1.4   The Proto-mitochondrion

Considering the explored diversity of mitochondria and evolutionarily related organelles, the picture that emerges is that of an extremely plastic organelle. As mentioned above, the only common functional ground for all diverse mitochondria is their involvement in the biosynthesis of Fe–S clusters. Besides, the presence of mitochondrially encoded genes for processes, such as oxidative phosphorylation, replication, transcription, and translation, indicates that these processes were also carried out by the proto-mitochondrial ancestor. As we will see below, this narrow set of activities informed the early models for explaining the initial endosymbiotic relationships between the alpha-proteobacterial symbiont and its host. To broaden the view about what metabolic capacities were present in the mitochondrial ancestor, Martijn Huynen and I followed a phylogenomic approach to identify proteins in modern eukaryotes whose origin could be traced back to the alpha-proteobacteria (Gabaldón and Huynen 2003). By reconstructing and examining thousands of phylogenies of families with alpha-proteobacterial and eukaryotic members, a minimal ancestral protomitochondrial proteome was reconstructed, based on the selection of trees whose topology was compatible with a vertical descent of the eukaryotic protein from an alpha-proteobacterial ancestor. This analysis yielded a total of 630 orthologous groups, a set that was later extended to 840 using more sophisticated phylogenetic techniques and a broader set of genomes (Gabaldón and Huynen 2007b). In any case, these numbers should be regarded as minimal estimates of the size of the proteome of the mitochondrial ancestor, since many genes would have been lost from the eukaryotic genomes considered, whereas others cannot be detected due to poor phylogenetic signal. In addition, processes such as horizontal gene transfer at the level of the alpha-proteobacterial ancestor may have confounded the phylogenetic affiliation of many proto-mitochondrial-derived genes (Esser et al. 2007). Indeed, our analyses could only recover an alpha-proteobacterial ancestry for roughly 65% of the genes encoded in the largest mitochondrial genome, that of *Reclinomonas americana* (Gabaldón and Huynen 2007b).

Nevertheless, although this reconstructed proteome is certainly incomplete, it can provide us with an overview of the metabolic properties of the mitochondrial ancestor. A putative metabolic map of the mitochondrial ancestor can be obtained by mapping the functions of the selected proteins onto known metabolic maps

(Fig. 1.2). The resulting picture is that of a (perhaps facultatively) aerobic alpha-proteobacterium able to catabolyze lipids, glycerol, and other substrates provided by the host. The presence of almost complete pathways for oxidative phosphorylation and beta-oxidation is indicative of an aerobic metabolism. However, it cannot be completely discarded that the proto-mitochondrion could have been a facultative anaerobe, as proposed by some hypotheses (Martin and Müller 1998), since no genome from a hydrogenosome-bearing organism was included in our analysis. Other pathways that can be reconstructed almost completely include lipid synthesis, biotin, vitamin B6, heme synthesis, and Fe–S clusters. In contrast, some mitochondrial pathways, such as the citric acid cycle, are incomplete, and others such as the urea cycle or glycolysis are totally absent. In addition, we did find partial presence of pathways from amino acid and nucleotide metabolism. The presence of pathways not directly involved in energy metabolism suggests a multi-faceted benefit for the eukaryotic host, rather than a symbiotic relationship based on the exchange of few molecules. This view is supported by the presence of a high number of metabolite transporters in the reconstruction, which suggests a host dependency of the proto-mitochondrion. The Fe^{2+} importer is particularly interesting because it could have provided the iron for the Fe–S cluster assembly pathway. There are several other cation transporters (Mg^{2+}/Co^{2+} and K^+) that could have been used either to maintain the ion homeostasis or to obtain the cofactors needed for the enzyme activities. Thus, altogether, this reconstructed metabolism provides a picture of a (facultatively) aerobic endosymbiont catabolizing lipids, glycerol, and amino acids provided by the eukaryotic host. From the host side, although energy conversion has been a dominant factor throughout mitochondrial evolution, this appears not to have been the sole benefit from the early symbiotic relationship. The comparison of this reconstructed proteome with that of modern mitochondria was made possible by the development of proteomics analysis of highly pure mitochondria from several organisms. In particular, the comparison of the reconstructed proto-mitochondrial proteome described above with comprehensive proteomics sets from yeast and human mitochondria was used to trace the transformations occurred during the transition from early symbiont to organelle and the subsequent specialization in two different lineages (Gabaldón and Huynen 2007b).

1.5 Scenarios for the Origin of Mitochondria

Regarding the metabolic rationale of the early symbiosis, various hypothetical scenarios have been put forward that differ in the metabolic properties assumed for the host and the endosymbiont (Fig. 1.3). First, the serial endosymbiotic theory, as suggested by Margulis (1970, 1981), proposed a exchange of ATP and glycolysis end products between the host and the endosymbiont that was mutually beneficial. This view was later abandoned once it was realized from phylogenetic analyses that the mitochondrial ADP/ATP transporter had a more recent origin. Two other

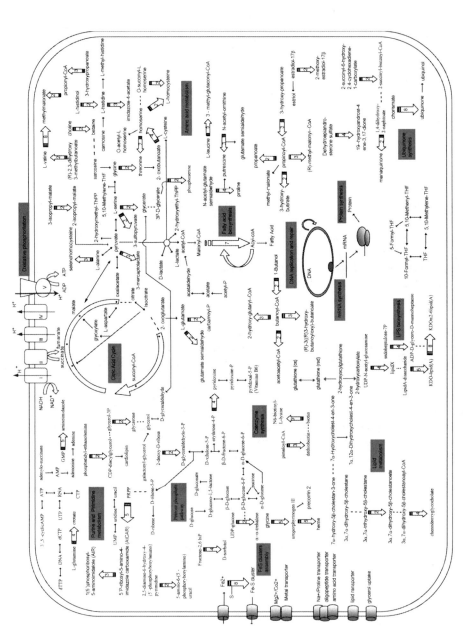

Fig. 1.2 Reconstructed metabolic map of the mitochondrion. Metabolic pathways and transport were deduced from the orthologous groups present in the estimated proteome, and their corresponding functions (Gabaldón and Huynen 2007b). *Boxes*, *arrows*, and *cylinders* indicate pathways, enzymes, and transporters, respectively. Several consecutive steps can be condensed into a *bigger arrow*, with a number indicating the steps included. Single missing steps connecting recovered pathways are indicated as *dashed lines*. Taken from Gabaldón and Huynen (2007b), with permission

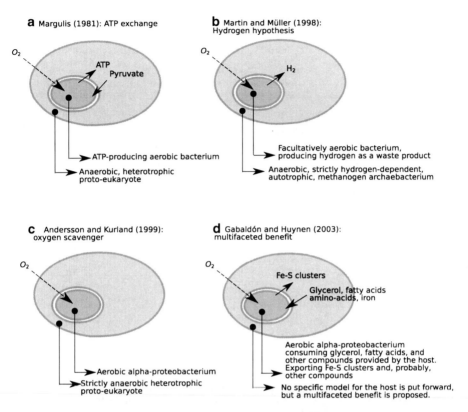

Fig. 1.3 Alternative metabolic scenarios for the ancestral symbiotic relationship between the proto-mitochondrion and its host, as proposed by several authors. *Large and small ovals* represent host and endosymbiont cells, respectively. *Small arrows* indicate inferred exchange of metabolites and other molecules. The ability of the endosymbiont to consume oxygen is indicated by a *dashed arrow*. Main properties of host and endosymbiont are discussed below the schemes. The different scenarios are discussed in the text. Modified from Gabaldón and Huynen (2007a)

hypotheses have been proposed more recently: one in which the primitive endo-symbiont is thought to be a hydrogen-producing, facultative anaerobe organism (Martin and Müller 1998), and another in which it would have acted as an oxygen scavenger (Kurland and Andersson 2000). The latter theory is based on the increase in oxygen in Earth's atmosphere supposedly happened approximately 2 billion years ago, with the oxygen tension going from 1% to more than 15% of the current levels in about 200 million years. According to this theory, the consumption of oxygen by the aerobic endosymbiont would have favored the survival of the anaerobic eukaryotic cells that established this symbiotic relationship. There are several problems to this theory. First, before oxygen reached the endosymbiont it would have diffused throughout the host cell, causing the damage that the endo-symbiont was supposed to avoid. Second, oxidative reactions taking place within the mitochondrion (most notably at the electron transport chain) are the source of many reactive oxygen species, which are more harmful to cellular structures such as lipids, proteins, or DNA, than O_2 itself. Finally, recent geological data and models suggest that, despite the increase of oxygen at atmospheric levels, the ocean waters, where early eukaryotic evolution took place, would have remained largely anoxic (Lyons and Reinhard 2009). Thus, the proposal by Martin and Müller of a facultative anaerobe endosymbiont (Martin and Müller 1998) seems to be more compatible with an anoxic environment. However, such model is not free of criticism. In particular, the assumption that the eukaryotic host was a methanogenic archaebacterium and not a proto-eukaryote has raised doubts about the possible mechanism of engulfment and possible membrane incompatibilities (de Duve 2007). Moreover, the capability of producing hydrogen by the proto-mitochondrion would imply facultative anaerobe ancestors in all lineages preceding the diversification of hydrogenosome-carrying eukaryotes and those with other types of mitochondria. This is in contrast with the finding that some hydrogenosomes appear to be rather the result of recent adaptations from aerobic mitochondria (Hackstein et al. 2006; van der Giezen 2009). In this respect, it is also important to note findings that suggest that at least some of the components for hydrogen-producing metabolism, such as iron hydrogenases, are also present in aerobic eukaryotes (Horner et al. 2002). The reconstruction of the proto-mitochondrial metabolism allowed further insight into the potential metabolic rationale of the initial endosymbiosis (Gabaldón and Huynen 2007b). Although in the absence of a reconstructed host proteome it is difficult to define a specific metabolic scenario, the prevalence of pathways and transporters in the reconstructed symbiont suggests that the metabolic exchange was more complex than previously proposed. Particularly, the presence of multiple transporters and pathways not directly related to energy metabolism indicate that the benefit for the host was rather multi-faceted.

1.6   Implications for the Origin of Eukaryotes

One of the most intriguing results of the reconstruction of the proto-mitochondrial proteome was the observation that a significant fraction (roughly 50%) of alpha-proteobacterial-derived proteins had been re-targeted to elsewhere in the host cell during the course of evolution (Gabaldón and Huynen 2003; Esser et al. 2004; Gabaldón and Huynen 2007b). This indicated that the contribution of the proto-mitochondrial ancestor to the eukaryotic cell went well beyond the mitochondrion itself. Proto-mitochondrial-derived proteins can be found almost everywhere in the eukaryotic cell, indicating that evolution has played around with the compartmentalization of proteins and pathways. In addition, the relative timing for the mitochondrial endosymbiosis has been revisited as mitochondria-related structures have been discovered in putative a-mitochondriate eukaryotes (Embley and Martin 2006). The current view is that mitochondrial endosymbiosis pre-dated the diversification of all known eukaryotic groups. Whether eukaryotes originated before or concomitantly to mitochondrial endosymbiosis constitutes a matter of heated debate (de Duve 2007; Koonin 2010). The classical view favors a proto-eukaryote with a developed cytoskeleton and thus is able to perform phagocytosis, the common mode of engulfing symbionts in modern eukaryotes. However, the finding of bacterial cells hosting other bacteria (von Dohlen et al. 2001; Davidov and Jurkevitch 2009) indicates that there exist alternative mechanisms for endosymbiosis. Some authors have proposed that mitochondrial endosymbiosis may have triggered the origin of key eukaryotic innovations such as introns and the cell nucleus (Martin and Koonin 2006), while paving the way for greater complexity (Lane and Martin 2010).

1.7 Concluding Remarks

During the last decade, thanks to the advent of new technologies such as whole genome sequencing and subcellular proteomics, a wealth of data from a broad range of eukaryotic organisms has been produced. This has facilitated advancing our understanding of the origin and evolution of the mitochondrion. The picture that is emerging provides different resolution for each of the two partners involved in the symbiotic event. Although our knowledge on the nature and metabolic properties of the proto-mitochondrion is slowly getting reasonably clear, the properties of the host cell remain fuzzy and highly controversial. One of the reasons for this is that, as new data have been gathered, the timing of the origin of mitochondria and that of the eukaryotic cell itself have come closer and closer. To the point that, since we lack any evidence for a supposed intermediate eukaryote – the "missing link" in the evolution of eukaryotes – it is reasonable to ask whether mitochondrial endosymbiosis itself triggered the origin of eukaryotes. Alternatively, the absence of any remainder of an intermediate stage does not necessarily mean that it did not exist. Deciding between these conflicting hypotheses is difficult, since nature shows that both an ancestral eukaryotic cell able to phagocytose and a simpler prokaryotic cell could have engulfed the primitive proto-mitochondrion. The answer lies perhaps in the discovery of new forms of eukaryotes, perhaps reminiscent of a pre-mitochondrial stage, or on inquisitive analysis of cellular, biochemical, or sequence data that may help us to discard one of the two scenarios. Alternatively, we may face the possibility that the exact relative timing of mitochondrial and eukaryotic origins remains uncertain, with many attractive scenarios being compatible with current data. One aspect is clear, in the years to come we will see how the universe of possible metabolic adaptations of mitochondria expands significantly, revealing an intricate history of evolutionary paths and astonishing adaptations to almost any inhabitable niche on earth.

Acknowledgments TG is funded in part by a grant from the Spanish ministry of science and innovation (BFU2009-09168).

References

Altmann R (1890) Die Elementarorganismen und ihre Beziehungen zu den Zellen. Veit & Comp, Liepzig

Atteia A, Adrait A, Brugiere S, Tardif M, van Lis R, Deusch O, Dagan T, Kuhn L, Gontero B, Martin W, Garin J, Joyard J, Rolland N (2009) A proteomic survey of *Chlamydomonas reinhardtii* mitochondria sheds new light on the metabolic plasticity of the organelle and on the nature of the alpha-proteobacterial mitochondrial ancestor. Mol Biol Evol 26:1533–1548

Au CE, Bell AW, Gilchrist A, Hiding J, Nilsson T, Bergeron JJ (2007) Organellar proteomics to create the cell map. Curr Opin Cell Biol 19:376–385

Boxma B, de Graaf RM, van der Staay GW, van Alen TA, Ricard G, Gabaldon T, van Hoek AH, Moon-van der Staay SY, Koopman WJ, van Hellemond JJ, Tielens AG, Friedrich T, Veenhuis M, Huynen MA, Hackstein JH (2005) An anaerobic mitochondrion that produces hydrogen. Nature 434:74–79

Brocks JJ, Logan GA, Buick R, Summons RE (1999) Archean molecular fossils and the early rise of eukaryotes. Science 285:1033–1036

Bui ET, Bradley PJ, Johnson PJ (1996) A common evolutionary origin for mitochondria and hydrogenosomes. Proc Natl Acad Sci U S A 93:9651–9656

Danovaro R, Dell'Anno A, Pusceddu A, Gambi C, Heiner I, Kristensen RM (2010) The first metazoa living in permanently anoxic conditions. BMC Biol 8:30

Davidov Y, Jurkevitch E (2009) Predation between prokaryotes and the origin of eukaryotes. Bioessays 31:748–757

de Duve C (2007) The origin of eukaryotes: a reappraisal. Nat Rev Genet 8:395–403

Dimmer KS, Rapaport D (2008) Proteomic view of mitochondrial function. Genome Biol 9:209

Embley TM, Martin W (2006) Eukaryotic evolution, changes and challenges. Nature 440:623–630

Emelyanov VV (2001) Evolutionary relationship of *Rickettsiae* and mitochondria. FEBS Lett 501:11–18

Esser C, Ahmadinejad N, Wiegand C, Rotte C, Sebastiani F, Gelius-Dietrich G, Henze K, Kretschmann E, Richly E, Leister D, Bryant D, Steel MA, Lockhart PJ, Penny D, Martin W (2004) A genome phylogeny for mitochondria among alpha-proteobacteria and a predominantly eubacterial ancestry of yeast nuclear genes. Mol Biol Evol 21:1643–1660

Esser C, Martin W, Dagan T (2007) The origin of mitochondria in light of a fluid prokaryotic chromosome model. Biol Lett 3:180–184

Gabaldón T, Huynen MA (2003) Reconstruction of the proto-mitochondrial metabolism. Science 301:609

Gabaldón T, Huynen MA (2004) Shaping the mitochondrial proteome. Biochim Biophys Acta 1659:212–220

Gabaldón T, Huynen M (2007a) Reconstruction of ancestral proteomes. In: Liberles DA (ed) Ancestral sequence reconstruction. Oxford University Press, Oxford, pp 128–138

Gabaldón T, Huynen MA (2007b) From endosymbiont to host-controlled organelle: the hijacking of mitochondrial protein synthesis and metabolism. PLoS Comput Biol 3:e219

Gabaldón T, Rainey D, Huynen MA (2005) Tracing the evolution of a large protein complex in the eukaryotes, NADH:ubiquinone oxidoreductase (complex I). J Mol Biol 348:857–870

Gray MW, Burger G, Lang BF (1999) Mitochondrial evolution. Science 283:1476–1481

Hackstein JH, Tjaden J, Huynen M (2006) Mitochondria, hydrogenosomes and mitosomes: products of evolutionary tinkering! Curr Genet 50:225–245

Hampl V, Hug L, Leigh JW, Dacks JB, Lang BF, Simpson AG, Roger AJ (2009) Phylogenomic analyses support the monophyly of Excavata and resolve relationships among eukaryotic "supergroups". Proc Natl Acad Sci U S A 106:3859–3864

Heazlewood JL, Tonti-Filippini JS, Gout AM, Day DA, Whelan J, Millar AH (2004) Experimental analysis of the *Arabidopsis* mitochondrial proteome highlights signaling and regulatory components, provides assessment of targeting prediction programs, and indicates plant-specific mitochondrial proteins. Plant Cell 16:241–256

Hjort K, Goldberg AV, Tsaousis AD, Hirt RP, Embley TM (2010) Diversity and reductive evolution of mitochondria among microbial eukaryotes. Philos Trans R Soc Lond B Biol Sci 365:713–727

Horner DS, Heil B, Happe T, Embley TM (2002) Iron hydrogenases – ancient enzymes in modern eukaryotes. Trends Biochem Sci 27:148–153

John P, Whatley FR (1975) *Paracoccus denitrificans* and the evolutionary origin of the mitochondrion. Nature 254:495–498

Keeling PJ, Burger G, Durnford DG, Lang BF, Lee RW, Pearlman RE, Roger AJ, Gray MW (2005) The tree of eukaryotes. Trends Ecol Evol 20:670–676

Koonin EV (2010) The origin and early evolution of eukaryotes in the light of phylogenomics. Genome Biol 11:209

Kurland CG, Andersson SG (2000) Origin and evolution of the mitochondrial proteome. Microbiol Mol Biol Rev 64:786–820

Lane N, Martin W (2010) The energetics of genome complexity. Nature 467:929–934

Li J, Cai T, Wu P, Cui Z, Chen X, Hou J, Xie Z, Xue P, Shi L, Liu P, Yates JR 3rd, Yang F (2009) Proteomic analysis of mitochondria from *Caenorhabditis elegans*. Proteomics 9:4539–4553

Lyons TW, Reinhard CT (2009) An early productive ocean unfit for aerobics. Proc Natl Acad Sci U S A 106:18045–18046

Mai Z, Ghosh S, Frisardi M, Rosenthal B, Rogers R, Samuelson J (1999) Hsp60 is targeted to a cryptic mitochondrion-derived organelle ("crypton") in the microaerophilic protozoan parasite *Entamoeba histolytica*. Mol Cell Biol 19:2198–2205

Marcet-Houben M, Marceddu G, Gabaldón T (2009) Phylogenomics of the oxidative phosphorylation in fungi reveals extensive gene duplication followed by functional divergence. BMC Evol Biol 9:295

Margulis L (1970) The origin of the eukaryotic cell. Yales University Press, New Haven

Margulis L (1981) Symbioses in cell evolution. W.H. Freeman, San Francisco

Martin W, Koonin EV (2006) Introns and the origin of nucleus-cytosol compartmentalization. Nature 440:41–45

Martin W, Müller M (1998) The hydrogen hypothesis for the first eukaryote. Nature 392:37–41

Mootha VK, Bunkenborg J, Olsen JV, Hjerrild M, Wisniewski JR, Stahl E, Bolouri MS, Ray HN, Sihag S, Kamal M, Patterson N, Lander ES, Mann M (2003) Integrated analysis of protein composition, tissue diversity, and gene regulation in mouse mitochondria. Cell 115:629–640

Pagliarini DJ, Calvo SE, Chang B, Sheth SA, Vafai SB, Ong SE, Walford GA, Sugiana C, Boneh A, Chen WK, Hill DE, Vidal M, Evans JG, Thorburn DR, Carr SA, Mootha VK (2008) A mitochondrial protein compendium elucidates complex I disease biology. Cell 134:112–123

Prokisch H, Andreoli C, Ahting U, Heiss K, Ruepp A, Scharfe C, Meitinger T (2006) MitoP2: the mitochondrial proteome database – now including mouse data. Nucleic Acids Res 34:D705–D711

Shutt TE, Gray MW (2006) Bacteriophage origins of mitochondrial replication and transcription proteins. Trends Genet 22:90–95

Smits P, Smeitink JA, van den Heuvel LP, Huynen MA, Ettema TJ (2007) Reconstructing the evolution of the mitochondrial ribosomal proteome. Nucleic Acids Res 35:4686–4703

Szklarczyk R, Huynen MA (2010) Mosaic origin of the mitochondrial proteome. Proteomics 10(22):4012–4024

Tielens AG, Rotte C, van Hellemond JJ, Martin W (2002) Mitochondria as we don't know them. Trends Biochem Sci 27:564–572

Tovar J, Fischer A, Clark CG (1999) The mitosome, a novel organelle related to mitochondria in the amitochondrial parasite *Entamoeba histolytica*. Mol Microbiol 32:1013–1021

Tovar J, Leon-Ávila G, Sanchez LB, Sutak R, Tachezy J, van der Giezen M, Hernandez M, Müller M, Lucocq JM (2003) Mitochondrial remnant organelles of *Giardia* function in iron-sulphur protein maturation. Nature 426:172–176 van der Giezen M (2009) Hydrogenosomes and mitosomes: conservation and evolution of functions. J Eukaryot Microbiol 56:221–231

Vo TD, Palsson BO (2007) Building the power house: recent advances in mitochondrial studies through proteomics and systems biology. Am J Physiol Cell Physiol 292:C164–C177

von Dohlen CD, Kohler S, Alsop ST, McManus WR (2001) Mealybug beta-proteobacterial endosymbionts contain gamma-proteobacterial symbionts. Nature 412:433–436

Chapter 2
Plastid Origins

John M. Archibald

2.1 Modern-Day Plastids and Their Genomes

The evolution of oxygenic photosynthesis in the ancestors of present-day cyanobacteria transformed the biosphere of our planet (Blankenship 1994; Reyes-Prieto et al. 2007). This landmark event was also an essential prerequisite for the evolution of photosynthetic eukaryotes. Plastids, or chloroplasts, are the light-gathering organelles of algae and plants whose origin can be traced back to cyanobacteria. Mereschkowsky (1905) is usually credited as being the first biologist to speculate on the possible evolutionary significance of similarities between cyanobacteria and plastids, and in the era of molecular biology and genomics, the evidence that plastids are derived from once free-living prokaryotes is now beyond refute. Modern-day plastids and the eukaryotes that harbor them are remarkably diverse in their morphology and biochemistry, but are nevertheless sufficiently similar to one another in their core features to be able to infer common ancestry. These include similarities in their plastid light-harvesting apparatus, the existence of protein import machinery with many cyanobacterial features, and an organellar genome of demonstrable cyanobacterial ancestry (Kim and Archibald 2009).

This chapter provides an overview of the origin and diversification of plastids across the eukaryotic tree of life, an area of basic research that has benefited tremendously from advances in genomics and molecular biology. Genome sequences from an evolutionarily diverse array of eukaryotic phototrophs are now available and have made it possible to sketch a general picture of how plastids evolved. Yet, while the evidence in support of a cyanobacterial origin for plastids is stronger than ever, other questions pertaining to the biology and evolution of

J.M. Archibald (✉)
Program in Integrated Microbial Biodiversity, Canadian Institute for Advanced Research,
Department of Biochemistry and Molecular Biology, Sir Charles Tupper Medical Building,
Dalhousie University, Halifax, NS, Canada B3H 4R2
e-mail: jmarchib@dal.ca

C.E. Bullerwell (ed.), *Organelle Genetics*,
DOI 10.1007/978-3-642-22380-8_2, © Springer-Verlag Berlin Heidelberg 2012

plastid-bearing organisms have become less clear in the light of more data. Most prominent among them is the issue of how – and how often – plastids have spread horizontally across the tree by endosymbioses involving two eukaryotes and the extent to which genes of algal/cyanobacterial ancestry in the genomes of plastid-lacking eukaryotes can be taken as evidence for plastid loss.

One of the major challenges associated with inferring the history of plastids is the vast amount of time that has transpired since they first evolved and the limited coding capacity of their genomes relative to those of cyanobacteria. Molecular clock-based analyses (e.g., Yoon et al. 2004) have suggested that plastids evolved >1 billion years ago, and the transition from cyanobacterial endosymbiont to fully integrated organelle is known to have involved the loss of many nonessential genes and the transfer of essential genes from the endosymbiont to the nuclear genome of its eukaryotic host. This process, referred to as endosymbiotic gene transfer (EGT), is of profound significance to the study of organelle evolution and is reviewed in detail by Timmis and colleagues in Chap. 7. Even the most gene-rich plastid genomes possess only ~200 protein genes; most possess far fewer than this (Martin et al. 1998; Kim and Archibald 2009). The nuclear genomes of algae and plants encode many hundreds of proteins of cyanobacterial/plastid ancestry, many (but not all) of which are translated on cytoplasmic ribosomes and targeted to the plastid post-translationally (Jarvis and Soll 2001; Gould et al. 2008). Interestingly, whole genome-scale analyses have revealed that genes of noncyanobacterial ancestry also contribute to the proteomes of modern-day plastids (e.g., Moustafa et al. 2008; Suzuki and Miyagishima 2010). Conversely, many of the genes donated to the nuclear genome by the cyanobacterial progenitor of the plastid subsequently acquired functions in the host eukaryote unrelated to the plastid and to photosynthesis (Martin et al. 2002; Archibald 2006; Reyes-Prieto et al. 2006). Photosynthetic eukaryotes are thus increasingly recognized as complex evolutionary "mosaics," with genes having been acquired via EGT as well as horizontal (or lateral) gene transfer (HGT) (Lane and Archibald 2008; Elias and Archibald 2009). Establishing the significance and relative contributions of these two sources of gene flow into algal nuclear genomes remains a major hurdle to overcome as the field moves toward a comprehensive understanding of the evolutionary history of plastid-bearing eukaryotes.

2.2   Primary Plastids

Plastids are typically classified as belonging to one of two types. "Primary" plastids are those considered to stem directly from the primordial endosymbiosis between a nonphotosynthetic eukaryote and the cyanobacterial plastid progenitor, while plastids that have spread indirectly from one eukaryote to another are designated "secondary" or "tertiary" organelles (Reyes-Prieto et al. 2007; Gould et al. 2008; Archibald 2009). Primary plastids are united in their shared possession of a two-membrane envelope, the leaflets of which are thought to correspond to the inner and outer membranes of the engulfed cyanobacterium (Reyes-Prieto et al. 2007; Gould

Table 2.1 Diversity and basic characteristics of plastids[a]

Lineage	Putative origin	Membranes	Pigmentation
Glaucophytes	1°	2[b]	Chl a + phycobiliproteins
Red algae	1°	2	Chl a + phycobiliproteins
Green algae + land plants	1°	2	Chl a + b
Cryptophytes[c]	2° (Red)	4	Chl a + c + phycobiliproteins
Haptophytes	2° (Red)	4	Chl a + c + fucoxanthin
Stramenopiles (Heterokonts)	2° (Red)	4	Chl a + c + fucoxanthin
Dinoflagellates[d]	2° (Red)	3	Chl a + c + peridinin
Perkinsids	2° (Red)	4	None (non-photosynthetic)
Apicomplexans	2° (Red)	4	None (non-photosynthetic)
Chromera	2° (Red)	4	Chl a
Euglenophytes	2° (Green)	3	Chl a + b
Chlorarachniophytes[c]	2° (Green)	4	Chl a + b

[a]Data taken primarily from Graham and Wilcox (2000) and Larkum et al. (2007). Numerous exceptions and additional complexities exist beyond the data presented in this table. Interested readers are referred to Kim and Archibald (2009) and references therein

[b]Glaucophyte plastids possess a layer of peptidoglycan between the inner and outer membranes, as in cyanobacteria

[c]The nucleus of the algal endosymbionts that gave rise to the cryptophyte and chlorarachniophyte plastids persists in a highly degenerate form called a nucleomorph. The nucleomorph is located in the periplastidial compartment, i.e., the space between the inner and outer pairs of plastid membranes

[d]Approximately 50% of known dinoflagellate species are photosynthetic. Plastid-bearing species usually possess a peridinin-pigmented plastid, although some dinoflagellates have also replaced this organelle with plastids acquired from haptophytes and diatoms (tertiary endosymbiosis) or green algae (serial secondary endosymbiosis). Plastid membrane number varies depending on plastid type. Refer to Hackett et al. (2004) for review

et al. 2008). In contrast, secondary and tertiary organelles possess additional membranes, with the precise number varying from lineage to lineage (Table 2.1).

The number of secondary and tertiary endosymbiotic events that have occurred during eukaryotic evolution is still very much an open question, but there is general agreement with regard to the origin of primary plastids and the lineages that harbor them: these are the red algae, glaucophyte (or glaucocystophyte) algae, and the green algae, the latter being the group from which land plants ultimately evolved (Delwiche et al. 2004; Reyes-Prieto et al. 2007). A single endosymbiotic capture of an ancestor of modern-day cyanobacteria by a full-blown eukaryotic host cell is believed by many researchers to have occurred in a common ancestor shared by the three lineages, followed by strict vertical inheritance thereafter (Palmer 2003; Reyes-Prieto et al. 2007) (Fig. 2.1).

A broad array of biochemical, molecular, and phylogenetic data has been brought to bear on the issue of whether primary plastids evolved once or more than once. For example, the plastids of red and green algae have been shown to possess light-harvesting complex (LHC) proteins that are not related to their functional equivalents in present-day cyanobacteria (Green and Durnford 1996; Durnford et al. 1999), the implication being that they represent singular eukaryote-

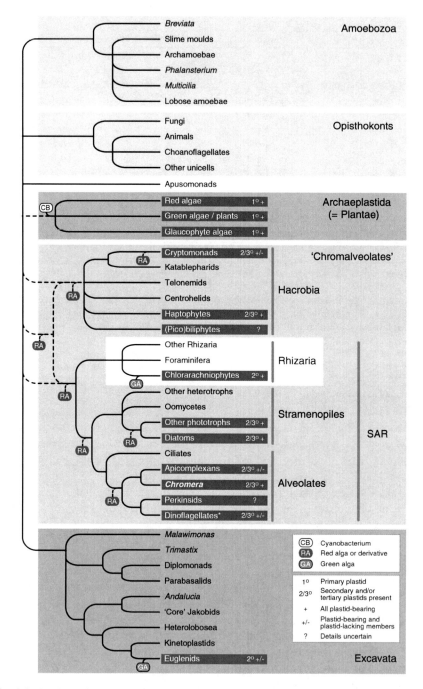

Fig. 2.1 A schematic representation of eukaryotic relationships with an emphasis on lineages containing one or more photosynthetic groups. The topology is a consensus of various nuclear multigene phylogenies (e.g., Burki et al. 2007, 2008; Hackett et al. 2007; Hampl et al. 2009; Parfrey et al. 2010). *Dashed lines* represent alternative topologies and/or areas of particular uncertainty. Primary and secondary/tertiary plastid-bearing lineages are indicated as described in the key. *Asterisks:* The dinoflagellates are known to harbor a wide diversity of plastids beyond those depicted in this figure (see main text)

specific evolutionary innovations occurring in a common ancestor of the two groups. A similar explanation has been proposed to account for the distribution of Tic110, a nucleus-encoded core subunit of the plastid protein import apparatus: Tic110 is found in red, green, and glaucophyte algae but is absent in all known cyanobacteria (McFadden and van Dooren 2004). The structure and coding capacity of primary plastid genomes is also suggestive of common ancestry. For example, the structure of the plastid *atpA* gene cluster is widely conserved among red, green, and glaucophyte algae (Stoebe and Kowallik 1999), as is the presence of ribosomal DNA (rDNA)-containing inverted repeats in the vast majority of primary plastid genomes (and those of their secondary derivatives). Inverted rDNA repeats are, however, also found in the genomes of some cyanobacteria, suggesting that this feature may have existed prior to the evolution of plastids by endosymbiosis (Glockner et al. 2000). The possibility of convergent evolution of basic plastid genome architecture is usually ignored but cannot be discounted (Palmer 2003; Stiller et al. 2003). Chapter 4 presents a detailed overview of plastid genomes and their structure.

Molecular phylogenetic analyses often, though by no means always, support the notion that red, green, and glaucophyte algae are each other's closest relatives. Such a relationship would be expected if primary plastids evolved only once in a common ancestor shared by the three lineages and if primary plastids have not been lost secondarily in plastid-lacking eukaryotic lineages. Early plastid 16S rDNA and elongation factor Tu (EF-Tu) sequence analyses supported the hypothesis of primary plastid monophyly (Delwiche et al. 1995; Helmchen et al. 1995), as have more recent multigene analyses in which dozens of plastid protein sequences are analyzed together as a single concatenate (e.g., Rodriguez-Ezpeleta et al. 2005). In contrast, phylogenies inferred from nuclear loci, such as the largest subunit of RNA polymerase II (RPB1) are sometimes much less clear (Stiller and Hall 1997; Longet et al. 2003). Tree topologies resulting from the latest "phylogenomic" analyses, including analyses of large 100+ protein data sets, have proven unexpectedly sensitive to taxon sampling and gene selection but often resolve the monophyly of "Archaeplastida" with high statistical support (e.g., Rodriguez-Ezpeleta et al. 2005, 2007a, b; Burki et al. 2007, 2008, 2009; Hackett et al. 2007). There is however no general consensus as to the branching order of red, green, and glaucophyte algae relative to one another. In cases where monophyly of these three main lines is not recovered, secondary plastid-bearing groups such as cryptophyte and haptophyte algae (see below) are often the intervening lineages (e.g., Hampl et al. 2009; Parfrey et al. 2010). Recent analysis of specific subsets of nuclear genes, in particular those deemed to be the most slowly evolving, appear to support the nonmonophyly of the primary plastid-bearing lineages and have been used as the basis for alternative scenarios in which primary plastids evolved in a common ancestor shared by red, green, and glaucophyte algae and other eukaryotic groups that currently lack a plastid (e.g., Nozaki et al. 2007). In sum, while it is widely held that primary plastids evolved only once, exactly when this occurred during eukaryotic evolution is unclear, and it is possible that primary plastid loss has occurred and/or that the double membrane-bound organelles in the three groups have not been inherited in a strictly vertical fashion (see Stiller and Hall 1997; Andersson and Roger 2002; Nozaki et al. 2003; Larkum et al. 2007; Stiller 2007; Kim and Graham 2008 and references therein for alternative scenarios). As will be underscored in the sections that follow, assessing the extent to which nuclear and organellar gene phylogenies can be judged to be congruent or incongruent with one another has become one of the most important issues in plastid evolution.

One final note on the evolution of primary plastids relates to the "chromatophores" of the testate amoeba *Paulinella chromatophora*. First discovered over 100 years ago (Lauterborn 1895; Melkonian and Mollenhauer 2005), *P. chromatophora*, a member of the Rhizaria (Fig. 2.1), harbors within its cytoplasm cyanobacteria belonging to the genus *Synechococcus* (Marin et al. 2005; Yoon et al. 2006). The chromatophore genome is somewhat reduced relative to its closest free-living relatives, although nowhere near that of a canonical plastid (Nowack et al. 2008; Yoon et al. 2009). Much has been written about whether these "recently" acquired photosynthetic intracellular inclusions should be considered endosymbionts or true cellular organelles (Theissen and Martin 2006; Bhattacharya and Archibald 2007; Bodyl et al. 2007). Chromatophore-to-host-nucleus gene transfer has been documented (Nakayama and Ishida 2009; Nowack et al. 2011) and a mechanism for protein import into the chromatophore has even been proposed (Bodyl et al. 2010; Mackiewicz and Bodyl 2010). It will thus be interesting to see how much of what we learn from studying *P. chromatophora* and its enigmatic "organelles" will shed light on the primary endosymbiotic origin of canonical plastids, which appears to have been a singular, truly ancient event in eukaryotic evolution.

2.3   Secondary and Tertiary Plastids: Origins and Evolution

As fundamental as they are, the uncertainties surrounding the evolution of plastids within red, green, and glaucophyte algae pale in comparison to those of phototrophic eukaryotes as a whole. Researchers have long recognized the extraordinary structural and biochemical diversity of plastids, particularly in the realm of microscopic algae, and in the 1970s, the possibility that plastids had moved from one eukaryotic lineage to another began to be taken seriously (Taylor 1974; Gibbs 1978). Definitive evidence of what is referred to as secondary endosymbiosis eventually came from the study of algae that possess two bona fide nuclei, a host nucleus and a plastid-associated, eukaryotic endosymbiont-derived nucleus, the latter being referred to as a "nucleomorph" (Greenwood 1974; Greenwood et al. 1977; McFadden et al. 1994). Nucleomorphs have been shown to harbor the smallest nuclear genomes known to science and represent interesting systems with which to study the processes of genome reduction and compaction (Archibald and Lane 2009; Moore and Archibald 2009).

Two distinct nucleomorph-bearing lineages are currently recognized. The cryptophytes and chlorarachniophytes are significant not only in their shared possession of the "smoking gun" of secondary endosymbiosis but also because they acquired photosynthesis independently. Cryptophyte nucleomorphs and plastids are derived from a red algal endosymbiont (Douglas et al. 1991, 2001; Douglas and Penny 1999; Graham and Wilcox 2000; Lane et al. 2007; Kim and Archibald 2009), while in chlorarachniophytes these organelles evolved from an endosymbiotic green alga (Gilson and McFadden 1996; Ishida et al. 1997, 1999; Gilson et al. 2006; Rogers et al. 2007). The closest relatives of their respective secondary endosymbionts within modern-day red and green algae are still uncertain, particularly in the case of cryptophytes, where comparative sequence data from diverse red algae are lacking. Nevertheless, karyotype and molecular sequence data have revealed that the cryptophyte and chlorarachniophyte nucleomorph genomes constitute a remarkable example of convergent evolution. Both harbor highly reduced, A+T-rich genomes less than 1 megabase-pair (Mbp) in size and partitioned into three chromosomes, each capped with subtelomeric ribosomal DNA (rDNA) loci (Moore and Archibald 2009). The evolutionary pressures responsible for and the biological significance of these similarities are for the most part not known.

The euglenophytes constitute a second lineage harboring green algal-derived secondary plastids. However, unlike chlorarachniophytes, whose plastid is surrounded by four membranes, the euglenophyte plastid has three membranes and lacks a plastid-associated nucleomorph (Table 2.1). Euglenophytes are classified as euglenids (or "euglenoids"), which in addition to phototrophs such as *Euglena gracilis*, include plastid-lacking heterotrophs capable of ingesting bacteria (bacteriovores) and eukaryotes (eukaryovores) (Leander et al. 2001). Together with the exclusively plastid-lacking kinetoplastids (e.g., parasites such as *Trypanosoma* and *Leishmania*), the euglenophytes reside within the "supergroup" Excavata (Hampl et al. 2009) (Fig. 2.1). In contrast, the chlorarachniophytes constitute the sole plastid-bearing lineage within the supergroup Rhizaria, a diverse collection of predominantly amoeboid, unicellular eukaryotes that include the foraminiferans and radiolarians (Nikolaev et al. 2004). The existence of green algal secondary plastids in both chlorarachniophytes and euglenophytes thus represents a case of discordant host–endosymbiont evolutionary histories and, at face value, is most consistent with the notion of independent secondary acquisitions of green algal plastids. Indeed, not only do the host components of chlorarachniophytes and euglenophytes belong to completely different supergroups, the latest plastid genome sequence comparisons suggest that their plastids evolved from distinct lines of green algae (e.g., Rogers et al. 2007; Turmel et al. 2009). Current data do not support the hypothesis of a single ancient green algal secondary endosymbiosis in a common ancestor shared by chlorarachniophytes and euglenophytes (Cavalier-Smith 1999).

An even broader array of eukaryotes harbors red algal-derived secondary plastids. In addition to the nucleomorph-bearing cryptophytes, these include the stramenopiles (e.g., diatoms and giant kelp), haptophytes (e.g., *Emiliania huxleyi*), some dinoflagellates, some apicomplexans such as the malaria parasite *Plasmodium falciparum*, as well as the newly discovered chromerids (Keeling 2009). Dinoflagellates are particularly impressive in their diversity of plastids, although it should be noted that only ~50% of known species actually possess a photosynthetic organelle (Taylor 1980). Of those that do, most harbor a peridinin-pigmented, red algal-derived secondary (or tertiary) plastid, while others have tertiary plastids specifically derived from haptophytes (Tengs et al. 2000), cryptophytes (Schnepf and Elbrächter 1988; Hackett et al. 2003), and diatoms (Dodge 1969; Inagaki et al. 2000) (see Hackett et al. 2004 and Archibald 2005 and references therein for detailed review). The dinoflagellate *Lepidodinium* possesses a recently acquired, green algal-derived plastid of serial secondary origin (Watanabe et al. 1990), and members of the genus *Kryptoperidinium* have an "unreduced" diatom plastid with a nucleus and mitochondria still associated with it (Chesnick et al. 1997; McEwan and Keeling 2004; Imanian et al. 2007; Imanian and Keeling 2007). In addition, some dinoflagellates possess transient plastids and carry out "acquired phototrophy." For instance, the heterotrophic dinoflagellate *Dinophysis acuminata* harbors a cryptophyte plastid that it obtains indirectly by regularly feeding on the ciliate *Myrionecta rubra*, which itself ingests cryptophytes of the *Teleaulax/Geminigera* clade (Park et al. 2008). Wisecaver and Hackett have recently shown that the nuclear genome of *D. acuminata* does not appear to be stocked with genes for plastid-targeted proteins, as is invariably the case for photosynthetic eukaryotes, and the few that have been found come primarily from algae other than cryptophytes (Wisecaver and Hackett 2010). The implication is that the *D. acuminata* plastid is truly temporary and incapable of being perpetuated to any great extent within the dinoflagellate cell.

Recent plastid acquisitions in dinoflagellates aside, how do the various red algal secondary plastids relate to one another? As for euglenids and chlorarachniophytes, demonstrating incongruent host and plastid phylogenies would support the notion of independent secondary endosymbioses involving distinct red algae and/or secondary hosts, but for various reasons, this has proven difficult to determine. For example, the highly derived nature of the plastid genomes of apicomplexans and dinoflagellates do not lend themselves to accurate phylogenetic reconstruction. The apicomplexans are nonphotosynthetic and appear to retain their plastids solely to carry out core metabolic processes, such as the synthesis of isoprenoids and fatty acids (Ralph et al. 2004). Consequently, the coding capacity of the "apicoplast" is quite limited and the genes that remain are typically highly divergent. It was in fact initially unclear whether the four-membrane-bound plastids of apicomplexans were of green or red algal ancestry. Data have been presented in support of both hypotheses (e.g., Köhler et al. 1997; Blanchard and Hicks 1999; Funes et al. 2002; Waller et al. 2003), but on balance, the evidence rests decidedly in favor of a red algal origin (Waller and McFadden 2005; Janouskovec et al. 2010; Lim and McFadden 2010). Dinoflagellates are even more problematic, as their peridinin plastid genomes are made up of single-gene minicircles encoding extraordinarily rapidly evolving genes (Zhang et al. 1999, 2000; Sanchez-Puerta et al. 2007a; Howe et al. 2008). The exciting discovery of *Chromera velia* (Moore et al. 2008), an alga with a relatively gene-rich plastid genome (Janouskovec et al. 2010), has made it possible to link both host- and plastid-associated features of dinoflagellates with those of apicomplexans (Keeling 2008; Janouskovec et al. 2010). Combined with the discovery of cryptic plastids in dinoflagellates that were previously assumed to be lacking plastids (Sanchez-Puerta et al. 2007b) and their close relatives, such as perkinsids (Stelter et al. 2007; Teles-Grilo et al. 2007) and *Oxyrrhis* (Slamovits and Keeling 2008), it now seems likely that the common ancestor of dinoflagellates and apicomplexans possessed a red algal-derived secondary plastid.

What about the other red secondary plastid-containing lineages? A long-standing and controversial idea in the field is that the plastids of apicomplexans and dinoflagellates are truly ancient, sharing a common endosymbiotic origin with *all* other known red secondary plastids (Table 2.1, Fig. 2.1). Cavalier-Smith's "chromalveolate" hypothesis unites the "chromists" (plastid-bearing cryptomonads (i.e., cryptophytes), stramenopiles, and haptophytes) with the alveolates (dinoflagellates, apicomplexans, and ciliates) and rests on the principle that secondary endosymbiosis is a complex process and should be invoked sparingly (Cavalier-Smith 1999). With each such event, hundreds of nuclear genes for plastid-targeted proteins must be transferred from the red or green algal nucleus to the secondary host nucleus, and a mechanism for importing such proteins must evolve "from scratch" (Cavalier-Smith 1999; McFadden 1999; Cavalier-Smith 2000; Gould et al. 2008). Critics of the chromalveolate hypothesis acknowledge these difficulties but point to the existence of many plastid-lacking "chromalveolate" taxa: *Goniomonas* (a basal cryptomonad), Hacrobia such as katablepharids and telonemids, heterotrophic stramenopiles such as oomycetes, and ciliates, a huge, diverse, and entirely plastid-lacking alveolate lineage (Fig. 2.1). If the chromalveolate hypothesis were true, then plastids would have had to be lost secondarily in each of these lineages. An alternative hypothesis is that red algal-derived secondary plastids have spread by one or more cryptic tertiary endosymbioses, as is known to have occurred in the case of dinoflagellates (Hackett et al. 2004). The chromalveolate hypothesis has proven to be something of a moving target: new eukaryotic lineages continue to be discovered and evolutionary relationships must continuously be retested as genomic sequence data accumulate.

Early single-locus analyses of plastid genes, such as for 16S rDNA and ribulose-1,5-bisphosphate carboxylase/oxygenase (RuBisCO), were inconsistent with the chromalveolate hypothesis, seeming to favor the notion that the different chromist lineages, i.e., cryptophytes, stramenopiles, and haptophytes, had acquired their plastids from different red algae (Daugbjerg and Andersen 1997; Oliveira and Bhattacharya 2000; Müller et al. 2001). With time and more sequence data, however, the balance tipped in favor of chromist plastid monophyly, albeit with varying levels of confidence (e.g., Yoon et al. 2002; Khan et al. 2007). Rare genomic characters, such as endosymbiotic gene replacements involving the plastid-associated genes for glyceraldehyde-3-phosphate dehydrogenase (GAPDH) (Fast et al. 2001) and fructose-1,6-bisphosphate (FBA) (Patron et al. 2004), were also presented as evidence to support plastid monophyly of subsets of chromalveolate taxa, and a rare lateral gene transfer involving the ribosomal protein gene *rpL36* suggested a specific relationship between haptophyte and cryptophyte plastids (Rice and Palmer 2006). However, a common origin of chromalveolate plastids is, as argued by Bodyl and others (Bodyl 2005, 2006; Sanchez-Puerta and Delwiche 2008; Bodyl et al. 2009), also consistent with evolutionary scenarios involving multiple tertiary endosymbioses. The "final frontier" has thus been examination of the phylogenetic signal contained in as many different nuclear loci as possible in an effort to confirm or refute the notion that the histories of the chromalveolate hosts and endosymbionts are congruent with one another.

The answer seems to be that they are not. Phylogenomic analyses of dozens to hundreds of nucleus-encoded proteins concatenated together have revealed that the chromalveolates, as originally defined (Cavalier-Smith 1999), do not form a monophyletic group to the exclusion of other eukaryotes, even though subsets of chromalveolate lineages are clearly related to one another, such as cryptophytes and haptophytes (Hackett et al. 2007; Patron et al. 2007) and stramenopiles and alveolates (Burki et al. 2007; Burki et al. 2008; Hampl et al. 2009; Parfrey et al. 2010). One recent twist has been the realization that the supergroup Rhizaria, to which the green algal secondary plastid-containing chlorarachniophytes belong, is robustly allied with stramenopiles and alveolates. This tripartite grouping has been dubbed SAR (Burki et al. 2007, 2008). The cryptophyte–haptophyte pair does *not* branch with the other chromalveolate groups and has been expanded to include the plastid-lacking telonemids, centrohelids, and katablepharids under the term Hacrobia (Burki et al. 2009; Okamoto et al. 2009). A recent analysis using smaller data sets than the above-mentioned studies but with expanded taxonomic sampling supported some but not all of these relationships, and found no evidence for chromalveolate monophyly (Parfrey et al. 2010). Finally, a rigorous test of the phylogenetic signal contained in the nuclear, mitochondrial, and plastid genomes of the CASH group (*c*ryptophytes, *a*lveolates, *s*tramenopiles, and *h*aptophytes) led Baurain et al. to "*…reject the chromalveolate hypothesis as falsified in favor of more complex evolutionary scenarios involving multiple higher order eukaryote-eukaryote endosymbioses*" (Baurain et al. 2010). The basis for their falsification is that the CASH lineage plastid genomes appear to have diverged from one another much more recently than their respective mitochondrial and nuclear genomes. Various alternative scenarios involving "higher-order" endosymbioses have recently been explored in the literature (e.g., Bodyl 2005, 2006; Sanchez-Puerta and Delwiche 2008; Archibald 2009; Bodyl et al. 2009), but unfortunately there is as yet little information with which to distinguish between them. The original recipient lineage(s) of the red algal secondary plastid is not obvious, nor is the number (and directionality) of subsequent tertiary endosymbioses needed to account for the apparent incongruence between chromalveolate hosts and plastids.

2.4   Genome Mosaicism: Evidence for Past Endosymbioses or "You Are What You Eat"?

One of the most unexpected developments in recent years has been the extent to which the nuclear genomes of secondary plastid-containing algae, and indeed all eukaryotes, harbor genes of mixed ancestry. HGT is now a well-established factor in the evolution of eukaryotic genomes (Keeling and Palmer 2008), but in the case of phototrophs it is often difficult to tell whether any given "foreign" gene is the product of an endosymbiotic gene transfer or was acquired in a endosymbiosis-independent fashion before, during, or after plastid acquisition. This is not a trivial distinction. Numerous studies invoking secondary plastid loss in eukaryotic groups are based entirely on the presence of algal/cyanobacterial genes in the genome (e.g., Huang et al. 2004; Tyler et al. 2006; Reyes-Prieto et al. 2008). Assessing the significance of genome mosaicism in algae thus has important implications for modeling the pattern and process of plastid evolution.

A 2003 study of the chlorarachniophyte *Bigelowiella natans*, which has a green algal secondary plastid (Table 2.1 and Fig. 2.1), provided some of the first comprehensive evidence for genome mosaicism in eukaryotes. As expected, most of the examined nucleus-encoded, plastid-targeted proteins in this organism were found to be of green algal ancestry, but red algal-derived genes, and even those from bacteria were also found (Archibald et al. 2003). The chlorarachniophytes are known to be capable of ingesting other algae and bacteria (Hibberd and Norris 1984), and the mosaic *B. natans* plastid proteome was deemed to be the product of both endosymbiotic and horizontal gene transfers, the latter related to its phagotrophic lifestyle. A more recent study of the dinoflagellate *Lepidodinium chlorophorum* revealed a similar pattern. This organism currently has a green algal plastid of serial secondary endosymbiotic origin and, not surprisingly, green algal-derived, plastid-associated genes reside in its nucleus (Minge et al. 2010). However, *L. chlorophorum* also harbors genes of red algal secondary endosymbiotic origin and, in this case, it seems reasonable to conclude that at least some of the red algal-type genes are "holdovers" from the ancestral peridinin-type plastid this dinoflagellate is believed to have harbored (Minge et al. 2010). Even though the patterns of plastid-associated gene mosaicism in *B. natans* and *L. chlorophorum* are similar, our interpretation of the underlying causes is different, invoking predominantly HGT in the former and EGT in the latter. Or is it different?

Considering that the supergroup Rhizaria, to which the chlorarachniophytes belong, now appears to be nested within traditional chromalveolate taxa (Burki et al. 2008; Parfrey et al. 2010) (Fig. 2.1), one could argue that at least some of the red algal genes in the *B. natans* genome (Archibald et al. 2003) are derived from ancient endosymbiotic gene transfer rather than HGT. The picture becomes even more complex when one considers a provocative hypothesis put forth by Moustafa et al. (2009). These authors showed that in chromalveolate taxa such as diatoms and haptophytes, genes of apparent green algal ancestry outnumber red algal genes by more than 3-to-1. Preliminary evidence for a green algal "footprint" in chromalveolates possessing a red algal-derived secondary plastid had in fact been observed previously (Frommolt et al. 2008), and Moustafa et al. interpret it as evidence for a cryptic green algal endosymbiont present in an ancient chromalveolate ancestor prior to the hypothesized red algal endosymbiotic event (Moustafa et al. 2009). Under such a model, the composition of the chlorarachniophyte nuclear genome would have conceivably been impacted by no fewer than three secondary endosymbionts at different times (green, red, then green again; Elias and Archibald 2009). Against an increasingly supported backdrop of HGT in chlorarachniophytes and other eukaryotes (Keeling and Palmer 2008; Takishita et al. 2009), it is not clear how such a hypothesis can be rigorously tested.

A similar challenge exists when probing the nuclear genomes of plastid-lacking eukaryotes for the "footprint" of past endosymbioses. Consistent with the chromalveolate hypothesis, it was proposed that the genome of the stramenopile *Phytophthora* contains hundreds of genes of algal/cyanobacterial ancestry, evidence for a plastid-bearing phase in its history (Tyler et al. 2006). However, reanalysis of the data by Stiller et al. (2009) indicates that the number of algal-like genes in *Phytophthora* does not in fact rise above "background," i.e., the number of algal genes found in the genomes of amoebozoans (Fig. 2.1), which would *not* be expected to possess an endosymbiotic footprint (Elias and Archibald 2009; Stiller et al. 2009). Similar concerns exist for the putative algal/cyanobacterial footprints in the genomes of ciliates (Archibald 2008; Reyes-Prieto et al. 2008), the apicomplexan *Cryptosporidium* (Huang et al. 2004), and other plastid-lacking eukaryotes (e.g., Maruyama et al. 2008, 2009).

2.5   Future Directions

The amount of genomic data with which to test hypotheses about the origin and evolution of plastids has increased tremendously. However, if the past decade of research in this area has revealed anything, it is that more data does not always lead to increased clarity. Detailed analyses of complete algal nuclear genome sequences have uncovered an unexpected degree of genome mosaicism, and there is as yet no clear consensus as to what it means. Distinguishing between *bona fide* past endosymbioses versus HGT-derived genomic footprints is a formidable challenge that will require a combination of further refinement and implementation of a priori testing procedures of the sort used by Stiller et al. (2009) and even more data from diverse primary and secondary plastid-bearing lineages. Fortunately, continued technological advances in DNA sequencing mean that virtually *any* eukaryote, no matter how large its genome, will become a viable target for whole genome and/or near-complete transcriptome sequencing in the very near future. Particularly important organisms and lineages include (a) the red algae, for which there is still only a single complete genome available (Matsuzaki et al. 2004), (b) the photosynthetic alveolate *Chromera* (Moore et al. 2008), (c) the plastid-lacking cryptomonad *Goniomonas* and other phagotrophs currently classified as Hacrobia (Okamoto et al. 2009), (d) various photosynthetic and nonphotosynthetic dinoflagellates, and (e) plastid-lacking lineages within the stramenopiles (e.g., bicosoecids). With so much new data on the horizon, it is difficult to predict which hypotheses will still be "in play" even a few years from now, especially considering that organisms are being discovered on a regular basis. One recent such example is the "rappemonads," an as-yet uncultured lineage defined solely on the basis of environmental plastid rDNA operon sequencing and fluorescence in situ hybridization (Kim et al. 2010). Rappemonads are most closely related to, but are clearly distinct from, the haptophytes, and constitute a genetically diverse lineage found in both marine and freshwater environments. It is sobering to consider that organisms that represent potentially important pieces of the endosymbiosis puzzle have escaped detection for decades.

Acknowledgments I thank members of the Archibald Laboratory for stimulating discussions on plastid evolution, endosymbiosis, and comparative genomics. Dr. Eunsoo Kim is thanked for helpful comments on an earlier version of this chapter. I gratefully acknowledge the Natural Science and Engineering Research Council of Canada and the Canadian Institutes of Health Research (CIHR) for funding, the CIHR New Investigator Program for salary support, and the Canadian Institute for Advanced Research, Program in Integrated Microbial Biodiversity, for Fellowship support.

References

Andersson JO, Roger AJ (2002) A cyanobacterial gene in nonphotosynthetic protists – an early chloroplast acquisition in eukaryotes? Curr Biol 12:115–119

Archibald JM (2005) Jumping genes and shrinking genomes – probing the evolution of eukaryotic photosynthesis using genomics. IUBMB Life 57:539–547

Archibald JM (2006) Algal genomics: examining the imprint of endosymbiosis. Curr Biol 16:R1033–R1035

Archibald JM (2008) Plastid evolution: remnant algal genes in ciliates. Curr Biol 18:R663–R665

Archibald JM (2009) The puzzle of plastid evolution. Curr Biol 19:R81–R88

Archibald JM, Lane CE (2009) Going, going, not quite gone: nucleomorphs as a case study in nuclear genome reduction. J Hered 100:582–590

Archibald JM, Rogers MB, Toop M, Ishida K, Keeling PJ (2003) Lateral gene transfer and the evolution of plastid-targeted proteins in the secondary plastid-containing alga *Bigelowiella natans*. Proc Natl Acad Sci U S A 100:7678–7683

Baurain D, Brinkmann H, Petersen J, Rodriguez-Ezpeleta N, Stechmann A, Demoulin V, Roger AJ, Burger G, Lang BF, Philippe H (2010) Phylogenomic evidence for separate acquisition of plastids in cryptophytes, haptophytes, and stramenopiles. Mol Biol Evol 27:1698–1709

Bhattacharya D, Archibald JM (2007) Response to Theissen and Martin: "the difference between endosymbionts and organelles". Curr Biol 16:R1017–R1018

Blanchard JL, Hicks JS (1999) The non-photosynthetic plastid in malarial parasites and other apicomplexans is derived from outside the green plastid lineage. J Eukaryot Microbiol 46:367–375

Blankenship RE (1994) Protein structure, electron transfer and evolution of prokaryotic photosynthetic reaction centers. Antonie van Leeuwenhoek 65:311–329

Bodyl A (2005) Do plastid-related characters support the chromalveolate hypothesis? J Phycol 41:712–719

Bodyl A (2006) Did the peridinin plastid evolve through tertiary endosymbiosis? A hypothesis. Eur J Phycol 41:435–448

Bodyl A, Mackiewicz P, Stiller JW (2007) The intracellular cyanobacteria of *Paulinella chromatophora*: endosymbionts or organelles? Trends Microbiol 15:295–296

Bodyl A, Stiller JW, Mackiewicz P (2009) Chromalveolate plastids: direct descent or multiple endosymbioses? Trends Ecol Evol 24:119–121

Bodyl A, Mackiewicz P, Stiller JW (2010) Comparative genomic studies suggest that the cyanobacterial endosymbionts of the amoeba *Paulinella chromatophora* possess an import apparatus for nuclear-encoded proteins. Plant Biol (Stuttg) 12:639–649

Burki F, Shalchian-Tabrizi K, Minge M, Skjaeveland Å, Nikolaev SI, Jakobsen KS, Pawlowski J (2007) Phylogenomics reshuffles the eukaryotic supergroups. PLoS One 8:e790

Burki F, Shalchian-Tabrizi K, Pawlowski J (2008) Phylogenomics reveals a new 'megagroup' including most photosynthetic eukaryotes. Biol Lett 4:366–369

Burki F, Inagaki Y, Brate J, Archibald JM, Keeling PJ, Cavalier-Smith T, Sakaguchi M, Hashimoto T, Horak A, Kuma K, Klaveness D, Jakobsen KS, Pawlowski J, Shalchian-Tabrizi K (2009) Large-scale phylogenomic analyses reveal that two enigmatic protist lineages, telonemia and centroheliozoa, are related to photosynthetic chromalveolates. Genome Biol Evol 1:231–238

Cavalier-Smith T (1999) Principles of protein and lipid targeting in secondary symbiogenesis: euglenoid, dinoflagellate, and sporozoan plastid origins and the eukaryote family tree. J Eukaryot Microbiol 46:347–366

Cavalier-Smith T (2000) Membrane heredity and early chloroplast evolution. Trends Plant Sci 5:174–182

Chesnick JM, Hooistra WH, Wellbrock U, Medlin LK (1997) Ribosomal RNA analysis indicates a benthic pennate diatom ancestry for the endosymbionts of the dinoflagellates *Peridinium foliaceum* and *Peridinium balticum* (Pyrrhophyta). J Eukaryot Microbiol 44:314–320

Daugbjerg N, Andersen RA (1997) Phylogenetic analyses of the *rbcL* sequences from haptophytes and heterokont algae suggest their chloroplasts are unrelated. Mol Biol Evol 14:1242–1251

Delwiche CF, Kuhsel M, Palmer JD (1995) Phylogenetic analysis of *tufA* sequences indicates a cyanobacterial origin of all plastids. Mol Phylogenet Evol 4:110–128

Delwiche C, Andersen RA, Bhattacharya D, Mishler BD (2004) Algal evolution and the early radiation of green plants. In: Cracraft J, Donoghue MJ (eds) Assembling the tree of life. Oxford University Press, New York, pp 121–137

Dodge JD (1969) Observations on the fine structure of the eyespot and associated organelles in the dinoflagellate *Glenodinium foliaceum*. J Cell Sci 5:479–493

Douglas SE, Penny SL (1999) The plastid genome of the cryptophyte alga, *Guillardia theta*: complete sequence and conserved synteny groups confirm its common ancestry with red algae. J Mol Evol 48:236–244

Douglas SE, Murphy CA, Spencer DF, Gray MW (1991) Cryptomonad algae are evolutionary chimaeras of two phylogenetically distinct unicellular eukaryotes. Nature 350:148–151

Douglas SE, Zauner S, Fraunholz M, Beaton M, Penny S, Deng L, Wu X, Reith M, Cavalier-Smith T, Maier U-G (2001) The highly reduced genome of an enslaved algal nucleus. Nature 410:1091–1096

Durnford DG, Deane JA, Tan S, McFadden GI, Gantt E, Green BR (1999) A phylogenetic assessment of the eukaryotic light-harvesting antenna proteins, with implications for plastid evolution. J Mol Evol 48:59–68

Elias M, Archibald JM (2009) Sizing up the genomic footprint of endosymbiosis. Bioessays 31:1273–1279

Fast NM, Kissinger JC, Roos DS, Keeling PJ (2001) Nuclear-encoded, plastid-targeted genes suggest a single common origin for apicomplexan and dinoflagellate plastids. Mol Biol Evol 18:418–426

Frommolt R, Werner S, Paulsen H, Goss R, Wilhelm C, Zauner S, Maier UG, Grossman AR, Bhattacharya D, Lohr M (2008) Ancient recruitment by chromists of green algal genes encoding enzymes for carotenoid biosynthesis. Mol Biol Evol 25:2653–2667

Funes S, Davidson E, Reyes-Prieto A, Magallón S, Herion P, King MP, Gonzalez-Halphen D (2002) A green algal apicoplast ancestor. Science 298:2155

Gibbs SP (1978) The chloroplasts of *Euglena* may have evolved from symbiotic green algae. Can J Bot 56:2883–2889

Gilson PR, McFadden GI (1996) The miniaturized nuclear genome of a eukaryotic endosymbiont contains genes that overlap, genes that are cotranscribed, and the smallest known spliceosomal introns. Proc Natl Acad Sci U S A 93:7737–7742

Gilson PR, Su V, Slamovits CH, Reith ME, Keeling PJ, McFadden GI (2006) Complete nucleotide sequence of the chlorarachniophyte nucleomorph: nature's smallest nucleus. Proc Natl Acad Sci U S A 103:9566–9571

Glockner G, Rosenthal A, Valentin K (2000) The structure and gene repertoire of an ancient red algal plastid genome. J Mol Evol 51:382–390

Gould SB, Waller RF, McFadden GI (2008) Plastid evolution. Annu Rev Plant Biol 59:491–517

Graham LE, Wilcox LW (2000) Algae. Prentice-Hall, Upper Saddle River, NJ

Green BR, Durnford DG (1996) The chlorophyll-carotenoid proteins of oxygenic photosynthesis. Annu Rev Plant Physiol Plant Mol Biol 47:685–714

Greenwood AD (1974) The Cryptophyta in relation to phylogeny and photosynthesis. In: Sanders JV, Goodchild DJ (eds) Proceedings of the eighth international congress on electron microscopy, vol 2, Canberra, Australia, pp 566–567

Greenwood AD, Griffiths HB, Santore UJ (1977) Chloroplasts and cell compartments in Cryptophyceae. Br Phycol J 12:119

Hackett JD, Maranda L, Yoon HS, Bhattacharya D (2003) Phylogenetic evidence for the cryptophyte origin of the plastid of Dinophysis (Dinophysiales, Dinophyceae). J Phycol 39:440–448

Hackett JD, Anderson DM, Erdner DL, Bhattacharya D (2004) Dinoflagellates: a remarkable evolutionary experiment. Am J Bot 91:1523–1534

Hackett JD, Yoon HS, Li S, Reyes-Prieto A, Rummele SE, Bhattacharya D (2007) Phylogenomic analysis supports the monophyly of cryptophytes and haptophytes and the association of rhizaria with chromalveolates. Mol Biol Evol 24:1702–1713

Hampl V, Hug L, Leigh JW, Dacks JB, Lang BF, Simpson AG, Roger AJ (2009) Phylogenomic analyses support the monophyly of Excavata and resolve relationships among eukaryotic "supergroups". Proc Natl Acad Sci U S A 106:3859–3864

Helmchen TA, Bhattacharya D, Melkonian M (1995) Analyses of ribosomal RNA sequences from glaucocystophyte cyanelles provide new insights into the evolutionary relationships of plastids. J Mol Evol 41:203–210

Hibberd DJ, Norris RE (1984) Cytology and ultrastructure of *Chlorarachnion reptans* (Chlorarachniophyta divisio nova, Chlorarachniophyceae classis nova). J Phycol 20:310–330

Howe CJ, Barbrook AC, Nisbet RE, Lockhart PJ, Larkum AW (2008) The origin of plastids. Philos Trans R Soc Lond B Biol Sci 363:2675–2685

Huang J, Mullapudi N, Lancto CA, Scott M, Abrahamsen MS, Kissinger JC (2004) Phylogenomic evidence supports past endosymbiosis, intracellular and horizontal gene transfer in *Cryptosporidium parvum*. Genome Biol 5:R88

Imanian B, Keeling PJ (2007) The dinoflagellates *Durinskia baltica* and *Kryptoperidinium foliaceum* retain functionally overlapping mitochondria from two evolutionarily distinct lineages. BMC Evol Biol 7:172

Imanian B, Carpenter KJ, Keeling PJ (2007) Mitochondrial genome of a tertiary endosymbiont retains genes for electron transport proteins. J Eukaryot Microbiol 54:146–153

Inagaki Y, Dacks JB, Doolittle WF, Watanabe KI, Ohama T (2000) Evolutionary relationship between dinoflagellates bearing obligate diatom endosymbionts: insight into tertiary endosymbiosis. Int J Syst Evol Microbiol 50(Pt 6):2075–2081

Ishida K, Cao Y, Hasegawa M, Okada N, Hara Y (1997) The origin of chlorarachniophyte plastids, as inferred from phylogenetic comparisons of amino acid sequences of EF-Tu. J Mol Evol 45:682–687

Ishida K, Green BR, Cavalier-Smith T (1999) Diversification of a chimaeric algal group, the chlorarachniophytes: phylogeny of nuclear and nucleomorph small-subunit rRNA genes. Mol Biol Evol 16:321–331

Janouskovec J, Horak A, Obornik M, Lukes J, Keeling PJ (2010) A common red algal origin of the apicomplexan, dinoflagellate, and heterokont plastids. Proc Natl Acad Sci USA 107:10949–10954

Jarvis P, Soll J (2001) Toc, Tic, and chloroplast protein import. Biochim Biophys Acta 1541:64–79

Keeling PJ (2008) Evolutionary biology: bridge over troublesome plastids. Nature 451:896–897

Keeling PJ (2009) Chromalveolates and the evolution of plastids by secondary endosymbiosis. J Eukaryot Microbiol 56:1–8

Keeling PJ, Palmer JD (2008) Horizontal gene transfer in eukaryotic evolution. Nat Rev Genet 9:605–618

Khan H, Parks N, Kozera C, Curtis BA, Parsons BJ, Bowman S, Archibald JM (2007) Plastid genome sequence of the cryptophyte alga *Rhodomonas salina* CCMP1319: lateral transfer of putative DNA replication machinery and a test of chromist plastid phylogeny. Mol Biol Evol 24:1832–1842

Kim E, Archibald JM (2009) Diversity and evolution of plastids and their genomes. In: Aronsson H, Sandelius AS (eds) The chloroplast-interactions with the environment. Springer, Berlin, pp 1–39

Kim E, Graham LE (2008) EEF2 analysis challenges the monophyly of Archaeplastida and Chromalveolata. PLoS One 3:e2621

Kim E, Harrison J, Sudek S, Jones MDM, Wilcox HM, Richards TA, Worden AZ, Archibald JM (2010) A new and diverse plastid-bearing branch on the eukaryotic tree of life. Proc Natl Acad Sci U S A 108:1496–1500

Köhler S, Delwiche CF, Denny PW, Tilney LG, Webster P, Wilson RJM, Palmer JD, Roos DS (1997) A plastid of probable green algal origin in apicomplexan parasites. Science 275:1485–1489

Lane CE, Archibald JM (2008) The eukaryotic tree of life: endosymbiosis takes its TOL. Trends Ecol Evol 23:268–275

Lane CE, van den Heuvel K, Kozera C, Curtis BA, Parsons B, Bowman S, Archibald JM (2007) Nucleomorph genome of *Hemiselmis andersenii* reveals complete intron loss and compaction as a driver of protein structure and function. Proc Natl Acad Sci U S A 104:19908–19913

Larkum AW, Lockhart PJ, Howe CJ (2007) Shopping for plastids. Trends Plant Sci 12:189–195

Lauterborn R (1895) Protozoenstudien II. Paulinella chromatophora nov. gen., nov. spec., ein beschalter Rhizopode des Subwassers mit blaugrunen chromatophorenartigen Einschlussen. Z Wiss Zool 59:537–544

Leander BS, Triemer RE, Farmer MA (2001) Character evolution in heterotrophic euglenids. Eur J Protistol 37:337–356

Lim L, McFadden GI (2010) The evolution, metabolism and functions of the apicoplast. Philos Trans R Soc Lond B Biol Sci 365:749–763

Longet D, Archibald JM, Keeling PJ, Pawlowski J (2003) Foraminifera and Cercozoa share a common origin according to RNA polymerase II hylogenies. Int J Syst Evol Microbiol 53:1735–1739

Mackiewicz P, Bodyl A (2010) A hypothesis for import of the nuclear-encoded PsaE protein of *Paulinella chromatophora* (Cercozoa, Rhizaria) into its cyanobacterial endosymbionts/plastids via the endomembrane system. J Phycol 46:847–859

Marin B, Nowack ECM, Melkonian M (2005) A plastid in the making: evidence for a second primary endosymbiosis. Protist 156:425–432

Martin W, Stoebe B, Goremykin V, Hansmann S, Hasegawa M, Kowallik KV (1998) Gene transfer to the nucleus and the evolution of chloroplasts. Nature 393:162–165

Martin W, Rujan T, Richly E, Hansen A, Cornelsen S, Lins T, Leister D, Stoebe B, Hasegawa M, Penny D (2002) Evolutionary analysis of *Arabidopsis*, cyanobacterial, and chloroplast genomes reveals plastid phylogeny and thousands of cyanobacterial genes in the nucleus. Proc Natl Acad Sci U S A 99:12246–12251

Maruyama S, Misawa K, Iseki M, Watanabe M, Nozaki H (2008) Origins of a cyanobacterial 6-phosphogluconate dehydrogenase in plastid-lacking eukaryotes. BMC Evol Biol 8:151

Maruyama S, Matsuzaki M, Misawa K, Nozaki H (2009) Cyanobacterial contribution to the genomes of the plastid-lacking protists. BMC Evol Biol 9:197

Matsuzaki M, Misumi O, Shin IT, Maruyama S, Takahara M, Miyagishima SY, Mori T, Nishida K, Yagisawa F, Nishida K, Yoshida Y, Nishimura Y, Nakao S, Kobayashi T, Momoyama Y, Higashiyama T, Minoda A, Sano M, Nomoto H, Oishi K, Hayashi H, Ohta

F, Nishizaka S, Haga S, Miura S, Morishita T, Kabeya Y, Terasawa K, Suzuki Y, Ishii Y, Asakawa S, Takano H, Ohta N, Kuroiwa H, Tanaka K, Shimizu N, Sugano S, Sato N, Nozaki H, Ogasawara N, Kohara Y, Kuroiwa T (2004) Genome sequence of the ultrasmall unicellular red alga *Cyanidioschyzon merolae* 10D. Nature 428:653–657

McEwan ML, Keeling PJ (2004) HSP90, tubulin and actin are retained in the tertiary endosymbiont genome of *Kryptoperidinium foliaceum*. J Eukaryot Microbiol 51:651–659

McFadden GI (1999) Plastids and protein targeting. J Eukaryot Microbiol 46:339–346

McFadden GI, van Dooren GG (2004) Evolution: red algal genome affirms a common origin of all plastids. Curr Biol 14:R514–R516

McFadden GI, Gilson PR, Hofmann CJ, Adcock GJ, Maier UG (1994) Evidence that an amoeba acquired a chloroplast by retaining part of an engulfed eukaryotic alga. Proc Natl Acad Sci USA 91:3690–3694

Melkonian M, Mollenhauer D (2005) Robert Lauterborn (1869–1952) and his *Paulinella chromatophora*. Protist 156:253–262

Mereschkowsky C (1905) Über Natur und Ursprung der Chromatophoren im Pflanzenreiche. Biol Centralbl 25:593–604

Minge MA, Shalchian-Tabrizi K, Torresen OK, Takishita K, Probert I, Inagaki Y, Klaveness D, Jakobsen KS (2010) A phylogenetic mosaic plastid proteome and unusual plastid-targeting signals in the green-colored dinoflagellate *Lepidodinium chlorophorum*. BMC Evol Biol 10:191

Moore CE, Archibald JM (2009) Nucleomorph genomes. Annu Rev Genet 43:251–264

Moore RB, Obornik M, Janouskovec J, Chrudimsky T, Vancova M, Green DH, Wright SW, Davies NW, Bolch CJ, Heimann K, Slapeta J, Hoegh-Guldberg O, Logsdon JM, Carter DA (2008) A photosynthetic alveolate closely related to apicomplexan parasites. Nature 452:900

Moustafa A, Reyes-Prieto A, Bhattacharya D (2008) Chlamydiae has contributed at least 55 genes to Plantae with predominantly plastid functions. PLoS One 3:e2205

Moustafa A, Beszteri B, Maier UG, Bowler C, Valentin K, Bhattacharya D (2009) Genomic footprints of a cryptic plastid endosymbiosis in diatoms. Science 324:1724–1726

Müller KM, Oliveira MC, Sheath RG, Bhattacharya D (2001) Ribosomal DNA phylogeny of the Bangiophycidae (Rhodophyta) and the origin of secondary plastids. Am J Bot 88:1390–1400

Nakayama T, Ishida K (2009) Another acquisition of a primary photosynthetic organelle is underway in *Paulinella chromatophora*. Curr Biol 19:R284–R285

Nikolaev SI, Berney C, Fahrni JF, Bolivar I, Polet S, Mylnikov AP, Aleshin VV, Petrov NB, Pawlowski J (2004) The twilight of Heliozoa and rise of Rhizaria, an emerging supergroup of amoeboid eukaryotes. Proc Natl Acad Sci U S A 101:8066–8071

Nowack ECM, Melkonian M, Glöckner G (2008) Chromatophore genome sequence of *Paulinella* sheds light on acquisition of photosynthesis by eukaryotes. Curr Biol 18:410–418

Nowack EC, Vogel H, Groth M, Grossman AR, Melkonian M, Glöckner G (2011) Endosymbiotic gene transfer and transcriptional regulation of transferred genes in *Paulinella chromatophora*. Mol Biol Evol 28:407–422

Nozaki H, Matsuzaki M, Takahara M, Misumi O, Kuroiwa H, Hasegawa M, Shin-i T, Kohara Y, Ogasawara N, Kuroiwa T (2003) The phylogenetic position of red algae revealed by multiple nuclear genes from mitochondria-containing eukaryotes and an alternative hypothesis on the origin of plastids. J Mol Evol 56:485–497

Nozaki H, Takano H, Misumi O, Terasawa K, Matsuzaki M, Maruyama S, Nishida K, Yagisawa F, Yoshida Y, Fujiwara T, Takio S, Tamura K, Chung SJ, Nakamura S, Kuroiwa H, Tanaka K, Sato N, Kuroiwa T (2007) A 100%-complete sequence reveals unusually simple genomic features in the hot-spring red alga *Cyanidioschyzon merolae*. BMC Biol 5:28

Okamoto N, Chantangsi C, Horak A, Leander BS, Keeling PJ (2009) Molecular phylogeny and description of the novel katablepharid *Roombia truncata* gen. et sp. nov., and establishment of the Hacrobia taxon nov. PLoS One 4(9):e7080

Oliveira MC, Bhattacharya D (2000) Phylogeny of the Bangiophycidae (Rhodophyta) and the secondary endosymbiotic origin of algal plastids. Am J Bot 87:482–492

Palmer JD (2003) The symbiotic birth and spread of plastids: how many times and whodunnit? J Phycol 39:4–11

Parfrey LW, Grant J, Tekle YI, Lasek-Nesselquist E, Morrison HG, Sogin ML, Patterson DJ, Katz LA (2010) Broadly sampled multigene analyses yield a well-resolved eukaryotic tree of life. Syst Biol 59:518–533

Park MG, Park JS, Kim M, Yih W (2008) Plastid dynamics during survival of *Dinophysis caudata* without its ciliate prey. J Phycol 44:1154–1163

Patron NJ, Rogers MB, Keeling PJ (2004) Gene replacement of fructose-1,6-bisphosphate aldolase supports the hypothesis of a single photosynthetic ancestor of chromalveolates. Eukaryot Cell 3:1169–1175

Patron NJ, Inagaki Y, Keeling PJ (2007) Multiple gene phylogenies support the monophyly of cryptomonad and haptophyte host lineages. Curr Biol 17:887–891

Ralph SA, van Dooren GG, Waller RF, Crawford MJ, Fraunholz MJ, Foth BJ, Tonkin CJ, Roos DS, McFadden GI (2004) Tropical infectious diseases: metabolic maps and functions of the *Plasmodium falciparum* apicoplast. Nat Rev Microbiol 2:203–216

Reyes-Prieto A, Hackett JD, Soares MB, Bonaldo MF, Bhattacharya D (2006) Cyanobacterial contribution to algal nuclear genomes is primarily limited to plastid functions. Curr Biol 16:2320–2325

Reyes-Prieto A, Weber AP, Bhattacharya D (2007) The origin and establishment of the plastid in algae and plants. Annu Rev Genet 41:147–168

Reyes-Prieto A, Moustafa A, Bhattacharya D (2008) Multiple genes of apparent algal origin suggest ciliates may once have been photosynthetic. Curr Biol 18:956–962

Rice DW, Palmer JD (2006) An exceptional horizontal gene transfer in plastids: gene replacement by a distant bacterial paralog and evidence that haptophyte and cryptophyte plastids are sisters. BMC Biol 4:31

Rodriguez-Ezpeleta N, Brinkmann H, Burey SC, Roure B, Burger G, Löffelhardt W, Bohnert HJ, Philippe H, Lang BF (2005) Monophyly of primary photosynthetic eukaryotes: green plants, red algae, and glaucophytes. Curr Biol 15:1325–1330

Rodriguez-Ezpeleta N, Brinkmann H, Burger G, Roger AJ, Gray MW, Philippe H, Lang BF (2007a) Toward resolving the eukaryotic tree: the phylogenetic positions of jakobids and cercozoans. Curr Biol 17:1420–1425

Rodriguez-Ezpeleta N, Brinkmann H, Roure B, Lartillot N, Lang BF, Philippe H (2007b) Detecting and overcoming systematic errors in genome-scale phylogenies. Syst Biol 56:389–399

Rogers MB, Gilson PR, Su V, McFadden GI, Keeling PJ (2007) The complete chloroplast genome of the chlorarachniophyte *Bigelowiella natans*: evidence for independent origins of chlorarachniophyte and euglenid secondary endosymbionts. Mol Biol Evol 24:54–62

Sanchez-Puerta MV, Delwiche CF (2008) A hypothesis for plastid evolution in chromalveolates. J Phycol 44:1097–1107

Sanchez-Puerta MV, Bachvaroff TR, Delwiche CF (2007a) Sorting wheat from chaff in multigene analyses of chlorophyll c-containing plastids. Mol Phylogenet Evol 44:885–897

Sanchez-Puerta MV, Lippmeier JC, Apt KE, Delwiche CF (2007b) Plastid genes in a nonphotosynthetic dinoflagellate. Protist 158:105–117

Schnepf E, Elbrächter M (1988) Cryptophycean-like double membrane-bound plastid chloroplast in the dinoflagellate, *Dinophysis* Ehrenb.: evolutionary, phylogenetic and toxicological implications. Bot Acta 101:196–203

Slamovits CH, Keeling PJ (2008) Plastid-derived genes in the non-photosynthetic alveolate *Oxyrrhis marina*. Mol Biol Evol 25:1297–1306

Stelter K, El-Sayed NM, Seeber F (2007) The expression of a plant-type ferredoxin redox system provides molecular evidence for a plastid in the early dinoflagellate *Perkinsus marinus*. Protist 158:119–130

Stiller JW (2007) Plastid endosymbiosis, genome evolution and the origin of green plants. Trends Plant Sci 12:391–396

Stiller JW, Hall BD (1997) The origin of red algae: implications for plastid evolution. Proc Natl Acad Sci U S A 94:4520–4525

Stiller JW, Reel DC, Johnson JC (2003) A single origin of plastids revisited: convergent evolution in organellar genome content. J Phycol 39:95–105

Stiller JW, Huang J, Ding Q, Tian J, Goodwillie C (2009) Are algal genes in nonphotosynthetic protists evidence of historical plastid endosymbioses? BMC Genomics 10:484

Stoebe B, Kowallik KV (1999) Gene-cluster analysis in chloroplast genomics. Trends Genet 15:344–347

Suzuki K, Miyagishima SY (2010) Eukaryotic and eubacterial contributions to the establishment of plastid proteome estimated by large-scale phylogenetic analyses. Mol Biol Evol 27:581–590

Takishita K, Yamaguchi H, Maruyama T, Inagaki Y (2009) A hypothesis for the evolution of nuclear-encoded, plastid-targeted glyceraldehyde-3-phosphate dehydrogenase genes in "chromalveolate" members. PLoS One 4:e4737

Taylor FJR (1974) Implications and extensions of the serial endosymbiosis theory of the origin of eukaryotes. Taxon 23:229–258

Taylor FJR (1980) On dinoflagellate evolution. Biosystems 13:65–108

Teles-Grilo ML, Tato-Costa J, Duarte SM, Maia A, Casal G, Azevedo C (2007) Is there a plastid in *Perkinsus atlanticus* (Phylum Perkinsozoa)? Eur J Protistol 43:163–167

Tengs T, Dahlberg OJ, Shalchian-Tabrizi K, Klaveness D, Rudi K, Delwiche CF, Jakobsen KS (2000) Phylogenetic analyses indicate that the 19′hexanoyloxy-fucoxanthin- containing dinoflagellates have tertiary plastids of haptophyte origin. Mol Biol Evol 17:718–729

Theissen U, Martin W (2006) The difference between organelles and endosymbionts. Curr Biol 16:R1016–R1017; author reply R1017–R1018

Turmel M, Gagnon MC, O'Kelly CJ, Otis C, Lemieux C (2009) The chloroplast genomes of the green algae *Pyramimonas*, *Monomastix*, and *Pycnococcus* shed new light on the evolutionary history of prasinophytes and the origin of the secondary chloroplasts of euglenids. Mol Biol Evol 26:631–648

Tyler BM, Tripathy S, Zhang X, Dehal P, Jiang RH, Aerts A, Arredondo FD, Baxter L, Bensasson D, Beynon JL, Chapman J, Damasceno CM, Dorrance AE, Dou D, Dickerman AW, Dubchak IL, Garbelotto M, Gijzen M, Gordon SG, Govers F, Grunwald NJ, Huang W, Ivors KL, Jones RW, Kamoun S, Krampis K, Lamour KH, Lee MK, McDonald WH, Medina M, Meijer HJ, Nordberg EK, Maclean DJ, Ospina-Giraldo MD, Morris PF, Phuntumart V, Putnam NH, Rash S, Rose JK, Sakihama Y, Salamov AA, Savidor A, Scheuring CF, Smith BM, Sobral BW, Terry A, Torto-Alalibo TA, Win J, Xu Z, Zhang H, Grigoriev IV, Rokhsar DS, Boore JL (2006) *Phytophthora* genome sequences uncover evolutionary origins and mechanisms of pathogenesis. Science 313:1261–1266

Waller RF, McFadden GI (2005) The apicoplast: a review of the derived plastid of apicomplexan parasites. Curr Issues Mol Biol 7:57–79

Waller RF, Keeling PJ, van Dooren GG, McFadden GI (2003) Comment on "A green algal apicoplast ancestor". Science 301:49a

Watanabe MM, Suda S, Inouye I, Sawaguchi I, Chihara M (1990) *Lepidodinium viride* gen et sp. nov. (Gymnodiniales, Dinophyta), a green dinoflagellate with a chlorophyll *a*- and *b*-containing endosymbiont. J Phycol 26:741–751

Wisecaver JH, Hackett JD (2010) Transcriptome analysis reveals nuclear-encoded proteins for the maintenance of temporary plastids in the dinoflagellate *Dinophysis acuminata*. BMC Genomics 11:366

Yoon HS, Hackett JD, Pinto G, Bhattacharya D (2002) The single, ancient origin of chromist plastids. Proc Natl Acad Sci U S A 99:15507–15512

Yoon HS, Hackett JD, Ciniglia C, Pinto G, Bhattacharya D (2004) A molecular timeline for the origin of photosynthetic eukaryotes. Mol Biol Evol 21:809–818

Yoon HS, Reyes-Prieto A, Melkonian M, Bhattacharya D (2006) Minimal plastid genome evolution in the *Paulinella* endosymbiont. Curr Biol 16:R670–R672

Yoon HS, Nakayama T, Reyes-Prieto A, Andersen RA, Boo SM, Ishida K, Bhattacharya D (2009) A single origin of the photosynthetic organelle in different *Paulinella lineages*. BMC Evol Biol 9:98

Zhang Z, Green BR, Cavalier-Smith T (1999) Single gene circles in dinoflagellate chloroplast genomes. Nature 400:155–159

Zhang Z, Green BR, Cavalier-Smith T (2000) Phylogeny of ultra-rapidly evolving dinoflagellate chloroplast genes: a possible common origin for sporozoan and dinoflagellate plastids. J Mol Evol 51:26–40

Part II
Organelle Genome Evolution

Chapter 3
Unusual Mitochondrial Genomes and Genes

Gertraud Burger, Chris J. Jackson, and Ross F. Waller

3.1 Introduction

This chapter summarizes current knowledge on the diversity of mitochondrial DNAs (mtDNAs) with a focus on unusual genomes discovered in the last decade. For broader reviews on mtDNAs in general and for specialized reviews on the intriguing mitochondrial genomes of kinetoplastids, we refer the reader to earlier publications (Shapiro and Englund 1995; Lang et al. 1999; Burger et al. 2003b; Lukes et al. 2005).

In the context of this chapter, the term "unusual" combines various meanings. One is synonymous with *departing from the traditional view* on mtDNAs that was minted by the first published mitochondrial genomes in the 1980s and is still surviving in many textbooks. For example, "small is beautiful" was the title of a Nature news and views article (Borst and Grivell 1981) featuring the first report of a complete mitochondrial genome, that of human (Anderson et al. 1981). Not long after that, a number of other mammalian mtDNAs were sequenced cementing the impression that this genome is commonly a small circle of ~16 kbp including a dozen protein-coding and two dozen structural RNA genes. Another meaning of "unusual" denotes *difference compared to the majority* of mtDNAs. According to the compilation in NCBI's Genome section (subdivision "Organelles"), the large majority of mtDNAs are 15–17 kbp long. This is obviously due to the much biased taxonomic sampling, with more than 3,000 sequences from animals, but only 450 from the other (50 or so) eukaryotic groups. A third meaning of "unusual" is

G. Burger (✉)
Department of Biochemistry, Robert-Cedergren Centre for Bioinformatics and Genomics, Université de Montréal, 2900 Edouard-Montpetit, Montreal, QC, H3T 1J4, Canada
e-mail: gertraud.burger@umontreal.ca

C.J. Jackson • R.F. Waller
School of Botany, University of Melbourne, Melbourne, VIC 3010, Australia
e-mail: chrisjackson1245@gmail.com; r.waller@unimelb.edu.au

C.E. Bullerwell (ed.), *Organelle Genetics*, 41
DOI 10.1007/978-3-642-22380-8_3, © Springer-Verlag Berlin Heidelberg 2012

deviation from the ancestral state. Mitochondria originated from an endosymbiotic α-Proteobacterium, and the ancestral genome was likely a bacteria-like circular molecule of a few million base pair with a thousand or so genes.

Here, we focus on mitochondrial genome size, genome architecture, and gene structure, touching upon posttranscriptional gene expression only insofar as it relates to exceptional gene structures. Otherwise, expression mechanisms of mitochondrial genes are dealt with more broadly in Chaps. 10–12. We also compiled Internet-accessible data sources on mitochondrial genomes, listed in the Appendix. For bioinformatics tools used in mitochondrial genome annotation, we refer the reader to Chap. 17.

3.2 Taxonomic Background

Recently reported most exceptional mitochondrial genomes are from three different groups of unicellular eukaryotes, ichthyosporeans, diplonemids, and dinoflagellates. Before describing their mtDNAs, we will briefly introduce what these organisms look like, how and where they make their living, and where they belong in the eukaryotic phylogenetic tree.

Traditional classification has subdivided eukaryotes into animals, fungi, and plants, and termed the leftover "protists" – a panoply of obscure creatures belonging to none of the three former divisions. More recent phylogeny-based classifications have abandoned the notion of a protist *lineage*, while animals and fungi, for instance, are now united together with certain protists from which the former two groups emerged. This new clade known as opisthokonts is one of the groups relevant to the present chapter. The other clades of interest are euglenozoans and alveolates, and all three are phylogenetically extremely distant from one another (Fig. 3.1).

Amoebidium (Opisthokonta) is the best-described genus of ichthyosporeans and has been recognized not too long ago as a close unicellular relative of animals (Lang et al. 2002). This protist group has an amazing lifestyle and morphology. *Amoebidium* species populate the armor of freshwater crustaceans and insect larvae, hitch-hiking (epibionts) rather than parasitizing (Fig. 3.2). In nature, an *Amoebidium* cell grows as a tiny bush with a thick cell wall (filamentous microthallus) and contains multiple nuclei. For asexual reproduction, microthalli produce uninuclear, naked amoeboid cells (hence the genus name) or walled spores [Fig. 3.2b–d; (Lichtwardt 1986)]. Under rich culture conditions in the laboratory, we observe large spheres, inside which form dozens of small daughter cells that are eventually released [Fig. 3.2a (Jostensen et al. 2002; Ruiz-Trillo et al. 2007)]. Ichthyosporea are specifically related to *Capsaspora*, choanoflagellates and animals that together form the Holozoa (Lang et al. 2002; Ruiz-Trillo et al. 2008). The only ichthyosporean for which mitochondrial genome information is available is *Amoebidium parasiticum* (for references, see below).

Diplonemids (Euglenozoa) are the poorly known sister group of the notorious kinetoplastids. Species of the two diplonemid genera, *Diplonema* and *Rhynchopus*,

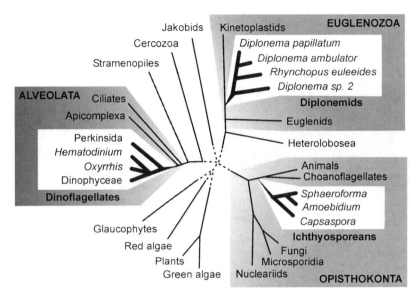

Fig. 3.1 Phylogenetic positions of ichthyosporeans, diplonemids, and dinoflagellates. *Dashed lines* indicate uncertain relationships

are abundant in marine habitats, but some occur also in freshwater. In contrast to kinetoplastids, which include numerous obligatory parasitic species pathogenic to humans, life stock and plants, diplonemids are free-living. They only occasionally parasitize lobsters, clams, and diatoms or cause the sudden decay of aquarium plants (Kent et al. 1987). Diplonemids in the "feeder" stage (under optimal growth conditions) are oval to pear-shaped with two flagella at one tip that are clearly visible in *Diplonema*, but concealed in *Rhynchopus* species [Fig. 3.3a–c; (Roy et al. 2007a)]. *Rhynchopus* has two morphologies. In addition to the moderately fast-moving, short-flagellated "feeder" under conditions when food is abundant, transformation is seen to a leaner, rapidly swimming "swarmer" with long flagella when nutrients are in short supply [Fig. 3.13d; (Vickerman 2000; Roy et al. 2007a)]. Mitochondrial genome information is available for three *Diplonema* species (*D. papillatum*, *D. ambulator*, *D.* sp. 2), and one from *Rhynchopus* (*R. euleeides*; for references, see below). The best characterized mtDNA is that of *D. papillatum*.

Dinoflagellates (Alveolata) are important unicellular organisms of the marine and aquatic ecosystems. This taxon is extremely diverse morphologically and biologically, including photosynthetic and heterotrophic species, predators, and symbionts. They not only are notorious for toxic red tides, but also engage in beneficial partnerships with reef-building corals. Photosynthetic dinoflagellates are major contributors to ocean carbon fixation. The great diversity of cell morphologies is achieved by flattened membrane sacs (alveoli) beneath the plasma membrane. These sacs may be rigid due to polysaccharides and form armors of most ludicrous shapes (Fig. 3.4). All dinoflagellates have two flagella that are

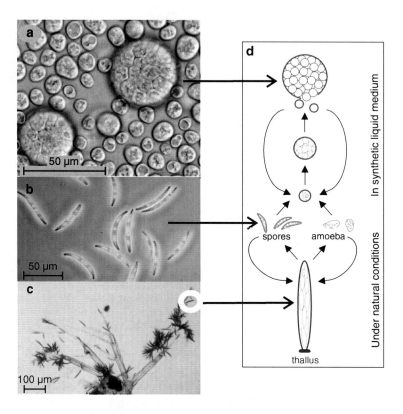

Fig. 3.2 *Amoebidium* cell morphology. (**a–c**) Light microscopy. (**a**) Spheric growth in synthetic medium (courtesy of Inaki Ruiz-Trillo). Shown is a relative of *Amoebidium*, *Spheroforma arctica*, which under these conditions, has the same morphology as *A. parasiticum*. (**b**) Sporangiospores released from an *A. parasiticum* thallus that grew on a waterflea. (**c**) *A. parasiticum* thalli sitting on the antennae of a waterflea (**b**, **c**, courtesy of Robert Lichtwardt). (**d**) *Amoebidium* lifecycle adapted from Lichtwardt (1973) by adding spheric forms that occur only in synthetic medium

inserted at the same point (in heterotrophic taxa at the "mouth"). One flagellum wraps around the cell, while the other is oriented perpendicularly to the first, and their combined action generates a whirling swimming pattern sometimes of prodigious speed. Investigation of dinoflagellate mitochondrial genomes was initiated in M.W. Gray's group (Norman and Gray 1997). Today, mtDNA data are available for about 15 different dinoflagellates species, and the ones with most genomic sequence available are *Amphidium carterae* (33 kbp), *Alexandrium catenella* (27 kbp), and *Karlodinium micrum* (25 kbp). Further mtDNA and/or EST data is available from *Crypthecodinium cohnii*, *Gonyaulax polyedra*, *Oxyrrhis marina*, *Prorocentrum micans*, *Katodinium rotundatum*, *Lingulodinium polyedrum*, *Heterocapsa triquetra*, *Pfiesteria piscicida*, *Karenia brevis*, *Symbiodinium*, and *Noctiluca scintillans* (for references see below).

Fig. 3.3 Diplonemid cell
morphology. (**a–c**) Light
microscopy of *Rhynchopus
euleeides*. (**a, b**) Feeder cells
dynamically changing
shape. (**c**) Swarmer cell.
(**d**) Scanning microscopy of
Diplonema papillatum
(courtesy of Brian Leander)

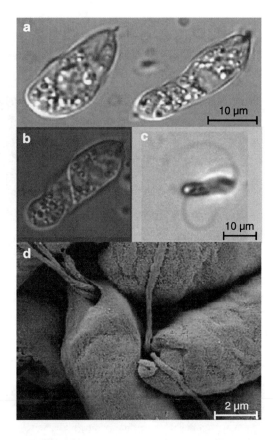

3.3 Approaches to Studying mtDNAs

When a newly described mtDNA is "usual," few will question whether the genome
is truly mitochondrial. Indeed, if the DNA has a "normal" circular-mapping shape
(see Sect. 3.3.2), and if there is really only one chromosome, then there would be
little reason to doubt it as the mitochondrial genome. Reservations only arise when
an apparent mtDNA does not conform. To substantiate nonconformity, a number of
biochemical methods are being employed.

3.3.1 Genome Localization

A major concern vis-à-vis an unusual mtDNA is whether the genome indeed resides
in mitochondria of the particular organism. Large chunks of mtDNA inserted in
nuclear genomes are relatively commonplace [e.g., *Arabidopsis* (Bensasson et al.
2001) and human (Richly and Leister 2004); for more details, see Chap. 7], and

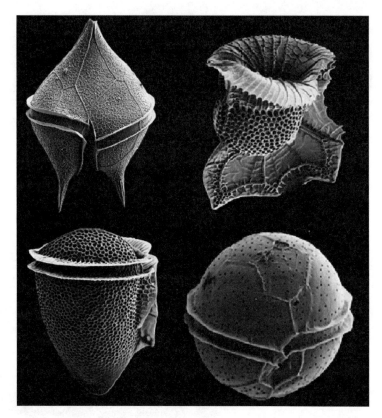

Fig. 3.4 Dinoflagellate cell morphology. Scanning electron microscopy of (clockwise from *top left*) *Protoperidinium claudicans*; *Ornithocercus* sp.; *Goniodema sphaericum*; *Phalachroma cuneus*. Images courtesy of David Hill

might be mistaken for the genuine organelle genome. Often used for separating mitochondrial DNA from nuclear DNA is elevated A + T content (i.e., in combination with dyes such as bisbenzimide or Hoechst dye that bind preferentially to A + T-rich DNA), or the propensity to form supercoils in the presence of intercalating dyes. Both alter the physical density of molecules, and CsCl equilibrium density centrifugation is able to distinguish these DNAs based on these differences. But these features will not stand critical perusal, because not all mtDNAs are richer in A + T than the nuclear DNA of a given organism, and only few mtDNAs can form supercoils as detailed in the next paragraph. A deviant mtDNA should be confirmed to come from isolated, pure mitochondria, or at least shown that it can be co-enriched with mitochondrial particles. For experimental methods, see (Lang and Burger 2007).

3.3.2   Mitochondrial Genome Shape

The topology of a genome can be inferred by various techniques, but on their own, each method has its particular limitations. For example, in DNA sequencing and restriction mapping, both rings and the more common linear tandem-arranged molecules appear as circular. Apparent circles determined by these two techniques should be referred to as circular-*mapping*. Circular and linear-tandem shapes are easily discernable by pulsed-field electrophoresis (PFE), but this technique cannot resolve more complicated shapes such as branched molecules. These latter forms are more readily discernable by electron microscopy (EM) or fluorescence microscopy [see e.g., (Bendich 1993)].

3.3.3   Number of Mitochondrial Chromosomes

The number of chromosomes is often inferred from PFE and EM, but these techniques determine the number of size *classes* and not of distinct chromosomes. To provide a comprehensive picture of a mitochondrial genome's architecture, it is necessary to combine the two methods with DNA sequencing. It should be noted that mitochondrial chromosome numbers reported in the literature are not always comparable, because they can be counted in different ways. Here, we define a mitochondrial chromosome as a (assumed) self-replicating unit, not considering recombination products. For instance, the multiple mtDNA molecules observed in angiosperms are recombination-generated subcircles of a single large mastercircle (Fauron et al. 1995); see discussion in Sugiyama et al. (2005), so the number of mitochondrial chromosome types in these plants is one. Another special case is *Dicyema*, a mesozoan animal of uncertain phylogenetic affiliation. Its mtDNA is a collection of small circles each carrying a single gene (Watanabe et al. 1999). Analysis of different tissues showed that the germ line of these animals contains one type of a larger master molecule, while the reported mitochondrial minicircles reside only in differentiated somatic cells that no longer replicate mtDNA (Awata et al. 2005). Again, under the above premise, the mitochondrial genome of *Dicyema* consists of a single chromosome.

Furthermore, we distinguish between the number of distinct chromosome types and the number of identical copies. For example, a trypanosome mtDNA is sometimes described as being composed of ~25 mitochondrial maxicircles (the molecule carrying the genes encoding proteins and structural RNAs) and ~5,000 minicircles [e.g., (Lukes et al. 1998)]. However, the figure for maxicircles refers to the number of identical copies, while that for minicircles, which carry gRNA genes involved in RNA editing, represents the total count of molecules including a hundred or so distinct types (Simpson 1997).

3.4   Mitochondrial Genome Architecture and Genome Size

Prevailing throughout eukaryotes is a single type of mitochondrial chromosome (in multiple copies), and this is presumably the ancestral state inherited from the mitochondrion's bacterial predecessor. There are, however, several exceptions to this rule in animals, fungi, and green algae, where canonical mitochondrial genes are distributed over a few, and in one case 18, chromosomes (Table 3.1). As we will see below, the mitochondrial genes from *Amoebidium*, diplonemids, and dinoflagellates are spread out over many more chromosomes and in most peculiar ways.

The majority of mtDNAs are circular-mapping with a physical shape of linear, head-to-tail concatenated molecules (Bendich 1993). Concatemers are likely the product of rolling-circle DNA replication as demonstrated in yeast (Ling and Shibata 2004) and the liverwort *Marchantia* (Oldenburg and Bendich 2001). Less frequent are truly circular mitochondrial chromosomes and monomeric linear molecules that occur sporadically across the eukaryotic tree [(Valach et al. 2011) and references therein; for multipartite genomes, see Table 3.1].

A size of 15–20 kbp – as seen in many animals – has long been regarded typical for mitochondrial genomes, but more recent data have changed this view dramatically. In animals alone, mtDNAs range in size from 11 kbp [(*Sagitta decipiens* (chaetognath) (Miyamoto et al. 2010)] to 43 kbp [(*Trichoplax* (placozoan) (Dellaporta et al. 2006; Burger et al. 2009)]. For eukaryotes as a whole, the smallest mtDNAs are found in *Plasmodium* and relatives, with 6 kbp only (Feagin et al. 1991), and the largest in the cucumber family with ~3,000 kbp (Ward et al. 1981). Currently, the largest fully sequenced mtDNA is that of pumpkin (*Cucurbita pepo*) with ~1,000 kbp (Alverson et al. 2010). In the face of such diversity, there is little meaning in sustaining the notion of a "usual" mitochondrial genome architecture and size.

3.4.1   Multiparite Genome Architectures in **Amoebidium**, *Diplonemids, and Dinoflagellates*

In the following, we will describe the makeup of the mitochondrial genomes in *Amoebidium*, diplonemids, and dinoflagellates. In all three cases, a novel, deviant mtDNA architecture has been observed, and the genome has been demonstrated to be located indeed inside mitochondria, with experimental evidence available for *Amoebidium parasiticum* (Burger et al. 2003a), the diplonemid *D. papillatum* (Marande et al. 2005), and the dinoflagellate *C. cohnii* (Jackson et al. 2007).

The mitochondrial genome of *Amoebidium* is extremely large. As of today, ~170 kbp of the genome have been sequenced, but it is still incomplete and the total size is unknown. There are probably several hundred different types of chromosomes that are all (monomeric) linear ranging in size from ~7 to 0.2 kbp (Fig. 3.5, left panel). Agarose gel electrophoresis visibly separates only the ten or so

Table 3.1 Multipartite mitochondrial genomes

Species (taxonomic group)	Chromosome		Shape	Total mtDNA size	References
	Number	Size			
Hydra attenuata, H. litteralis, H. magnipapillata (cnidarians, animals)	2	~8 kbp	Linear	~16 kbp	Warrior and Gall (1985); Voigt et al. (2008)
Globodera pallida (nematodes, animals)	>6	6.3–9.5 kbp	Circular	>46 kbp	Armstrong et al. (2000)
Brachionus plicatilis (rotifers, animals)	2	12.7 kbp, 11.2 kbp	Circular	11,153 kbp	Suga et al. (2008)
Pediculus humanus (arthropods, animals)	18	3–4 kbp	Circular	50–60 kbp	Shao et al. (2009)
Spizellomyces punctatus (Spizellomycetes, Fungi)	3	1.1 kbp, 1.4 kbp, 55.8 kbp	Circular	61.4 kbp	Lang et al. (1999)
Polytomella parva (Volvocales, Chlorophyceae)	2	3.5 kbp, 13.5 kbp	Linear	17 kbp	Fan (2002); Smith (2010)
Amoebidium parasiticum (Ichthyosporea, opisthokonts)	Several hundred	0.2–7 kbp	Linear	Very large	Burger et al. (2003a)
Diplonemids (Euglenozoa)	~100	5–12 kbp	Circular	~600 kbp	Vlcek et al. (2011); Kiethega et al. (2011)
Dinoflagellates (alveolates)	Numerous "DNA elements"	0.5–30 kbp	Linear	Very large	Norman and Gray (2001); Jackson et al. (2007); Nash et al. (2007)

Fig. 3.5 Appearance of
mitochondrial DNA.
Electrophoretic separation on
agarose gel of mtDNAs from
Amoebidium parasiticum
(Amo), *Diplonema
papillatum* (Dip), and the
dinoflagellate *A. carterae*
[(Din); courtesy of Ellen
Nisbet]

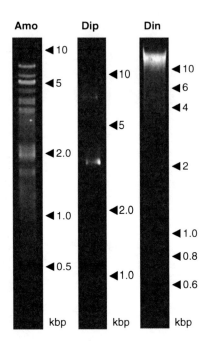

largest chromosomes (7–3.5 kbp), while the smaller ones (<3.5 kbp) are so
numerous that they appear as a contiguous smear. Physical size determination of
the linear chromosomes and size estimation based on DNA sequencing are in full
agreement (Burger et al. 2003a).

In *D. papillatum*, ~220 kbp of mtDNA have been sequenced and the estimated
genome size is in the order of 600 kbp. This genome appears to possess somewhat
fewer distinct mitochondrial chromosomes than that of *Amoebidium*, with probably
about one hundred in total. But in contrast to *Amoebidium*, *Diplonema* molecules
are circular, and furthermore, they fall in only two different size classes of 6 and
7 kbp (Fig. 3.5, middle panel). Size and shape were determined by gel electrophoresis and EM, and distinctness of chromosomes within size classes was established
by DNA sequencing. Three other diplonemid species, *D. ambulator*, *D. sp. 2*, and
R. euleeides also have circular mitochondrial chromosomes of unequal size classes,
but sizes range between 5 and 10 kbp (Roy et al. 2007b; Kiethega et al. 2011).

Dinoflagellate mtDNAs have confused us from the outset. Southern hybridization
of uncut mtDNA reveals a continuum of molecule sizes from ~15 kbp (upper size
limit of resolution under the particular experimental conditions) down to 0.5 kbp
[Fig. 3.5, right panel; (Norman and Gray 2001; Jackson et al. 2007)], whereas
preliminary PFE separation in *A. cartera* indicates ~30-kbp molecules (Nash et al.
2007). In any case, DNA sequencing shows that genes occur as multiple copies all
in different contexts, probably reflecting a high level of recombination. Apparently,
this has resulted in inflated genome sizes, and sequencing of upward of 30 kbp in
some taxa continues to find novel genomic combinations (Nash et al. 2008; Waller
and Jackson 2009). Whether these hundreds of mitochondrial DNA molecules

detected in electrophoresis and Southern analyses are self-replicating units or recombination products of one or several master molecules remains obscure. Therefore, the notion of chromosomes is not applied in this system; instead, "mtDNA elements" is being used (Norman and Gray 2001; Jackson et al. 2007).

3.5 Noncoding Regions of mtDNA

Genomic sequences are subdivided into coding regions – which are the stretches occupied by genes specifying proteins and structural RNAs and may or may not contain introns – and noncoding regions. These latter harbor replication and transcription origins (known for only few mtDNAs), telomeric repeats in linear molecules, and then what is sometimes referred to as "junk" DNA, i.e., genome regions of unknown biological role. Note that experimental determination of essential and nonessential mtDNA regions is not readily tractable, because reverse genetics techniques are not available for the large majority of mitochondrial systems; and where established [yeast, *Chlamydomonas* (Butow and Fox 1990)], the methodology is extremely cumbersome.

Some lineages such as vertebrate animals, *Chlamydomonas*, and apicomplexans have streamlined their mtDNAs to a degree that genes are cramped with virtually no space in between. In other lineages, mtDNAs expand through accumulation not only of introns inserted in genes, but also of repeats and other untranscribed sequences between genes. Angiosperm mtDNAs also accumulate large quantities of mostly inactive chloroplast sequences [for a review, see (Kubo and Newton 2008)]. Noncoding regions are the major contributor to size differences of mtDNAs from closely related species. For example, the mtDNA of different *Schizosaccharomyces* species varies in size by a factor of four despite the nearly identical gene content (Bullerwell et al. 2003).

3.5.1 Conspicuous Noncoding mtDNA Regions of Amoebidium, Diplonemids, and Dinoflagellates

In all three taxa, *Amoebidium*, diplonemids, and dinoflagellates, it is the noncoding part of mtDNA that predominates.

In *Amoebidium*, the overall noncoding portion of its mtDNA is estimated at 80%, ranging between 20 and 100% for individual chromosomes. In fact, the majority of small chromosomes (>2 kbp) lacks recognizable coding sequence. The noncoding regions are structured in an amazingly regular pattern [Fig. 3.6a; (Burger et al. 2003a)]. At the very ends of linear mitochondrial chromosomes sits one copy each of a ~45-nt motif (termed repx) in inverted orientation that has the propensity to form a guanine quadruplex structure known to stabilize ends of nuclear

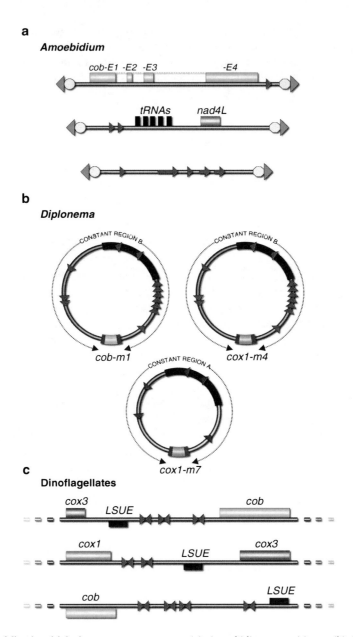

Fig. 3.6 Mitochondrial chromosome structures. (**a**) *Amoebidium parasiticum*; (**b**) *Diplonema papillatum*; (**c**) Dinoflagellates. Protein-coding sequence is shown as *shaded cylinders*, structural RNAs as *black boxes*. Repeats are shown by *arrows/arrowheads*. In *Amoebidium,* terminal repeats and subterminal motifs (large *triangles* and *circles/octagons*, respectively) are found on each chromosome. In diplonemid *circular* chromosomes the *black* region is common to all chromosomes (irrespective of their class), and the constant regions A and B are common to class A and B chromosomes, respectively. In dinoflagellates, head-to-head *arrows* indicate inverted repeats that are abundant in intergenic regions. *cob*-E1 to E4, exons of the *cob* gene. *LSUE*, fragment number 5 of the large subunit rRNA gene

chromosomes [Figs. 3.5, left panel and 3.6a, large arrowheads; (Bartoszewski et al. 2004)]. Terminal inverted repeats have been implicated in replication initiation (Pritchard and Cummings 1981) and in circularization of linear molecules prior to replication (Rycovska et al. 2004). The terminal repeats in *Amoebidium* mtDNA are flanked by subterminal motifs that are specific for the "left" (repa) and "right" (repb) ends of chromosomes (Fig. 3.6a, circles and octagons). These motifs are most likely involved in transcription initiation (repa) and termination (repb), as all genes are oriented in the direction repa → repb. In addition, *Amoebidium* mitochondrial chromosomes enclose in their central region at least 20 further repeat motifs (>50 bp) that all have the same orientation and are arranged either as dispersed single units or in tandem arrays (Fig. 3.6a, small arrows/arrowheads).

Diplonemid mtDNA contains ~95% noncoding sequence on each of its numerous circular chromosomes. These noncoding regions display a highly regular structure that is diametrically opposite from what is seen in *Amoebidium* [Fig. 3.6b; (Vlcek et al. 2011)]. In *D. papillatum*, the two size classes of circular chromosomes (6 kbp, Class A and 7 kbp, Class B) contain a contiguous stretch of approximately 5.7 and 6.7 kbp noncoding sequence that is nearly identical in molecules of the same class. These sequences include a moderate number (>10) of distinct repeat motifs (40 bp or longer), all oriented in the same direction, some dispersed others in tandem (Fig. 3.6b, arrowheads). Most conspicuous is a 2.2 kbp-long tandem array of a ~70-bp motif in class B chromosomes. Shared between Class A and B chromosomes is a stretch of 2.6 kbp, which likely includes the replication origin [Fig. 3.6b, black bars; (Vlcek et al. 2011)]. Noncoding sequence that is unique to individual chromosomes is on average only ~150 bp long and flanks the coding region on both sides (see below).

In dinoflagellates, noncoding regions of mtDNA are roughly estimated at 85% for *A. carterae* [the only taxon for which mtDNA has been purified and randomly sequenced; (Nash et al. 2008)]. A conspicuous feature of dinoflagellate noncoding mtDNA (noted from this species and *C. cohnii* and *K. micrum*) is the presence of numerous distinct repeat motifs, most in inverted orientation, with the propensity to form densely packed arrays of stem-loop structures that occasionally overlap genes [Fig. 3.6c; (Norman and Gray 2001; Jackson et al. 2007; Nash et al. 2007)]. Palindromic sequences have been reported before in mtDNAs from various other taxa and have been implicated in different biological processes. For example, the double-hairpin elements present in several fungal mtDNAs are believed to be mobile, spreading across considerable phylogenetic distances and mediating lateral gene transfer (Paquin et al. 2000; Bullerwell et al. 2003). In other organisms, palindromes have been proposed to play roles in mitochondrial recombination (Bartoszewski et al. 2004), replication (Kornberg and Baker 1992), chromosomal rearrangements (Lewis and Cote 2006), and transcript stability (Kuhn et al. 2001). Which role(s) palindromic sequences play in dinoflagellates is yet to be unraveled.

3.6 Mitochondrion-Encoded Genes

The noncoding regions substantially define the architecture and regulation of mitochondrial genomes, but it is the gene coding regions that define genome functionality. While the biological processes and gene sets residing on mtDNAs are generally confined to a handful of common categories, there is still considerable variation seen throughout the eukaryotic groups. Moreover, the structure of genes themselves can be subject to further modification.

3.6.1 Pathways and Biological Processes Involving mtDNA-Encoded Genes

Broad sampling of mtDNAs from throughout the eukaryotic tree shows that the pathways and biological processes involving mtDNA-encoded genes are confined to a very select set (Table 3.2). This is in spite of mitochondria performing numerous biological functions, the majority of which rely entirely on nucleus-encoded genes. Two processes that universally depend on at least some mtDNA-encoded genes are electron transport plus oxidative phosphorylation (often referred to collectively as OXPHOS), and mitochondrial translation. Whereas the first of these processes necessitates protein-encoding mitochondrial genes, the latter may be only represented by mitochondrial genes encoding structural RNAs, particularly the ribosomal RNAs (rRNAs) and often, but not always, tRNAs. The basic requirement of the mtDNA to service these two functions has been attributed to the necessity for fast, organelle-based regulation of the redox state of mitochondria through control of oxidative phosphorylation [see Chap. 5 on the CoRR hypothesis (Allen 2003)], and for the presumed difficulty in importing large structural RNAs such as the rRNAs. Mitochondrial genomes that encode only genes for OXPHOS and translation are broadly found among the "usual" mtDNAs of most animals and fungi, but also in more disparate groups including apicomplexans and chlorophycean algae. It is common, however, for mtDNAs to encode molecules involved in a small number of other processes (see Table 3.2). These generally include not only the transmembrane protein transport via the twin-arginine translocase, but also, rarely, the SecY-type transport system. In many lineages, the process of cytochrome c maturation, namely heme transport into the inner-membrane space and its covalent linkage to cytochrome c, are controlled by mitochondrial genes. An RNase P RNA for tRNA processing, and a cytochrome oxidase assembly protein, are specified by mtDNA in select lineages. Finally, mitochondrial genes for transcription have been detected, but in only a single lineage, the jakobids [for a review, see (Gray et al. 2004)].

All of the mitochondrial processes outlined above are of bacterial origin and are directly derived from the organelle's bacterial progenitor. The breadth of functions covered by mtDNA-encoded genes generally corresponds to their level of gene

Table 3.2 Gene content and biological processes encoded by mtDNAs across eukaryotic diversity[a]

Taxon	OxPhos[b]	rRNAs	Ribosomal proteins[c]	tRNAs[c]	TAT transport	SecY transport	Cytochrome c maturation	tRNA processing	CIV assembly	Transcription
Jakobida[d]	CI, II, III, IV, V	rns, rnl, rrn5	rps: (11–12), rpl: (11–15)	(25–26)	tatA, C	secY	ccmA-C, F	rnpB	cox11	rpoA-D
Heterolobosea[d]	CI, II, III, IV, V	rns, rnl	rps: (11), rpl: (6)	(20)	tatC	/	ccmC, F	/	cox11	/
Kinetoplastida	CI, III, IV, V	rns, rnl	rps: (1)	/	/	/	/	/	/	/
Apicomplexa	CIII, IV	rns, rnl	/	/	/	/	/	/	/	/
Ciliates	CI, III, IV, V	rns, rnl	rps: (4–5), rpl: (3–4)	4(–7)	/	/	ccmF	/	/	/
Oomycetes	CI, III, IV, V	rns, rnl	rps: (11), rpl: (5)	(23–25)	tatC	/	/	/	/	/
Bacillariophyta[e]	CI, III, IV, V	rns, rnl	rps: (11), rpl: (5)	(25)	tatA, C	/	/	/	/	/
Phaeophyceae[e]	CI, III, IV, V	rns, rnl, rrn5	rps: (11), rpl: (6)	(24–25)	tatC	/	/	/	/	/
Haptophyceae[e]	CI, III, IV, V	rns, rnl	rps: (4), rpl: (1)	(23)	/	/	/	/	/	/
Cryptophyta[e]	CI, II, III, IV, V	rns, rnl	rps: (10), rpl: (4)	(27)	tatA, C	/	/	/	/	/
Rhodophyta[e]	CI, II, III, IV, V	rns, rnl, rrn5	rps: (3–6), rpl: (1–5)	(23–25)	tatA, C	/	ccmA-C, F	/	/	/
Chlorophyceae[e]	CI, II, III, IV, (V)	rns, rnl	/	(1–26)	/	/	/	/	/	/
Prasinophyceae[e]	CI, III, IV, V	rns, rnl, rrn5	rps: (11), rpl: (4)	(25–26)	tatC	/	/	rnpB	/	/
Plants	CI, II, III, IV, V	rns, rnl, rrn5	rps: (4–12), rpl: (2–4)	(15–27)	tatC	/	ccmB, C, F	/	/	/
Amoebozoa[d]	CI, III, IV, V	rns, rnl, rrn5	rps: (9–10), rpl: (5–6)	(16–18)	/	/	/	/	/	/
Fungi	(CI), III, IV, V	rns, rnl	rps(0–1)	(7–26)	/	/	/	rnpB	/	/
Choanoflagellida	CI, III, IV, V	rns, rnl	rps: (7), rpl: (4)	(23)	tatC	/	/	/	/	/
Metazoa	CI, III, IV, (V)	rns, rnl	/	(0–25)	/	/	/	/	/	/

[a]Taxonomic groups have been chosen to represent major eukaryotic diversity but are not exhaustive. Gene and product names are rns, rnl, rrn5, small subunit, large subunit, and 5 S ribosomal rRNA; rps, rpl, small-subunit, and large-subunit ribosomal protein; tatA, C, sec-independent protein translocase components A and C; secY, sec-type transporter protein; ccmA-C, F, ABC transporter ATP-binding subunit, channel subunit, subunit C, and haem lyase; rpoA-D, RNA polymerase subunit alpha, beta, beta′ and sigma-like factor. Data are taken from http://gobase.bcm.umontreal.ca/searches/compliations.php and (Gray et al. 1998; Lang et al. 1999; Gray et al. 2004)

[b]The five oxidative phosphorylation ("OxPhos") complexes are indicated as CI-V. Bracketed complex numbers indicates variable presence in group members

[c]Figures in parentheses are the total number of distinct genes. Where these numbers are variable within an individual taxon, they are shown as a range

[d]Additional genes found in mtDNA are ssrA [tmRNA; Jakobida; (Jacob et al. 2004)] and, tufA [(elongation factor EFTu; Jakobida, Naegleria (Heterolobosea), and Hartmanella (Amoebozoa)]

[e]For a recent review on algal mtDNAs, see (Burger and Nedelcu 2011)

reduction, but not necessarily to their phylogenetic affinities. This is evidenced by several of the pathways being distributed seemingly randomly across disparate lineages (Table 3.2). Furthermore, it implies that the genes for these pathways persisted well after the radiation of most eukaryotic groups, and that loss from the mitochondrial genome has happened numerous times independently. Thus in terms of eukaryotic diversity there is really no such thing as a "usual" set of mitochondrion-encoded biological processes.

3.6.2 Gene Sets

The mtDNA-encoded genes representing the processes discussed above follow themselves a pattern where some are almost universally retained, and others show multiple instances of independent loss (Table 3.2). The large and small subunit rRNA genes (*rnl, rns*) are encoded on all known mtDNAs, whereas the gene for mitochondrial 5 S rRNA is found only sporadically [namely, in some plants, green, red, and brown algae, amoebozoans, and jakobids (Gray et al. 2004)]. The number of mtDNA-encoded tRNAs is quite variable. In many instances, the gene set serves all codons observed in mitochondrial protein-coding genes, but partial sets and complete absence is seen as well (see Chap. 17). Select genes for the respiratory chain complexes and oxidative phosphorylation are universally retained, while others show a hierarchically pattern of those frequently retained to those seldom retained on mtDNA. The genes for cytochrome b (*cob*) of Complex III, and cytochrome oxidase subunit 1 (*cox1*) of Complex IV reside in all, and additional genes for Complex IV (*cox2* and *cox3*) in most mtDNAs. Complex I subunits are mitochondrion-encoded in most eukaryotes, the basic gene set including *nad1, nad4,* and *nad5,* and in most cases also *nad2, nad3, nad4L,* and *nad6,* while further *nad* genes are less frequent. In some lineages such as apicomplexans and certain fungi in the Saccharomycetales, this complex has been entirely lost, and the function has been substituted by a nucleus-encoded single-subunit enzyme (van Dooren et al. 2006). Several genes for Complex V are typically located on mtDNA (notably *atp6, atp8,* and *atp9*), although mtDNA-encoded genes for this complex are completely absent from apicomplexans, some green alga and some animals. Finally, genes for Complex II are less common in mtDNAs, present only in a few lineages.

Genes for ribosomal proteins are the other major class of mtDNA-encoded genes (Table 3.2). While up to 27 such genes are found in jakobids, other lineages contain few or no such genes on their mtDNAs. When multiple ribosomal protein genes exist, genes for the small subunit are most common (notably *rps1-4, 7, 8, 11–14, 19*) with typically fewer genes for the large subunit (notably *rpl2, 5, 6, 11, 14, 16*). Analyses of various plant mtDNAs demonstrate well that mtDNA gene loss has taken place frequently and independently. Some ribosomal protein genes were lost over 40 different times within plants alone (Adams et al. 2002; Bergthorsson et al. 2003).

A small set of rarely occurring mitochondrial genes were probably gained secondarily through lateral gene transfer. These specify DNA and RNA polymerases (*dpo* and *rpo*) and maturase (*matR*) and reverse transcriptase (*rtl*) that are involved in intron propagation and splicing, and are typically comprised in introns, but sometimes free-standing. Rarely, mtDNAs encode DNA mismatch repair protein (*mutS*) and adenine methyl transferase (*dam* or *mtf*) (Gray et al. 2004).

ORFs (unidentified open reading frames) constitute the final gene class, and these are potential protein-encoding genes. They might be either unrecognized, highly divergent versions of common genes [e.g., the former *orfB* (*atp8*) (Gray et al. 1998), *ymf39* (*atp4*) (Burger et al. 2003c) and *murf1* (*nad2*) (Kannan and Burger 2008)], or open reading frames by chance, neither transcribed nor translated. The highest numbers of ORFs (up to 100 and more) are correlated with the inflated genome sizes seen in plants; the majority of these ORFs most likely occur by chance.

The mtDNAs with the largest gene count belong to the Jakobida, with 66 identified protein-encoding genes and a further 31 genes for structural RNAs (Lang et al. 1997). This genome contains all of the genes represented on any other mtDNA, in addition to unique ones such as *ssrA*, which specifies tmRNA that releases stalled ribosomes from "stop-less" mRNAs (Jacob et al. 2004). This most ancestral gene complement is restricted to the jakobids. At the other end of the spectrum are the more common minimal mtDNA with 4–9 genes: *rns, rnl, cob,* and *cox1,* and often also *cox2, cox3, nad1, nad4,* and *nad5.*

3.6.3 Gene Structure

The "prototype" gene structure is a contiguous coding sequence that corresponds to a single contiguous product. In mitochondria, the coding sequence may be interrupted by one or more Group I or Group II introns that are removed posttranscriptionally. Introns are most abundant in plants and fungi, and a few protist lineages [for a recent review, see (Lang et al. 2007)], but are completely lacking in other protists, e.g., apicomplexans, ciliates, and kinetoplastids.

Several mitochondrial genes have broken the basic convention of gene structure in a number of creative ways. Perhaps, the simplest of these is that genes have become discontinuous, resulting in multiple gene products instead of one. This is the case with rRNA genes that can be split into two pieces as for *rnl* in ciliates, green algae, fungi, and animals (Heinonen et al. 1987; Boer and Gray 1988; Nedelcu et al. 2000; Forget et al. 2002; Dellaporta et al. 2006), to more than 20 pieces as for the apicomplexan *rns* and *rnl* (Feagin et al. 1997). Gene pieces are usually arranged on mtDNAs in an unordered fashion, and the corresponding transcripts assemble in the ribosome through intermolecular interactions.

Protein-encoding genes are also known to fragment, with two possible outcomes for the protein product: either discontinuous or contiguous. For example, the *nad1* gene of ciliates is split into two coding sequences that apparently result in a split protein (Edqvist et al. 2000). A gene split can also be accompanied by relocation of one or both parts to the nucleus. This has happened several times independently in green algae, plants, amoebozoans, and apicomplexans (Nedelcu et al. 2000;

Adams et al. 2001; Funes et al. 2002a, b; Waller and Keeling 2006; Gawryluk and Gray 2010). The second outcome for fragmented genes is restoration of a single gene product at the RNA level, and the processes that enable this are discussed in Sect. 3.7.1.

A further variation on gene structure seen in mtDNAs is gene fusion. The Complex IV genes *cox1* and *cox2* are found in the same reading frame and produce a single mRNA in some amoebozoans (Burger et al. 1995; Ogawa et al. 2000). In *Acanthamoeba castellanii*, there is evidence that the two proteins are separate, but it is not known whether this is due to an unusual translation termination of the *cox1/cox2* mRNA or to posttranslational processing (Lonergan and Gray 1996).

3.6.3.1 Fragmented, Recombined, and Intact Gene Versions in *Amoebidium* mtDNA

In the currently sequenced portion of *Amoebidium* mtDNA, we find all genes known from animals, plus three ribosomal protein genes that are also present in their choanoflagellate neighbors (Burger et al. 2003a) (Table 3.3). Unlike mito-chondrial genes from the other Holozoa, those of *Amoebidium* enclose more than 20 introns that are predominantly of the Group I type. Surprisingly, *Amoebidium* mtDNA contains many gene fragments in addition to complete gene versions. For four *nad* genes (see Table 3.3), complete gene versions are missing and are thought to be located on as-yet-unsequenced chromosomes. It is also conceivable that that these genes are in the process of migrating to the nucleus leaving incomplete pseudogenes behind. This idea will be testable when more nuclear genome data become available.

The most abundant gene class in *Amoebidium* mtDNA is tRNA genes, existing in astounding numbers. Among the 85 tRNA-like sequences [identified by tRNAScan (Lowe and Eddy 1997)], 54 are *bona fide* functional genes and 31 are pseudo genes (Burger unpublished). The large majority (80%) of all tRNA genes (functional plus pseudo) reside on only three chromosomes in clusters of 20–25 genes (while protein-coding genes are single or grouped by two at most per chromosome). Functional tRNA genes occur in up to four almost identical copies and are often arranged in tandem. Pseudo tRNA genes appear to be mostly recombination products of two or more functional tRNAs. The explosion of tRNA-related sequences indicates ongoing and frequent recombination among and within chromosomes, likely facilitated by similar sequence motifs in tRNA genes as well as intergenic repeat elements.

3.6.3.2 Systematically Fragmented Mitochondrial Genes in Diplonemids

As in *Amoebidium*, the mtDNA of *D. papillatum* is not fully sequenced. The gene content in the currently known portion of *Diplonema* mtDNA (Vlcek et al. 2011) is not much different from that of its sister group, the kinetoplastids (Table 3.3), and

Table 3.3 Mitochondrial genes from amoebidium, diplonema, dinoflagellates, and their relatives[a]

Taxon	Genes contained in mtDNA					Fragmented protein genes	Number of mitochondrial chromosomes	mtDNA size
	Electron transport + oxidative phosphorylation	rRNA	Translation / Ribosomal proteins	tRNAs	Other			
Amoebidium parasiticum	*atp6,8,9; cob; cox1,2,3; nad3,4L,5,6*	*rnl, rns*	*rps3,4,13*	*trnA-trnY (18)*		Yes	Several hundred	Very large
Animals	*atp6, 8,[9]; cob; cox1,2,3; nad1,2,3,4,4L,5,6*	*rnl, rns*	/	*trnA-trnY (~22)*		No	1 (~16)	12 (~60) kbp
Diplonema papillatum	*atp6; cob; cox1,2,3; nad1,4,5,7,8*	*rnl, rns?*	/	/		Yes	About hundred	Very large
Heterolobosea	*atp1,3,6,8,9; cob; cox1,2,3; nad1,2,3,4,4L,5,6,7,8,9*	*rnl, rns*	*rpl2,5,6,11,14,16; rps2,3,4,7,8,10, 11,12,13,14,19*	*trnD-trnY (20)*	*cox11; sdh2*	No	1	50 kbp
Dinoflagellates[b]	*cob; cox1,3*	*rnl, rns*	/	/		Yes	Unknown	Very large
Ciliates	*atp9; cob; cox1,2,3; nad1,2,3,4,4L,5,6,7,9,11*	*rnl, rns*	*rpl2,6,14,16; rps3,12,13,14,19*	*trnE-trnY (7)*	*ccmF*	No	1	40–47 kbp

[a]Number of distinct genes (excluding ORFs). Note that mtDNAs from *Amoebidium*, diplonemids, and dinoflagellates have not been sequenced completely. For references, see text and Table 3.2. A *question mark* indicates that the gene is expected, but has not yet been detected

[b]Species studied are *A. carterae, A. catenella, K. micrum, C. cohnii, G. polyedra, O. marina, P. micans, K. rotundatum, L. polyedrum, H. triquetra, P. piscicida, K. brevis, Symbiodinium,* and *Noctiluca scintillans.* For references, see text

includes eleven protein-encoding genes with the common *atp6, cob, cox1-3*, and five *nad* genes. The situation for structural RNA genes is less clear. For *rnl*, only the 3' terminal portion of the gene has been detected. Apparently, LSU rRNA is fragmented, but the number of pieces is uncertain. The otherwise omnipresent and well conserved *rns* remains elusive. Transfer RNA genes seem to be missing from mtDNA of diplonemids, as is the case in kinetoplastids (Simpson et al. 1989).

Gene identification is most difficult in *Diplonema* mtDNA, not only because the sequences are highly divergent, but also because of a most startling dispersed-fragmented gene structure. In fact, all genes in *D. papillatum* mtDNA seem to be broken up into multiple pieces. But in contrast to LSU rRNA, protein-coding regions rejoin at the RNA level. The absence of complete protein gene versions from both the mitochondrial and nuclear genome has been confirmed by PCR on total cellular DNA.

Gene fragmentation in *D. papillatum* mtDNA is surprisingly regular, with pieces (also referred to as modules) of relatively constant size (on average 170 nt, +/−100), so that genes consist of a total of four (e.g., the small *atp9* gene) to twelve (the large *nad7* gene) parts. Even more surprising, each gene module resides on a separate chromosome, rationalizing the large number of distinct chromosomes in *Diplonema* mitochondria. But exuberance is paired with parsimony: no chromosome has been detected that does not contain a (potential) short coding region (in contrast to *Amoebidium* mtDNA, where the majority of chromosomes appears to be noncoding; see above). The peculiar, systematically fragmented structure of mitochondrial genes not only occurs in *D. papillatum*, but also is shared by all diplonemids as we conclude from a survey of the *cox1* gene in three additional species from both diplonemid genera (Kiethega et al. 2011).

3.6.3.3 Not So Systematically Fragmented Mitochondrial Genes in Dinoflagellates

The gene content of dinoflagellate mtDNA most likely reflects that of the sister phylum Apicomplexa, containing only *cob, cox1*, and *cox3*, along with heavily fragmented *rns* and *rnl*, and no tRNAs (Norman and Gray 2001; Jackson et al. 2007; Kamikawa et al. 2007, 2009; Nash et al. 2007; Slamovits et al. 2007). Although the complex structure of dinoflagellate mtDNAs has prevented a complete genome survey, broad sampling of this genome has been conducted in several taxa, and there is no evidence of further mitochondrial genes. For example, *cox2* has relocated to the nucleus in split form (Waller and Keeling 2006), and EST data suggest that dinoflagellates share the loss of Complex I (neither mitochondrial nor nucleus-encoded genes are found) with apicomplexans (Waller and Jackson 2009), fission yeasts (Bullerwell et al. 2003), and a subgroup of budding yeasts including *Saccharomyces cerevisiae* (Foury et al. 1998; Su et al. 2011).

The complete set of *rns* and *rnl* fragments is yet to be identified in dinoflagellate mtDNA, but based on available information, they appear to closely resemble those in apicomplexans in terms of size, fragmentation pattern, and sequence boundaries

(Jackson et al. 2007). Unlike in apicomplexans, protein genes exist in numerous and varyingly sized pieces, and together all coding sequences are present in many copies and genomic arrangements (Figs. 3.5c and 3.6c) (Norman and Gray 2001; Jackson et al. 2007; Nash et al. 2007; Slamovits et al. 2007; Kamikawa et al. 2009). In addition to these gene fragments, full-length coding sequences are found for *cob* and *cox1*, and it is currently unknown whether the gene fragments serve any function. The *cox3* gene is an exception in that complete gene sequences were not detected in dinoflagellate mtDNA, although full-length mRNAs were observed (see Sect 3.7.1.2) (Jackson et al. 2007; Waller and Jackson 2009). Altogether, mitochondrial gene structure in dinoflagellates is somewhat reminiscent of that in *Amoebidium*.

The above-described characteristics of dinoflagellate mtDNA genes prevail in a broad taxonomic sample, yet basal dinoflagellate lineages display some variation. *Oxyrrhis marina* does encode a complete *cox3*, but this is merged with the upstream *cob* united in a contiguous ORF (Slamovits et al. 2007). It is unknown whether a fused protein is generated, but given that *cob* and *cox3* contribute to different complexes, they likely form two proteins. Fragmented versions of all genes are also seen in *O. marina*. Curiously, fragments of genes are only found linked to complete copies or fragments of the same gene (for example, coding sequences of *cox1* are only linked to other coding sequences of *cox1*, but never to those of other genes). This suggests a particular, short-range-restricted form of recombination in this taxon. *Hematodinium* sp. is another basal dinoflagellate that shares most mtDNA features with other dinoflagellates. In this taxon, however, full-length coding sequences of all three protein-encoding genes (*cox1*, *cox3,* and *cob*) exist, along side copious numbers of gene fragments (Jackson and Waller unpublished).

3.7 Expression of Mitochondrial Genes

In general, mitochondrial gene expression is relatively poorly understood across eukaryotic diversity (particularly at the level of regulation, see Chaps. 11–13 and 18). Several observations can be made, however, that pertain to "usual" versus "unusual." The machinery for transcription in mitochondria was apparently inherited initially from the progenitor α-proteobacterium, evident by the persistence of genes *rpoA-C* for bacterial-type RNA-polymerase in the jakobid, *Reclinomonas americana* (Lang et al. 1997). This state, however, is very unusual and replacement of this polymerase with a bacteriophage type RNA-polymerase in all other eukaryotic groups suggests an early move to this phage-type system (Shutt and Gray 2006). Transcription in mitochondria from many eukaryotic lineages (e.g., ciliates, apicomplexans, green and red algae, stramenopiles, amoebozoans) is polycistronic with a small number of transcription initiation sites employed (Gray and Boer 1988; Wolff and Kuck 1996; Richard et al. 1998; Edqvist et al. 2000; Rehkopf et al. 2000). Individual gene transcripts are then generated by precise processing between the often closely spaced genes [e.g., (Wolff and Kuck 1996; Rehkopf et al. 2000)]. An implication of this system is that much of the regulation of gene expression must be posttranscriptional given that large banks of genes are initially expressed as one. This relatively simple mode of mitochondrial transcription might be common to many eukaryotes, but is unlikely to apply to mtDNAs that are much less gene-dense (e.g., plant mtDNAs) or in those with coding elements dispersed across separate molecules as, for example, in diplonemids.

Polyadenylation of transcripts is known from several mitochondrial systems including animals, apicomplexans, and trypanosomes (Anderson et al. 1981; Gillespie et al. 1999), but also diplonemids and dinoflagellates. The length of the poly(A) tail can contribute to translation control [(Etheridge et al. 2008); for a review, see (Gagliardi et al. 2004)], and we will discuss below that nucleotides of this tail can also contribute to the coding information. Two further posttranscriptional processes can have profound impacts on the expression of mitochondrial genes, notably (a) trans-splicing and (b) RNA recoding. These processes are able to rescue effectively fragmented and/or cryptic genes.

3.7.1   Trans-splicing

Trans-splicing produces complete RNAs from transcribed pieces of fragmented genes. In mitochondria, this process was first described in plants (Bonen 1993) where trans-splicing of mRNAs is mediated by Group II intron structures. Trans-splicing takes place for several of the Complex I genes (*nad1-3*, *nad5*) and requires cofactor molecules, some encoded in the nucleus [reviewed in (Bonen and Vogel 2001; Glanz and Kuck 2009)]. Recently, trans-splicing mediated by discontinuous Group I introns has been reported in early branching animals (placozoans), a lycophyte plant, and a green alga (Burger et al. 2009; Grewe et al. 2009; Pombert and Keeling 2010). In either case, initial evidence for trans-splicing was gathered by modeling complete Group I intron structures from the partial intron sequences that flank gene fragments. In addition, cDNA or RT-PCR data have provided experimental confirmation that trans-splicing takes place in vivo (For a recent review on trans-splicing of all intron types, see Moreira et al. 2011). However, not all trans-splicing in mitochondria is mediated by Group I or II introns, as we will discuss in the following sections.

3.7.1.1   Trans-splicing in Diplonemid Mitochondria

The systematically fragmented mitochondrial gene sequences of diplonemids are joined at the RNA level. The process of gene module trans-splicing has been investigated in detail for *cox1* of *D. papillatum* by employing various experimental techniques. These include Northern hybridization using individual gene modules as probes, demonstrating that intermediates of trans-splicing are abundant in the cell. Furthermore, sequencing of a cDNA library and of amplicons generated by targeted

RT-PCR (reverse transcription followed by polymerase chain reaction) detected module transcripts with noncoding 5′ and 3′ extensions, partially processed transcripts, and various intermediates of the module joining process.

The diverse processing intermediates allow reconstruction of the steps involved in the biogenesis of the *cox1* mRNA (Fig. 3.7). First, gene modules are transcribed individually, together with several hundred nucleotides of the constant region

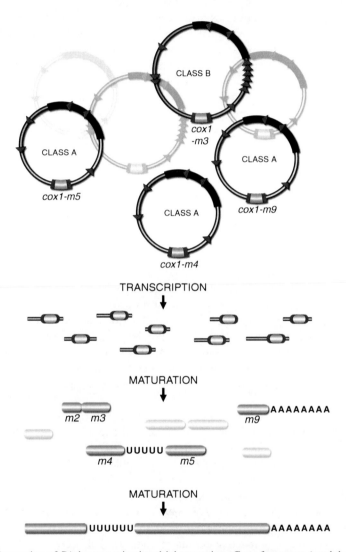

Fig. 3.7 Processing of *Diplonema* mitochondrial transcripts. Gene fragments (modules) together with noncoding flanking regions are transcribed as separate RNA fragments. Noncoding sequence is then removed and the last module is poly-adenylated. Gene modules (here indicated as m1 to m9) are joined in no particular directionality. In the case of *cox1* shown here, six Us are inserted between Modules 4 and 5

upstream and downstream of the module. Subsequently, noncoding regions are clipped off, leaving only the module RNAs. Finally, processed modules engage in trans-splicing, apparently in no specific directionality, to yield a mature mRNA (Marande and Burger 2007). Module joining is a most accurate process, since misassembled RNA species (e.g., Module 1 linked to Module 3) were not encountered.

The key question is how neighbor modules recognize each other in trans-splicing given a population of one hundred or so different gene pieces to be assembled into a dozen mRNAs. Initial hypotheses postulated discontinuous introns of Group I or Group II type, or alternatively introns of the archaeal type. Yet, no signatures of these introns or spliceosomal introns were detected at module junctions, nor reverse-complementary motifs in adjacent modules or their corresponding flanking regions. Not even single residues are conserved across the various *cox1* module boundaries from the same diplonemid species or in the same *cox1* module boundary across different species (Kiethega et al. 2011). The lack of significant motifs in cis suggests that module matchmaking is achieved by a third party, for example, guide RNAs similar to those that mediate RNA editing in kinetoplastid mitochondria. Equally possible are guiding proteins. Work is in progress to identify the nature of matchmaking molecules.

3.7.1.2   Trans-splicing in Dinoflagellate Mitochondria

Unlike in diplonemids, the majority of gene fragments in dinoflagellate mtDNA probably do not contribute to functional transcripts via splicing events. For example, fragmented rRNAs can be observed as discrete poly-adenylated transcripts both by RT-PCR-based techniques and Northern hybridization analysis, with no evidence of larger species being generated by fragment ligation (Jackson and Waller unpublished). Poly(A) tails are typically ~10–20 nt in length and are presumably tolerated in the assembled ribosome by complementary base pairing (Jackson et al. 2007; Slamovits et al. 2007). For the protein-encoding genes *cob* and *cox1*, only transcripts corresponding to complete gene sequences are seen, and the detected shorter transcripts do not match gene fragments (Jackson et al. 2007; Nash et al. 2007). Although long polycistronic transcripts containing gene fragments are occasionally found in EST data, these are not sufficiently abundant for Northern detection, and their fate and utility is unknown.

Expression of *Karlodinium micrum cox3* is unlike that of *cob* and *cox1* in that trans-splicing is required to generate a complete transcript. In this species, partial *cox3* transcripts correspond precisely in length to two *cox3* gene fragments encoding nucleotides 1–712, and 718–839 of "ordinary" *cox3* [Fig. 3.8a (Jackson et al. 2007; Waller and Jackson 2009)]. Both of these transcripts lack additional 5′ sequence beyond their respective coding regions, are poly-adenylated, and readily detectable by Northern hybridization (Jackson and Waller unpublished). It is conceivable that a split Cox3 protein is generated from these two transcripts. Yet, a third transcript with roughly equal copy number as the two partial ones represents a full-length Cox3 coding sequence (Jackson et al. 2007). It likely arises by

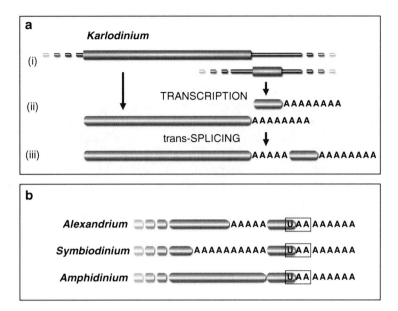

Fig. 3.8 Trans-splicing of *cox3* transcripts in dinoflagellates. (**a**) Gene fragments (i) are transcribed as separate, poly-adenylated RNA fragments (ii) that are then trans-spliced to form a continuous *cox3* mRNA (iii). The junction between the two spliced fragments likely inherits adenosine nucleotides from the poly-adenylation tail of the upstream fragment. (**b**) The length of the internal adenosine stretch varies across dinoflagellate taxa in order to maintain the correct length and reading frame of the encoded product. Poly-adenylation is used in all dinoflagellate *cox3* transcripts to generate a UAA termination codon (*boxes*)

trans-splicing of the two shorter RNAs because no single gene corresponds to it, and because the extremities coincide perfectly with those of the two shorter transcripts. The lack of any flanking sequences resembling introns, both in the coding sequences and in the transcripts, suggests that a process other than discontinuous intron splicing takes place in this system. The only *cox3* region that seems not encoded by mtDNA is from nucleotides 713–717, between the two fragments, and this is filled with five As in the full-length *K. micrum cox3* transcript. Given that the upstream *cox3* fragment is poly-adenylated after nucleotide 712, splicing could retain some of these adenosines (As) in the splice product. This event must be controlled precisely to avoid generating a frameshift.

Several dinoflagellate taxa for which EST data are available show equally truncated *cox3* transcripts in addition to full-length transcripts with several As bridging the two fragments. The number of As varies between taxa according to the gap in the *cox3* coding sequence, and thus the reading frame and protein length are generally maintained (Fig. 3.8b). Precise trans-splicing within the poly(A) tail likely requires some form of a guide molecule, but as in *Diplonema* mitochondria, no candidates for this role have yet been identified.

3.7.2   RNA Recoding: Alternative Genetic Codes and Editing

Cryptic genes are those where the sequence of a gene does not correspond to that of the gene's product. Discrepancies might be amino acid substitutions or interruptions to the reading frame by stop codons or frameshifts. Some cryptic genes can be so obscured as to be completely unrecognizable. Two scenarios can account for such cryptic genes. One is the use of an alternative genetic code to what is expected. Recoding of this type is quite frequent in mitochondria amounting to 16 deviations from the "universal" translation code scattered across plant and animal (including human) mitochondria alone [(Lekomtsev et al. 2007); see also Chap. 17]. Such changes can be identified in multiple protein alignments where otherwise conserved residues are consistently exchanged for another residue. Mitochondrial genomes with very few and divergent gene sequences can present challenges to identifying code changes, and it is conceivable that some have gone unnoticed. A particularly difficult case was *cox1* of the dinoflagellate *Perkinsus marinus* (Masuda et al. 2010), which contains numerous frameshifts in the gene sequence. One hypothesis brought forward by the authors invokes quadruplet and quintuplet codons that are recognized by special tRNAs, and another proposes programmed ribosome frameshifting. Both modes would constitute a radical way of recoding during the translation process.

A second type of cryptic genes are those recoded by RNA editing. There are several forms of RNA editing that are both mechanistically and evolutionarily distinct (Gray 2003). In plant mitochondria (and chloroplasts), C-to-U substitutions are found in most genes, with U-to-C changes less frequent in vascular plants but abundant in some ferns and mosses. Enzymatic base conversion by cytidine deaminases (that is without cleaving the RNA backbone) is presumed for C-to-U substitution, although the exact biochemistry is unknown (Takenaka et al. 2008). The specificity of changes in plant mitochondria is directed by a large suite of RNA-binding proteins that are thought to interact with sequence motifs in the region of the edited nucleotide. Similar substitution editing is also seen in some animal and protist groups, but again, the mechanisms are unknown [reviewed in (Gray 2003, 2009)]. More drastic editing takes the form of insertion and deletion editing that corrects frameshifts obscuring even a gene's identity. This is best investigated in trypanosomatid mitochondria where extensive insertion and/or deletion of single to multiple Us is accomplished by elaborate "editosome" complexes. As mentioned earlier, this editing is directed by short RNA molecules known as guide RNAs that bind to the transcript through base-paring [reviewed in (Stuart et al. 2005)]. Insertion editing is also known from myxomycete protists (slime moulds) where all nucleotide types can be inserted, but unlike in trypanosomatids, editing takes place during transcription and most likely without the participation of guide RNAs (Gott and Emeson 2000; Gray 2003).

The above discussed editing systems can act on all gene types – those encoding proteins, rRNAs and tRNAs – and can have profound effects on the gene product. In protein genes, substitutional changes predominate in the first two codon positions where they usually specify an amino acid change [e.g., (Lin et al. 2002)], whereas insertions/deletions mostly restore open reading frames [e.g., (Liu and Bundschuh 2005)]. Editing of tRNAs can even reengineer the anti-codon, and in trypanosomatids this recodes the UGA stop as a tryptophan codon to achieve an alternative genetic code (Alfonzo et al. 1999).

3.7.2.1 Rare RNA Editing in Diplonemid Mitochondria

In diplonemid mitochondria recoding is rare, but oddly, it seems to always occur exactly at gene module boundaries. Most noticeable is the addition of six nonencoded uridines (Us) at the junction between Modules 4 and 5 of the *cox1* transcript, and this is the case in all diplonemids investigated (Kiethega et al. 2011). The U residues are added in frame 3, contributing position 3 of a first codon, a complete second codon, and positions 1 and 2 of a third codon. This editing event has an important consequence for the protein, because the corresponding three amino acids in the Cox1 protein are invariably present, albeit not highly conserved, across eukaryotes. A lack of these Us would drastically change not only the protein sequence, but also its secondary and tertiary structure and make the protein non-functional (Kiethega et al. 2011). Three scenarios of U addition are conceivable. First, the nucleotides could originate from a gene module, encoded by mtDNA and transcribed and processed as described above. But this is unlikely, because no cassette has been found in *D. papillatum* mtDNA that encloses six contiguous Us, and known gene pieces are all significantly longer. Alternatively, the additional nucleotides could be inserted in a *cox1*-precursor transcript at the junction of Modules 4 and 5 via concerted action of an endonuclease and a nucleotide transfer-ase. Yet, no RNA species has been detected where the junction 4/5 lacks these nucleotides. Another possibility is that these Us are appended to one of the unjoined modules prior to trans-splicing, and indeed, we have experimental evidence for this latter scenario (Yan and Burger unpublished). The editing event of diplonemid *cox1* is conspicuously similar to the much more frequent RNA editing in kinetoplastid mitochondria, which is also limited to Us, but includes nucleotide deletions [for a review, see e.g., (Stuart et al. 2005)]. The main difference is that kinetoplastid mitochondria have contiguous primary transcripts that, prior to nucleotide addition or removal, need to be cleaved by an endonuclease that is integral part of the editosome. In diplonemids, however, U residues are simply appended to the free end of a module transcript.

A few less obvious and less consistent recoding events appear to take place in *Diplonema* at the ends of mitochondrial mRNAs. These events affect the termination signal for translation and will be discussed in Sect. 3.7.3.1.

3.7.2.2 Abundant and Assorted RNA Editing in Dinoflagellate Mitochondria

RNA editing in dinoflagellate mitochondria has a major impact on gene expression. Both protein-encoding and rRNA transcripts are edited with up to 6% of nucleotides affected per gene (Jackson et al. 2007; Nash et al. 2007; Waller and Jackson 2009; Kamikawa et al. 2007; Lin et al. 2002; Zhang et al. 2008). This editing is exclusively substitutional. Dinoflagellates are exceptional in that at least nine out of 12 possible forms of nucleotide substitution are observed including both transitions and transversions. Base conversion could not account for all of these changes and therefore, excision-replacement has to take place. However, the details of the underlying mechanism are unknown including how changes are specified.

Conservation of some editing sites across dinoflagellate taxa indicates a certain evolutionary stability of this process, yet frequent emergence of new editing sites in a lineage-specific manner also demonstrates that it is adaptable (Lin et al. 2002; Zhang et al. 2008). The vast majority of editing events in the protein-encoding genes occur at the first and second codon positions that typically lead to changes in the amino acid specified, including the removal of internal stop codons in several instances (Zhang et al. 2008; Waller and Jackson 2009). Thus, this editing process can recover cryptic genes from mtDNA sequences.

3.7.3 Start and Stop Codons of Mitochondrial Reading Frames

In the nucleus and organelles likewise, the coding region of protein-coding genes is framed by a start codon at the 5' end (an ATG) that initiates translation, and a stop codon at the 3' end (in mitochondria usually TAA or TAG) that signals termination of polypeptide elongation and release of the protein and the mRNA from the ribosome. The use of alternative initiation codons such as GTG, ATT, ATA, etc. is seen in bacterial systems (Kozak 1983) and also sporadically in mitochondria from diverse eukaryotic groups (Feagin 1992; Bock et al. 1994; Edqvist et al. 2000). The evidence for alternative start codons in mitochondria is generally indirect and based on multiple protein alignments of close relatives, where the probable beginning of the coding region lacking an ATG includes instead one of the possible alternatives. However, when sequence information from closely related species is unavailable, the inferred protein sequence is divergent, and protein sequence data are unavailable, its N terminus can be placed only tentatively.

Alternative stop codons are rather rare and have been reported, for example, in a green alga (Nedelcu et al. 2000), and proposed (Jukes and Osawa 1990), but recently refuted (Temperley et al. 2010), in humans (see discussion in Chap. 16). Sometimes, the stop codon is incomplete at the gene level and becomes completed at the transcript level by attachment of the poly(A) tail as for instance in animal mitochondria (Anderson et al. 1981).

Apparently, a stop codon is not always required in the mitochondrial system. One example is "nonstop" mRNAs in plant mitochondria that give rise to functional polypeptides (Raczynska et al 2006). Another is a human mitochondrial transcript that lost its stop via a deletion (Chrzanowska-Lightowlers et al. 2004). Here, the RNA is polyadenylated and translated normally, i.e., without read-through into the poly(A) tail that would otherwise generate a polylysine extension of the protein. This is thought to be achieved by poly(A) binding proteins that stall the ribosome, RNases that subsequently trim the A-tail, and specific release factors that allow ribosome detachment at the transcripts 3' end in the absence of a stop codon.

3.7.3.1 Unusual Start and Stop Codons in *Diplonema* Mitochondria

Information on mitochondrial start and stop codons in *D. papillatum* has been inferred from 11 genes [five complete (GenBank acc. nos. HQ288820-22; EU123538), and six incomplete ones lacking the 5' portion (Burger et al. unpublished]. Canonical initiation codons (in the gene and transcript sequence) are found for half of the genes, while GTG appears to serve as a start codon in the other half (Vlcek et al. 2011). The picture for stop codons is confusing. EST data indicate that the most C-terminal modules of four out of five genes (*cob cox2, cox3* and *nad7*) lack a conventional stop signal. For *cob*, U addition in the transcript seems not only to complete the last amino-acid codon to specify Phe, but also to supply the first position of the stop codon that, in turn, is apparently completed by addition of poly(A) (Kiethega and Burger unpublished). For *cox2, cox3*, and *nad7*, the situation is similar, except that the added nucleotide positions are polymorphic so that a sizable number of transcripts lack canonical stop codons. These observations need to be validated by experiments not relying on a reverse transcriptase reaction primed with an anchored oligo-d(T) primer, which might introduce artifacts [for methods, see (Rodriguez-Ezpeleta et al. 2009)]. If confirmed, this raises several questions. What directs nucleotide addition at the end of the last modules prior to polyadenylation? Are transcript versions without stop codons translated like the nonstop mRNAs mentioned above?

The *cox1* gene of *Diplonema* does have a stop codon encoded by mtDNA, notably an in-frame TAG at a position of the gene where the *cox1* reading frame of most other eukaryotes ends. However, this codon is followed by a T (or U) in the genomic and transcript sequence of *cox1* Module 9, and this nucleotide is completed upon poly-adenylation to a UAA codon, thus adding a second termination signal. Perhaps, the upstream nucleotide context makes the UAG stop codon less effective (Mottagui-Tabar and Isaksson 1998), which may have required recruitment of a second one.

3.7.3.2 Unusual Start and Stop Codons in Dinoflagellate Mitochondria

Transcripts of *cox1* and *cox3* from several dinoflagellate taxa consistently lack an AUG in the 5' region (Jackson et al. 2007; Nash et al. 2007; Slamovits et al. 2007;

Kamikawa et al. 2009). While some *cob* transcripts do contain an AUG codon toward the 5′ end of the transcript, multiple protein alignments suggest that these are downstream of the N terminus. Given the high A + T content of dinoflagellates mtDNAs, any of the many AUA or AUU codons might serve as alternative start codons. Nonstandard initiation codons seem also to be used in most mitochondrial genes of apicomplexans, and the predominance of this trait appears to unite the two groups.

In dinoflagellates, neither the *cox1* or *cob* genes, nor their edited transcripts, contain a stop codon, and the poly(A) tail starts immediately after the predicted 3′ coding sequence (Jackson et al. 2007; Nash et al. 2007; Slamovits et al. 2007; Kamikawa et al. 2009). A potential alternative stop codon is not observed at the 3′ end of either of these sequences. It is unknown which mechanism enacts translation termination and ribosome detachment.

Dinoflagellate *cox3* transcripts are consistently distinct in that a UAA termination codon is present, although not in the gene sequence but in the transcript upon poly-adenylation immediately after an in-frame U [Fig. 3.8; (Jackson et al. 2007; Slamovits et al. 2007)]. This suggests that more conventional translation termination could apply to Cox3. It remains puzzling why in both mitochondrial systems, in *Diplonema* and dinoflagellates, only a minority of genes would retain a standard stop codon, and also why this codon is generated through poly-adenylation rather than coding for it in the gene sequence.

3.8   Convergent Evolution of Highly Derived mtDNAs

As detailed above, the mtDNAs of *Amoebidium*, diplonemids, and dinoflagellates share an extraordinary large genome size, multi-chromosome genome structure, and fragmented genes. The two latter groups share also poly-adenylation, trans-splicing, and RNA editing of mitochondrial transcripts, as well as certain peculiar features of nuclear gene expression and subcellular organization [for a review, see (Lukes et al. 2009)]. The resemblances are startling because these taxa belong to three completely different eukaryotic lineages, opisthokonts, euglenozoans, and alveolates (see Fig. 3.1). In fact, as illustrated in Table 3.3, there is considerably more similarity between mtDNAs of *Amoebidium*, diplonemids, and dinoflagellates than with those of their phylogenetic neighbors, for example, between *Amoebidium* and animals, diplonemids and heteroloboseans, and dinoflagellates and ciliates. These neighbors possess all relatively traditional mtDNAs, which implies that the shared, deviant characters of *Amoebidium*, diplonemids, and dinoflagellates have emerged independently and represent spectacular cases of convergent evolution.

3.9 Which Forces Shape the Evolution of Organelle Genomes?

Mitochondrial DNAs of *Amoebidium*, diplonemids, and dinoflagellates are indeed eccentric, each in its own way. It is even not evident that these genomes stem from one common ancestor and that this ancestor is an α-proteobacterial genome with a single large, compact chromosome that encodes a thousand or more genes.

Commonly, evolution is perceived as a force seeking innovative solutions toward a selective advantage for the species (adaptive evolution). But novelty can also emerge in other ways. In the cases of the three protist groups discussed here, their respective ancestors may have been faced with deteriorating mitochondrial replication or gone-wild recombination leading to massive genome and gene fragmentation. Instead of restitution of the original state or extinction, the damage may have been countered by diverse and quite complex compensatory "measures." For example, the ancestor of diplonemid mitochondria may have coped with gene fragmentation by adapting an existing RNA ligation process to enable trans-splicing of fragmented genes. Furthermore, some DNA repair machinery may have been recruited and tailored to fix defective gene fragments at the RNA level. As pointed out earlier in the explanation of how RNA editing may have emerged, the compensatory system must have pre-existed (Covello and Gray 1993). Obviously, innovation is not driven alone by natural selection, and we have to consider as well what is called nonadaptive or neutral evolution (Jacob 1977; Lynch et al. 2006). In this light, as Gray and co-workers have put it, highly complex phenomena "generally regarded as evidence of 'fine tuning' or 'sophistication,' ...might be better interpreted as the consequences of runaway bureaucracy – as biological parallel of nonsensically complex Rube Goldberg machines" (Gray et al. 2010).

Appendix

Public, Internet-accessible data sources on mitochondrial genomes of all eukaryotes:

1. NCBI's complete organelle genome section. (http://www.ncbi.nlm.nih.gov/genomes/genlist.cgi?taxid=2759&type=4&name=Eukaryotae Organelles)
2. GOBASE is a taxonomically broad database on genomes from mitochondria and chloroplasts as well as selected bacteria belonging to groups from which these organelles originated. GOBASE integrates DNA and protein sequences, RNA secondary structures, and information on RNA editing, taxonomy and human mitochondrial DNA mutations and associated diseases. Data are drawn from various sources including NCBI's GenBank, and curated diligently. The last update is from June 2010. The database is being maintained, but further updates are not anticipated due to termination of funding.

References

Adams KL, Ong HC, Palmer JD (2001) Mitochondrial gene transfer in pieces: fission of the ribosomal protein gene *rpl2* and partial or complete gene transfer to the nucleus. Mol Biol Evol 18:2289–2297

Adams KL, Qiu YL, Stoutemyer M, Palmer JD (2002) Punctuated evolution of mitochondrial gene content: high and variable rates of mitochondrial gene loss and transfer to the nucleus during angiosperm evolution. Proc Natl Acad Sci U S A 99:9905–9912

Alfonzo JD, Blanc V, Estevez AM, Rubio MA, Simpson L (1999) C to U editing of the anticodon of imported mitochondrial tRNA(Trp) allows decoding of the UGA stop codon in *Leishmania tarentolae*. EMBO J 18:7056–7062

Allen JF (2003) The function of genomes in bioenergetic organelles. Philos Trans R Soc Lond B Biol Sci 358:19–37; discussion 37–18

Alverson AJ, Wei X, Rice DW, Stern DB, Barry K, Palmer JD (2010) Insights into the evolution of mitochondrial genome size from complete sequences of *Citrullus lanatus* and *Cucurbita pepo* (Cucurbitaceae). Mol Biol Evol 27:1436–1448

Anderson S et al (1981) Sequence and organization of the human mitochondrial genome. Nature 290:457–465

Armstrong MR, Blok VC, Phillips MS (2000) A multipartite mitochondrial genome in the potato cyst nematode *Globodera pallida*. Genetics 154:181–192

Awata H, Noto T, Endoh H (2005) Differentiation of somatic mitochondria and the structural changes in mtDNA during development of the dicyemid *Dicyema japonicum* (Mesozoa). Mol Genet Genomics 273:441–449

Bartoszewski G, Katzir N, Havey MJ (2004) Organization of repetitive DNAs and the genomic regions carrying ribosomal RNA, *cob*, and *atp9* genes in the cucurbit mitochondrial genomes. Theor Appl Genet 108:982–992

Bendich AJ (1993) Reaching for the ring: the study of mitochondrial genome structure. Curr Genet 24:279–290

Bensasson D, Zhang D, Hartl DL, Hewitt GM (2001) Mitochondrial pseudogenes: evolution's misplaced witnesses. Trends Ecol Evol 16:314–321

Bergthorsson U, Adams KL, Thomason B, Palmer JD (2003) Widespread horizontal transfer of mitochondrial genes in flowering plants. Nature 424:197–201

Bock H, Brennicke A, Schuster W (1994) *Rps3* and *rpl16* genes do not overlap in *Oenothera* mitochondria: GTG as a potential translation initiation codon in plant mitochondria? Plant Mol Biol 24:811–818

Boer PH, Gray MW (1988) Scrambled ribosomal RNA gene pieces in *Chlamydomonas reinhardtii* mitochondrial DNA. Cell 55:399–411

Bonen L (1993) Trans-splicing of pre-mRNA in plants, animals, and protists. FASEB J 7:40–46

Bonen L, Vogel J (2001) The ins and outs of group II introns. Trends Genet 17:322–331

Borst P, Grivell LA (1981) Small is beautiful–portrait of a mitochondrial genome. Nature 290:443–444

Bullerwell CE, Leigh J, Forget L, Lang BF (2003) A comparison of three fission yeast mitochondrial genomes. Nucleic Acids Res 31:759–768

Burger G, Plante I, Lonergan KM, Gray MW (1995) The mitochondrial DNA of the amoeboid protozoon, *Acanthamoeba castellanii*: complete sequence, gene content and genome organization. J Mol Biol 245:522–537

Burger G, Forget L, Zhu Y, Gray MW, Lang BF (2003a) Unique mitochondrial genome architecture in unicellular relatives of animals. Proc Natl Acad Sci U S A 100:892–897

Burger G, Gray MW, Lang BF (2003b) Mitochondrial genomes - anything goes. Trends Genet 19:709–716

Burger G, Lang BF, Braun HP, Marx S (2003c) The enigmatic mitochondrial ORF ymf39 codes for ATP synthase chain b. Nucleic Acids Res 31:2353–2360

Burger G, Nedelcu A (2011) Mitochondrial genomes of algae. In: Bock R, Knoop V (eds) Advances in Photosynthesis and Respiration. Springer, Berlin

Burger G, Yan Y, Javadi P, Lang BF (2009) Group I-intron trans-splicing and mRNA editing in the mitochondria of placozoan animals. Trends Genet 25:381–386

Butow RA, Fox TD (1990) Organelle transformation: shoot first, ask questions later. Trends Biochem Sci 15:465–468

Chrzanowska-Lightowlers ZM, Temperley RJ, Smith PM, Seneca SH, Lightowlers RN (2004) Functional polypeptides can be synthesized from human mitochondrial transcripts lacking termination codons. Biochem J 377:725–731

Covello PS, Gray MW (1993) On the evolution of RNA editing. Trends Genet 9:265–268

Dellaporta SL et al (2006) Mitochondrial genome of *Trichoplax adhaerens* supports placozoa as the basal lower metazoan phylum. Proc Natl Acad Sci U S A 103:8751–8756

Edqvist J, Burger G, Gray MW (2000) Expression of mitochondrial protein-coding genes in *Tetrahymena pyriformis*. J Mol Biol 297:381–393

Etheridge RD, Aphasizheva I, Gershon PD, Aphasizhev R (2008) 3′ adenylation determines mRNA abundance and monitors completion of RNA editing in *T. brucei* mitochondria. EMBO J 27:1596–1608

Fan J, Lee RW (2002) Mitochondrial genome of the colorless green alga *Polytomella parva*: two linear DNA molecules with homologous inverted repeat Termini. Mol Biol Evol 19:999–1007

Fauron C, Casper M, Gao Y, Moore B (1995) The maize mitochondrial genome: dynamic, yet functional. Trends Genet 11:228–235

Feagin JE (1992) The 6-kb element of *Plasmodium falciparum* encodes mitochondrial cytochrome genes. Mol Biochem Parasitol 52:145–148

Feagin JE, Gardner MJ, Williamson DH, Wilson RJ (1991) The putative mitochondrial genome of *Plasmodium falciparum*. J Protozool 38:243–245

Feagin JE, Mericle BL, Werner E, Morris M (1997) Identification of additional rRNA fragments encoded by the *Plasmodium falciparum* 6 kb element. Nucleic Acids Res 25:438–446

Forget L, Ustinova J, Wang Z, Huss VA, Lang BF (2002) *Hyaloraphidium curvatum*: a linear mitochondrial genome, tRNA editing, and an evolutionary link to lower fungi. Mol Biol Evol 19:310–319

Foury F, Roganti T, Lecrenier N, Purnelle B (1998) The complete sequence of the mitochondrial genome of *Saccharomyces cerevisiae*. FEBS Lett 440:325–331

Funes S et al (2002a) The typically mitochondrial DNA-encoded ATP6 subunit of the F_1F_0-ATPase is encoded by a nuclear gene in *Chlamydomonas reinhardtii*. J Biol Chem 277:6051–6058

Funes S et al (2002b) A green algal apicoplast ancestor. Science 298:2155

Gagliardi D, Stepien PP, Temperley RJ, Lightowlers RN, Chrzanowska-Lightowlers ZM (2004) Messenger RNA stability in mitochondria: different means to an end. Trends Genet 20:260–267

Gawryluk RM, Gray MW (2010) An ancient fission of mitochondrial Cox1. Mol Biol Evol 27:7–10

Gillespie DE, Salazar NA, Rehkopf DH, Feagin JE (1999) The fragmented mitochondrial ribosomal RNAs of *Plasmodium falciparum* have short A tails. Nucleic Acids Res 27:2416–2422

Glanz S, Kuck U (2009) Trans-splicing of organelle introns–a detour to continuous RNAs. Bioessays 31:921–934

Gott JM, Emeson RB (2000) Functions and mechanisms of RNA editing. Annu Rev Genet 34:499–531

Gray MW (2003) Diversity and evolution of mitochondrial RNA editing systems. IUBMB Life 55:227–233

Gray MW (2009) The path to RNA editing in plant mitochondria: the Halifax chapter. IUBMB Life 61:1114–1117

Gray MW, Boer PH (1988) Organization and expression of algal (*Chlamydomonas reinhardtii*) mitochondrial DNA. Philos Trans R Soc Lond B Biol Sci 319:135–147

Gray MW et al (1998) Genome structure and gene content in protist mitochondrial DNAs. Nucleic Acids Res 26:865–878

Gray MW, Lang BF, Burger G (2004) Mitochondria of protists. Annu Rev Genet 38:477–524

Gray MW, Lukes J, Archibald JM, Keeling PJ, Doolittle WF (2010) Cell biology. Irremediable complexity? Science 330:920–921

Grewe F, Viehoever P, Weisshaar B, Knoop V (2009) A trans-splicing group I intron and tRNA-hyperediting in the mitochondrial genome of the lycophyte *Isoetes engelmannii*. Nucleic Acids Res 37:5093–5104

Heinonen TY, Schnare MN, Young PG, Gray MW (1987) Rearranged coding segments, separated by a transfer RNA gene, specify the two parts of a discontinuous large subunit ribosomal RNA in *Tetrahymena pyriformis* mitochondria. J Biol Chem 262:2879–2887

Jackson CJ, Norman JE, Schnare MN, Gray MW, Keeling PJ, Waller RF (2007) Broad genomic and transcriptional analysis reveals a highly derived genome in dinoflagellate mitochondria. BMC Biol 5:41

Jacob F (1977) Evolution and tinkering. Science 196:1161–1166

Jacob Y, Seif E, Paquet PO, Lang BF (2004) Loss of the mRNA-like region in mitochondrial tmRNAs of jakobids. RNA 10:605–614

Jostensen J-P, Sperstad S, Johansen S, Landfald B (2002) Molecular-phylogenetic, structural and biochemical features of a cold-adapted, marine ichthyosporean near the animal-fungal divergence, described from *in vitro* cultures. Eur J Protistol 38:93–104

Jukes TH, Osawa S (1990) The genetic code in mitochondria and chloroplasts. Experientia 46:1117–1126

Kamikawa R, Inagaki Y, Sako Y (2007) Fragmentation of mitochondrial large subunit rRNA in the dinoflagellate *Alexandrium catenella* and the evolution of rRNA structure in alveolate mitochondria. Protist 158:239–245

Kamikawa R, Nishimura H, Sako Y (2009) Analysis of the mitochondrial genome, transcripts, and electron transport activity in the dinoflagellate *Alexandrium catenella* (Gonyaulacales, Dinophyceae). Phycol Res 57:1–11

Kannan S, Burger G (2008) Unassigned MURF1 of kinetoplastids codes for NADH dehydrogenase subunit 2. BMC Genomics 9:455

Kent ML, Elston RA, Nerad TA, Sawer TK (1987) An *Isonema*-like flagellate (Protozoa: Mastigophora) infection in larval Geoduck clams, *Panope abrupta*. J Invertebr Pathol 50:221–229

Kiethega G, Turcotte M, Burger G (2011) Evolutionary conserved trans-splicing without cis-motifs. Mol Biol Evol. doi:10.1093/molbev/msr075

Kornberg A, Baker J (1992) DNA replication, 2nd edn. W.H. Freeman, & Co, New-York

Kozak M (1983) Comparison of initiation of protein synthesis in procaryotes, eucaryotes, and organelles. Microbiol Rev 47:1–45

Kubo T, Newton KJ (2008) Angiosperm mitochondrial genomes and mutations. Mitochondrion 8:5–14

Kuhn J, Tengler U, Binder S (2001) Transcript lifetime is balanced between stabilizing stem-loop structures and degradation-promoting polyadenylation in plant mitochondria. Mol Cell Biol 21:731–742

Lang BF, Burger G (2007) Purification of mitochondrial and plastid DNA. Nat Protoc 2:652–660

Lang BF et al (1997) An ancestral mitochondrial DNA resembling a eubacterial genome in miniature. Nature 387:493–497

Lang BF, Gray MW, Burger G (1999) Mitochondrial genome evolution and the origin of eukaryotes. Annu Rev Genet 33:351–397

Lang BF, O'Kelly C, Nerad T, Gray MW, Burger G (2002) The closest unicellular relatives of animals. Curr Biol 12:1773–1778

Lang BF, Laforest MJ, Burger G (2007) Mitochondrial introns: a critical view. Trends Genet 23:119–125

Lekomtsev S, Kolosov P, Bidou L, Frolova L, Rousset JP, Kisselev L (2007) Different modes of stop codon restriction by the *Stylonychia* and *Paramecium* eRF1 translation termination factors. Proc Natl Acad Sci U S A 104:10824–10829

Lewis SM, Cote AG (2006) Palindromes and genomic stress fractures: bracing and repairing the damage. DNA Repair (Amst) 5:1146–1160

Lichtwardt RW (1973) Trichomycetes. In: Ainsworth GC, Sparrow FK, Sussman AS (eds) The fungi. An advances treatise. Academic press, New York, pp 237–243

Lichtwardt RW (1986) The trichomycetes: fungal associates of arthropods. Springer, New York

Lin S, Zhang H, Spencer DF, Norman JE, Gray MW (2002) Widespread and extensive editing of mitochondrial mRNAs in dinoflagellates. J Mol Biol 320:727–739

Ling F, Shibata T (2004) Mhr1p-dependent concatemeric mitochondrial DNA formation for generating yeast mitochondrial homoplasmic cells. Mol Biol Cell 15:310–322

Liu T, Bundschuh R (2005) Model for codon position bias in RNA editing. Phys Rev Lett 95:088101

Lonergan KM, Gray MW (1996) Expression of a continuous open reading frame encoding subunits 1 and 2 of cytochrome c oxidase in the mitochondrial DNA of *Acanthamoeba castellanii*. J Mol Biol 257:1019–1030

Lowe TM, Eddy SR (1997) tRNAscan-SE: a program for improved detection of transfer RNA genes in genomic sequence. Nucleic Acids Res 25:955–964

Lukes J, Jirku M, Avliyakulov N, Benada O (1998) Pankinetoplast DNA structure in a primitive bodonid flagellate, *Cryptobia helicis*. EMBO J 17:838–846

Lukes J, Hashimi H, Zikova A (2005) Unexplained complexity of the mitochondrial genome and transcriptome in kinetoplastid flagellates. Curr Genet 48:277–299

Lukes J, Leander BS, Keeling PJ (2009) Cascades of convergent evolution: the corresponding evolutionary histories of euglenozoans and dinoflagellates. Proc Natl Acad Sci U S A 106 (Suppl 1):9963–9970

Lynch M, Koskella B, Schaack S (2006) Mutation pressure and the evolution of organelle genomic architecture. Science 311:1727–1730

Marande W, Burger G (2007) Mitochondrial DNA as a genomic jigsaw puzzle. Science 318:415

Marande W, Lukeš J, Burger G (2005) Unique mitochondrial genome structure in diplonemids, the sister group of kinetoplastids. Eukaryot Cell 4:1137–1146

Masuda I, Matsuzaki M, Kita K (2010) Extensive frameshift at all AGG and CCC codons in the mitochondrial cytochrome c oxidase subunit 1 gene of *Perkinsus marinus* (Alveolata; Dinoflagellata). Nucleic Acids Res 38:6186–6194

Miyamoto H, Machida RJ, Nishida S (2010) Complete mitochondrial genome sequences of the three pelagic chaetognaths *Sagitta nagae*, *Sagitta decipiens* and *Sagitta enflata*. Comp Biochem Physiol Part D Genomics Proteomics 5:65–72

Mottagui-Tabar S, Isaksson LA (1998) The influence of the 5' codon context on translation termination in *Bacillus subtilis* and *Escherichia coli* is similar but different from *Salmonella typhimurium*. Gene 212:189–196

Moreira S, Breton S, Burger G (2011) Wiley Interdiscip Rev RNA, in press

Nash EA, Barbrook AC, Edwards-Stuart RK, Bernhardt K, Howe CJ, Nisbet RE (2007) Organization of the mitochondrial genome in the dinoflagellate *Amphidinium carterae*. Mol Biol Evol 24:1528–1536

Nash EA, Nisbet RE, Barbrook AC, Howe CJ (2008) Dinoflagellates: a mitochondrial genome all at sea. Trends Genet 24:328–335

Nedelcu AM, Lee RW, Lemieux C, Gray MW, Burger G (2000) The complete mitochondrial DNA sequence of *Scenedesmus obliquus* reflects an intermediate stage in the evolution of the green algal mitochondrial genome. Genome Res 10:819–831

Norman JE, Gray MW (1997) The cytochrome oxidase subunit 1 gene (cox1) from the dinoflagellate, *Crypthecodinium cohnii*. FEBS Lett 413:333–338

Norman JE, Gray MW (2001) A complex organization of the gene encoding cytochrome oxidase subunit 1 in the mitochondrial genome of the dinoflagellate, *Crypthecodinium cohnii*: homologous recombination generates two different *cox1* open reading frames. J Mol Evol 53:351–363

Ogawa S et al (2000) The mitochondrial DNA of *Dictyostelium discoideum*: complete sequence, gene content and genome organization. Mol Gen Genet 263:514–519

Oldenburg DJ, Bendich AJ (2001) Mitochondrial DNA from the liverwort *Marchantia polymorpha*: circularly permuted linear molecules, head-to-tail concatemers, and a 5′ protein. J Mol Biol 310:549–562

Paquin B, Laforest MJ, Lang BF (2000) Double-hairpin elements in the mitochondrial DNA of allomyces: evidence for mobility. Mol Biol Evol 17:1760–1768

Pombert JF, Keeling PJ (2010) The mitochondrial genome of the entomoparasitic green alga *Helicosporidium*. PLoS One 5:e8954

Pritchard AE, Cummings DJ (1981) Replication of linear mitochondrial DNA from *Paramecium*: sequence and structure of the initiation-end crosslink. Proc Natl Acad Sci U S A 78:7341–7345

Raczynska KD, Le Ret M, Rurek M, Bonnard G, Augustyniak H, Gualberto JM (2006) Plant mitochondrial genes can be expressed from mRNAs lacking stop codons. FEBS Lett 580:5641–5646

Rehkopf DH, Gillespie DE, Harrell MI, Feagin JE (2000) Transcriptional mapping and RNA processing of the *Plasmodium falciparum* mitochondrial mRNAs. Mol Biochem Parasitol 105:91–103

Richard O, Bonnard G, Grienenberger JM, Kloareg B, Boyen C (1998) Transcription initiation and RNA processing in the mitochondria of the red alga *Chondrus crispus*: convergence in the evolution of transcription mechanisms in mitochondria. J Mol Biol 283:549–557

Richly E, Leister D (2004) NUMTs in sequenced eukaryotic genomes. Mol Biol Evol 21:1081–1084

Rodriguez-Ezpeleta N, Teijeiro S, Forget L, Burger G, Lang BF (2009) 3. Generation of cDNA libraries: Protists and Fungi. In: Parkinson J (ed) Methods in Molecular Biology: Expressed Sequence Tags (ESTs). Humana Press, Totowa, NJ

Roy J, Faktorova D, Benada O, Lukes J, Burger G (2007a) Description of *Rhynchopus euleeides* n. sp. (Diplonemea), a free-living marine euglenozoan. J Eukaryot Microbiol 54:137–145

Roy J, Faktorova D, Lukes J, Burger G (2007b) Unusual mitochondrial genome structures throughout the Euglenozoa. Protist 158:385–396

Ruiz-Trillo I et al (2007) The origins of multicellularity: a multi-taxon genome initiative. Trends Genet 23:113–118

Ruiz-Trillo I, Roger AJ, Burger G, Gray MW, Lang BF (2008) A phylogenomic investigation into the origin of metazoa. Mol Biol Evol 25:664–672

Rycovska A, Valach M, Tomaska L, Bolotin-Fukuhara M, Nosek J (2004) Linear versus circular mitochondrial genomes: intraspecies variability of mitochondrial genome architecture in *Candida parapsilosis*. Microbiology 150:1571–1580

Shao R, Kirkness EF, Barker SC (2009) The single mitochondrial chromosome typical of animals has evolved into 18 minichromosomes in the human body louse, *Pediculus humanus*. Genome Res 19:904–912

Shapiro TA, Englund PT (1995) The structure and replication of kinetoplast DNA. Annu Rev Microbiol 49:117–143

Shutt TE, Gray MW (2006) Homologs of mitochondrial transcription factor B, sparsely distributed within the eukaryotic radiation, are likely derived from the dimethyladenosine methyltransferase of the mitochondrial endosymbiont. Mol Biol Evol 23:1169–1179

Simpson L (1997) The genomic organization of guide RNA genes in kinetoplastid protozoa: several conundrums and their solutions. Mol Biochem Parasitol 86:133–141

Simpson AM, Suyama Y, Dewes H, Campbell DA, Simpson L (1989) Kinetoplastid mitochondria contain functional tRNAs which are encoded in nuclear DNA and also contain small minicircle and maxicircle transcripts of unknown function. Nucleic Acids Res 17:5427–5445

Slamovits CH, Saldarriaga JF, Larocque A, Keeling PJ (2007) The highly reduced and fragmented mitochondrial genome of the early-branching dinoflagellate *Oxyrrhis marina* shares characteristics with both apicomplexan and dinoflagellate mitochondrial genomes. J Mol Biol 372:356–368

Smith DR, Hua J, Lee RW (2010) Evolution of linear mitochondrial DNA in three known lineages of *Polytomella*. Curr Genet 56:427–438

Stuart KD, Schnaufer A, Ernst NL, Panigrahi AK (2005) Complex management: RNA editing in trypanosomes. Trends Biochem Sci 30:97–105

Su D, Lieberman A, Lang BF, Simonovic M, Soll D, Ling J (2011) An unusual tRNA^{Thr} derived from tRNA^{His} reassigns in yeast mitochondria the CUN codons to threonine. Nucleic Acids Res 39(11):4866–4874

Suga K, Mark Welch DB, Tanaka Y, Sakakura Y, Hagiwara A (2008) Two circular chromosomes of unequal copy number make up the mitochondrial genome of the rotifer *Brachionus plicatilis*. Mol Biol Evol 25:1129–1137

Sugiyama Y et al (2005) The complete nucleotide sequence and multipartite organization of the tobacco mitochondrial genome: comparative analysis of mitochondrial genomes in higher plants. Mol Genet Genomics 272:603–615

Takenaka M, Verbitskiy D, van der Merwe JA, Zehrmann A, Brennicke A (2008) The process of RNA editing in plant mitochondria. Mitochondrion 8:35–46

Temperley R, Richter R, Dennerlein S, Lightowlers RN, Chrzanowska-Lightowlers ZM (2010) Hungry codons promote frameshifting in human mitochondrial ribosomes. Science 327:301

Valach M et al (2011) Evolution of linear chromosomes and multipartite genomes in yeast mitochondria. Nucleic Acids Res 39(10):4202–4219

van Dooren GG, Stimmler LM, McFadden GI (2006) Metabolic maps and functions of the *Plasmodium* mitochondrion. FEMS Microbiol Rev 30:596–630

Vickerman K (2000) Diplonemids. In: Lee JJ, Leedale GF, Bradbury P (eds) An illustrated guide to the protozoa. Allen Press, Lawrence, Kansas, pp 1157–1159

Vlcek C, Marande W, Teijeiro S, Lukes J, Burger G (2011) Systematically fragmented genes in a multipartite mitochondrial genome. Nucleic Acids Res 39(3):979–988

Voigt O, Erpenbeck D, Worheide G (2008) A fragmented metazoan organellar genome: the two mitochondrial chromosomes of *Hydra magnipapillata*. BMC Genomics 9:350

Waller RF, Jackson CJ (2009) Dinoflagellate mitochondrial genomes: stretching the rules of molecular biology. Bioessays 31:237–245

Waller RF, Keeling PJ (2006) Alveolate and chlorophycean mitochondrial *cox2* genes split twice independently. Gene 383:33–37

Ward BL, Anderson RS, Bendich AJ (1981) The mitochondrial genome is large and variable in a family of plants (Cucurbitaceae). Cell 25:793–803

Warrior R, Gall J (1985) The mitochondrial DNA of *Hydra attenuata* and *Hydra littoralis* consists of two linear molecules. Arch Sci Geneva 38:439–445

Watanabe KI, Bessho Y, Kawasaki M, Hori H (1999) Mitochondrial genes are found on minicircle DNA molecules in the mesozoan animal *Dicyema*. J Mol Biol 286:645–650

Wolff G, Kuck U (1996) Transcript mapping and processing of mitochondrial RNA in the chlorophyte alga *Prototheca wickerhamii*. Plant Mol Biol 30:577–595

Zhang H, Bhattacharya D, Maranda L, Lin S (2008) Mitochondrial *cob* and *cox1* genes and editing of the corresponding mRNAs in *Dinophysis acuminata* from Narragansett Bay, with special reference to the phylogenetic position of the genus *Dinophysis*. Appl Environ Microbiol 74:1546–1554

Chapter 4
Plastid Genomes of Parasitic Plants: A Trail of Reductions and Losses

Kirsten Krause

Abbreviations

bp	Base pairs
NEP	Nuclear-encoded RNA polymerase
PEP	Plastid encoded RNA polymerase
Rubisco	Ribulose-1,5-bisphosphate carboxylase/oxygenase

4.1 Introduction

The view that plastids have evolved from cyanobacteria by endosymbiosis and that this event took place originally only once is nowadays widely accepted (see Chap. 2). This primary endosymbiosis gave rise to the common ancestor of today's three primary plastid-containing lineages (together known as Archaeplastida, Table 4.1): (1) green algae and plants, (2) red algae, and (3) glaucophyte algae (Reyes-Prieto and Bhattacharya 2007; Lane and Archibald 2008). From these lineages, plastids have spread laterally by secondary endosymbiosis to form further lineages (see Chap. 2). Thus, two unrelated lineages containing plastids of green algal ancestry (euglenoids and chlorarachniophytes) and several lineages containing plastids of red algal ancestry (haptophytes, cryptophytes, stramenopiles, dinoflagellates, and apicomplexa) have evolved (Cavalier-Smith 1999, 2002; Lane and Archibald 2008). In addition, the occurrence of serial secondary endosymbiosis and tertiary plastids that replaced the original secondary plastid (Yoon et al. 2005) is under discussion (Lane and Archibald 2008; Janouskovec et al. 2010). Most of the resulting species gain their energy through photoautotrophic carbon fixation. However, in probably every land plant and algal lineage, parasitic species have evolved that have abandoned

K. Krause (✉)
Department for Arctic and Marine Biology, University of Tromsø, Dramsvegen 201,
N-9037 Tromsø, Norway
e-mail: kirsten.krause@uit.no

C.E. Bullerwell (ed.), *Organelle Genetics*,
DOI 10.1007/978-3-642-22380-8_4, © Springer-Verlag Berlin Heidelberg 2012

Table 4.1 Plastid genomes sequenced from different lineages of the plant kingdom

Group[a]	Number of completely sequenced genomes[b]	Example species shown in Fig. 4.2
ARCHAEPLASTIDA		
Streptophytes	179	*Nicotiana tabacum*
Chlorophytes	22	*Chlamydomonas rheinhardtii*
Rhodophytes	5	*Cyanidium caldarium*
Glaucocystophytes	1	*Cyanophora paradoxa*
EXCAVATA		
Euglenids	2	*Euglena gracilis*
RHIZARIA		
Chlorarachniophytes	1	*Bigelowiella natans*
CHROMALVEOLATA		
Dinoflagellates	2	–
Cryptophytes	3	*Guillardia theta*
Stramenopiles	10	*Vaucheria litorea*
Apicomplexa	5	*Toxoplasma gondii*
Haptophytes	1	*Emiliania huxleyi*

[a]The classification was adapted after Lane and Archibald (2008)
[b]The numbers are based on the sequences published in the NCBI genomes database (http://www.ncbi.nlm.nih.gov/genomes/genlist.cgi?taxid=2759&type=4&name=Eukaryotae%20Organelles)

photoautotrophic growth and rather live by parasitizing on other plants, fungi, or even animals.

Species of colorless unicellular algae with reduced plastid genomes have been described among the green algae (de Koning and Keeling 2004; Tartar and Boucias 2004; Borza et al. 2005; de Koning and Keeling 2006), the euglenoid algae (Gockel and Hachtel 2000), and the dinoflagellates (Sanchez-Puerta et al. 2007; Matsuzaki et al. 2008). The apicomplexan group of unicellular parasitic organisms, last but not least, provides the most prominent example that plastid-containing parasites have been able to utilize a very wide host range (Lim and McFadden 2010). More recently, claims of a photosynthetic/red algal (and therefore, possibly, plastid) history of the oomycete *Phytophthora* (Tyler et al. 2006) and of ciliates (Reyes-Prieto et al. 2008) have been made (Janouskovec et al. 2010), taking the discussion around adaptations to a nonphotosynthetic life style another step ahead.

Parasitic land plants are – in contrast to their algal counterparts – quite eye catching (Fig. 4.1) and have received much attention not only due to their special lifestyle but also due to the damage they can inflict on agricultural land use. It is estimated that approximately 1% of all angiosperm species from at least 11 different lineages have resorted to a parasitic lifestyle (Barkman et al. 2007). Based on their attachment sites on their hosts, root parasites and shoot parasites are being distinguished (Fig. 4.1). The liverwort *Aneura mirabilis* (formerly known as *Cryptothallus mirabilis*) is one example – and as a matter of fact, to date, the only known example – of a nonvascular parasitic land plant that has evolved into a completely nonphotosynthetic lifestyle (Bidartondo et al. 2003).

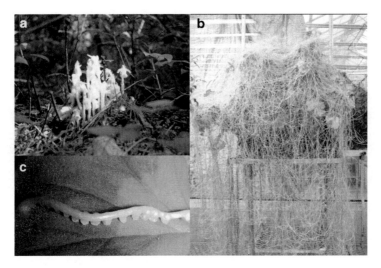

Fig. 4.1 Parasitic land plants. Parasitic plants can attach to the roots of their hosts, as in *Monotropa uniflora* (Indian pipe, **a**), or to the stems or leaves as in species of the genus *Cuscuta* (**b, c**). The attachment organs, or haustoria, are visible in (**c**)

The following chapters will summarize our current knowledge on the structure and function of plastid genomes from algae to higher plants and will focus on the degeneration of plastid genomes in species that have resorted to a parasitic lifestyle.

4.2 The Plastid Genome

4.2.1 The Organization of Plastid DNA

Every plastid possesses its own genetic information that was inherited from the cyanobacterial ancestors of these organelles. Reflecting this ancestry, the plastid genome has retained many features of prokaryotic genomes, including the overall structure, physical properties, gene organization, and regulatory features necessary for gene expression (Bock 2007). Plastid DNA is present in multiple copies per plastid and is compacted by DNA-binding proteins (Kuroiwa 1991). Microscopic studies in combination with fluorescent staining of these high-molecular-weight DNA–protein complexes generally known as nucleoids, revealed that their number, shape, size, and distribution can vary significantly between species and depending on the developmental stage (Kuroiwa 1991). Plastid genome maps of all plant lineages traditionally show the organelle's genetic information as circular molecules. The only striking exception to this seem to be the dinoflagellates where the normal single circle has been replaced by minicircles, which contain one or a few genes each (Zhang et al. 1999; Barbrook and Howe 2000). Electron microscopic pictures later revealed that the circular conformation is only one of many that the plastid DNA of higher plants can appear in. In addition to circular monomeric molecules, a variety of linear and branched monomers and oligomers (Deng et al. 1989; Lilly et al. 2001; Oldenburg and Bendich 2004; Scharff and Koop 2006) and possibly also shorter fragments (Kolodner and Tewari 1972) have been observed.

In contrast to the structural variability of the nucleoids, plastid coding capacity and gene organization have remained remarkably conserved across the plant kingdom, with higher plants, in particular, showing very little variation in their gene content and plastid genome sizes (Jansen et al. 2007). The selective pressure on the maintenance of photosynthesis-related genes and genes for subunits of the plastid gene-expression machinery seems to have been instrumental for maintaining the plastid genome and conserving a core set of plastid genes (Bock 2007). It is this selective pressure exerted by photosynthesis or, rather opposite, the obvious lack of it in parasitic plants that has drawn the attention to parasitic plant plastid genomes and their particular evolution.

4.2.2 Structure and Coding Capacity of Plastid Genomes

The first complete plastid genome sequences, namely that of the bryophyte *Marchantia polymorpha* and of the seed plant *Nicotiana tabacum*, were reported as early as 1986 (Ohyama et al. 1986; Shinozaki et al. 1986). Since then, more than 230 plastid genomes have been completely sequenced. The vast majority of these plastid genome sequences are from land plants and green algae, while other big groups, such as the red algae, dinoflagellates, or stramenopiles, are still strongly underrepresented (Table 4.1).

A common feature of plastid genomes are two inverted repeat regions (IR_A and IR_B) that split the remainder of the chromosome into a large and a small single-copy region (LSC and SSC, respectively). The inverted repeats can vary significantly in size, from 75 kbp in *Pelargonium* (Chumley et al. 2006) to 0.5 kbp in *Pinus* (Wakasugi et al. 1994). The functional role of this tetrapartite structure is unclear and examples of species without this organization are known from green (Hallick et al. 1993; Wakasugi et al. 1997; Jansen et al. 2008) and red plastid lineages (Glockner et al. 2000; Ohta et al. 2003; Hagopian et al. 2004), suggesting that it could be dispensable.

Compared to the ancestral prokaryotic genome, the plastome of all plants and algae is more or less drastically reduced. The genetic information for the majority of plastid-localized proteins either has been lost altogether in the course of evolution or has been transferred to the nucleus from where the gene products are imported back into the organelle (see Chap. 7). With a few exceptions, plastid genomes of the green lineage range between 120 and 160 kbp in size and code for 100 $\pm$ 20 proteins as well as about 40 genes encoding stable RNA species (rRNAs and tRNAs) (Palmer 1990). Even the plastome of the hypothetical ancestor to all green lineages was estimated to have contained a total of only 137 protein-coding genes (Turmel et al.

1999). Free-living descendents of the red plastid lineage have, on average, retained larger plastid genomes compared to the "standard plastome" of the green lineage. Nevertheless, gene numbers and sizes in these genomes are still in the same order of magnitude and feature around 200 protein-coding genes on approximately 160–180 kbp (Reith and Munholland 1993; Ohta et al. 2003). Extreme deviations from the average sizes, on the other hand, do exist in single cases and have been reported, for example, for *Acetabularia* species whose plastid genomes can be up to 1.5 Mbp in size (Simpson and Stern 2002), while probably the smallest plastid genome of a photosynthetic organism with only 72 kbp is that of the green alga *Ostreococcus tauri* (Robbens et al. 2007).

The information content of the plastid genome can be roughly divided into three large groups (see also Table 4.2): (1) genetic system genes comprising the RNA and protein components of the transcription and translation machineries as well as a few proteins involved in post-transcriptional and post-translational steps, (2) photosynthesis genes for subunits of the light and dark reactions and the ATPase, and (3) conserved open reading frames and genes with miscellaneous functions.

Genetic system genes required for transcription, translation, and processing steps represent a large portion of all plastid genomes and also represent the major fraction of strongly reduced plastid genomes, such as that of the apicomplexan parasite *Toxoplasma gondii* (Fig. 4.2). The number of photosynthesis and genetic system genes is slightly higher in red algae and most lineages derived from them, indicating specific losses in the green plastids, whereas a constant number of

Table 4.2 Core set of genes encoded by the plastid genome of higher plants

Protein complex or functional category	# genes	Gene designation
GROUP I: Genetic system genes		
RNA polymerase	4	*rpo* genes
Intron maturase	1	*matK*
Ribosomal small subunit	14	*rps* genes
Ribosomal large subunit	11	*rpl* genes
Ribosomal RNAs	4	*rrn* genes
Transfer RNAs	30	*trn* genes
GROUP II: Photosynthesis and energy production		
Photosystem I	5	*psa* genes
Photosystem II	14	*psb* genes
Cytochrome b6f complex	6	*pet* genes
NAD(P)H dehydrogenase	11	*ndh* genes
ATPase	6	*atp* genes
Rubisco	1	*rbcL*
GROUP III: Conserved hypothetical reading frames and other genes		
Lipid metabolism	1	*accD*
Chaperone and protease	1	*clpP*
Conserved hypothetical reading frames[a]	8	*ycf* genes

[a]The list contains the *ycf* genes under their original designation, instead of under the newer gene designations that exist for some *ycf*s (see text)

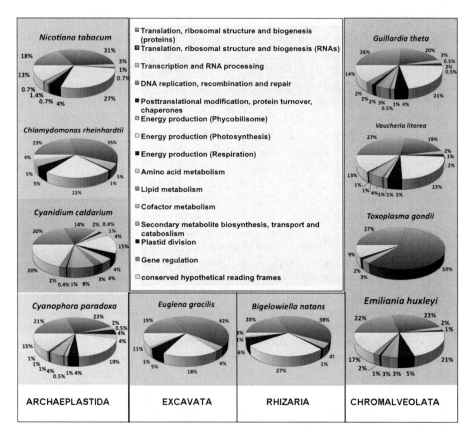

Fig. 4.2 Plastome coding capacity expressed in proportion of the various functional categories in the different lineages of the plant kingdom. Genomes used as examples were published as follows: *N. tabacum* (Shinozaki et al. 1986; Wakasugi et al. 1998); *C. rheinhardtii* (Maul et al. 2002); *Cyanidium caldarium* (Glockner et al. 2000); *Cyanophora paradoxa* (Genbank Acc # U30821); *Euglena gracilis* (Hallick et al. 1993); *Bigelowiella natans* (Rogers et al. 2007); *Emiliania huxleyi* (Sanchez Puerta et al. 2005); *Toxoplasma gondii* (Genbank Acc # U87145); *Vaucheria litorea* (Rumpho et al. 2008); and *Guillardia theta* (Douglas and Penny 1999). *Background colors* behind the pies indicate whether the plastids are of green algal origin (*green*), red algal origin (*pink*) or of glaucophyte origin (*blue*)

six genes (equaling 4–5% of the coding potential) on almost all plastid genomes is dedicated to the subunits of the plastid ATPase. In green plastid-derived lineages, the third group that comprises genes with other functions makes up around 20% of the genes, the majority of which are conserved reading frames of unknown function (YCFs) (Fig. 4.2). In most rhodophytes, glaucophytes, and chromalveolates, this gene group is considerably expanded and contains genes for amino acid, lipid, pigment, and cofactor metabolism (Fig. 4.2) that are absent from land plants and green algae. The only exception is the chromalveolate group of apicomplexans, where many typical gene groups of the red plastid lineage have been lost (Wilson et al. 1996). This reduction in genome-coding capacity is related partly to the parasitic life that members of this group are leading and partly to losses that must have occurred independent of the parasitic lifestyle.

4.3   Plastid Genomes from Parasitic Species

The angiosperm holoparasite *Epifagus virginiana* (Beechdrops) has been one of the first parasites to be thoroughly investigated with respect to its plastid genome sequence (Wolfe et al. 1992b); and these analyses have revealed a number of drastic changes that involve mainly gene losses and pseudogenizations. The significant reductions of the coding potential that embrace, among others, all photosynthesis genes in *E. virginiana* (Table 4.3 and Fig. 4.4), have been explained with the relaxation of the selective pressure exerted otherwise by photosynthesis. The discovery that, among the more than 150 species that are assembled within the holoparasitic genus *Cuscuta*, not all exhibit the same severe physiological reductions found in *Epifagus* but that some have, in fact, retained some basal photosynthetic activity (Hibberd et al. 1998; van der Kooij et al. 2000) has more recently enabled glimpses into the transition from photoautotrophy via intermediate mixotrophic states to complete heterotrophy (Krause 2008). A likewise gradient has, so far, not been found anywhere else.

4.3.1   Structural Changes

The typical organization of plastid chromosomes with a large single-copy region (LSC) and a small single-copy region (SSC) separated by two inverted repeat regions (IR_A and IR_B) has been retained by all parasitic angiosperms (Wolfe et al. 1992b; Funk et al. 2007; McNeal et al. 2007) (Fig. 4.3). For two *Cuscuta* species, *C. reflexa* and *C. gronovii*, overlapping PCR products have indicated the existence of a circular form of the plastid chromosomes (Funk et al. 2007), suggesting overall structural similarities between parasitic and nonparasitic plastid genomes. The same holds true for parasitic algae. Divergences from the standard pattern, for example, in *Euglena longa* are also present in the photosynthetic relative *E. gracilis* and are, therefore, unrelated to parasitism (Hallick et al. 1993; Gockel and Hachtel 2000).

   The inverted repeats have been assigned a role as a stabilizing factor that limits genome rearrangements in chloroplasts. Nevertheless, the boundaries of inverted repeats were found to be hot spots for gene duplications or deletions (Yue et al. 2008). In line with this, the IR_A–LSC junction (JLA) in *C. reflexa* and *C. exaltata* was found within the *ycf2* gene, leaving one copy of this gene truncated. As a result of this reduction in IR size, there is only one copy each of *rpl2* and *trnI-CAU* and one complete *ycf2* gene (Funk et al. 2007; McNeal et al. 2007). Generally, the size of the inverted repeats in *Cuscuta* was reduced proportionally to the size of the

Table 4.3 Plastid gene losses in parasitic versus nonparasitic higher plants

	Genes missing from parasitic plant genomes[a]	Genes missing from photosynthetic (nonparasitic) species[b]
PHOTOSYNTHESIS AND CHLORORESPIRATORY GENES		
ndhA-K	Cr (ndhB:Ψ), Ce (ndhB:Ψ), Cg, Co, Ev (ndhB:Ψ)	Gymnosperms Phalaenopsis
psaI	Cg, Co, Ev	–
psaA,B,C, and J	Ev	–
psbA,B,C,D,E,F,H, I,J, K,L,M,N, and T	Ev (psbA, B:Ψ)	–
petA,B,D,G L, and N	Ev	–
atpA,B,E,F,H, and I	Ev (atpA, B:Ψ)	–
rbcL	Ev (Ψ)	–
RNA POLYMERASE AND MATURASE GENES		
rpoA	Cg, Co, Ev (Ψ)	Pelargonium Passiflora
rpoB,C1, and C2	Cg, Co, Ev	–
matK	Cg, Co	–
RIBOSOMAL PROTEIN AND INITIATION FACTOR GENES		
infA	Cr, Ce, Cg, and Co	e.g., Arabidopsis, Brassica, Citrus, Cucumis, Glycine, Gossypium, Manihot, Medicago, Nicotiana, Oenothera, Solanum, etc.
rpl23	Cr (Ψ), Ce (Ψ), Cg, Co, Ev (Ψ)	Trachelium Spinacia
rpl32	Cg, Co, Ev	Populus Yucca
rps16	Cr (Ψ), Ce (Ψ), Cg, Co, Ev	Medicago, Populus, Passiflora Pinus
rpl14, rpl22, and rps15	Ev (rpl14: Ψ)	–
OTHER PROTEIN GENES		
Ycf3, 4, 5, 9, 10, and 15	Cr, Ce, Cg, Co, Ev (ycf15: Ψ)	–

Genes that were reported missing from the plastid genomes of *Cuscuta* or *Epifagus* are listed according to their functional categories alongside reported losses from the plastid genomes of nonparasitic species as reported by Jansen et al. (2007)

[a] Ψ = Gene is an unfunctional pseudogene. *Cr C. reflexa* (Funk et al. 2007); *Ce C. exaltata* (McNeal et al. 2007); *Cg C. gronovii* (Funk et al. 2007); *Co C. obtusiflora* (McNeal et al. 2007); *Ev E. virginiana* (Wolfe et al. 1992b)

[b] – = No reported losses

plastid genome (Fig. 4.3). In *E. virginiana*, in contrast, the inverted repeats have suffered much less reductions so that their sizes relative to the single-copy regions are much larger. The main reason for this is that the ribosomal genes that are located on the IRs have been retained while many genes that are normally part of the single-copy regions (such as the photosynthesis genes) have been lost. However, differences between IR length and IR gene content are fairly common in higher plants, anyway, as exemplified by the comparison between tobacco and *Pelargonium* (Fig. 4.3), and even between species of the same genus (Goulding et al. 1996), so inverted repeat sizes cannot be correlated with a particular lifestyle.

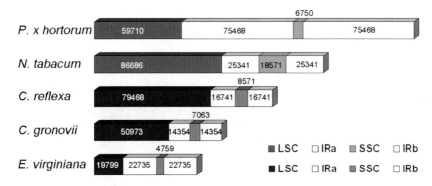

Fig. 4.3 Sizes of the large and small single-copy regions and the inverted repeats. Two photosynthetic angiosperms, *Pelargonium* and tobacco, and three parasitic angiosperms, *Cuscuta reflexa*, *Cuscuta gronovii*, and *Epifagus virginiana*, are shown. The size of each respective region is shown in basepairs

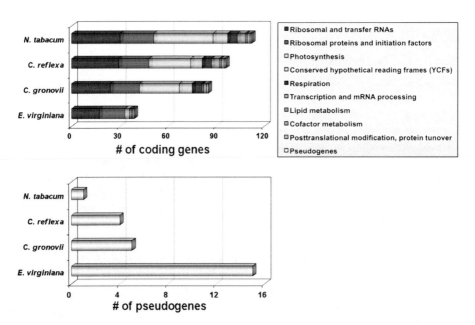

Fig. 4.4 Amount of functional genes and pseudogenes in three parasitic and one nonparasitic angiosperm species

Insertions and deletions ("indels") of larger fragments and inversions that can affect the order of genes on the plastid genome are considered to be almost as important for the evolution of genomes as nucleotide substitutions (Yue et al. 2008). Compared to tobacco and other angiosperm plastid genomes, species of the *Cuscuta* subgenus *Monogyna* exhibit three typical sequence inversions within the plastid chromosome, two in the large single-copy region ~2 kb and ~13 kb in length and one of ~1.5 kb length in the small single-copy region (Stefanovic and Olmstead 2005;

Funk et al. 2007). Otherwise, the gene order in parasitic plants was not found to be different. Whether the three inversions are of any functional significance is, however, unclear.

A high ratio of coding versus noncoding sequences was found for all *Cuscuta* species and some parasitic algae, such as *Helicosporidium* and *Euglena longa*, as well as for aplicoplast genomes. In contrast, *Epifagus* exhibits almost the same coding to noncoding sequence ratio as photosynthetic plants (Krause 2008). It has been speculated that an early reaction of the plastid genome to the parasitic lifestyle was the condensation of the genome by loss of many noncoding and possibly unimportant parts of the plastid DNA (Funk et al. 2007). As the adaptations to parasitism became more pronounced, pseudogenization and loss of functional reading frames occurred (Fig. 4.4), with the result that the relative amount of noncoding areas have increased again, as observed in *Epifagus*. The observation of highly compact genomes in parasitic algae and in *Cuscuta* might indicate, however, that the *Epifagus* plastid genome represents an exception.

4.3.2   Gene Losses in Parasitic Plant Plastomes

A recent study of 64 nonparasitic higher plant plastid genomes has revealed that 66 individual losses of genes have occurred in different species during evolution (Jansen et al. 2007). As many as 62 of these losses were confined to the more derived monocot and eudicot clades, and only four genes (*chlB, L,* and *N* as well as *trnP-GGG*) have disappeared from the plastid genome at a very early stage (Jansen et al. 2007). Two genes were lost particularly often: *infA* for which 11 independent losses have been recorded in eudicots and that is missing, among others from all Solanaceae (Table 4.3), and *accD* with a total of six independent losses in monocots and some eudicots. The reported losses seem to follow no specific pattern.

The adaptation to parasitism, on the other hand, has resulted in a loss of genes that is the more pronounced, the lesser the photosynthetic capacity of the parasite has become. Here, functional correlations between gene losses and metabolic activities can be drawn more easily.

4.3.2.1   *Ndh* Genes

Eleven *ndh* genes on a "normal" plastid genome code for a chloroplast NAD(P)H dehydrogenase (Table 4.2). These genes were reported missing in several photosynthetic genera, such as *Pinus*, *Phalaenopsis*, and *Chlamydomonas* (Wakasugi et al. 1994; Maul et al. 2002; Chang et al. 2006), which was interpreted as evidence that these genes are nonessential. This question seems, however, far from being resolved, as a recent report defends the view that the Ndh complex plays an essential role in photosynthesis and discusses evidence for a possible transfer of *ndh* genes to the nucleus in conifers and Gnetales (Martin and Sabater 2010).

A loss of *ndh* genes was also reported for all sequenced plastid genomes of parasitic dicots [*C. reflexa, C. exaltata, C. gronovii, C. obtusiflora*, and *E. virginiana* (Wolfe et al. 1992b; Funk et al. 2007; McNeal et al. 2007)] (Table 4.3) and the recently published underground orchid *Rhizantella gardneri* (Delannoy et al. 2011) and besides also for the parasitic liverwort *Aneura mirabilis* (Wickett et al. 2008a, b). Like in all other cases where these genes were reported missing, the entire set consisting of 11 genes has been lost without exception, since these species do no longer rely on photosynthesis.

In *E. longa*, the *ndh* genes are also reported missing. However, in this case, the loss is shared with the photosynthetic species *E. gracilis*.

4.3.2.2 Photosystem Genes

Genes for photosystem I and photosystem II as well as the cytochrome b6f complex are encoded on all nonparasitic species. Most photosystem genes were retained in the four species of *Cuscuta* whose plastid genome sequences are available. *C. reflexa* exhibits no losses, which coincides with comparatively mild reductions in the rate of photosynthesis (van der Kooij et al. 2000), while photosynthesis rates are more severely compromised in *C. gronovii*. Since all genes were found to be transcribed (Berg et al. 2003) in both species, the finding that the *psaI* gene was lost in *C. gronovii* (Funk et al. 2007) could be of significance in that context. The retention of photosynthesis genes is the biggest difference to the plastid genome of *E. virginiana* (Table 4.3). All genes associated with the bioenergetic processes of photosynthetic electron transport and ATP synthesis were either lost or pseudogenized (Table 4.3 and Fig. 4.4). The condensation grade of the plastid genome of *E. virginiana* with only 70 kbp and a coding capacity of just 42 genes (Wolfe et al. 1992b) is in higher plants only outmatched by the underground orchid *Rhizantella* (Delannoy et al. 2011).

Analyses of nucleotide substitution rates showed that the *psaI* gene that is missing in, for example, *C. gronovii*, showed significantly increased K_A/K_S values in *C. reflexa* and *C. exaltata* (Krause 2010). PsaI is a small subunit of photosystem I that has only one transmembrane domain and is involved in the docking of the PsaL subunit to this photosystem (Yu et al. 2008; Vanselow et al. 2009). The high K_A/K_S values suggest that this protein is obviously not evolving under selective constraint, and that this lack of selective pressure is what presumably leads to its eventual loss. It has, however, not yet been determined whether a copy of this gene has been transferred to the nucleus and can functionally replace the lost plastid gene.

4.3.2.3 The *rbcL* Gene

The large subunit of Rubisco, *rbcL*, is encoded by the plastid genome. In accordance with the loss of photosynthetic activity, this gene was lost in some aphotosynthetic species, such as *E. virginiana* (Wolfe et al. 1992b), *C. odorata* (van der Kooij et al. 2000), and *R. gardneri* (Delannoy et al. 2011). However, many reports on other parasitic plant families, among them holoparasitic Scrophulariaceae, showed that *rbcL* was surprisingly conserved as a functional plastid gene independent of whether the photosynthetic capacity was retained or not. Open reading frames for *rbcL* were detected, for example, in *Lathrea clandestina*, a parasite of alder (*Alnus glutinosa*), where it seems to be expressed, despite the fact that this plant lacks chlorophyll (Lusson et al. 1998). Similar situations have been described for other holoparasites (Thalouarn et al. 1994; Wolfe and dePamphilis 1997, 1998; Delavault and Thalouarn 2002).

Likewise, the parasitic liverwort *Aneura mirabilis* and the euglenoid alga *E. longa* have retained seemingly functional *rbcL* genes, while all photosystem genes were deleted (Gockel and Hachtel 2000; Wickett et al. 2008a). The fact that *rbcL* has been retained in many but not all parasitic plant plastomes makes it difficult to associate a particular meaning to this, but it has been discussed that Rubisco could have a separate metabolic function independent of photosynthesis (Krause 2008).

4.3.2.4 Ribosomal Protein Genes

A total of 25 genes of the ribosomal small and large subunits are encoded by most higher plant plastomes (Table 4.2). Although no tendency toward enhanced nucleotide substitution rates in ribosomal protein genes was observed in *Cuscuta* species (Krause 2010), parasitic plant genomes exhibit several losses of *rpl* and *rps* genes (Morden et al. 1991; Funk et al. 2007). Like with the photosynthesis genes and other gene groups, the number of losses roughly follows the gradient of dependency on heterotrophic growth. In *C. reflexa*, only two genes, *rpl23* and *rps16*, have nonfunctional reading frames and behave as pseudogenes. Along with a third gene, *rpl32*, both genes have been completely lost in *C. gronovii*. *E. virginiana* has even suffered four losses (*rpl22*, *rpl32*, *rps15*, and *rps16*) in addition to two pseudogenizations (*rpl14* and *rps23*) (Table 4.3).

The *rpl32* gene is not only missing from some parasitic plant plastid genomes but was also lost in a number of photosynthetic angiosperms (Jansen et al. 2007), among them *Populus alba* (Table 4.3). In *P. alba*, the corresponding chloroplast gene appears to have been transferred to the nucleus. The transfer of chloroplast genes to the nucleus is a process that requires many steps, including the removal of possible introns, the gain of suitable regulatory elements, as well as the acquisition of a transit peptide that can direct the nuclear gene product back to the plastids (Bungard 2004; Ravi et al. 2008). In case of the *P. alba rpl32* gene, it could be shown that it acquired the transit peptide from another plastid targeted gene, cp *sod-1* (Ueda et al. 2007), thereby paving the way for the Rpl32 protein's return into the chloroplast.

Another example for a ribosomal gene whose loss has also been observed in *P. alba* is *rps16* (Ueda et al. 2008). In this case, however, the gene was not simply transferred to the nucleus. Rather, the original plastid *rps16* gene has been lost, and this loss has been compensated for by import of the mitochondrial *rps16* gene. Apparently, the nuclear gene for the mitochondrial *rps16* gene has acquired a dual-targeting signal that is able to direct it to the plastids in addition to mitochondria, rendering the plastid's own gene dispensable.

This functional replacement of a ribosomal gene from one organelle by a dually targeted counterpart from the other organelle is not a unique case. A recent report has shown that, for example, the *rpl10* gene in several plant mitochondrial genomes has been replaced by a dually targeted copy of the original "chloroplast-only" targeted *rpl10* isoform (Kubo and Arimura 2010). It is possible that some gene losses from parasitic plant plastomes have been compensated for in a similar manner.

4.3.2.5 *Rpo* Genes

Higher plants possess two RNA polymerases, PEP and NEP, which share the responsibility of transcribing the plastid genetic information (Hess and Börner 1999). The PEP is a multi-subunit enzyme and is encoded by four plastid genes: *rpoA*, *rpoB*, *rpoC1*, and *rpoC2* (Table 4.2). The encoded subunits display a similarity to the eubacterial multisubunit RNA polymerase, and PEP recognizes promoters with a structure similar to bacterial promoters (Hess and Börner 1999). *PEP is* predominantly responsible for the transcription of photosynthesis-related genes. The generation of homoplastomic *rpo* knock-out mutants in tobacco, consequently, leads to a loss of photosynthetic capacity that is accompanied by an off-white phenotype (DeSantis-Maciossek et al. 1999) and a reduction in the amount of transcripts of photosynthesis genes (Krause et al. 2000; Legen et al. 2002).

The entire set of *rpo* genes was reported missing in *E. virginiana*, which has lost any photosynthetic activity and with that also any of the PEP-dependent genes. Losses of the *rpoA* subunit were both previously and subsequently reported for photosynthetic species, such as *Euglena gracilis* (Hallick et al. 1993), *Pelargonium* (Chumley et al. 2006), and *Passiflora* (Jansen et al. 2007) (Table 4.3). The operon containing *rpoB*, *C1*, and *C2*, on the other hand, has generally been retained on plastid genomes. To date, the only exceptions are the plastid genomes of *C. gronovii* and *C. obtusiflora*, which were the first and only plastid genomes with a confirmed loss of the entire *rpo* gene set in a plastid genome where photosynthesis genes are not only present (Funk et al. 2007) but, moreover, actively expressed (Berg et al. 2003). It has been shown that transcription of these genes was taken over by the nuclear-encoded RNA polymerase, NEP, which is highly homologous to the mitochondrial phage-type RNA polymerase and might have evolved from it by gene duplication (see Chap. 12).

Unlike parasitic angiosperms, apicomplexans, *Helicosporidium* and *E. longa* have retained the *rpo* genes in their reduced plastid genomes (Wilson et al. 1996; Gockel and Hachtel 2000; Cai et al. 2003; de Koning and Keeling 2006; Janouskovec et al. 2010). In contrast to higher plants, algae appear to possess only a single nuclear gene for a phage-type RNA polymerase and its gene product seems to be exclusively localized to the mitochondria. A second nuclear-encoded but plastid-localized RNA polymerase, NEP, seems to be missing in algae (Smith and Purton 2002). A loss of the PEP subunits encoded by the plastid *rpo* genes would therefore deprive the plastids of the possibility to transcribe their genetic information, making these genes essential.

4.3.2.6 Hypothetical Conserved Reading Frames (*Ycfs*)

Plastid genomes contain a number of conserved open reading frames of unknown function. The conservation of some of these sequences within plastomes from higher plants down to algae or even cyanobacteria is interpreted as strong indication for their functional importance (Ravi et al. 2008). Attempts to uncover the function of the *ycf* gene products have been successful in some cases, and the use of aliases to the *ycf* designations is becoming more common. For example, *ycf5* has been renamed *ccsA* [for *c*-type *c*ytochrome *s*ynthesis, (Xie and Merchant 1996)], *ycf9* is *psbZ* (Swiatek et al. 2001), and *ycf10* is now called *cemA* [for *c*hloroplast *e*nvelope *m*embrane protein *A*, (Rolland et al. 1997)]. For ease of reference, the older *ycf* designations will be used here in this chapter.

Nicotiana species contain nine conserved reading frames in their plastid genome sequence (*ycf1-5*, *ycf9,10*, and *15* and *orf350*) (Wakasugi et al. 1998; Yukawa et al. 2006), six of which have been lost in *E. virginiana* (*ycf3*, *4*, *5*, *9*, *10*, and *15*). Only the pseudogenization of *ycf15* is shared with the *Cuscuta* species (Table 4.3). The retention of *ycf1* and *ycf2* on all of the parasitic plant plastid genomes indicates that a function of these genes in photosynthesis can be most likely excluded and that rather, a possible role in gene expression or a photosynthesis-unrelated metabolic function can be envisaged. This observation corroborates previous findings that knockout mutants of *ycf1* and *ycf2* never yielded viable homoplastomic lines and that these genes are therefore essential for the survival of higher plants (Drescher et al. 2000).

4.3.2.7 tRNA Genes

The set of tRNA genes encoded by plastomes of photosynthetic plants encompasses around 30 genes and it has been argued that a transfer of tRNA genes to the nucleus and re-import of the RNAs is impossible (Barbrook et al. 2006; Howe and Purton 2007). Nevertheless, some parasitic species do exhibit extensive losses of tRNA genes.

In *C. reflexa*, only a single tRNA, that for lysine (*trnK-UUU*) is missing. More extended losses of tRNA genes were observed in *C. gronovii*, where in addition to *trnK-UUU* the sequence for *trnV-UAC* was completely eliminated from the plastid DNA, and four tRNA genes (*trnA-UGC*, *trnG-UCC*, *trnI-GAU*, and *trnR-AGC*) have been reduced to pseudogenes. In *E. virginiana*, a total of five tRNAs (*trnA-UGC*, *trnC-GCA*, *trnI-GAU*, *trnR-UCU*, and *trnS-GGA*) were pseudogenized and eight tRNAs (*trnG-GCC*, trnG-UCC, *trnK-UUU*, *trnL-UAA*, *trnT-GGU*, *trnT-UGU*, *trnV-GAC*, and *trnV-UAC*) lost.

The extensive loss of tRNA genes from parasitic plastids has, therefore, raised the question whether the missing tRNAs have been replaced by nuclear-encoded ones or whether the codon usage was adapted to these losses. An analysis of codon usages in two *Cuscuta* species has shown that all 61 sense codons were found in the coding regions of plastid genes in a similar proportion as in nonparasitic plants that possess a "full" plastid tRNA set (Funk et al. 2007). A similar picture emerges from the *Epifagus* plastid genome. This was interpreted as circumstantial evidence for an import of cytosolic tRNAs into the chloroplasts (Wolfe et al. 1992a; Funk et al. 2007). It is, however, also possible that an "extended wobble" behavior of the remaining tRNAs partly compensates for the losses.

4.3.2.8   *trnK*-UUU and *matK*

Among the tRNA genes, *trnK-UUU* has a special role as this tRNA gene harbors in its intron the only RNA maturase gene found on the plastid genome, *matK*. The *trnK-UUU* gene was found missing on all parasitic angiosperm plastomes, but neither the gene nor its intron and the *matK* gene contained within have been reported missing in any nonparasitic plants. In *C. reflexa*, *C. exaltata*, and *E. virginiana*, surprisingly, *matK* has been retained as a free-standing gene, confirming the probably essential function in plastid intron maturation that has been attributed earlier to its gene product (Hess et al. 1994; Hubschmann et al. 1996; Jenkins et al. 1997; Vogel et al. 1997). Unprecedented in higher plants, however, was the complete loss of *matK* from *C. gronovii* and *C. obtusiflora* (Funk et al. 2007; McNeal et al. 2007).

The exceptional loss of *matK* from the plastid genomes of *C. gronovii* and *C. obtusiflora* is probably closely related to the loss of group IIa introns from a whole suite of genes in parasitic plant plastids. Group IIa introns have been discussed as targets of MatK (Liere and Link 1995; Hubschmann et al. 1996; Jenkins et al. 1997; Vogel et al. 1997). Out of originally eight group IIa introns, only a single group IIa intron, namely intron 2 of *clpP*, is found in *C. gronovii* and *C. obtusiflora*, where the *matK* gene was deleted. This intron was shown to be faithfully spliced and might therefore not be a target of MatK action (Funk et al. 2007). This conclusion was supported by recent biochemical and molecular evidence linking MatK to all group IIa introns, except the *clpP* intron 2 (Zoschke et al. 2010).

In Apicomplexa and in *E. longa*, an absence of the gene *matK* along with all group II introns was observed as well but this does not seem to be the result of a parasitic lifestyle here. The loss of *matK* in these species is shared with their photosynthetic relatives since *matK* was reported to be already missing in *Chromera velia* (Janouskovec et al. 2010) and *E. gracilis* (Gockel and Hachtel 2000).

4.3.3   Reduction of RNA-Editing Sites

The loss of introns was not the only posttranscriptional processing step that has experienced a reduction. A surprise was the reduction of RNA-editing sites that was not only the result of the loss of genes that are typically richer in editing sites. Of the 30–40 editing sites of photosynthetic seed plants (Tsudzuki et al. 2001), only 17 potential sites remain in *C. reflexa*, while 12 have been found in *C. gronovii*. Several sites are only partially edited or not edited at all. The loss of editing competence as well as the reduction in the number of introns has been discussed in a very recent review (Tillich and Krause 2010) and the reader is referred to this review for further details.

4.3.4   Loss of PEP Promoters

Each of the two enzymes sharing the responsibility for plastid transcription (PEP and NEP) differs with respect to the promoters they bind to (Hess and Börner 1999). PEP promoters resemble prokaryotic promoters and occur mainly upstream of the photosynthetic genes, whereas the phage-type NEP promoters can be found upstream of genes for the genetic system. Many genes and operons, such as the gene cluster for the ribosomal RNAs, even possess both promoter types.

The loss of the *rpo* genes from two *Cuscuta* plastid genomes where photosynthesis genes are still present (Krause et al. 2003; Berg et al. 2004) raised the question of how the corresponding promoters have developed. The analysis of promoter motifs upstream of photosynthesis genes revealed that the consensus -10 and -35 boxes of PEP promoters have been so severely changed that they must be considered to be nonfunctional (Funk et al. 2007). For the *rrn* operon and the *rbcL* gene it could be, moreover, demonstrated that the start sites of transcription have been shifted relative to those of tobacco and that the $5'$ region of the novel transcription start site revealed striking similarities to the consensus sequence recognized by the NEP polymerase, indicating a shift from PEP- to NEP-driven transcription of these genes. This shift obviously enables the plastids to transcribe the previous PEP genes with high enough efficiency to allow for low photosynthetic activity.

4.4   Gene Retentions and the "Raison d'être" of Reduced Plastid Genomes

Overall, many of the changes that were seen in connection with a parasitic lifestyle seem to be shared between higher plant and algal parasites. The postulated forces that must exist, according to deKoning and Keeling, for algal parasites and that shape plastid genomes even after relaxation of photosynthetic selection pressure (de Koning and Keeling 2006) do seemingly also apply to higher plants. The most reduced plastid genomes in both groups (i.e., *Epifagus* for seed plants and Apicomplexa for algae) are characterized by a domination of genetic system genes and only two to four genes with functions outside of gene expression are present.

One question that has repeatedly been asked but so far never satisfactorily been answered is the mystery why plastid genomes were retained in nonphotosynthetic organisms. In this context, genes that encode for subunits of the gene-expression machinery, such as ribosomal RNAs and proteins, are hardly of much interest, since they are presumed to only secure the expression of the "key" plastid gene(s). Consequently, the answers why the plastid genome has not been lost altogether have been sought in the other retained genes.

In many plastid genomes of parasitic plant species, intact reading frames of the *rbcL* gene that codes for the photosynthesis protein Rubisco have been retained, despite the loss of photosynthetic activity (Krause 2008). It has been suggested that Rubisco could assume a second function in lipid biosynthesis (Schwender et al. 2004). An argument that strengthens the "essential lipid biosynthesis" hypothesis is the retention of the *accD* gene even in strongly reduced genomes, such as that from *E. virginiana*. Tobacco plastome mutants, where *accD* was interrupted, did never reach a homoplastomic state, underlining the essentiality of this gene for plants (Kode et al. 2005). However, in Apicomplexa there is no *accD* gene and *rbcL* is missing in apicomplexan parasites as well as in *Epifagus* and some *Cuscuta* species, just to name some (see Sect. 4.3.2).

Another set of genes that appear essential for plastid development independent of photosynthetic capacity are those that encode the subunits of the Clp Protease, *clpP* and *clpC*. Clp most likely performs chaperone functions and is engaged in protein import into the plastid. While in seed plants, the *clpP* gene was retained even in reduced plastomes, such as that of *Epifagus*, Apicomplexa have retained the gene *clpC*. However, exceptions are found also here (e.g., *Helicosporidium*), weakening any hypothesis that was tentatively built up around the *clp* genes.

Among the protein-coding genes, the two hypothetical reading frames *ycf1* and *ycf2* remain. Knockouts of each gene resulted persistently in heteroplasmy (Drescher et al. 2000; Shikanai et al. 2001; Kuroda and Maliga 2003) and both belong to the reduced gene set of extremely reduced plastid genomes, such as that of *E. virginiana*. However, their function might well be found to be associated with gene expression (Bock 2007), which would eliminate also these genes from being candidates for the "raison d'être" of plastid genomes.

Some recent alternative attempts at an explanation for the retention of plastid genomes circle around two transfer RNA genes. The first is the gene for the glutamyl-tRNA (*trnE*). This (*trnE*) gene fulfills three tasks in plastids: besides its role in protein biosynthesis, it plays a well-known role in the synthesis of δ-aminolaevulinic acid and, thereby, in heme biosynthesis (Jahn et al. 1992), and may regulate transcriptional activity of the NEP (Hanaoka et al. 2005), although this last function has been later challenged (Bohne et al. 2009). The essentiality of heme biosynthesis and the belief that a functional transfer of tRNA genes from the plastid to the nuclear genome is unlikely if not impossible (Barbrook et al. 2006) has been used as an argument for why a plastid genome has been retained, however much reduced it is. It has even been predicted some years ago that *trnE* may be the only gene that is found in all genomes, regardless of the degree of reductions (Barbrook et al. 2006). So far, all sequenced plastomes fulfill this prediction. However, also this hypothesis has a caveat, since in *Plasmodium*, at least, heme was found to be synthesized by an exclusively mitochondrial-located pathway, and it is therefore independent of plastid *trnE*. A second hypothesis was brought up recently, where the formylmethionyl-tRNA ($tRNA^{fM}$) plays the main role (Howe and Purton 2007). $tRNA^{fM}$ is needed for translation initiation in plastids and mitochondria, but the only $tRNA^{fM}$ gene in *Plasmodium* is the one that is located in the apicoplast. Therefore, the formylmethionyl-tRNA pool of the plastids was proposed to be shared by the mitochondria, rendering this particular tRNA indispensable (Howe and Purton 2007). Whether this holds true for further parasitic species still awaits confirmation.

4.5   From Loss of Photosynthesis to Loss of Plastids?

As described in the previous chapters, all nonphotosynthetic land plants without exception have retained a more or less cryptic plastid with a plastome that exhibits a set of typical losses and also of typical gene retentions. A number of nonphoto-synthetically living algae and descendents thereof have likewise retained a cryptic plastid that apparently fulfills some essential functions for the parasites.

For a long time, the debate has been going on why these plastids with their "crippled" plastid genomes have remained so steadfast in the nonphotosynthetic species. The discovery that the apicomplexan species *Cryptosporidium* has lost its plastid (Huang et al. 2004) has given the debate a new spin. Evidence for plastid-derived genes in the nonphotosynthetic oomycete *Phytophthora* (Tyler et al. 2006), in Ciliates (Reyes-Prieto et al. 2008), and in trypanosomatid parasites (Bodyl et al. 2010) that has recently been presented, has nourished the discussion of whether these lineages have evolved from a photosynthetic, plastid-bearing ancestor (Janouskovec et al. 2010). This scenario would imply that the loss of photosynthesis genes and the reduction of other plastome features could be just one intermediate step and that there are no evolutionary restrictions that would preclude the total loss of the plastid genome or the entire plastid. It is surprising, however, that so far no species has been found where the plastid DNA but not the plastid compartment as a whole were lost.

4.6 Conclusion and Perspectives

Analyses of nucleotide substitution rates have revealed that the mutation rates in plastid genomes are considerably lower than in their nuclear counterparts (Wolfe et al. 1987). The hypothesis that the selective pressure exerted by the photoautotrophic lifestyle has contributed significantly to this conservation of the plastid genome can best be tested by analyzing species that have evolved under a different type or level of selective constraint. Such species are present in the various groups of parasitic plants and algae. In the "omics" age, tools for high-throughput analysis and annotation of genomes are not only available, they are, more importantly, also affordable and require very small amounts of plant material. Consequently, not only the number of published chloroplast genomes from agriculturally important plants have increased considerably over the last 5 years, but also the number of genomes from "cryptic" plastids of parasitic species. While this information has been instrumental in getting an insight into the evolution of plastid genomes, a number of questions await answers in the future. Among those is the extent of a possible nuclear transfer of genes that are regarded as essential and that are as of now reported missing from the plastid genome. Another question concerns the coordination of cryptic plastid and nuclear gene expression in a nonphotosynthetic setting. To answer these questions, it will be necessary to concentrate on the nuclear genomes of some of these species in the future.

Acknowledgments Critical comments by Prof. Karsten Fischer, University of Tromsø, during the preparation of this manuscript are gratefully acknowledged. Prof. Thomas Börner is thanked for reviewing the final version of this chapter.

References

Barbrook AC, Howe CJ (2000) Minicircular plastid DNA in the dinoflagellate *Amphidinium operculatum*. Mol Gen Genet 263:152–158

Barbrook AC, Howe CJ, Purton S (2006) Why are plastid genomes retained in non-photosynthetic organisms? Trends Plant Sci 11:101–108

Barkman TJ, McNeal JR, Lim SH, Coat G, Croom HB, Young ND, Depamphilis CW (2007) Mitochondrial DNA suggests at least 11 origins of parasitism in angiosperms and reveals genomic chimerism in parasitic plants. BMC Evol Biol 7:248

Berg S, Krupinska K, Krause K (2003) Plastids of three *Cuscuta* species differing in plastid coding capacity have a common parasite-specific RNA composition. Planta 218:135–142

Berg S, Krause K, Krupinska K (2004) The rbcL genes of two *Cuscuta* species, *C. gronovii* and *C. subinclusa*, are transcribed by the nuclear-encoded plastid RNA polymerase (NEP). Planta 219:541–546

Bidartondo MI, Bruns TD, Weiss M, Sergio C, Read DJ (2003) Specialized cheating of the ectomycorrhizal symbiosis by an epiparasitic liverwort. Proc Biol Sci 270:835–842

Bock R (2007) Structure, function, and inheritance of plastid genomes. In: Bock R (ed) Cell and molecular biology of plastids, vol 19, Topics in current genetics. Springer, Heidelberg

Bodyl A, Mackiewicz P, Milanowski R (2010) Did trypanosomatid parasites contain a eukaryotic alga-derived plastid in their evolutionary past? J Parasitol 96:465–475

Bohne A-V, Weihe A, Börner T (2009) Transfer RNAs inhibit *Arabidopsis* phage-type RNA polymerases. Endocytobiosis Cell Res 19:63–69

Borza T, Popescu CE, Lee RW (2005) Multiple metabolic roles for the nonphotosynthetic plastid of the green alga *Prototheca wickerhamii*. Eukaryot Cell 4:253–261

Bungard RA (2004) Photosynthetic evolution in parasitic plants: insight from the chloroplast genome. Bioessays 26:235–247

Cai X, Fuller AL, McDougald LR, Zhu G (2003) Apicoplast genome of the coccidian *Eimeria tenella*. Gene 321:39–46

Cavalier-Smith T (1999) Principles of protein and lipid targeting in secondary symbiogenesis: euglenoid, dinoflagellate, and sporozoan plastid origins and the eukaryote family tree. J Eukaryot Microbiol 46:347–366

Cavalier-Smith T (2002) The phagotrophic origin of eukaryotes and phylogenetic classification of Protozoa. Int J Syst Evol Microbiol 52:297–354

Chang CC, Lin HC, Lin IP, Chow TY, Chen HH, Chen WH, Cheng CH, Lin CY, Liu SM, Chaw SM (2006) The chloroplast genome of *Phalaenopsis aphrodite* (Orchidaceae): comparative analysis of evolutionary rate with that of grasses and its phylogenetic implications. Mol Biol Evol 23:279–291

Chumley TW, Palmer JD, Mower JP, Fourcade HM, Calie PJ, Boore JL, Jansen RK (2006) The complete chloroplast genome sequence of Pelargonium x hortorum: organization and evolution of the largest and most highly rearranged chloroplast genome of land plants. Mol Biol Evol 23:2175–2190

de Koning AP, Keeling PJ (2004) Nucleus-encoded genes for plastid-targeted proteins in *Helicosporidium*: functional diversity of a cryptic plastid in a parasitic alga. Eukaryot Cell 3:1198–1205

de Koning AP, Keeling PJ (2006) The complete plastid genome sequence of the parasitic green alga *Helicosporidium* sp. is highly reduced and structured. BMC Biol 4:12

Delannoy E, Fujii S, des Francs CC, Brundrett M, Small I (2011) Rampant gene loss in the underground orchid *Rhizanthella gardneri* highlights evolutionary constraints on plastid genomes. Mol Biol Evol 28(7):2077–2086

Delavault P, Thalouarn P (2002) The obligate root parasite *Orobanche cumana* exhibits several rbcL sequences. Gene 297:85–92

Deng XW, Wing RA, Gruissem W (1989) The chloroplast genome exists in multimeric forms. Proc Natl Acad Sci U S A 86:4156–4160

DeSantis-Maciossek G, Kofer W, Bock A, Schoch S, Maier RM, Wanner G, Rüdiger W, Koop HU, Herrmann RG (1999) Targeted disruption of the plastid RNA polymerase genes *rpoA*, *B* and *C1*: molecular biology, biochemistry and ultrastructure. Plant J 18:477–489

Douglas SE, Penny SL (1999) The plastid genome of the cryptophyte alga, *Guillardia theta*: complete sequence and conserved synteny groups confirm its common ancestry with red algae. J Mol Evol 48:236–244

Drescher A, Ruf S, Calsa T Jr, Carrer H, Bock R (2000) The two largest chloroplast genome-encoded open reading frames of higher plants are essential genes. Plant J 22:97–104

Funk HT, Berg S, Krupinska K, Maier UG, Krause K (2007) Complete DNA sequences of the plastid genomes of two parasitic flowering plant species, *Cuscuta reflexa* and *Cuscuta gronovii*. BMC Plant Biol 7:45

Glockner G, Rosenthal A, Valentin K (2000) The structure and gene repertoire of an ancient red algal plastid genome. J Mol Evol 51:382–390

Gockel G, Hachtel W (2000) Complete gene map of the plastid genome of the nonphotosynthetic euglenoid flagellate *Astasia longa*. Protist 151:347–351

Goulding SE, Olmstead RG, Morden CW, Wolfe KH (1996) Ebb and flow of the chloroplast inverted repeat. Mol Gen Genet 252:195–206

Hagopian JC, Reis M, Kitajima JP, Bhattacharya D, de Oliveira MC (2004) Comparative analysis of the complete plastid genome sequence of the red alga *Gracilaria tenuistipitata* var. liui provides insights into the evolution of rhodoplasts and their relationship to other plastids. J Mol Evol 59:464–477

Hallick RB, Hong L, Drager RG, Favreau MR, Monfort A, Orsat B, Spielmann A, Stutz E (1993) Complete sequence of *Euglena gracilis* chloroplast DNA. Nucleic Acids Res 21:3537–3544

Hanaoka H, Kanamaru K, Fujiwara M, Takahashi H, Tanaka K (2005) Glutamyl-tRNA mediates a switch in RNA polymerase use during chloroplast biogenesis. EMBO Rep 6:545–550

Hess W, Börner T (1999) Organellar RNA polymerases of higher plants. Int Rev Cytol 190:1–59

Hess W, Müller A, Nagy F, Börner T (1994) Ribosome-deficient plastids affect transcription of light-induced nuclear genes: genetic evidence for a plastid-derived signal. Mol Gen Genet 242:305–312

Hibberd JM, Bungard RA, Press MC, Jeschke WD, Scholes JD, Quick WP (1998) Localization of photosynthetic metabolism in the parasitic angiosperm *Cuscuta reflexa*. Planta 205:506–513

Howe CJ, Purton S (2007) The little genome of apicomplexan plastids: its raison d'etre and a possible explanation for the 'delayed death' phenomenon. Protist 158:121–133

Huang J, Mullapudi N, Lancto CA, Scott M, Abrahamsen MS, Kissinger JC (2004) Phylogenomic evidence supports past endosymbiosis, intracellular and horizontal gene transfer in *Cryptosporidium parvum*. Genome Biol 5:R88

Hubschmann T, Hess WR, Borner T (1996) Impaired splicing of the rps12 transcript in ribosome-deficient plastids. Plant Mol Biol 30:109–123

Jahn D, Verkamp E, Soll D (1992) Glutamyl-transfer RNA: a precursor of heme and chlorophyll biosynthesis. Trends Biochem Sci 17:215–218

Janouskovec J, Horak A, Obornik M, Lukes J, Keeling PJ (2010) A common red algal origin of the apicomplexan, dinoflagellate, and heterokont plastids. Proc Natl Acad Sci U S A 107:10949–10954

Jansen RK, Cai Z, Raubeson LA, Daniell H, Depamphilis CW, Leebens-Mack J, Muller KF, Guisinger-Bellian M, Haberle RC, Hansen AK, Chumley TW, Lee SB, Peery R, McNeal JR, Kuehl JV, Boore JL (2007) Analysis of 81 genes from 64 plastid genomes resolves relationships in angiosperms and identifies genome-scale evolutionary patterns. Proc Natl Acad Sci U S A 104:19369–19374

Jansen RK, Wojciechowski MF, Sanniyasi E, Lee SB, Daniell H (2008) Complete plastid genome sequence of the chickpea (*Cicer arietinum*) and the phylogenetic distribution of rps12 and clpP intron losses among legumes (Leguminosae). Mol Phylogenet Evol 48:1204–1217

Jenkins BD, Kulhanek DJ, Barkan A (1997) Nuclear mutations that block group II RNA splicing in maize chloroplasts reveal several intron classes with distinct requirements for splicing factors. Plant Cell 9:283–296

Kode V, Mudd EA, Iamtham S, Day A (2005) The tobacco plastid accD gene is essential and is required for leaf development. Plant J 44:237–244

Kolodner R, Tewari KK (1972) Molecular size and conformation of chloroplast deoxyribonucleic acid from pea leaves. J Biol Chem 247:6355–6364

Krause K (2008) From chloroplasts to "cryptic" plastids: evolution of plastid genomes in parasitic plants. Curr Genet 54:111–121

Krause K (2010) Plastid genome variation and selective constraint in species of the parasitic plant genus *Cuscuta*. In: Urbano KV (ed) Advances in genetics research, vol. 2, chapter 15. Nova Science Publishers, New York, pp 1–15

Krause K, Maier RM, Kofer W, Krupinska K, Herrmann RG (2000) Disruption of plastid-encoded RNA polymerase genes in tobacco: expression of only a distinct set of genes is not based on selective transcription of the plastid chromosome. Mol Gen Genet 263:1022–1030

Krause K, Berg S, Krupinska K (2003) Plastid transcription in the holoparasitic plant genus *Cuscuta*: parallel loss of the rrn16 PEP-promoter and of the rpoA and rpoB genes coding for the plastid-encoded RNA polymerase. Planta 216:815–823

Kubo N, Arimura S (2010) Discovery of the rpl10 gene in diverse plant mitochondrial genomes and its probable replacement by the nuclear gene for chloroplast RPL10 in two lineages of angiosperms. DNA Res 17:1–9

Kuroda H, Maliga P (2003) The plastid *clpP1* protease gene is essential for plant development. Nature 425:86–89

Kuroiwa T (1991) The replication, differentiation, and inheritance of plastids with emphasis on the concept of organelle nuclei. Int Rev Cytol 128:1–62

Lane CE, Archibald JM (2008) The eukaryotic tree of life: endosymbiosis takes its TOL. Trends Ecol Evol 23:268–275

Legen J, Kemp S, Krause K, Profanter B, Herrmann RG, Maier RM (2002) Comparative analysis of plastid transcription profiles of entire plastid chromosomes from tobacco attributed to wild-type and PEP-deficient transcription machineries. Plant J 31:171–188

Liere K, Link G (1995) RNA-binding activity of the matK protein encoded by the chloroplast trnK intron from mustard (*Sinapis alba* L.). Nucleic Acids Res 23:917–921

Lilly JW, Havey MJ, Jackson SA, Jiang J (2001) Cytogenomic analyses reveal the structural plasticity of the chloroplast genome in higher plants. Plant Cell 13:245–254

Lim L, McFadden GI (2010) The evolution, metabolism and functions of the apicoplast. Philos Trans R Soc Lond B Biol Sci 365:749–763

Lusson NA, Delavault PM, Thalouarn PA (1998) The rbcL gene from the non-photosynthetic parasite *Lathraea clandestina* is not transcribed by a plastid-encoded RNA polymerase. Curr Genet 34:212–215

Martin M, Sabater B (2010) Plastid ndh genes in plant evolution. Plant Physiol Biochem 48:636–645

Matsuzaki M, Kuroiwa H, Kuroiwa T, Kita K, Nozaki H (2008) A cryptic algal group unveiled: a plastid biosynthesis pathway in the oyster parasite *Perkinsus marinus*. Mol Biol Evol 25:1167–1179

Maul JE, Lilly JW, Cui L, dePamphilis CW, Miller W, Harris EH, Stern DB (2002) The *Chlamydomonas reinhardtii* plastid chromosome: islands of genes in a sea of repeats. Plant Cell 14:2659–2679

McNeal JR, Kuehl JV, Boore JL, de Pamphilis CW (2007) Complete plastid genome sequences suggest strong selection for retention of photosynthetic genes in the parasitic plant genus *Cuscuta*. BMC Plant Biol 7:57

Morden CW, Wolfe KH, dePamphilis CW, Palmer JD (1991) Plastid translation and transcription genes in a non-photosynthetic plant: intact, missing and pseudo genes. EMBO J 10:3281–3288

Ohta N, Matsuzaki M, Misumi O, Miyagishima SY, Nozaki H, Tanaka K, Shin IT, Kohara Y, Kuroiwa T (2003) Complete sequence and analysis of the plastid genome of the unicellular red alga *Cyanidioschyzon merolae*. DNA Res 10:67–77

Ohyama K, Fukuzawa H, Kohchi T, Shirai H, Sano T, Sano S, Takeuchi M, Chang Z, Aota S-I, Inokuchi H, Ozeki H (1986) Chloroplast gene organization deduced form complete sequence of liverwort *Marchantia polymorpha* chloroplast DNA. Nature 322:572–574

Oldenburg DJ, Bendich AJ (2004) Most chloroplast DNA of maize seedlings in linear molecules with defined ends and branched forms. J Mol Biol 335:953–970

Palmer JD (1990) Contrasting modes and tempos of genome evolution in land plant organelles. Trends Genet 6:115–120

Ravi V, Khurana JP, Tyagi AK, Khurana P (2008) An update on chloroplast genomes. Plant Syst Evol 271:101–122

Reith M, Munholland J (1993) A high-resolution gene map of the chloroplast genome of the red alga *Porphyra purpurea*. Plant Cell 5:465–475

Reyes-Prieto A, Bhattacharya D (2007) Phylogeny of nuclear-encoded plastid-targeted proteins supports an early divergence of glaucophytes within Plantae. Mol Biol Evol 24:2358–2361

Reyes-Prieto A, Moustafa A, Bhattacharya D (2008) Multiple genes of apparent algal origin suggest ciliates may once have been photosynthetic. Curr Biol 18:956–962

Robbens S, Derelle E, Ferraz C, Wuyts J, Moreau H, Van de Peer Y (2007) The complete chloroplast and mitochondrial DNA sequence of *Ostreococcus tauri*: organelle genomes of the smallest eukaryote are examples of compaction. Mol Biol Evol 24:956–968

Rogers MB, Gilson PR, Su V, McFadden GI, Keeling PJ (2007) The complete chloroplast genome of the chlorarachniophyte *Bigelowiella natans*: evidence for independent origins of chlorarachniophyte and euglenid secondary endosymbionts. Mol Biol Evol 24:54–62

Rolland N, Dorne AJ, Amoroso G, Sultemeyer DF, Joyard J, Rochaix JD (1997) Disruption of the plastid ycf10 open reading frame affects uptake of inorganic carbon in the chloroplast of Chlamydomonas. EMBO J 16:6713–6726

Rumpho ME, Worful JM, Lee J, Kannan K, Tyler MS, Bhattacharya D, Moustafa A, Manhart JR (2008) Horizontal gene transfer of the algal nuclear gene psbO to the photosynthetic sea slug *Elysia chlorotica*. Proc Natl Acad Sci U S A 105:17867–17871

Sanchez Puerta MV, Bachvaroff TR, Delwiche CF (2005) The complete plastid genome sequence of the haptophyte *Emiliania huxleyi*: a comparison to other plastid genomes. DNA Res 12:151–156

Sanchez-Puerta MV, Lippmeier JC, Apt KE, Delwiche CF (2007) Plastid genes in a non-photosynthetic dinoflagellate. Protist 158:105–117

Scharff LB, Koop HU (2006) Linear molecules of tobacco ptDNA end at known replication origins and additional loci. Plant Mol Biol 62:611–621

Schwender J, Goffman F, Ohlrogge JB, Shachar-Hill Y (2004) Rubisco without the Calvin cycle improves the carbon efficiency of developing green seeds. Nature 432:779–782

Shikanai T, Shimizu K, Ueda K, Nishimura Y, Kuroiwa T, Hashimoto T (2001) The chloroplast *clp* Gene, encoding a proteolytic subunit of ATP-dependent protease, is indispensable for chloroplast development in tobacco. Plant and Cell Physiol 42:264–273

Shinozaki K, Ohme M, Tanaka M, Wakasugi T, Hayashida N, Matsubayashi T, Zaita N, Chunwongse J, Obokata J, Yamaguchi-Shinozaki K, Ohto C, Torazawa K, Meng BY, Sugita M, Deno H, Kamogashira T, Yamada K, Kusuda J, Takaiwa F, Kato A, Tohdoh N, Shimada H, Sugiura M (1986) The complete nucleotide sequence of the tobacco chloroplast genome: its gene organization and expression. EMBO J 5:2043–2049

Simpson CL, Stern DB (2002) The treasure trove of algal chloroplast genomes. Surprises in architecture and gene content, and their functional implications. Plant Physiol 129:957–966

Smith AC, Purton S (2002) The transcriptional apparatus of algal plastids. Eur J Phycol 37:301–311

Stefanovic S, Olmstead RG (2005) Down the slippery slope: plastid genome evolution in Convolvulaceae. J Mol Evol 61:292–305

Swiatek M, Kuras R, Sokolenko A, Higgs D, Olive J, Cinque G, Muller B, Eichacker LA, Stern DB, Bassi R, Herrmann RG, Wollman FA (2001) The chloroplast gene ycf9 encodes a photosystem II (PSII) core subunit, PsbZ, that participates in PSII supramolecular architecture. Plant Cell 13:1347–1367

Tartar A, Boucias DG (2004) The non-photosynthetic, pathogenic green alga Helicosporidium sp. has retained a modified, functional plastid genome. FEMS Microbiol Lett 233:153–157

Thalouarn PA, Theodet C, Russo NM, Delavault P (1994) The reduced plastid genome of a non-photosynthetic angiosperm *Orobanche hederae* has retained the *rbcL* gene. Plant Physiol Biochem 32:233–242

Tillich M, Krause K (2010) The ins and outs of editing and splicing of plastid RNAs: lessons from parasitic plants. N Biotechnol 27:256–266

Tsudzuki T, Wakasugi T, Sugiura M (2001) Comparative analysis of RNA editing sites in higher plant chloroplasts. J Mol Evol 53:327–332

Turmel M, Otis C, Lemieux C (1999) The complete chloroplast DNA sequence of the green alga *Nephroselmis olivacea*: insights into the architecture of ancestral chloroplast genomes. Proc Natl Acad Sci U S A 96:10248–10253

Tyler BM, Tripathy S, Zhang X, Dehal P, Jiang RH, Aerts A, Arredondo FD, Baxter L, Bensasson D, Beynon JL, Chapman J, Damasceno CM, Dorrance AE, Dou D, Dickerman AW, Dubchak IL,

Garbelotto M, Gijzen M, Gordon SG, Govers F, Grunwald NJ, Huang W, Ivors KL, Jones RW, Kamoun S, Krampis K, Lamour KH, Lee MK, McDonald WH, Medina M, Meijer HJ, Nordberg EK, Maclean DJ, Ospina-Giraldo MD, Morris PF, Phuntumart V, Putnam NH, Rash S, Rose JK, Sakihama Y, Salamov AA, Savidor A, Scheuring CF, Smith BM, Sobral BW, Terry A, Torto-Alalibo TA, Win J, Xu Z, Zhang H, Grigoriev IV, Rokhsar DS, Boore JL (2006) Phytophthora genome sequences uncover evolutionary origins and mechanisms of pathogenesis. Science 313:1261–1266

Ueda M, Fujimoto M, Arimura S, Murata J, Tsutsumi N, Kadowaki K (2007) Loss of the rpl32 gene from the chloroplast genome and subsequent acquisition of a preexisting transit peptide within the nuclear gene in *Populus*. Gene 402:51–56

Ueda M, Nishikawa T, Fujimoto M, Takanashi H, Arimura S, Tsutsumi N, Kadowaki K (2008) Substitution of the gene for chloroplast RPS16 was assisted by generation of a dual targeting signal. Mol Biol Evol 25:1566–1575

van der Kooij TA, Krause K, Dorr I, Krupinska K (2000) Molecular, functional and ultrastructural characterisation of plastids from six species of the parasitic flowering plant genus *Cuscuta*. Planta 210:701–707

Vanselow C, Weber AP, Krause K, Fromme P (2009) Genetic analysis of the Photosystem I subunits from the red alga, *Galdieria sulphuraria*. Biochim Biophys Acta 1787:46–59

Vogel J, Hubschmann T, Borner T, Hess WR (1997) Splicing and intron-internal RNA editing of trnK-matK transcripts in barley plastids: support for MatK as an essential splice factor. J Mol Biol 270:179–187

Wakasugi T, Tsudzuki J, Ito S, Nakashima K, Tsudzuki T, Sugiura M (1994) Loss of all ndh genes as determined by sequencing the entire chloroplast genome of the black pine *Pinus thunbergii*. Proc Natl Acad Sci U S A 91:9794–9798

Wakasugi T, Nagai T, Kapoor M, Sugita M, Ito M, Ito S, Tsudzuki J, Nakashima K, Tsudzuki T, Suzuki Y, Hamada A, Ohta T, Inamura A, Yoshinaga K, Sugiura M (1997) Complete nucleotide sequence of the chloroplast genome from the green alga *Chlorella vulgaris*: the existence of genes possibly involved in chloroplast division. Proc Natl Acad Sci U S A 94:5967–5972

Wakasugi T, Sugita M, Tsudzuki T, Sugiura M (1998) Updated map of tobacco chloroplast DNA. Plant Mol Biol Rep 16:231–241

Wickett NJ, Fan Y, Lewis PO, Goffinet B (2008a) Distribution and evolution of pseudogenes, gene losses, and a gene rearrangement in the plastid genome of the nonphotosynthetic liverwort, *Aneura mirabilis* (Metzgeriales, Jungermanniopsida). J Mol Evol 67:111–122

Wickett NJ, Zhang Y, Hansen SK, Roper JM, Kuehl JV, Plock SA, Wolf PG, DePamphilis CW, Boore JL, Goffinet B (2008b) Functional gene losses occur with minimal size reduction in the plastid genome of the parasitic liverwort *Aneura mirabilis*. Mol Biol Evol 25:393–401

Wilson RJ, Denny PW, Preiser PR, Rangachari K, Roberts K, Roy A, Whyte A, Strath M, Moore DJ, Moore PW, Williamson DH (1996) Complete gene map of the plastid-like DNA of the malaria parasite *Plasmodium falciparum*. J Mol Biol 261:155–172

Wolfe AD, dePamphilis CW (1997) Alternate paths of evolution for the photosynthetic gene rbcL in four nonphotosynthetic species of Orobanche. Plant Mol Biol 33:965–977

Wolfe AD, dePamphilis CW (1998) The effect of relaxed functional constraints on the photosynthetic gene rbcL in photosynthetic and nonphotosynthetic parasitic plants. Mol Biol Evol 15:1243–1258

Wolfe KH, Li WH, Sharp PM (1987) Rates of nucleotide substitution vary greatly among plant mitochondrial, chloroplast, and nuclear DNAs. Proc Natl Acad Sci U S A 84:9054–9058

Wolfe KH, Morden CW, Ems SC, Palmer JD (1992a) Rapid evolution of the plastid translational apparatus in a nonphotosynthetic plant: loss or accelerated sequence evolution of tRNA and ribosomal protein genes. J Mol Evol 35:304–317

Wolfe KH, Morden CW, Palmer JD (1992b) Function and evolution of a minimal plastid genome from a nonphotosynthetic parasitic plant. Proc Natl Acad Sci U S A 89:10648–10652

Xie Z, Merchant S (1996) The plastid-encoded ccsA gene is required for heme attachment to chloroplast c-type cytochromes. J Biol Chem 271:4632–4639

Yoon HS, Hackett JD, Van Dolah FM, Nosenko T, Lidie KL, Bhattacharya D (2005) Tertiary endosymbiosis driven genome evolution in dinoflagellate algae. Mol Biol Evol 22:1299–1308

Yu J, Ma PJ, Shi DJ, Li SM, Wang CL (2008) Homologous comparisons of photosynthetic system I genes among cyanobacteria and chloroplasts. J Integr Plant Biol 50:929–940

Yue F, Cui L, dePamphilis CW, Moret BM, Tang J (2008) Gene rearrangement analysis and ancestral order inference from chloroplast genomes with inverted repeat. BMC Genomics 9 (Suppl 1):S25

Yukawa M, Tsudzuki T, Sugiura M (2006) The chloroplast genome of *Nicotiana sylvestris* and *Nicotiana tomentosiformis*: complete sequencing confirms that the *Nicotiana sylvestris* progenitor is the maternal genome donor of *Nicotiana tabacum*. Mol Genet Genomics 275:367–373

Zhang Z, Green BR, Cavalier-Smith T (1999) Single gene circles in dinoflagellate chloroplast genomes. Nature 400:155–159

Zoschke R, Nakamura M, Liere K, Sugiura M, Borner T, Schmitz-Linneweber C (2010) An organellar maturase associates with multiple group II introns. Proc Natl Acad Sci U S A 107:3245–3250

Chapter 5
Mitochondria, Hydrogenosomes and Mitosomes in Relation to the CoRR Hypothesis for Genome Function and Evolution

Wilson B.M. de Paula, John F. Allen, and Mark van der Giezen

5.1 Introduction

Mitochondria and chloroplasts are energy-converting organelles in the cytoplasm of eukaryotic cells. Chloroplasts perform photosynthesis; the capture and conversion of the energy of sunlight. Mitochondria perform respiration; the release of this stored energy when work is done. Mitochondria and chloroplasts also contain small, specialised genetic systems to make a few of their own proteins. Both the genetic and the energy-converting machineries of mitochondria and chloroplasts are descended, with little modification, from those of the free-living bacteria that these organelles once were. Today, almost all genes for proteins of chloroplasts and mitochondria are found on chromosomes in the nuclei of eukaryotic cells. There they code for protein precursors that are made in the cytosol for import into these two bioenergetic organelles, there to be trimmed down into their mature, functional forms. So why are any characters at all still inherited through the cytoplasm? Why do just a few genes remain steadfastly within chloroplasts and mitochondria as vestiges of ancestral, bacterial DNA?

In 1925, the American cytologist Edmund B. Wilson wrote as follows in "The cell in development and heredity," third edition, (Wilson 1925).

> ...much interest has been aroused in recent years by cytological studies on the mitochondria and chondriosomes, cytoplasmic structural elements now widely believed to play an important part in chemical activities of cells and also in differentiation; by some

W.B.M. de Paula • J.F. Allen
Centre for Eukaryotic Evolutionary Microbiology, Biosciences, College of Life and Environmental Sciences, University of Exeter, Devon EX4 4QJ, UK
e-mail: w.depaula@qmul.ac.uk; j.f.allen@qmul.ac.uk

M. van der Giezen (✉)
School of Biological and Chemical Sciences, Queen Mary, University of London, Mile End Road, London E1 4NS, UK
e-mail: m.vandergiezen@exeter.ac.uk

C.E. Bullerwell (ed.), *Organelle Genetics*,
DOI 10.1007/978-3-642-22380-8_5, © Springer-Verlag Berlin Heidelberg 2012

authors, accordingly (Benda, Meves) they have been regarded as representing a mechanism of "cytoplasmic heredity" comparable in importance with that represented by the chromosomes. This view, still very far from substantiation, remains a subject of controversy and must be taken with proper scepticism; but in spite of its doubtful status it should be kept clearly in view in all cytological discussions of these problems.

Today, there can be little doubt that mitochondria, as they are universally now known, indeed play a pivotal role in biochemistry, development, cytoplasmic inheritance and evolution (Lane 2005). In particular, the evolutionary origin of mitochondria from endosymbiotic bacteria is widely accepted (Allen et al. 2007; Gray et al. 1999). Until the discovery of mitochondrial DNA (mtDNA), however, pioneer scientists in endosymbiosis were disregarded and their conclusions displaced by other theories. A predominant theory at the time was that all structural elements of the eukaryotic cell evolved sequentially, in one lineage. Also, it was generally assumed that mitochondria are synthesised de novo as differentiated compartments within a wholly autogenous eukaryotic cell, as reviewed critically by Margulis (1970, 1981). In 1905, a Russian scientist named Constantin Mereschkowsky published a theory describing our contemporary concept of endosymbiosis (Martin and Kowallik 1999; Mereschkowsky 1905). Mereschkowsky made prodigious assumptions for his time, such as that chloroplasts, which he termed *chromatophores* (colour bearers), and autotrophy descend from cyanobacteria (then known as unicellular algae), and that nuclei have originated from an invasion of a small "micrococcus" into a larger heterotrophic amoeba-like host cell. Chloroplasts are coloured and thus conspicuous by light microscopy as intracellular organelles in plant and algal cells. Mitochondria, also termed "chondriosomes," were less well characterised and little was known of their function. Priority for the proposal of an endosymbiotic origin of mitochondria may be unclear (Martin 2007), but Ivan Wallin (1923) questioned the interpretation of previous experiments that appeared to support the nuclear-cytosolic origin of mitochondria. Wallin's experiments provided evidence in support of his assertion of a bacterial origin of this organelle (Martin 2007; Wallin 1923). The scientific community generally seems to have remained sceptical until the pioneering synthesis in 1967 by Lynn Sagan (later adopting the name Lynn Margulis) who reviewed observations supporting the endosymbiotic origin of mitochondria and chloroplasts (Sagan 1967). A milestone, this work provided an alternative to "direct filiation" (Margulis 1970) by reviewing properties that mitochondria and chloroplasts have in common with modern bacteria, such as 70 S ribosomes, circular DNA and reproduction by binary fission. Undoubtedly, this period was a scientific revolution and the beginning of a new paradigm in organelle biology.

Following evidence that mitochondria contained DNA (Luck and Reich 1964; Nass and Nass 1963), the first mitochondrial genomes were sequenced (Anderson et al. 1981). Initial analysis of mtDNA sequence and expression found unmistakable evidence that contemporary mitochondria were once free-living bacteria (Gray and Doolittle 1982). Furthermore, it has also been agreed that mitochondria are closely related to the contemporary free-living α-proteobacteria (Yang et al. 1985). However, even though studies on mtDNA have significantly increased our understanding on the evolutionary aspects of this organelle, many other questions have arisen within and alongside the endosymbiosis theory. Which α-proteobacterial species is the closest candidate to the proposed mitochondrial ancestor? What was the evolutionary driving force for this endosymbiotic event to happen and to be maintained over billions of years? Even though the answers to these questions remain uncertain, several compelling studies have formulated assumptions worth describing.

Studies in the 1990s led to the suggestion that the α-proteobacterial order Rickettsiales contains most characteristics shared with the putative mitochondrial ancestor (Gray et al. 1998). An "Ox-Tox" model (Kurland and Andersson 2000), whereby the proto-mitochondrion served to quench oxygen free radicals fitted nicely with the strict aerobic nature of *Rickettsia* species. The main thrust of this theory was that the acquisition of oxygen tolerance, at a time when atmospheric oxygen concentration was rising due to photosynthetic activity, was the most valuable advantage for a strict anaerobic host cell to acquire from an aerobic symbiotic organism (such as *Rickettsia prowazekii*). Additionally, the "Ox-Tox" theory propounds the importance of the origin of an ADP/ATP mitochondrial translocator, which made the exchange of energetic currencies between the symbiont and the host cell possible. *Rickettsia*'s ADP/ATP translocator, however, is unrelated to the mitochondrial one (Winkler and Neuhaus 1999), partly undermining the theory that *Rickettsia* is related to the mitochondrial endosymbiont. A large-scale analysis to assess the contribution of the mitochondrial endosymbiont to eukaryote nuclear genomes indicates the massive effect of endosymbiotic gene transfer on overall eukaryotic evolution (Esser et al. 2004). This work also indicated the likelihood of other, more biochemically versatile, α-proteobacteria being good candidates for the mitochondrial endosymbiont. However, exactly pinpointing the mitochondrial endosymbiont to an extant α-proteobacterium is complicated by the dynamic nature of prokaryotic genomes due to lateral gene transfer and gene loss (Esser et al. 2007). Nonetheless, biochemically more versatile α-proteobacteria such as *Rhodobacter* do seem more often come to the fore in more intense studies (Atteia et al. 2009).

Recently, a hypothesis has been put forward elucidating the role of bioenergetics in the prokaryote to eukaryote transition. Mitochondria play an indispensable role in this hypothesis (Lane and Martin 2010). According to this hypothesis, the ATP produced by mitochondrial oxidative phosphorylation has provided the energy necessary for the expression of an immensely larger number of genes than would have been possible without such a powerhouse. The endosymbiotic event that resulted in the establishment of the mitochondrion was therefore a crucial event for the evolution of the eukaryotes as a whole. The opposite, which the evolution of a complex eukaryote enabled the endosymbiotic event that lead to the establishment of the mitochondrion, is untenable according to the rules of bioenergetics. In addition, there is currently no evidence that mitochondria-free eukaryotes ever existed. Supposedly "primitive" eukaryotes that would be devoid of mitochondria (Cavalier-Smith 1983) have been shown to be secondarily derived. Eukaryotes such as microsporidia, *Giardia*, *Trichomonas* and *Entamoeba* were put forward as related to the putative host to the mitochondrial endosymbiont due to their simple cell structures. Initial molecular phylogenies indeed placed these organisms (except

Entamoeba) at the base of the eukaryotes (Sogin et al. 1989; Vossbrinck et al. 1987). Subsequent work showed that these early phylogenetic reconstructions were fraught with methodological artefacts such as long-branch attraction (Brinkmann et al. 2005; Embley and Hirt 1998) and the true relationships within the eukaryotes are currently not certain (Simpson and Roger 2004). More importantly, the assumption that these eukaryotes were devoid of mitochondria proved to be unfounded. They were, however, devoid of "classic" 2 μm oval cristate mitochondria as their mitochondria turned out to be very small non-descript vesicles. The notion that these "primitive" eukaryotes were not that primitive after all became apparent when genes encoding typical mitochondrial proteins, such as chaperonins, were found in these organisms (Clark and Roger 1995; Horner et al. 1996; Roger et al. 1996). For *Entamoeba*, antibodies raised against these proteins localised in a punctuate pattern throughout the cytoplasm suggestive of an organellar localisation (Mai et al. 1999; Tovar ct al. 1999). In addition, immunogold electron microscopy clearly labelled small organelles that had two membranes for *Giardia* and microsporidia (Tovar et al. 2003; Williams et al. 2002). The presence of two surrounding membranes is a defining feature of organelles of endosymbiotic origin (Henze and Martin 2003). These organelles of *Entamoeba*, *Giardia* and microsporidia were termed mitosomes. Subsequent genome projects and large-scale proteomics attempts to elucidate the nature of these elusive mitochondria have not been able to provide much information about the role these organelles play. A common feature seems to be the production of iron–sulphur clusters as in other mitochondria (Lill and Mühlenhoff 2005). Mitosomes do not seem to play a role in ATP production and are devoid of components of the mitochondrial electron transport chain. No organellar genome has been detected, either directly (León-Avila and Tovar 2004) or indirectly from the genome projects for these organisms (Clark et al. 2007; Loftus et al. 2005; Morrison et al. 2007). In the case of the trichomonads, the situation was slightly different as an unusual organelle was known to be present for quite some time (Cerkasovová et al. 1973; Lindmark and Müller 1973). This hydrogenosome had been shown to play a role in cellular energetics but unusually produced molecular hydrogen as a metabolic end-product (Müller 1993). Despite some initial claims (Cerkasovová et al. 1976), no organellar genome could be detected in hydrogenosomes (Turner and Müller 1983). Several mitochondrial proteins do, however, localise to these organelles (Bui et al. 1996; Horner et al. 1996; Lahti et al. 1992, 1994) and are targeted there by means of cleavable mitochondrial-like targeting signals (Bradley et al. 1997). More recently, many more variations of hydrogenosomes and mitosomes have been discovered (see for an overview van der Giezen 2009). Relevant for this chapter are the hydrogenosomes from *Nyctotherus ovalis* (Boxma et al. 2005) and *Blastocystis* (Stechmann et al. 2008). Both these organisms contain hydrogenosomes that are less derived than other hydrogenosomes. These organelles are able to take up active dyes such as Rhodamine123 or MitoTracker, which require an electrochemical gradient across the mitochondrial membrane to be actively taken up. This suggests that a proton-pumping activity would be present in these organelles and indeed, molecular evidence has been found suggesting that both organisms contain parts of Complex

I in their hydrogenosomes. Both *N. ovalis* and *Blastocystis* have been shown to contain a hydrogenosomal genome (Boxma et al. 2005; de Graaf et al. 2011; Pérez-Brocal and Clark 2008; Stechmann et al. 2008; Wawrzyniak et al. 2008). The presence of hydrogenosomes and mitosomes as part of the mitochondrial family of organelles indicates a clear spread from simple metabolic organelles such as mitosomes that have lost their organellar genomes to classic textbook aerobic mitochondria. These novel organelles do also fit with the above-mentioned bioenergetic theory of eukaryotic origins (Lane and Martin 2010; Martin and Müller 1998). However, despite numerous attempts to determine the nature of the mitochondrial ancestor and the evolutionary driving force behind endosymbiosis, this topic is still open to debate.

Since high-throughput DNA sequencing became available, and several mitochondrial genomes were sequenced, it has become obvious how variable mitochondrial genome size and gene content is among different eukaryotes (Burger et al. 2003; Lukeš et al. 2002; Martin and Müller 1998). This is the case not only between distantly related organisms but also between closely related species (see, for example, Pérez-Brocal et al. 2010). From here, another intriguing question emerges: Why is mitochondrial genome size and gene content so variable among species?

At present, it is clear that mitochondria possess their own genome which has been derived from an endosymbiotic bacterial ancestor. Many mitochondrial genomes have been sequenced by now. In addition, many α-proteobacterial genomes have been sequenced as well. Comparative genomic analysis clearly shows that modern mitochondrial genomes are severely reduced compared to those from α-proteobacteria. This has been caused by gene loss, but most importantly, because the endosymbiont's genes have been functionally transferred to the nucleus over time (Adams and Palmer 2003; Race et al. 1999; Timmis et al. 2004). This process was named endosymbiotic gene transfer (Martin et al. 2001) and is a special case of lateral (or horizontal) gene transfer. As this was not an instantaneous event but something that happened over time, each and every lineage has transferred and lost genes at his or her own pace. As a result of this, a large variety of mitochondrial and chloroplast genome sizes and genome contents can be found. An interesting example is the causative agent of malaria *Plasmodium falciparum,* which is known to possess one of the smallest mitochondrial genomes with only 5,967 base pairs (bp). This very small genome, nonetheless, still contains three core genes encoding the cytochrome *b* and cytochrome oxidase subunits I and III of the respiratory electron transport chain (Omori et al. 2007). On the other hand, the gene-rich mitochondrial genome of *Reclinomonas americana* is comprised of 69,034 bp and 67 protein-coding genes (Lang et al. 1997). Although there are plenty of notable studies unveiling the mechanisms and forces driving the lateral gene transfer to happen, this chapter aims to canvass the other side of the coin: Why is a small subset of genes always kept in the organellar genome of contemporary eukaryotes?

There are several hypotheses attempting to explain the selective pressure that maintains genomes in mitochondria and chloroplasts. The hydrophobicity hypothesis suggests that certain organellar genes encode hydrophobic proteins which may be problematic for cellular targeting systems (see, for example von Heijne 1986). Moreover, it also suggests that hydrophobic proteins may be mistargeted to the endoplasmic reticulum (von Heijne and Segrest 1987). Support for this hypothesis comes from the observation that coxI and cob genes are present in every mitochondrial genome sequenced so far (but not on hydrogenosomal genomes from *Blastocystis* and *N. ovalis*), and the respective proteins encoded by these genes are classified as typically hydrophobic peptides (Claros et al. 1995). Moreover, experimental analysis has shown that cytosolic synthesised apocytochrome b in yeast is not properly imported into mitochondria (Daley et al. 2002). In plants, a similar experiment reported that the in vitro synthesised COX2 protein is unable to be imported into soybean mitochondria, unless one of the transmembrane domains is removed and a few critical amino acid changes are made (Daley et al. 2002). Although these data have upheld the hydrophobicity hypothesis for some genes, this hypothesis remains unable to explain various other cases. For example, using the recently published structure of *Thermus thermophilus* NADH ubiquinone oxireductase (Efremov et al. 2010) as template, we built homology models for *Arabidopsis thaliana* and *Caenorhabditis elegans* (see Fig. 5.1). From this figure, it becomes obvious that the hydrophobicity hypothesis would predict that for both species the genes encoding some of these hydrophobic proteins need to be mitochondrially located and that some can be transferred to the nucleus. The hydrophilic nad7 and nad9 subunits have indeed been transferred to the nucleus

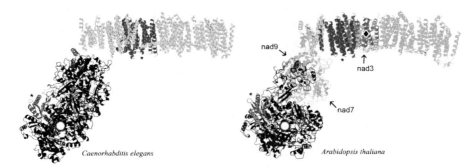

Fig. 5.1 Respiratory complex I structure of the bacterium *Thermus thermophillus* coded according to homologous gene locations in *Arabidopsis thaliana* and *Caenorhabdidis elegans*. The subunits indicated by a *white dot* are nuclear encoded, those in *light grey* are mitochondrial encoded and the subunit indicated by a *filled diamond* is present on both genomes. Subunits where no similarity at the amino acid level is found between *T. thermophillus* and the other two species are indicated by an *asterisk*. Nad 7 and nad 9 are hydrophilic polypeptides that would be encoded in the nucleus according to the hydrophobicity hypothesis. In fact they are mitochondrially encoded in *A. thaliana*, which is consistent with the CoRR hypothesis since these subunits contain the site of ubiquinone reduction by the iron–sulphur centre N2 and the initial site of chemical, redox-driven vectorial proton translocation (Efremov et al. 2010) that initiates conformationally driven proton translocation in the hydrophobic, membrane-intrinsic domain. Nad 3 is a polypeptide with a sequence of amino acid residues predicted by the nucleotide sequence of both a nuclear and a mitochondrial gene in *A. thaliana*

in *C. elegans*. However, both genes are still present on the mitochondrial genome for *A. thaliana* while there would be no hydrophobicity barrier for them to be transferred to the nucleus as well. Another problem for the hydrophobicity hypothesis is the *nad3* gene, which encodes a transmembrane H^+ pump subunit – a highly hydrophobic peptide. Intriguingly, *nad3* has two identical copies in *A. thaliana*, one located in its nuclear genome, possibly as an unexpressed and recent transformation, and the other in the mitochondrial genome. The earlier mentioned mitochondrial ADP/ATP translocator is another problem for the hydrophobicity hypothesis as these carrier proteins are almost completely buried in the mitochondrial inner membrane and are very hydrophobic but are, nonetheless, always nuclear encoded. It might be argued that these are eukaryotic inventions and would therefore never have been mitochondrially encoded in the first place. This is true but, nonetheless, the cell has no problem targeting this highly hydrophobic protein containing six membrane spanning domains to the correct cellular compartment.

Another theory seeking to explain the presence of organellar genomes states that some organellar genes have an idiosyncratic codon usage, which would preclude their nuclear expression, therefore locking them into the organelles (Doolittle 1998). In contrast, it has been shown in tobacco that the chloroplast gene encoding the large subunit of the Rubisco (rbcL) can be expressed in the nuclear genome of tobacco if relocated (Kanevski and Maliga 1994), therefore exposing a drawback on the idiosyncratic codon usage hypothesis.

Based on previous studies that had shown that gene expression was controlled by the redox state in bacteria, a hypothesis was published by one of us (Allen 1993a). This hypothesis suggests that mitochondria and chloroplasts have kept some specific genes in their genomes in order to enable an in situ redox regulation of their expression. These specific genes are thought to encode either the respiratory core subunits in the mitochondria, or the photosynthetic apparatus core subunits in the chloroplasts. In other words, if they were relocated to the nucleus, transcriptional regulation of these genes by the organellar redox state would not be possible. This hypothesis was named the CoRR hypothesis, which stands for Co-location for Redox Regulation (Allen 2003a, b). According to this hypothesis, there are two main players participating in the regulation of redox-driven gene expression: a redox sensor and a redox response regulator. The redox sensor is thought to be an electron-carrier that initiates control of gene expression upon oxidation or reduction. The redox response regulator, on the other hand, is proposed to be a DNA-binding protein that modifies gene expression as a result of the action of the redox sensor (Fig. 5.2).

Reactive oxygen species (ROS) result from excitation or incomplete reduction of molecular oxygen, and are unwelcome harmful by-products of normal cellular metabolism in aerobic organisms (Chance et al. 1979; Chen et al. 2003). In plants, mitochondrial and chloroplast electron transport is the major generator of ROS (Møller and Sweetlove 2010). It is extremely important for the plant cell to keep the ROS levels under control to avoid cell damage (Jo et al. 2001) as this can eventually lead to the initiation of programmed cell death (PCD) (Gechev et al. 2006). The CoRR hypothesis suggests that by regulating the individual expression of genes

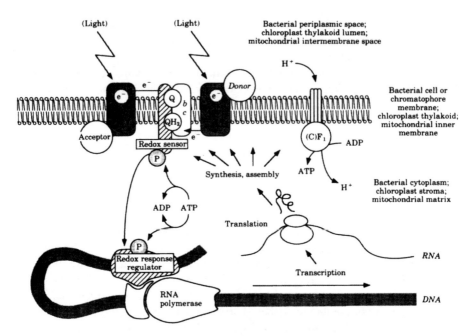

Fig. 5.2 Two-component redox regulation of transcription in bacteria, chloroplasts and mitochondria (after Allen 1993a). Vectorial electron and proton transfer exerts regulatory control over expression of genes encoding proteins directly involved in, or affecting, redox poise. This regulatory coupling requires co-location of such genes with their gene products. *CoRR – Co-location for Redox Regulation* – predicts that this regulatory coupling operated continuously before, during, and after the transition from prokaryote to eukaryotic organelle

encoding core subunits of the electron transport chain in mitochondria, and the photosystem I and II core subunits in chloroplasts, ROS levels can be controlled.

Light quality and quantity are known to influence the plastoquinone pool, and hence the redox state of chloroplasts (Allen et al. 1995b). Indeed, it has been shown that changes in the redox status of the plastoquinone pool by variation of light quality do control the rate of transcription of genes encoding reaction-centre apoproteins of photosystem I and II (Pfannschmidt et al. 1999). It was predicted that a redox sensor protein would control this switch. In 2008, this sensor protein was identified in *Arabidopsis thaliana* (Puthiyaveetil et al. 2008). It was termed chloroplast sensor kinase or CSK. Phylogenetic analysis shows that the plant CSK shares common ancestry with cyanobacterial histidine kinases, suggesting that photosynthetic CoRR regulation has been present since chloroplasts were still free-living cyanobacteria. Furthermore, recent studies on *A. thaliana* CSK have also indicated that specific cysteine residues are well conserved between cyanobacteria and higher plants. These are thought to be crucial for sensing the redox state of the chloroplast plastoquinone pool (Ibrahim 2009; Puthiyaveetil et al. 2010). Thiol-based regulatory switches involving cysteine residues are known to play central roles in cellular responses to oxidative stress (Paget and Buttner 2003).

For example, *Escherichia coli* ArcB is a sensor kinase that contains redox-active cysteine residues, and, upon changes in redox states of the quinone and menaquinone pools, regulates the transcription of aerobic genes (Bekker et al. 2010).

Few relevant studies have been carried out unravelling the role of the mitochondrial redox state on the regulation of organellar gene expression. One such study describes the incorporation of ^{35}S-methionine into newly synthesised mitochondrial proteins in relation to the redox status of the electron transport chain (ETC) ubiquinone pool (Allen et al. 1995a). By the use of inhibitors of specific sites of the ETC, it was reported that protein synthesis was precluded by inhibitors of ubiquinone reduction, but not by inhibitors of ubiquinol oxidation. Furthermore, it was found that electron transport through succinate:ubiquinone oxidoreductase (Complex II) was specifically required for protein synthesis, strongly suggesting that a subunit of complex II, or a component closely associated with this complex, is involved in a regulatory system that couples electron transport to protein synthesis (Escobar Galvis et al. 1998). Another study using *Solanum tuberosum* (potato) mitochondria investigated the role of a variety of electron transport inhibitors on organellar RNA synthesis. It was found that the redox state of the Rieske iron–sulphur protein was the major determinant of organellar RNA synthesis (Wilson et al. 1996). RNA synthesis was positively affected by inhibitors that act on the substrate side of the Rieske iron–sulphur protein. These inhibitors cause oxidation on the oxygen side of their site of action. On the other hand, if inhibitors were used that reduce the substrate site, then RNA synthesis was decreased. Redox regulation of plant mitochondrial glutamate dehydrogenase (Tarasenko et al. 2009) and DNA topoisomerase (Konstantinov et al. 2001) has also been reported. It has been suggested that this plays a role in coupling respiratory electron transport with mitochondrial gene expression.

In *Arabidopsis thaliana*, a large family of cysteine-rich receptor-like kinases (CRKs) have been described (Wrzaczek et al. 2010). Although only a few of these have been functionally characterised, it has been suggested that CRKs play an important role in the regulation of pathogen defence and programmed cell death, which are mainly driven by changes in ROS levels. Moreover, Wrzaczek et al. have also shown that several CRK mutants altered in hormone biosynthesis or signalling showed changes in basal and O_3-induced transcriptional responses (2010). In addition, a thorough survey of the *A. thaliana* mitochondrial proteome detected the presence of a nuclear encoded CRK protein among other kinases (Heazlewood et al. 2004). Much like CSK in *A. thaliana* and ArcB in *E. coli*, the presence of a cysteine-rich kinase such as CRK in the mitochondrial proteome strongly suggests that it might be involved in redox sensing activity and perhaps gene expression regulation of mitochondrial genes. Others studies also suggest the presence of a CoRR-like regulatory system in plant mitochondria. Transcription of mitochondrial genes in animals, fungi and plants relies on the T3/T7 phage-type RNA polymerases (RPOT). Two types of RPOTs are found in Eudicotyledonous plants. Whereas both types are nuclear encoded, one (RPOTm) is exclusively targeted to the mitochondria while the other (RPOTmp) is targeted to both mitochondria and chloroplasts. Transcriptional profiling of *A. thaliana* RPOTmp mutants has indicated that RPOTmp is able to transcribe a small subset of genes of the mitochondrial genome. These include *nad2* and *nad6*, which encode Complex I subunits, and *coxI*, which encodes a subunit of Complex IV (Kuhn et al. 2009). In contrast, the *A. thaliana* RPOTm mutant shows a lethal phenotype indicating that RPOTm is responsible for the transcription of essential mitochondrial biogenesis and maintenance genes. It is intriguing that the only ETC complexes regulated by RPOTmp in mitochondria are complexes I and IV, which are two antagonistic redox protein complexes. In addition to these mitochondrial studies, chloroplast transcriptional regulation has been studied in *A. thaliana*. Here, the synthesis of a protein called NIP (NEP Interacting Protein) is triggered by light, subsequently activating the RPOTmp in chloroplasts (Azevedo et al. 2008). It has been shown that there are two nuclear genes encoding NIPs in *A. thaliana*. One is targeted to chloroplasts while computer algorithms predict that the other is targeted to mitochondria. Interestingly, the fact that NIP protein synthesis is up-regulated under illumination implies that NIP proteins might arise due to necessity of transcriptional regulation upon redox changes in the organelles. No further studies have been carried out to study the possible interactions between the putative mitochondrial NIP and RPOTmp in mitochondria.

The CoRR hypothesis for the function and evolutionary persistence of organellar genetic systems predicts that organelles homologous with mitochondria, such as hydrogenosomes and mitosomes, would lose their genomes when the redox and proton-motive machinery of oxidative phosphorylation are lost. In these cases, there would be no requirement for direct, local control of gene expression. It is therefore gratifying to see that mitosomes, organelles which have lost their complete electron transport chains, do not contain organellar genomes. In addition, those hydrogenosomes that have kept the ability to maintain an active proton-motive force across their organellar membranes, have indeed kept an organellar genome. Organisms such as *N. ovalis* (Boxma et al. 2005) and *Blastocystis* (Stechmann et al. 2009) offer additional support for the CoRR hypothesis and it is interesting to note that their unusual organelles, and mitosomes as a whole, were not known when CoRR was first put forward.

When taken together with the hydrogen hypothesis for the first eukaryote (Martin and Müller 1998), CoRR makes it possible to understand the distribution of cytoplasmic genomes among the full range of mitochondrial organelles, including hydrogenosomes and mitosomes, of eukaryotic cells.

A focus of interest for future research will be identification of the predicted mitochondrial sensor kinase (MSK) and mitochondrial response regulator (MRR) (Fig. 5.2). Alternatively, mitochondrial redox signalling might take the form of a single-component, iron–sulphur protein based signalling mechanism involving a mitochondrial repressor (MRP) or activator protein (MAP) (Allen 1993b).

There is now no doubt that mitochondria are agents of cytoplasmic inheritance. Wilson's words of caution (Wilson 1925) may now apply to a proposed explanation, since the CoRR hypothesis "...*remains a subject of controversy and must be taken with proper skepticism...*". Amongst proposed explanations of the function and significance of organellar genomes, we suggest that CoRR "...*should be kept clearly in view in all cytological discussions of these problems"*. Co-location for Redox Regulation (CoRR) is consistent with available evidence and remains testable in making clear predictions concerning the nature and distribution of redox regulatory systems controlling cytoplasmic gene expression in all eukaryotic cells.

Acknowledgements JFA gratefully acknowledges a research grant from The Leverhulme Trust and MvdG is grateful for the continuous support from the University of Exeter.

References

Adams KL, Palmer JD (2003) Evolution of mitochondrial gene content: gene loss and transfer to the nucleus. Mol Phylogenet Evol 29:380–395

Allen JF (1993a) Control of gene expression by redox potential and the requirement for chloroplast and mitochondrial genomes. J Theor Biol 165:609–631

Allen JF (1993b) Redox control of transcription – sensors, response regulators, activators and repressors. FEBS Lett 332:203–207

Allen JF (2003a) The function of genomes in bioenergetic organelles. Philos Trans R Soc Lond B Biol Sci 358:19–37

Allen JF (2003b) Why chloroplasts and mitochondria contain genomes. Comp Funct Genomics 4:31–36

Allen CA, Hakansson G, Allen JF (1995a) Redox conditions specify the proteins synthesized by isolated chloroplasts and mitochondria. Redox Rep 1:119–123

Allen JF, Alexciev K, Hakansson G (1995b) Photosynthesis. Regulation by redox signalling. Curr Biol 5:869–872

Allen CA, van der Giezen M, Allen JF (2007) Origin, function, and transmission of mitochondria. In: Martin W, Müller M (eds) Origins of mitochondria and hydrogenosomes. Springer, Berlin, pp 39–56

Anderson S, Bankier AT, Barrell BG, de Bruijn MH, Coulson AR, Drouin J, Eperon IC, Nierlich DP, Roe BA, Sanger F, Schreier PH, Smith AJ, Staden R, Young IG (1981) Sequence and organization of the human mitochondrial genome. Nature 290:457–465

Atteia A, Adrait A, Brugiere S, Tardif M, van Lis R, Deusch O, Dagan T, Kuhn L, Gontero B, Martin W, Garin J, Joyard J, Rolland N (2009) A proteomic survey of *Chlamydomonas reinhardtii* mitochondria sheds new light on the metabolic plasticity of the organelle and on the nature of the alpha-proteobacterial mitochondrial ancestor. Mol Biol Evol 26:1533–1548

Azevedo J, Courtois F, Hakimi MA, Demarsy E, Lagrange T, Alcaraz JP, Jaiswal P, Marechal-Drouard L, Lerbs-Mache S (2008) Intraplastidial trafficking of a phage-type RNA polymerase is mediated by a thylakoid RING-H2 protein. Proc Natl Acad Sci USA 105:9123–9128

Bekker M, Alexeeva S, Laan W, Sawers G, Teixeira de Mattos J, Hellingwerf K (2010) The ArcBA two-component system of *Escherichia coli* is regulated by the redox state of both the ubiquinone and the menaquinone pool. J Bacteriol 192:746–754

Boxma B, de Graaf RM, van der Staay GW, van Alen TA, Ricard G, Gabaldon T, van Hoek AH, Moon-van der Staay SY, Koopman WJ, van Hellemond JJ, Tielens AG, Friedrich T, Veenhuis M, Huynen MA, Hackstein JH (2005) An anaerobic mitochondrion that produces hydrogen. Nature 434:74–79

Bradley PJ, Lahti CJ, Plümper E, Johnson PJ (1997) Targeting and translocation of proteins into the hydrogenosome of the protist *Trichomonas*: similarities with mitochondrial protein import. EMBO J 16:3484–3493

Brinkmann H, van der Giezen M, Zhou Y, Poncelin de Raucourt G, Philippe H (2005) An empirical assessment of long-branch attraction artefacts in deep eukaryotic phylogenomics. Syst Biol 54:743–757

Bui ETN, Bradley PJ, Johnson PJ (1996) A common evolutionary origin for mitochondria and hydrogenosomes. Proc Natl Acad Sci USA 93:9651–9656

Burger G, Gray MW, Franz Lang B (2003) Mitochondrial genomes: anything goes. Trends Genet 19:709–716

Cavalier-Smith T (1983) A 6 kingdom classification and a unified phylogeny. In: Schwemmler W, Schenk HEA (eds) Endocytobiology II. De Gruyter, Berlin, pp 1027–1034

Cerkasovová A, Lukasová G, Cerkasòv J, Kulda J (1973) Biochemical characterization of large granule fraction of *Tritrichomonas foetus* (strain KV1). J Protozool 20:525

Cerkasovová A, Cerkasòv J, Kulda J, Reischig J (1976) Circular DNA and cardiolipin in hydrogenosomes, microbody-like organelles in *Trichomonas*. Folia Parasitol 23:33–37

Chance B, Sies H, Boveris A (1979) Hydroperoxide metabolism in mammalian organs. Physiol Rev 59:527–605

Chen Q, Vazquez EJ, Moghaddas S, Hoppel CL, Lesnefsky EJ (2003) Production of reactive oxygen species by mitochondria – central role of complex III. J Biol Chem 278:36027–36031

Clark CG, Roger AJ (1995) Direct evidence for secondary loss of mitochondria in *Entamoeba histolytica*. Proc Natl Acad Sci USA 92:6518–6521

Clark CG, Alsmark UC, Tazreiter M, Saito-Nakano Y, Ali V, Marion S, Weber C, Mukherjee C, Bruchhaus I, Tannich E, Leippe M, Sicheritz-Ponten T, Foster PG, Samuelson J, Noel CJ, Hirt RP, Embley TM, Gilchrist CA, Mann BJ, Singh U, Ackers JP, Bhattacharya S, Bhattacharya A, Lohia A, Guillen N, Duchene M, Nozaki T, Hall N (2007) Structure and content of the *Entamoeba histolytica* genome. Adv Parasitol 65:51–190

Claros MG, Perea J, Shu Y, Samatey FA, Popot JL, Jacq C (1995) Limitations to in vivo import of hydrophobic proteins into yeast mitochondria. The case of a cytoplasmically synthesized apocytochrome b. Eur J Biochem 228:762–771

Daley DO, Clifton R, Whelan J (2002) Intracellular gene transfer: reduced hydrophobicity facilitates gene transfer for subunit 2 of cytochrome c oxidase. Proc Natl Acad Sci USA 99:10510–10515

de Graaf RM, Ricard G, van Alen TA, Duarte I, Dutilh BE, Burgtorf C, Kuiper JW, van der Staay GW, Tielens AG, Huynen MA, Hackstein JH (2011) The organellar genome and metabolic potential of the hydrogen-producing mitochondrion of *Nyctotherus ovalis*. Mol Biol Evol. doi:10.1093/molbev/msr1059

Doolittle WF (1998) You are what you eat: a gene transfer ratchet could account for bacterial genes in eukaryotic nuclear genomes. Trends Genet 14:307–311

Efremov RG, Baradaran R, Sazanov LA (2010) The architecture of respiratory complex I. Nature 465:441–445

Embley TM, Hirt RP (1998) Early branching eukaryotes? Curr Opin Genet Dev 8:624–629

Escobar Galvis ML, Allen JF, Hakansson G (1998) Protein synthesis by isolated pea mitochondria is dependent on the activity of respiratory complex II. Curr Genet 33:320–329

Esser C, Ahmadinejad N, Wiegand C, Rotte C, Sebastiani F, Gelius-Dietrich G, Henze K, Kretschmann E, Richly E, Leister D, Bryant D, Steel MA, Lockhart PJ, Penny D, Martin W (2004) A genome phylogeny for mitochondria among alpha-proteobacteria and a predominantly eubacterial ancestry of yeast nuclear genes. Mol Biol Evol 21:1643–1660

Esser C, Martin W, Dagan T (2007) The origin of mitochondria in light of a fluid prokaryotic chromosome model. Biol Lett 3:180–184

Gechev TS, Van Breusegem F, Stone JM, Denev I, Laloi C (2006) Reactive oxygen species as signals that modulate plant stress responses and programmed cell death. Bioessays 28:1091–1101

Gray MW, Doolittle WF (1982) Has the endosymbiont hypothesis been proven? Microbiol Rev 46:1–42

Gray MW, Lang BF, Cedergren R, Golding GB, Lemieux C, Sankoff D, Turmel M, Brossard N, Delage E, Littlejohn TG, Plante I, Rioux P, Saint-Louis D, Zhu Y, Burger G (1998) Genome structure and gene content in protist mitochondrial DNAs. Nucleic Acids Res 26:865–878

Gray MW, Burger G, Lang BF (1999) Mitochondrial evolution. Science 283:1476–1481

Heazlewood JL, Tonti-Filippini JS, Gout AM, Day DA, Whelan J, Millar AH (2004) Experimental analysis of the Arabidopsis mitochondrial proteome highlights signaling and regulatory components, provides assessment of targeting prediction programs, and indicates plant-specific mitochondrial proteins. Plant Cell 16:241–256

Henze K, Martin W (2003) Essence of mitochondria. Nature 426:127–128

Horner DS, Hirt RP, Kilvington S, Lloyd D, Embley TM (1996) Molecular data suggest an early acquisition of the mitochondrion endosymbiont. Proc R Soc Lond B Biol Sci 263:1053–1059

Ibrahim IM (2009) Characterizing chloroplast sensor kinase. Biosci Horizons 2:191–196

Jo SH, Son MK, Koh HJ, Lee SM, Song IH, Kim YO, Lee YS, Jeong KS, Kim WB, Park JW, Song BJ, Huh TL (2001) Control of mitochondrial redox balance and cellular defense against oxidative damage by mitochondrial NADP$^+$-dependent isocitrate dehydrogenase. J Biol Chem 276:16168–16176

Kanevski I, Maliga P (1994) Relocation of the plastid rbcL gene to the nucleus yields functional ribulose-1,5-bisphosphate carboxylase in tobacco chloroplasts. Proc Natl Acad Sci USA 91:1969–1973

Konstantinov YM, Tarasenko VI, Rogozin IB (2001) Redox modulation of the activity of DNA topoisomerase I from carrot (Daucus carota) mitochondria. Dokl Biochem Biophys 377:263–265

Kuhn K, Richter U, Meyer EH, Delannoy E, de Longevialle AF, O'Toole N, Borner T, Millar AH, Small ID, Whelan J (2009) Phage-type RNA polymerase RPOTmp performs gene-specific transcription in mitochondria of Arabidopsis thaliana. Plant Cell 21:2762–2779

Kurland CG, Andersson SG (2000) Origin and evolution of the mitochondrial proteome. Microbiol Mol Biol Rev 64:786–820

Lahti CJ, D'Oliveira CE, Johnson PJ (1992) β-Succinyl-coenzyme A synthetase from Trichomonas vaginalis is a soluble hydrogenosomal protein with an amino-terminal sequence that resembles mitochondrial presequences. J Bacteriol 174:6822–6830

Lahti CJ, Bradley PJ, Johnson PJ (1994) Molecular characterization of the α-subunit of Trichomonas vaginalis hydrogenosomal succinyl CoA synthetase. Mol Biochem Parasitol 66:309–318

Lane N (2005) Power, sex, suicide: mitochondria and the meaning of life. Oxford Press, Oxford

Lane N, Martin W (2010) The energetics of genome complexity. Nature 467:929–934

Lang BF, Burger G, O'Kelly CJ, Cedergren R, Golding GB, Lemieux C, Sankoff D, Turmel M, Gray MW (1997) An ancestral mitochondrial DNA resembling a eubacterial genome in miniature. Nature 387:493–497

León-Avila G, Tovar J (2004) Mitosomes of Entamoeba histolytica are abundant mitochondrion-related remnant organelles that lack a detectable organellar genome. Microbiology 150:1245–1250

Lill R, Mühlenhoff U (2005) Iron-sulfur-protein biogenesis in eukaryotes. Trends Biochem Sci 30:133–141

Lindmark DG, Müller M (1973) Hydrogenosome, a cytoplasmic organelle of the anaerobic flagellate Tritrichomonas foetus, and its role in pyruvate metabolism. J Biol Chem 248:7724–7728

Loftus B, Anderson I, Davies R, Alsmark UC, Samuelson J, Amedeo P, Roncaglia P, Berriman M, Hirt RP, Mann BJ, Nozaki T, Suh B, Pop M, Duchene M, Ackers J, Tannich E, Leippe M, Hofer M, Bruchhaus I, Willhoeft U, Bhattacharya A, Chillingworth T, Churcher C, Hance Z, Harris B, Harris D, Jagels K, Moule S, Mungall K, Ormond D, Squares R, Whitehead S, Quail MA, Rabbinowitsch E, Norbertczak H, Price C, Wang Z, Guillen N, Gilchrist C, Stroup SE, Bhattacharya S, Lohia A, Foster PG, Sicheritz-Ponten T, Weber C, Singh U, Mukherjee C,

El-Sayed NM, Petri WA, Clark CG, Embley TM, Barrell B, Fraser CM, Hall N (2005) The genome of the protist parasite *Entamoeba histolytica*. Nature 433:865–868

Luck DJL, Reich E (1964) DNA in mitochondria of *Neurospora crassa*. Proc Natl Acad Sci USA 52:931–938

Lukeš J, Guilbride DL, Votýpka J, Zíková A, Benne R, Englund PT (2002) Kinetoplast DNA network: evolution of an improbable structure. Eukaryot Cell 1:495–502

Mai Z, Ghosh S, Frisardi M, Rosenthal B, Rogers R, Samuelson J (1999) Hsp60 is targeted to a cryptic mitochondrion-derived organelle ("crypton") in the microaerophilic protozoan parasite *Entamoeba histolytica*. Mol Cell Biol 19:2198–2205

Margulis L (1970) Origin of eukaryotic cells. Yale University Press, New Haven, USA

Margulis L (1981) Symbiosis in cell evolution. W.H. Freeman, New York

Martin W (2007) Eukaryote and mitochondrial origins: two sides of the same coin and too much ado about oxygen. In: Falkowski P, Knoll A (eds) Primary producers of the sea. Academic Press, New York, pp 55–73

Martin W, Kowallik KV (1999) Annotated english translation of Mereschkowsky's 1905 paper 'Uber Natur und Ursprung der Chromatophoren im Pflanzenreiche'. Eur J Phycol 34:287–295

Martin W, Müller M (1998) The hydrogen hypothesis for the first eukaryote. Nature 392:37–41

Martin W, Hoffmeister M, Rotte C, Henze K (2001) An overview of endosymbiotic models for the origins of eukaryotes, their ATP-producing organelles (mitochondria and hydrogenosomes), and their heterotrophic lifestyle. Biol Chem 382:1521–1539

Mereschkowsky C (1905) Über Natur und Ursprung der Chromatophoren im Pflanzenreiche. Biol Centralbl 25:593–604

Møller IM, Sweetlove LJ (2010) ROS signalling – specificity is required. Trends Plant Sci 15:370–374

Morrison HG, McArthur AG, Gillin FD, Aley SB, Adam RD, Olsen GJ, Best AA, Cande WZ, Chen F, Cipriano MJ, Davids BJ, Dawson SC, Elmendorf HG, Hehl AB, Holder ME, Huse SM, Kim UU, Lasek-Nesselquist E, Manning G, Nigam A, Nixon JE, Palm D, Passamaneck NE, Prabhu A, Reich CI, Reiner DS, Samuelson J, Svard SG, Sogin ML (2007) Genomic minimalism in the early diverging intestinal parasite *Giardia lamblia*. Science 317:1921–1926

Müller M (1993) The hydrogenosome. J Gen Microbiol 139:2879–2889

Nass MM, Nass S (1963) Intramitochondrial fibers with DNA characteristics. I. Fixation and electron staining reactions. J Cell Biol 19:593–611

Omori S, Sato Y, Isobe T, Yukawa M, Murata K (2007) Complete nucleotide sequences of the mitochondrial genomes of two avian malaria protozoa, *Plasmodium gallinaceum* and *Plasmodium juxtanucleare*. Parasitol Res 100:661–664

Paget MS, Buttner MJ (2003) Thiol-based regulatory switches. Annu Rev Genet 37:91–121

Pérez-Brocal V, Clark CG (2008) Analysis of two genomes form the mitochondrion-like organelle of the intestinal parasite *Blastocystis*: complete sequences, gene content and genome organization. Mol Biol Evol 25:2475–2482

Pérez-Brocal V, Shahar-Golan R, Clark CG (2010) A linear molecule with two large inverted repeats: the mitochondrial genome of the stramenopile *Proteromonas lacertae*. Genome Biol Evol 2:257–266

Pfannschmidt T, Nilsson A, Allen JF (1999) Photosynthetic control of chloroplast gene expression. Nature 397:625–628

Puthiyaveetil S, Kavanagh TA, Cain P, Sullivan JA, Newell CA, Gray JC, Robinson C, van der Giezen M, Rogers MB, Allen JF (2008) The ancestral symbiont sensor kinase CSK links photosynthesis with gene expression in chloroplasts. Proc Natl Acad Sci USA 105:10061–10066

Puthiyaveetil S, Ibrahim IM, Jelicic B, Tomasic A, Fulgosi H, Allen JF (2010) Transcriptional control of photosynthesis genes: the evolutionarily conserved regulatory mechanism in plastid genome function. Genome Biol Evol

Race HL, Herrmann RG, Martin W (1999) Why have organelles retained genomes? Trends Genet 15:364–370

Roger AJ, Clark CG, Doolittle WF (1996) A possible mitochondrial gene in the early-branching amitochondriate protist *Trichomonas vaginalis*. Proc Natl Acad Sci USA 93:14618–14622

Sagan L (1967) On the origin of mitosing cells. J Theor Biol 14:225–274

Simpson AG, Roger AJ (2004) The real 'kingdoms' of eukaryotes. Curr Biol 14:R693–R696

Sogin ML, Gunderson JH, Elwood HJ, Alonso RA, Peattie DA (1989) Phylogenetic meaning of the kingdom concept: an unusual ribosomal RNA from *Giardia lamblia*. Science 243:75–77

Stechmann A, Hamblin K, Pérez-Brocal V, Gaston D, Richmond GS, van der Giezen M, Clark CG, Roger AJ (2008) Organelles in *Blastocystis* that blur the distinction between mitochondria and hydrogenosomes. Curr Biol 18:580–585

Stechmann A, Tsaousis AD, Hamblin KA, van der Giezen M, Pérez-Brocal V, Clark CG (2009) The *Blastocystis* mitochondrion-like organelles. In: Clark CG, Adam RD, Johnson PJ (eds) Anaerobic parasitic protozoa: genomics and molecular biology. Horizon Scientific Press, Norwich, pp 205–219

Tarasenko VI, Garnik EY, Shmakov VN, Konstantinov YM (2009) Induction of Arabidopsis gdh2 gene expression during changes in redox state of the mitochondrial respiratory chain. Biochemistry (Mosc) 74:47–53

Timmis JN, Ayliffe MA, Huang CY, Martin W (2004) Endosymbiotic gene transfer: organelle genomes forge eukaryotic chromosomes. Nat Rev Genet 5:123–135

Tovar J, Fischer A, Clark CG (1999) The mitosome, a novel organelle related to mitochondria in the amitochondrial parasite *Entamoeba histolytica*. Mol Microbiol 32:1013–1021

Tovar J, León-Avila G, Sánchez L, Sutak R, Tachezy J, van der Giezen M, Hernández M, Müller M, Lucocq JM (2003) Mitochondrial remnant organelles of *Giardia* function in iron-sulphur protein maturation. Nature 426:172–176

Turner G, Müller M (1983) Failure to detect extranuclear DNA in *Trichomonas vaginalis* and *Tritrichomonas foetus*. J Parasitol 69:234–236

van der Giezen M (2009) Hydrogenosomes and mitosomes: conservation and evolution of functions. J Eukaryot Microbiol 56:221–231

von Heijne G (1986) Why mitochondria need a genome. FEBS Lett 198:1–4

von Heijne G, Segrest JP (1987) The leader peptides from bacteriorhodopsin and halorhodopsin are potential membrane-spanning amphipathic helices. FEBS Lett 213:238–240

Vossbrinck CR, Maddox TJ, Friedman S, Debrunner-Vossbrinck BA, Woese CR (1987) Ribosomal RNA sequence suggests microsporidia are extremely ancient eukaryotes. Nature 326:411–414

Wallin IE (1923) The mitochondria problem. Am Nat 57:255–261

Wawrzyniak I, Roussel M, Diogen M, Couloux A, Texier C, Tan KSW, Vivarès C, Delbac F, Winckler P, El Alaoui H (2008) Complete circular DNA in the mitochondria-like organelles of *Blastocystis hominis*. Int J Parasitol 38:1377–1382

Williams BAP, Hirt RP, Lucocq JM, Embley TM (2002) A mitochondrial remnant in the microsporidian *Trachipleistophora hominis*. Nature 418:865–869

Wilson EB (1925) The cell in development and heredity. The Macmillan Company, New York

Wilson SB, Davidson GS, Thomson LM, Pearson CK (1996) Redox control of RNA synthesis in potato mitochondria. Eur J Biochem 242:81–85

Winkler HH, Neuhaus HE (1999) Non-mitochondrial ATP transport. Trends Biochem Sci 24:64–68

Wrzaczek M, Brosche M, Salojarvi J, Kangasjarvi S, Idanheimo N, Mersmann S, Robatzek S, Karpinski S, Karpinska B, Kangasjarvi J (2010) Transcriptional regulation of the CRK/DUF26 group of receptor-like protein kinases by ozone and plant hormones in Arabidopsis. BMC Plant Biol 10:95

Yang D, Oyaizu Y, Oyaizu H, Olsen GJ, Woese CR (1985) Mitochondrial origins. Proc Natl Acad Sci USA 82:4443–4447

Part III
Mechanisms of Organelle Gene Loss

Chapter 6
Evolutionary Rate Variation in Organelle Genomes: The Role of Mutational Processes

Daniel B. Sloan and Douglas R. Taylor

6.1 Introduction

The field of molecular evolution makes extensive use of the comparative analysis of evolutionary rates. Rates of nucleotide substitution are often used as tools to study the history of selection, with corrections being made to account for underlying differences in the mutation rate. But there is also a broad interest in the mutation rate, per se, as an evolutionary force. For example, sex and recombination may be favored to enhance the clearance of deleterious mutations from the genome (Muller 1964), complex genomes may be streamlined (Lynch 2006, 2007), or genes may be moved to other compartments of the cell (i.e., the nucleus) (Brandvain and Wade 2009) to reduce the occurrence of deleterious mutations, and mutational biases may influence many aspects of genome size and structure (Mira et al. 2001; Petrov 2002). Organelle genomes are of particular interest in this context because they exhibit some of the greatest natural variation in mutation rate.

Mitochondria and plastids represent the oldest examples of a much larger group of endosymbiotic relationships in which formerly free-living organisms have evolved to live exclusively inside the cells of other organisms. A common characteristic uniting this diverse group of organelles and endosymbionts is an accelerated rate of DNA sequence evolution (Brown et al. 1979; Andersson and Kurland 1998; Moran et al. 2009). This acceleration can be partially explained by changes in the strength and efficacy of natural selection (and hence the probability of fixation of mutations) resulting from an intracellular lifestyle (Moran 1996; Lynch and Blanchard 1998). However, increases in the mutation rate (the probability of occurrence of mutations) are also responsible, as these genomes have lost large numbers of genes, including many involved in DNA replication and repair.

D.B. Sloan (✉) • D.R. Taylor
Department of Biology, University of Virginia, Charlottesville, VA 22904 USA
e-mail: dbs4a@virginia.edu; drt3b@virginia.edu

C.E. Bullerwell (ed.), *Organelle Genetics*,
DOI 10.1007/978-3-642-22380-8_6, © Springer-Verlag Berlin Heidelberg 2012

Mitochondrial and plastid genomes provide the most extreme examples of gene loss. In fact, most organelle genomes completely lack DNA replication and repair genes, as the control of organelle genome maintenance has been transferred to the nucleus (Timmis 2004). It is clear that eukaryotic lineages differ significantly in their nuclear-encoded replication and repair machinery, which may provide an explanation for dramatic variation in organelle mutation rates. In addition, differences in physiology and life history may produce further variation in mutational input across lineages. In this chapter, we review the causes and consequences of substitution rate variation in organelle genomes, focusing predominantly on mutational processes. Given the ample evidence for mitochondrial and plastid mutation rate variation, these systems represent a valuable model for investigating the role of mutational processes in shaping genome evolution.

6.2 Methods for Estimating Organelle Mutation Rates

In order to understand the patterns of mutation rate variation in organelle genomes, we must first consider how mutation rates are estimated. The vast majority of rate estimates have been inferred from phylogenetic studies that quantify the extent of DNA sequence divergence among organisms. However, with the recent advent of high-throughput DNA sequencing technology, a complementary approach has also become feasible. Whole genome resequencing of mutation accumulation (MA) lines allows for genome-wide identification of genetic changes that have accumulated over a defined number of generations in the lab. In this section, we review these methods, highlighting their relative strengths and weaknesses.

6.2.1 Phylogenetic Methods

Molecular phylogenetic analyses are often used to generate an estimate of the number of substitutions that have occurred on each branch in an evolutionary tree. When combined with estimates of the age of that branch, this information can be used to calculate absolute substitution rates. Furthermore, restricting such analyses to neutrally evolving sequences can provide an estimate of the underlying mutation rate (see below). The advantage of these methods is that they are generally easy to implement and can often be conducted with publicly available data, making them amenable to comparisons of large numbers of species. However, as discussed below, phylogenetic methods also suffer from a number of assumptions and technical challenges that potentially bias their mutation rate estimates (see also Lanfear et al. (2010) for a recent review of the use of phylogenetic methods to estimate evolutionary rates).

6.2.1.1 Synonymous Substitution Rates and the Assumption of Neutrality

One of the pillars of molecular evolution theory is that, if a sequence is not under selection, the nucleotide substitution rate is equal to its mutation rate (Kimura 1983). Phylogenetic methods rely on this expectation to infer mutation rates from DNA sequence data. Synonymous substitutions are changes in DNA sequence that do not affect the corresponding amino acid sequence because of the redundancy of the genetic code. These so-called silent sites are therefore expected to be relatively free from selection pressures. Thus, the rate of synonymous substitutions in protein genes is commonly used as an estimate of the mutation rate (Table 6.1).

The assumption of complete neutrality at synonymous sites, however, is unrealistic. There is ample evidence for selective forces influencing rates of synonymous substitution. These include selection for biased codon usage, conservation of regulatory motifs, and stability of RNA secondary structure (Chamary et al. 2006). Therefore, there is likely to be some degree of purifying selection acting on synonymous sites in organelle genomes, resulting in a downward bias on mutation rate estimates based on synonymous substitution rates. Data from MA experiments and pedigree analyses suggest that this bias may be substantial – perhaps as much as one or two orders of magnitude (Denver et al. 2000; Howell et al. 2003; Haag-Liautard et al. 2008; Lynch et al. 2008).

An additional concern with the use of synonymous sites to estimate mutation rates is that they offer a potentially nonrepresentative sample of the genome. For example, fourfold synonymous sites in the *Drosophila* mitochondrial genome have an extremely skewed nucleotide composition (94% A + T), which reflects a strong mutational bias toward A:T base pairs (Haag-Liautard et al. 2008). Nonsynonymous sites are also AT rich but not to the same extent (66% A + T), presumably because of functional constraint on amino acid sequence. MA experiments have shown that the

Table 6.1 Phylogenetic estimates of mitochondrial mutation rates

Taxon	Mitochondrial synonymous substitution rate ($\times 10^{-9}$ per site per year)	Sources
Mammals	18.2–54.5	Wolfe et al. (1987)[a]
	7.0–643.4	Nabholz et al. (2008)[b]
Birds	3.0–90.0	Nabholz et al. (2009)[b]
Amphibians	13.8–21.6	Lynch et al. (2006)
Insects	16.6–34.0	Lynch et al. (2006)
Seed Plants	0.2–1.1	Wolfe et al. (1987)[a]
	0.02–90.1	Mower et al. (2007)

[a] The early study of Wolfe et al. was limited to a small number of species comparisons within both mammals and seed plants. The wider ranges of rate estimates in subsequent studies reflect much more extensive sampling in these groups

[b] Based on the original authors' interpretation, the indicated ranges for each of the Nabholz et al. studies exclude the most extreme 5% of rate estimates as potentially unreliable outliers

Table 6.2 Estimates of mitochondrial mutation based on resequencing of laboratory mutation accumulation lines

Species	Mitochondrial mutation rate ($\times 10^{-9}$ per site per generation $\pm$ SEM)	Sources
Caenorhabditis briggsae (HK104)	110 ($\pm$38)	Howe et al. (2010)
Caenorhabditis briggsae (PB800)	72 ($\pm$34)	Howe et al. (2010)
Caenorhabditis elegans	97 ($\pm$27)	Denver et al. (2000)
Drosophila melanogaster (Florida)	43 ($\pm$19)	Haag-Liautard et al. (2008)[a]
Drosophila melanogaster (Madrid)	81 ($\pm$24)	Haag-Liautard et al. (2008)[a]
Saccharomyces cerevisiae	12.2 ($\pm$3.6)	Lynch et al. (2008)

[a]The standard errors indicated for the *Drosophila* lines were approximated as ¼ of the 95% confidence intervals reported in the original study

rate of mutation in *Drosophila* mitochondrial DNA (mtDNA) is higher at nonsynonymous than synonymous sites (Haag-Liautard et al. 2008). This result is not surprising given the differences in G + C content between synonymous and nonsynonymous sites and the fact that most mutations in *Drosophila* mtDNA convert G/C to A/T. Therefore, in the case of *Drosophila*, the synonymous mutation rate is not representative of the genome-wide mutation rate. Similarly, Morton (2003) found that, because of the constraints of the genetic code, synonymous sites in the plastid genomes of grasses are preferentially found adjacent to particular upstream and downstream nucleotides. Because the patterns of mutation at a given site depend on these flanking nucleotides, synonymous sites in plastid DNA can also be subject to different mutational pressures than the rest of the genome.

6.2.1.2 Saturation

One key to phylogenetic rate estimates is to accurately identify the number of nucleotide substitutions that have occurred along a particular branch. This can become challenging when there have been multiple substitutions at the same site. Models of sequence evolution are generally used to estimate the total number of substitutions that occurred in a lineage when only a fraction of those changes are directly observable by sequence comparison (Sullivan and Joyce 2005). In the extreme, however, when all sites have experienced one or more substitutions, such models cannot be effectively applied.

To circumvent this problem of "saturation," Nabholz et al. (2008, 2009) have employed a hierarchical strategy to analyze large datasets of mtDNA sequences in mammals and birds. Their method subdivides a dataset into smaller taxonomic groups, in which saturation is not a problem. The age of each group can be inferred based on more slowly evolving amino acid sequence data. Standard phylogenetic analyses are then performed separately on each subdivision. This approach represents a promising option that could be extended to other large datasets with levels of divergence that are too high for more standard phylogenetic analyses.

6.2.1.3 Phylogenetic Artifacts

Phylogenetic methods are subject to a number of biases that can potentially affect estimates of substitution rate. For example, although models of evolution attempt to account for the number of unobservable substitutions resulting from recurrent changes at the same site, there is often a bias toward underestimating the total number of substitutions in long branches. This is less of a problem in well-sampled clades with shorter internal branches. The discrepancy in rate estimates associated with sampling intensity is known as the node density effect (Venditti et al. 2006). Employing more phylogenetically balanced sampling strategies may minimize this effect. Alternatively, for imbalanced datasets, the magnitude of the node density effect can be assessed and possibly mitigated with statistical measures (Venditti et al. 2006, 2008).

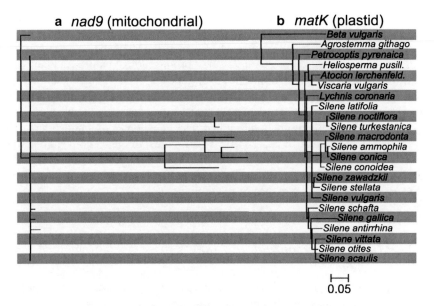

Fig. 6.1 An example of recent and extreme acceleration in mitochondrial substitution rates within the angiosperm genus *Silene*. (**a**) A phylogenetic tree with branch lengths representing the number of substitutions per synonymous site in the mitochondrial gene *nad9*. (**b**) For comparison, a tree based on a gene from the plastid genome (which does not exhibit comparable increases in substitution rate). Both trees are based on previously published data (Sloan et al. 2009)

A second source of bias in phylogenetic analysis is intraspecific polymorphism. When only one or a small number of individuals are sampled from each species, estimates of between-species divergence can be inflated by the existence of within-species polymorphism (Peterson and Masel 2009; Charlesworth 2010). This bias is particularly important for divergence at young nodes in a tree and when relatively ancient polymorphisms are maintained across species boundaries. It leads to higher branch length estimates at the tips of a tree as compared to internal branches near the root. However, this effect does not explain asymmetry in synonymous branch lengths between sister taxa, which is the classic signature of substitution rate variation across lineages (Fig. 6.1). One solution to this problem is to sample numerous individual per species to quantify the effect of polymorphism. For large-scale phylogenetic analyses, however, this solution may be impractical. A reasonable alternative may be to focus on deeper splits in the phylogenetic tree or on the relative rate differences between sister lineages.

A third issue with phylogenetic analyses is the basic assumption of a single tree-like model of evolution, which can be violated when organelle genomes undergo some forms of recombination. Organelle genomes are generally thought of as being uniparentally (usually maternally) inherited and, therefore, asexual. There is, however, enormous variation in the modes of organelle inheritance across eukaryotes with countless exceptions to the rule of uniparental inheritance (Barr et al. 2005). When it occurs, biparental inheritance generates the opportunity for sexual recombination in organelle genomes, which can result in different sections of the genome having different genealogies. When sequence data from recombining genomes are analyzed under the assumption of a single genealogy, distortions in branch lengths and substitution rate estimates can result. Cases of lateral gene transfer can have especially severe effects. For example, the "promiscuous" nature of plant mito-chondrial genomes has resulted in transfer of genes or gene fragments among very distant related lineages (Ellis 1982; Bergthorsson et al. 2003). In, perhaps, the most extreme examples, there is evidence of recent gene conversion between *anciently* homologous genes in the mitochondrial and plastid genomes of angiosperms (Hao and Palmer 2009; Sloan et al. 2010). These cases of gene conversion affect relatively small stretches of DNA sequence and can superficially appear to be the result of local increases in substitution rate. Therefore, it is important to test phylogenetic datasets for evidence of recombination.

6.2.2 Mutation Accumulation Lines

Whereas phylogenetic estimates of mutation rates focus on portions of the genome that are believed to be relatively free from selection, MA lines represent an alternative approach in which experimental manipulation is used to reduce or eliminate the effect of selection across the entire genome. The logic behind MA experiments is that the efficacy of natural selection is proportional to the effective population size (N_e). Therefore, by repeated bottlenecking of a population through one or a small number of individuals, N_e can be reduced to a point that only the most severe mutations will be removed by selection. Therefore, the genetic changes that accumulate over time in these experimental lines should very closely reflect the unfiltered mutational input. The ever-decreasing cost of DNA sequencing has made it feasible to resequence the entire genomes of individuals from MA lines. This approach has now been used to analyze mitochondrial mutation patterns in a handful of classic laboratory model systems, including yeast, *Drosophila*, and *Caenorhabditis*.

6.2.2.1 Findings from Mutation Accumulation Experiments

The number of MA studies on organelle genomes is still very limited, but there is a clear trend suggesting that these more direct measures of mutation rates produce much higher estimates than those inferred from synonymous substitution rates. As opposed to phylogenetic analyses, which produce absolute (i.e., per year) rate estimates, the results of MA studies are measured on a per generation basis (Table 6.1). In the first major sequencing experiment based on MA lines, Denver et al. (2000) reported a surprisingly high rate of point mutations in the mitochondrial genome of the nematode *Caenorhabditis elegans*. Assuming a 4-day generation time in *C. elegans*, this rate was roughly 2 orders of magnitude higher than previous estimates of mitochondrial mutation rates based on sequence divergence. A subsequent study of the related nematode *C. briggsae* found similarly high point mutation rates (although interestingly the mitochondrial genomes of these two closely related species differed substantially in their rates of large deletion mutations)(Howe et al. 2010). MA experiments have also yielded unexpectedly high mitochondrial mutation rate estimates in yeast and *Drosophila* (Haag-Liautard et al. 2008; Lynch et al. 2008).

It appears that the disagreement between mutation rate estimates based on phylogenetic methods and MA experiments is not limited to mitochondrial genomes. Nuclear mutation rate estimates from MA experiments have also been elevated (e.g., Denver et al. 2004). However, at least in the cases of *Drosophila* and yeast, the discrepancy between the two methods is more pronounced in the mitochondrial genome, resulting in higher estimates of the ratio of mitochondrial to nuclear mutation rates (Haag-Liautard et al. 2008; Lynch et al. 2008).

Although the discrepancy between mutation rate estimates derived from synonymous substitution rates and MA lines may reflect some of the shortcomings of phylogenetic analyses (Sect. 6.2.1), it is also important to consider that the substitution process for mitochondrial mutations is complex, reflecting the hierarchical organization of eukaryotic organisms. Below, we discuss these complexities in the context of MA experiments.

6.2.2.2 Heteroplasmy and Mutation Accumulation Experiments

Unlike the nuclear genome, which typically occurs as a single diploid copy in each cell, organelle genomes are highly polyploid often with many thousands of genome copies distributed across multiple organelles in each cell (Moraes 2001; Day and Madesis 2007). Therefore, to reach fixation, a novel mutation in an organelle genome must spread not only among individuals within a population but also among the many genome copies within a cell. The state of coexistence between different copies of an organelle genome within the same cell or individual is known as heteroplasmy.

The hierarchical organization of biological populations creates multiple levels at which selection can act. The bottlenecking approach employed in MA experiments is designed to reduce the efficacy of selection at the individual level. However, this approach does not necessarily reduce the effects of mechanism such as replication advantage and mitophagy, which may act at lower levels of selection to bias the fate of new organelle mutations (Taylor et al. 2002; Tolkovsky 2009). For example, it has been found in experimental yeast populations that reducing the population size results in the dominance of within-cell selection pressures favoring the spread of mtDNA deletions that increase the rate of genome replication but eliminate the ability of the cell to respire (Taylor et al. 2002). In addition, mouse lines engineered to have higher mitochondrial mutation rates resulting from a proofreading-deficient mtDNA polymerase show evidence of substantial selection against nonsynonymous changes in the mitochondrial genome within only a few generations (Stewart et al. 2008). The presumably small N_e and rapid response to selection in these lines suggest that there may be mechanisms of selection occurring below the individual level.

Interestingly, the fraction of mutations found in the heteroplasmic state has differed substantially among MA experiments with different organisms (Denver et al. 2000; Haag-Liautard et al. 2008; Howe et al. 2010). This may reflect differences in the severity of the "mitochondrial bottleneck" among species, i.e., the number of mitochondria (or mitochondrial genome copies) transmitted to the offspring. To what extent do the dynamics of mitochondria within cells affect the accumulation and selective filtering of mutations? How does the mitochondrial bottleneck within individuals act to reduce selection among these variants? A better understanding of the replication and transmission dynamics of organelle genomes in a heteroplasmic state will be an essential step toward accurately interpreting both phylogenetic estimates of evolutionary rates and the results obtained from MA experiments.

6.3 Phylogenetic Variation in Organelle Substitution Rates

The branching structure of phylogenetic trees is inherently fractal, and comparing rates of sequence evolution across different evolutionary timescales suggests that rate variation has a fractal nature as well. From some of the deepest splits in the eukaryotic phylogeny all the way down to the intraspecific level, there is evidence for evolutionary rate variation in organelle genomes, much of which can be explained by differences in the underlying mutation rate. As discussed above (Sect. 6.2), most evidence for organelle mutation rate variation has been inferred from phylogenetic patterns of sequence divergence. In this section, we summarize patterns of substitution rate variation that have been indentified across different phylogenetic scales.

6.3.1 Early Evidence for Mitochondrial Substitution Rate Differences Between Plants and Animals

In a classic study, Brown et al. (1979) showed that mtDNA from four primate species evolves approximately an order of magnitude faster than single-copy nuclear genes. As predicted by Brown et al., the rapid rate of mtDNA sequence evolution in most animals has made mtDNA a preferred tool for phylogenetic and population genetic studies. However, an elevated rate of sequence evolution in the mitochondrial genome (relative to the nucleus) is not a universal rule in eukaryotes. The opposite pattern generally occurs in plants. Wolfe et al. (1987) found that substitution rates in angiosperm mtDNA are approximately an order of magnitude slower than corresponding rates in the nucleus, while substitution rates in the plastid genome fall in between these two levels. By comparing rates of nucleotide substitution in absolute terms, Wolfe et al. (1987) established that, while rates of nuclear sequence evolution are comparable between plants and animals, rates of mtDNA evolution in the two lineages have evolved to opposite extremes, differing by 100-fold or more.

6.3.2 Limited Data from Other Eukaryotic Lineages

In contrast to the wealth of data available for plants and animals, estimates of organelle substitution rates are sorely lacking in other eukaryotic lineages including fungi and protists. Nevertheless, there is some evidence to suggest that most eukaryotic lineages have experienced rates of mitochondrial evolution that fall in between the extremes observed in multicellular plants and animals. For example, global phylogenies of slowly evolving rRNA genes exhibit intermediate branch lengths for protists and fungi (Yang et al. 1985; Gray et al. 1989). In addition, studies on a handful of protist and fungal lineages have found ratios of mitochondrial to nuclear divergence closer to 1:1, contrasting with biased ratios observed in plants and animals (Clark-Walker 1991; Lynch and Blanchard 1998; Lynch et al. 2006). However, there are at least some lineages, such as the mushroom order Boletales, that exhibit elevated ratios of mitochondrial to nuclear divergence (Bruns and Szaro 1992). Furthermore, data from MA lines in yeast show a very high ratio of mitochondrial to nuclear mutation rates (~37:1) (Lynch et al. 2008), which is in conflict with estimates derived from synonymous substitutions (Clark-Walker 1991; Lynch and Blanchard 1998; Lynch et al. 2006).

Given the ever-increasing availability of DNA sequence data, systematic analyses of mitochondrial rate variation – similar to those recently conducted in diverse groups such as seed plants (Mower et al. 2007), mammals (Nabholz et al. 2008), and birds (Nabholz et al. 2009) – would be a valuable contribution to the field. Even in cases where reliable divergence times cannot be estimated to calculate absolute substitution rates, comparisons of the relative rate of nuclear and mitochondrial substitution would be informative.

6.3.3 Mitochondrial Substitution Rate Variation Within Major Taxonomic Groups

While the long-standing generalization that animal mtDNA evolves rapidly and plant mtDNA evolves slowly has remained largely intact, more recent research has also shown that there is substantial rate variation within each of these groups. Notably, it is not clear that the high rate observed in most animal mitochondrial genomes is the ancestral state for all animals, because many nonbilaterians (including corals and sponges) have markedly slower rates (Shearer et al. 2002; Huang et al. 2008). Instead, it has been proposed that there was a mitochondrial rate acceleration in the ancestor of all bilaterians (Hellberg 2006; Huang et al. 2008).

There is also evidence for significant rate variation at lower taxonomic levels, even in vertebrate mtDNA, which for years was viewed as one the strongest cases for a molecular clock as evidenced by the famous "2% per million years" rule of thumb. For example, the rates of evolution in turtle and shark mitochondrial genomes are slow relative to other vertebrates (Avise et al. 1992; Martin et al. 1992), and recent in-depth studies of mitochondrial sequence divergence within both mammals and birds have revealed surprising levels of rate variation (Nabholz et al. 2008, 2009). Mammalian species in particular differ by 2 orders of magnitude in synonymous substitution rate, shattering the misconception of constant mutation rates even on relatively local phylogenetic scales (Galtier et al. 2009).

Research over the last decade on typically slow-evolving plant mtDNA has uncovered some of the most extreme examples of substitution rate variation ever identified. Flowering plants from multiple independent lineages (including the genera *Pelargonium*, *Plantago*, and *Silene*) exhibit massive accelerations in mito-chondrial synonymous substitution rate, sometimes in excess of 1,000-fold (Cho et al. 2004; Parkinson et al. 2005; Mower et al. 2007; Sloan et al. 2009). As a result, rates in these lineages often approach those of some of the fastest-evolving animal mitochondria. In contrast to the dramatic increases in mitochondrial substitution rates in these lineages, plastid and nuclear substitution rates appear generally unchanged, suggesting these species have experienced a mitochondrial-specific increase in mutation rate (Cho et al. 2004; Parkinson et al. 2005; Mower et al. 2007; Sloan et al. 2009; but see Erixon and Oxelman 2008; Guisinger et al. 2008). In some cases, particularly within the genus *Silene*, these changes have occurred quite recently (<10 Mya), resulting in closely related species with highly divergent mitochondrial rates (Fig. 6.1) (Mower et al. 2007; Sloan et al. 2009). Further variation has been generated by apparent rate reversions in a subset of the accelerated lineages (Parkinson et al. 2005).

It is not surprising that some of the most extreme examples of variation in mitochondrial substitution rates have been documented in mammals, birds, and seed plants (Table 6.1). This almost certainly reflects the greater intensity of study in these groups and highlights the need for improved sampling in other eukaryotic lineages. Given the evidence that phylogenetic patterns of rate variation may extend all the way to the intraspecific level (Sloan et al. 2008) and that organelle mutation rate can vary among genomic regions (Wolfe et al. 1987) and even among individual genes (Sloan et al. 2009), it appears that our ability to detect rate variation in organelle genomes may be limited only by how close we are willing to look.

The existence of rate variation across these diverse biological scales is of fundamental importance for analyses of sequence data. In particular, many population genetic tests based on sequence diversity make the assumption of a constant underlying mutation rate, at least at some phylogenetic scale. Frequent violations of these assumptions in organelle genomes highlight the importance of using a local measure of the neutral substitution rate to correct estimates of diversity (Barr et al. 2007; Nabholz et al. 2009).

6.4 DNA Replication and Repair in Organelle Genomes

The substantial substitution rate variation among organelle genomes and among taxa at every phylogenetic scale begs for both mechanistic and evolutionary explanations. With respect to mechanistic explanations of mutation rate variation, most attention has been focused on systems of DNA replication and repair.

For the most part, organelle genomes completely lack the genes necessary for their own DNA replication and repair. Although there are occasional exceptions (e.g., the plastid genomes of many nongreen algae contain a *dnaB*-like gene; Day and Madesis 2007), the genetic control of organelle genome replication resides entirely in the nucleus. Even in *Reclinomonas americana*, a protist with the most gene-rich mitochondrial genome identified to date, there are no mitochondrially encoded genes known to be involved in genome replication and repair [although, unlike most eukaryotes, *Reclinomonas* does maintain a mitochondrially encoded copy of a eubacterial-like RNA polymerase (Lang et al. 1997)]. Furthermore, in angiosperms in which plastid ribosomes have been artificially eliminated, plastid genome replication still occurs, indicating that the replication process is not strictly dependent on plastid-encoded proteins (Zubko and Day 2002).

In 1974, Clayton et al. showed that mammalian cells were incapable of repairing pyrimidine dimers in their mitochondrial DNA, indicating that mitochondria lacked the nucleotide excision repair pathways found in the nucleus. In some ways, this and other studies may have been overinterpreted to mean that mitochondria lack DNA repair mechanisms altogether – a notion that has been clearly refuted with subsequent research identifying a host of different mechanisms involved in preventing and repairing mutations in mtDNA (Bogenhagen 1999; Holt 2009). Nevertheless, these early studies were important in establishing that the machineries involved in nuclear and organelle genome maintenance are not always the same. Instead, DNA replication and repair in organelle genomes depends on numerous genes that are specifically targeted to the mitochondria and/or plastids. Understanding the functional roles of these genes and how they vary across eukaryotic lineages is essential to understanding variation in organelle mutation rates.

6.4.1 Origins of Organelle DNA Replication and Repair Genes

6.4.1.1 Endosymbiotic Gene Transfer

The dominant pattern in the evolution of organelle genomes since their endosymbiotic origin is one of gene loss. The genomes of free-living bacteria contain thousands of protein genes. In contrast, mitochondrial and plastid genomes contain fewer than 250, in most cases far fewer. Animal mitochondrial genomes contain a nearly universal complement of only 13 protein genes, and the mtDNA of *Plasmodium* species encodes only 3 proteins (Gray et al. 1999). Some of the reduction in organelle genome coding content can be explained by outright gene loss, but the history of eukaryotic evolution has also been characterized by a massive transfer of genes from the organelles to the nucleus – a process that remains active in many lineages (Timmis 2004).

Evidence of endosymbiotic gene transfer (EGT) can be found in the genes responsible for organelle DNA replication and repair. Genes of both proteobacterial and cyanobacterial origin (presumably reflecting the progenitors of mitochondria and plastids, respectively) have been identified as components of organelle DNA replication and repair machinery (Van Dyck et al. 1992; Eisen and Hanawalt 1999; Karlberg et al. 2000; Kimura et al. 2002; Wall et al. 2004; Lin et al. 2007; Shedge et al. 2007). These include some of the classic players in bacterial DNA repair such as *mutS*, *mutL*, and *recA* (Eisen and Hanawalt 1999; Lin et al. 2007; Shedge et al. 2007). In addition, some key processes in organelle DNA replication are apparently mediated by eubacterial-like proteins, including single-stranded DNA binding and (in some eukaryotes) DNA polymerization (Van Dyck et al. 1992; Ono et al. 2007; Moriyama et al. 2008).

Despite the role of EGT in the evolution of organelle DNA replication and repair, it is also clear that many genes involved in these processes did not originate with the bacterial progenitors of mitochondria and plastids (Karlberg et al. 2000;

Suzuki and Miyagishima 2010). Instead, many components of organelle DNA replication and repair machinery appear to have been co-opted or acquired from other sources, as we discuss below.

6.4.1.2 Viral Origins

Sequencing of the yeast mitochondrial RNA polymerase resulted in the surprising observation that it is not homologous to other known eukaryotic RNA polymerases or to the eubacterial RNA polymerase, as might be expected given the α-proteobacterial origins of mitochondria (Masters et al. 1987). This finding has since been extended to diverse eukaryotic lineages (Cermakian et al. 1996). Rather than being derived from a eubacterial ancestor, it appears that mitochondrial RNA polymerases are related to those encoded by T3/T7 bacteriophages, indicating that in some cases the genes controlling organelle genome function in eukaryotes may have been acquired from viruses.

Subsequent studies have found that key components controlling organelle DNA replication may also be of bacteriophage origin, raising the possibility that these genes were simultaneously acquired from a viral ancestor early on in eukaryotic evolution, perhaps in association with the bacterial endosymbiont that gave rise to mitochondria (Filee and Forterre 2005; Shutt and Gray 2006a). For example, there is evidence in diverse eukaryotic lineages for another T7 bacteriophage homolog (known as Twinkle in humans) that is responsible for helicase activity in mitochondrial genome replication (Spelbrink et al. 2001; Shutt and Gray 2006b). In addition, phylogenetic analysis suggests that DNA polymerase γ, the enzyme responsible for mitochondrial genome replication in animals and fungi, also has a T3/T7 bacteriophage homolog (Filee et al. 2002). Therefore, it is apparent that viral genes have played an important role in the evolution of organelle genome replication and expression.

6.4.1.3 Dual Targeting to Mitochondria and Plastids

The co-existence of two or more genomes within the eukaryotic cell creates the opportunity to share components of cellular machinery across genomic compartments. Nowhere is this more apparent than in the growing list of nuclear-encoded proteins that are targeted to both mitochondria and plastids. The set of known plant proteins that are dual targeted to the mitochondria and plastids is significantly enriched for genes involved in DNA synthesis and processing (Carrie et al. 2009a). Dual targeted proteins include DNA polymerases, helicases, topoisomerases and a RecA homolog (Wall et al. 2004; Christensen et al. 2005; Shedge et al. 2007; Carrie et al. 2009b). These examples clearly demonstrate the ability of the evolutionary process to co-opt existing genetic machinery for function in other organelles. The importance of this process is further illustrated by numerous genes shown to be involved in the repair of both nuclear and mitochondrial DNA, including DNA glycosylases, an apurinic/apyrimidinic endonuclease, and DNA ligase III (Larsen et al. 2005; Holt 2009; and references therein).

Even in cases where gene products are targeted to only the mitochondria or the plastids, there are often closely related paralogs that perform similar functions in the other organelle. For example, the plant-specific OSB gene family has been shown to be involved in the generation and maintenance of alternative conformations of the mitochondrial genome. This family includes paralogs that are targeted to the mitochondria, to the plastids, or possibly to both organelles (Zaegel et al. 2006). A history of gene duplication and replacement has been clearly demonstrated in other organelle processes including translation. In particular, multiple angiosperm lineages appear to have experienced replacement of mitochondrial-encoded ribosomal protein genes by duplicated copies of (nuclear-encoded) plastid homologs (Adams et al. 2002; Mower and Bonen 2009; Kubo and Arimura 2010). Collectively, these phenomena illustrate a history of co-opting and modifying existing genes to function in organelles, and they have played an especially important role in the evolution of organelle DNA replication and repair genes.

Although our understanding of organelle genome replication and repair remains limited in many respects, the available data illustrate that these processes depend on a complex and evolutionary labile assemblage of viral, bacterial, and eukaryotic genes. As we discuss in the next section, the flexibility of these systems has led to significant divergence in replication and repair across eukaryotic lineages.

6.4.2   Variation in Mitochondrial Genome Replication and Repair Machinery Across Eukaryotes

The best-characterized mitochondrial systems are probably those from mammals and yeast. Comparisons between the processes of mtDNA replication and repair in these two lineages have revealed important differences that likely reflect enormous variation across the diversity of eukaryotes. In some cases, differences in repair machinery may explain observed variation in mitochondrial substitution rates.

Yeast nuclear genomes contain a homolog of the bacterial DNA repair gene *mutS* (*MSH1*), which encodes a protein that functions in mitochondrial mismatch repair (Reenan and Kolodner 1992). In contrast, mammalian mitochondria lack a *mutS*-based mismatch repair system, which may at least partially explain the higher rates of point mutations in mammalian mtDNA (Foury et al. 2004). Interestingly, the Msh1 protein has been shown to preferentially recognize mismatches that would result in transitions (Chi and Kolodner 1994). Therefore, the lack of *mutS*-based mismatch repair in mammals is consistent with the extreme bias observed in transition:transversion ratios in animal mtDNA, which can be well in excess of 10:1 (Tamura and Nei 1993). In contrast, there is no significant excess of transitions observed in yeast mtDNA (Vanderstraeten et al. 1998; Lynch et al. 2008).

There are also components of mammalian mtDNA replication and repair that have not been identified in yeast. For example, a homolog of the helicase Twinkle has not been found in yeast, and it is unclear what gene is responsible for helicase activity during yeast mtDNA replication. Yeast also lack an identifiable mitochondrial DNA polymerase γ accessory factor, while such a factor is known to function in animal mitochondrial DNA synthesis and in the related T3/T7 bacteriophage system (Shutt and Gray 2006a).

These few examples likely represent the tip of the iceberg when it comes to variation among eukaryotes in organelle DNA replication and repair. Although our understanding of the organelle genetic machinery remains limited (note that even in the most well-characterized organelle systems there is ongoing uncertainty and controversy about the basic mechanisms of replication; Day and Madesis 2007; Holt 2009), there is evidence for distinct origins of the mitochondrial DNA polymerase in different eukaryotic lineages (Shutt and Gray 2006a; Ono et al. 2007; Moriyama et al. 2008). Such differences suggest that even the most central components of organelle DNA replication and repair machinery may fundamentally differ from one species to the next. Furthermore, mutation screens and directed mutagenesis have been effective at identifying/generating variants with altered organelle genome replication and repair machinery and corresponding changes in mutation rates (Foury et al. 2004; Trifunovic et al. 2004). A valuable next step would be to identify the genetic basis of organelle mutation rate variation in natural populations. Cases of recent and extreme increases in organelle mutation rates would be a good place to start (Mower et al. 2007; Sloan et al. 2009).

6.5   Evolutionary Explanations for Organelle Mutation Rate Variation

Up to this point, we have largely focused on mechanisms of organelle DNA repair as a potential cause of mutation rate variation. The observed mutation rate, however, depends not only on the efficacy of DNA repair, but also on the total amount of mutational input (Baer et al. 2007). Accordingly, extensive comparative work has been performed (particularly in vertebrates) to develop and test hypothesis about the causes of mitochondrial mutation rate variation, focusing on the role of physiology and life history. Historically, most hypotheses have treated organelle mutation rate variation as a byproduct of other biological differences (e.g., generation time and metabolic rate). In recent years, additional emphasis is being placed on the effects of mutation rate variation and the more direct role of natural selection in shaping the mutation rate.

6.5.1 Generation Time and Metabolic Rate Hypothesis

Comparative work exploring the causes of variation in mitochondrial substitution rates has focused predominantly on vertebrates and particularly mammals for which there are ample data on life history and physiology. In a now famous study, Martin and Palumbi (1993) noted the existence of a strong negative relationship between body size and the rate of DNA sequence evolution in mammals. Although it is unlikely that there is any direct effect of body size, this trait is strongly correlated with both metabolic rate and generation time, which are at the center of leading hypotheses about the sources of mutation rate variation in mitochondrial genomes.

6.5.1.1 Metabolic Rate Hypothesis

The basic operation of metabolic pathways in mitochondria is associated with the production of mutagenic byproducts including reactive oxygen species (Wallace 2005). Therefore, one natural prediction is that species with higher metabolic rates will experience higher rates of mutational damage to their mitochondrial genomes. This prediction is consistent with the negative correlation between evolutionary rate and body size in mammals, because smaller mammals tend to have higher mass-specific metabolic rates. Nevertheless, studies that have attempted to decouple effects of metabolic rate from confounded variables have found limited support for this hypothesis, particularly outside of mammals (Lanfear et al. 2007; Nabholz et al. 2009).

6.5.1.2 Generation Time Hypothesis

An alternative hypothesis is based on the expectation that many or most mutations are the result of DNA replication errors and that species with shorter generation times will undergo more rounds of germline DNA replication per year. This hypothesis is also consistent with the negative relationship between body size and substitution rate, because smaller species tend to have shorter generation times. It is not entirely clear how this hypothesis should extend to other eukaryotic lineages, particularly those that do not have a sequestered germline. Nevertheless, there is support for a generation time effect in invertebrates and plants, suggesting that it may have some generality outside of mammals (Smith and Donoghue 2008; Thomas et al. 2010).

6.5.2   Variation in the Efficacy and Intensity of Selection on Mutation Rates

Although the idea that mutation rates can be shaped by the forces of natural selection is not new (Sturtevant 1937), recent arguments have placed renewed emphasis on how variation in the selective environment may explain difference in organelle mutation rates. In general, selection is expected to favor reductions in the mutation rate because the vast majority of nonneutral mutations are deleterious. However, the strength of that selection and the ability of populations to respond to it may vary across species. For example, comparative analyses in vertebrates have found that mitochondrial synonymous substitution rates are correlated with lifespan, even when controlling for related life history traits including generation time. It has been proposed that these results reflect more intense selection in long-lived organisms for reduced mitochondrial mutation rates (Nabholz et al. 2008). This hypothesis represents an extension of the mitochondrial theory of aging, which posits that a positive feedback between the rate of mitochondrial mutations and the decline of mitochondrial function is responsible for the physiological signs of aging (Kujoth et al. 2007).

The outcome of selection on mutation rate may also depend on variation in the efficacy of selection. Based on the observation that species with low N_e tend to have higher (per generation) point mutation rates, Lynch (2010) has argued that small populations cannot effectively select against weakly deleterious alleles that increase the mutation rate. This hypothesis, however, is not specific to organelle genomes, and Lynch notes that it cannot explain some of the major patterns in mitochondrial mutation rate variation across eukaryotes (e.g., the combination of extremely low mitochondrial rates and small N_e in land plants).

The population genetic theory of mutation rate evolution in organelle genomes remains largely unexplored. In general, the magnitude of selection acting on mutation rate modifiers is dependent on genetic linkage between these modifiers and mutations throughout the genome (as well as any direct fitness effects of the modifier) (Sniegowski et al. 2000). Because organelle genomes generally experience little or no sexual recombination, a mutator located within the genome would remain tightly linked with resulting mutations and, therefore, be subject to strong selection. However, most of the genetic control of organelle mutation rates likely resides with nuclear-encoded DNA replication and repair machinery. Therefore, linkage between modifiers and mutation load should be dependent on the frequency of outcrossed sexual reproduction. A valuable area for theoretical and empirical population genetic research would be to investigate the evolutionary forces acting on nuclear-encoded modifiers of organelle mutation rates including the effects of mating system, N_e, and the relative frequency of deleterious and beneficial mutations.

6.6 Mutational Processes and the Evolution of Organelle Genome Architecture

Although mutational mechanisms are often viewed as a directionless player in evolution, generating "random" variation on which natural selection can act, they can also have clear directional effects. Based on a combination of empirical and theoretical arguments, it has been proposed that variation in mutational processes can explain some of the striking diversity of genome architecture found across living organisms, including differences in genome size, structure, and organization.

6.6.1 Biased Mutation as a Directional Force

Mutational patterns are often highly skewed, preferentially affecting certain portions of a genome and exhibiting a bias for certain types of nucleotide substitutions as well as disparities in the number and size of insertions vs. deletions (i.e., the indel spectrum). In the absence of counterbalancing selection, these biases represent a directional force in genome evolution. Directional mutation pressures have been linked with variation in nucleotide composition and genome size. In particular, differences in nuclear genome size in animals have been attributed to corresponding differences in the indel spectrum (Petrov 2002). Likewise, it has been proposed that high gene densities in bacteria result from a deletion bias (Mira et al. 2001; Kuo and Ochman 2009).

Organelle genomes (particularly mitochondrial genomes) exhibit dramatic variation in genome size and gene density (Gray et al. 1999), but the extent to which directional mutation pressures can explain these differences in unclear. There is recent evidence that rates of indels in mtDNA can vary even between very closely related species (Howe et al. 2010), but overall very little is known about variation in the mitochondrial indel spectrum. Comparative analyses determining the number and size of mitochondrial indels in diverse eukaryotic lineages would be a valuable contribution.

6.6.2 Mutation Pressure as a Selective Force

The mutational burden hypothesis presents another possible role for mutational pressures in shaping genome architecture (Lynch 2006, 2007). The idea is that complex genomic features experience a small selective cost associated with the probability that they will be disrupted by mutation. For example, a mutation altering the splice donor site of an intron can prevent proper splicing resulting in deleterious or lethal consequences depending on the functional importance of the gene. Lynch has argued that this form of selection acts as a general deterrent to the expansion of noncoding content. However, the selection coefficient on any single feature is expected to be quite small (proportional to the per nucleotide mutation rate). Therefore, the effects of this mechanism are predicted to vary across lineages, depending on both the efficacy and intensity of selection, which should be proportional to N_e and the mutation rate, respectively (Lynch 2007).

Consequently, the mutational burden hypothesis has been put forth as an explanation for some of the most dramatic differences in genome architecture observed among living organisms, e.g., prokaryotes (large N_e) vs. eukaryotes (small N_e) or animal mitochondria (high mutation rate) vs. plant mitochondria (low mutation rate). These comparisons, however, span enormous phylogenetic scales, which confound countless biological differences and raise alternative interpretations for observed variation in genome architecture. Given the growing evidence for organelle mutation rate variation among much more closely related species, we suggest that mitochondrial and chloroplast genomes represent an ideal model for dissecting the genomic consequences of mutation rate variation.

6.7   Conclusion

The organelle genomes of eukaryotes exhibit remarkable variation in nucleotide substitution rates. Despite the challenges in estimating spontaneous mutation rates, differences in evolutionary rates at sites under relatively weak selection points to substantial mutation rate variation in organelle genomes. In most cases, the underlying molecular mechanisms remain elusive, though select examples of organelle mutation rate variation may be attributed to documented differences in DNA replication and repair machinery. Mutation rate variation also reflects more ultimate evolutionary causes. Recent studies have placed renewed focus on differences across species in the efficacy and intensity of selection on mutation rate modifiers. Finally, mutation itself may be a powerful evolutionary force. It has been proposed that biased mutation may drive many aspects of genome structure and that selection exerted by deleterious mutations may favor reduced genome complexity. Since mutation rate variation arises repeatedly over small phylogenetic scales, organelle genomes represent potentially powerful systems for testing these hypotheses.

Acknowledgements Our research on mutation rates and the evolution of organelle genomes has been supported by the NSF (DEB-0808452 and MCB-1022128).

References

Adams KL, Daley DO, Whelan J, Palmer JD (2002) Genes for two mitochondrial ribosomal proteins in flowering plants are derived from their chloroplast or cytosolic counterparts. Plant Cell 14:931–943
Andersson SG, Kurland CG (1998) Reductive evolution of resident genomes. Trends Microbiol 6:263–268

Avise JC, Bowen BW, Lamb T, Meylan AB, Bermingham E (1992) Mitochondrial DNA evolution at a turtle's pace: evidence for low genetic variability and reduced microevolutionary rate in the testudines. Mol Biol Evol 9:457–473

Baer CF, Miyamoto MM, Denver DR (2007) Mutation rate variation in multicellular eukaryotes: causes and consequences. Nat Rev Genet 8:619–631

Barr CM, Neiman M, Taylor DR (2005) Inheritance and recombination of mitochondrial genomes in plants, fungi and animals. New Phytol 168:39–50

Barr CM, Keller SR, Ingvarsson PK, Sloan DB, Taylor DR (2007) Variation in mutation rate and polymorphism among mitochondrial genes in *Silene vulgaris*. Mol Biol Evol 24:1783–1791

Bergthorsson U, Adams KL, Thomason B (2003) Widespread horizontal transfer of mitochondrial genes in flowering plants. Nature 424:197–201

Bogenhagen DF (1999) Repair of mtDNA in vertebrates. Am J Hum Genet 64:1276–1281

Brandvain Y, Wade MJ (2009) The functional transfer of genes from the mitochondria to the nucleus: the effects of selection, mutation, population size and rate of self-fertilization. Genetics 182:1129–1139

Brown WM, George M, Wilson AC (1979) Rapid evolution of animal mitochondrial DNA. Proc Natl Acad Sci USA 76:1967–1971

Bruns TD, Szaro TM (1992) Rate and mode differences between nuclear and mitochondrial small-subunit rRNA genes in mushrooms. Mol Biol Evol 9:836–855

Carrie C, Giraud E, Whelan J (2009a) Protein transport in organelles: dual targeting of proteins to mitochondria and chloroplasts. FEBS J 276:1187–1195

Carrie C et al (2009b) Approaches to defining dual-targeted proteins in Arabidopsis. Plant J 57:1128–1139

Cermakian N, Ikeda TM, Cedergren R, Gray MW (1996) Sequences homologous to yeast mitochondrial and bacteriophage T3 and T7 RNA polymerases are widespread throughout the eukaryotic lineage. Nucleic Acids Res 24:648–654

Chamary JV, Parmley JL, Hurst LD (2006) Hearing silence: non-neutral evolution at synonymous sites in mammals. Nat Rev Genet 7:98–108

Charlesworth D (2010) Apparent recent elevation of mutation rate: don't forget the ancestral polymorphisms. Heredity 105:509–510

Chi NW, Kolodner RD (1994) Purification and characterization of MSH1, a yeast mitochondrial protein that binds to DNA mismatches. J Biol Chem 269:29984–29992

Cho Y, Mower JP, Qiu YL, Palmer JD (2004) Mitochondrial substitution rates are extraordinarily elevated and variable in a genus of flowering plants. Proc Natl Acad Sci 101:17741–17746

Christensen AC et al (2005) Dual-domain, dual-targeting organellar protein presequences in Arabidopsis can use non-AUG start codons. Plant Cell 17:2805–2816

Clark-Walker GD (1991) Contrasting mutation rates in mitochondrial and nuclear genes of yeasts versus mammals. Curr Genet 20:195–198

Clayton DA, Doda JN, Friedberg EC (1974) The absence of a pyrimidine dimer repair mechanism in mammalian mitochondria. Proc Natl Acad Sci USA 71:2777–2781

Day A, Madesis P (2007) DNA replication, recombination and repair in plastids. In: Bock R (ed) Cell and molecular biology of plastids, vol 19. Springer, Heidelberg, pp 65–119

Denver DR, Morris K, Lynch M, Vassilieva LL, Thomas WK (2000) High direct estimate of the mutation rate in the mitochondrial genome of *Caenorhabditis elegans*. Science 289:2342–2344

Denver DR, Morris K, Lynch M, Thomas WK (2004) High mutation rate and predominance of insertions in the *Caenorhabditis elegans* nuclear genome. Nature 430:679–682

Eisen JA, Hanawalt PC (1999) A phylogenomic study of DNA repair genes, proteins, and processes. Mutat Res 435:171–213

Ellis J (1982) Promiscuous DNA–chloroplast genes inside plant mitochondria. Nature 299:678–679

Erixon P, Oxelman B (2008) Whole-gene positive selection, elevated synonymous substitution rates, duplication, and indel evolution of the chloroplast clpP1 gene. PLoS One 3:e1386

Filee J, Forterre P (2005) Viral proteins functioning in organelles: a cryptic origin? Trends Microbiol 13:510–513

Filee J, Forterre P, Sen-Lin T, Laurent J (2002) Evolution of DNA polymerase families: evidences for multiple gene exchange between cellular and viral proteins. J Mol Evol 54:763–773

Foury F, Hu J, Vanderstraeten S (2004) Mitochondrial DNA mutators. Cell Mol Life Sci 61:2799–2811

Galtier N, Nabholz B, Glemin S, Hurst GD (2009) Mitochondrial DNA as a marker of molecular diversity: a reappraisal. Mol Ecol 18:4541–4550

Gray MW, Cedergren R, Abel Y, Sankoff D (1989) On the evolutionary origin of the plant mitochondrion and its genome. Proc Natl Acad Sci 86:2267–2271

Gray MW, Burger G, Lang BF (1999) Mitochondrial evolution. Science 283:1476–1481

Guisinger MM, Kuehl JV, Boore JL, Jansen RK (2008) Genome-wide analyses of geraniaceae plastid DNA reveal unprecedented patterns of increased nucleotide substitutions. Proc Natl Acad Sci USA 105:18424–18429

Haag-Liautard C et al (2008) Direct estimation of the mitochondrial DNA mutation rate in *Drosophila melanogaster*. PLoS Biol 6:1706–1714

Hao W, Palmer JD (2009) Fine-scale mergers of chloroplast and mitochondrial genes create functional, transcompartmentally chimeric mitochondrial genes. Proc Natl Acad Sci 106:16728–16733

Hellberg ME (2006) No variation and low synonymous substitution rates in coral mtDNA despite high nuclear variation. BMC Evol Biol 6:24

Holt IJ (2009) Mitochondrial DNA replication and repair: all a flap. Trends Biochem Sci 34:358–365

Howe DK, Baer CF, Denver DR (2010) High rate of large deletions in *Caenorhabditis briggsae* mitochondrial genome mutation processes. Genome Biol Evol 2:29–38

Howell N et al (2003) The pedigree rate of sequence divergence in the human mitochondrial genome: there is a difference between phylogenetic and pedigree rates. Am J Hum Genet 72:659–670

Huang D, Meier R, Todd PA, Chou LM (2008) Slow mitochondrial COI sequence evolution at the base of the metazoan tree and its implications for DNA barcoding. J Mol Evol 66:167–174

Karlberg O, Canback B, Kurland CG, Andersson SG (2000) The dual origin of the yeast mitochondrial proteome. Yeast 17:170–187

Kimura M (1983) The neutral theory of molecular evolution. Cambridge University Press, Cambridge

Kimura S et al (2002) A novel DNA polymerase homologous to *Escherichia coli* DNA polymerase I from a higher plant, rice (*Oryza sativa* L.). Nucleic Acids Res 30:1585–1592

Kubo N, Arimura S (2010) Discovery of the rpl10 gene in diverse plant mitochondrial genomes and its probable replacement by the nuclear gene for chloroplast RPL10 in two lineages of angiosperms. DNA Res 17:1–9

Kujoth GC, Bradshaw PC, Haroon S, Prolla TA (2007) The role of mitochondrial DNA mutations in mammalian aging. PLoS Genet 3:e24

Kuo CH, Ochman H (2009) Deletional bias across the three domains of life. Genome Biol Evol 1:145–152

Lanfear R, Thomas JA, Welch JJ, Brey T, Bromham L (2007) Metabolic rate does not calibrate the molecular clock. Proc Natl Acad Sci 104:15388–15393

Lanfear R, Welch JJ, Bromham L (2010) Watching the clock: studying variation in rates of molecular evolution between species. Trends Ecol Evol 25:495–503

Lang BF et al (1997) An ancestral mitochondrial DNA resembling a eubacterial genome in miniature. Nature 387:493–497

Larsen NB, Rasmussen M, Rasmussen LJ (2005) Nuclear and mitochondrial DNA repair: similar pathways? Mitochondrion 5:89–108

Lin Z, Nei M, Ma H (2007) The origins and early evolution of DNA mismatch repair genes–multiple horizontal gene transfers and co-evolution. Nucleic Acids Res 35:7591–7603

Lynch M (2006) Streamlining and simplification of microbial genome architecture. Annu Rev Microbiol 60:327–349

Lynch M (2007) The origins of genome architecture. Sinauer Associates, Sunderland, MA

Lynch M (2010) Evolution of the mutation rate. Trends Genet 26:345–352

Lynch M, Blanchard JL (1998) Deleterious mutation accumulation in organelle genomes. Genetica 103:29–39

Lynch M, Koskella B, Schaack S (2006) Mutation pressure and the evolution of organelle genomic architecture. Science 311:1727–1730

Lynch M et al (2008) A genome-wide view of the spectrum of spontaneous mutations in yeast. Proc Natl Acad Sci USA 105:9272–9277

Martin AP, Palumbi SR (1993) Body size, metabolic rate, generation time, and the molecular clock. Proc Natl Acad Sci 90:4087–4091

Martin AP, Naylor GJP, Palumbi SR (1992) Rates of mitochondrial DNA evolution in sharks are slow compared with mammals. Nature 357:153–155

Masters BS, Stohl LL, Clayton DA (1987) Yeast mitochondrial RNA polymerase is homologous to those encoded by bacteriophages T3 and T7. Cell 51:89–99

Mira A, Ochman H, Moran NA (2001) Deletional bias and the evolution of bacterial genomes. Trends Genet 17:589–596

Moraes CT (2001) What regulates mitochondrial DNA copy number in animal cells? Trends Genet 17:199–205

Moran NA (1996) Accelerated evolution and Muller's rachet in endosymbiotic bacteria. Proc Natl Acad Sci USA 93:2873–2878

Moran NA, McLaughlin HJ, Sorek R (2009) The dynamics and time scale of ongoing genomic erosion in symbiotic bacteria. Science 323:379–382

Moriyama T, Terasawa K, Fujiwara M, Sato N (2008) Purification and characterization of organellar DNA polymerases in the red alga Cyanidioschyzon merolae. FEBS J 275:2899–2918

Morton BR (2003) The role of context-dependent mutations in generating compositional and codon usage bias in grass chloroplast DNA. J Mol Evol 56:616–629

Mower JP, Bonen L (2009) Ribosomal protein L10 is encoded in the mitochondrial genome of many land plants and green algae. BMC Evol Biol 9:265

Mower JP, Touzet P, Gummow JS, Delph LF, Palmer JD (2007) Extensive variation in synonymous substitution rates in mitochondrial genes of seed plants. BMC Evol Biol 7:135

Muller HJ (1964) The relation of recombination to mutational advance. Mutat Res 106:2–9

Nabholz B, Glemin S, Galtier N (2008) Strong variations of mitochondrial mutation rate across mammals–the longevity hypothesis. Mol Biol Evol 25:120–130

Nabholz B, Glémin S, Galtier N (2009) The erratic mitochondrial clock: variations of mutation rate, not population size, affect mtDNA diversity across birds and mammals. BMC Evol Biol 9:54

Ono Y et al (2007) NtPolI-like1 and NtPolI-like2, bacterial DNA polymerase I homologs isolated from BY-2 cultured tobacco cells, encode DNA polymerases engaged in DNA replication in both plastids and mitochondria. Plant Cell Physiol 48:1679–1692

Parkinson CL et al (2005) Multiple major increases and decreases in mitochondrial substitution rates in the plant family Geraniaceae. BMC Evol Biol 5:73

Peterson GI, Masel J (2009) Quantitative prediction of molecular clock and Ka/Ks at short timescales. Mol Biol Evol 26:2595–2603

Petrov DA (2002) Mutational equilibrium model of genome size evolution. Theor Popul Biol 61:531–544

Reenan RA, Kolodner RD (1992) Characterization of insertion mutations in the Saccharomyces cerevisiae MSH1 and MSH2 genes: evidence for separate mitochondrial and nuclear functions. Genetics 132:975–985

Shearer TL, Van Oppen MJH, Romano SL, Wörheide G (2002) Slow mitochondrial DNA sequence evolution in the anthozoa (cnidaria). Mol Ecol 11:2475–2487

Shedge V, Arrieta-Montiel M, Christensen AC, Mackenzie SA (2007) Plant mitochondrial recombination surveillance requires unusual RecA and MutS homologs. Plant Cell 19:1251–1264

Shutt TE, Gray MW (2006a) Bacteriophage origins of mitochondrial replication and transcription proteins. Trends Genet 22:90–95

Shutt TE, Gray MW (2006b) Twinkle, the mitochondrial replicative DNA helicase, is widespread in the eukaryotic radiation and may also be the mitochondrial DNA primase in most eukaryotes. J Mol Evol 62:588–599

Sloan DB, Barr CM, Olson MS, Keller SR, Taylor DR (2008) Evolutionary rate variation at multiple levels of biological organization in plant mitochondrial DNA. Mol Biol Evol 25:243–246

Sloan DB, Oxelman B, Rautenberg A, Taylor DR (2009) Phylogenetic analysis of mitochondrial substitution rate variation in the angiosperm tribe Dileneae (Caryophyllaceae). BMC Evol Biol 9:260

Sloan DB, Alverson AJ, Storchova H, Palmer JD, Taylor DR (2010) Extensive loss of translational genes in the structurally dynamic mitochondrial genome of the angiosperm *Silene latifolia*. BMC Evol Biol 10:274

Smith SA, Donoghue MJ (2008) Rates of molecular evolution are linked to life history in flowering plants. Science 322:86–89

Sniegowski PD, Gerrish PJ, Johnson T, Shaver A (2000) The evolution of mutation rates: separating causes from consequences. Bioessays 22:1057–1066

Spelbrink JN et al (2001) Human mitochondrial DNA deletions associated with mutations in the gene encoding twinkle, a phage T7 gene 4-like protein localized in mitochondria. Nat Genet 28:223–231

Stewart JB et al (2008) Strong purifying selection in transmission of mammalian mitochondrial DNA. PLoS Biol 6:e10

Sturtevant AH (1937) Essays on evolution. I. On the effects of selection on mutation rate. Q Rev Biol 12:464

Sullivan J, Joyce P (2005) Model selection in phylogenetics. Annu Rev Ecol Evol Syst 36:445–466

Suzuki K, Miyagishima SY (2010) Eukaryotic and eubacterial contributions to the establishment of plastid proteome estimated by large-scale phylogenetic analyses. Mol Biol Evol 27:581–590

Tamura K, Nei M (1993) Estimation of the number of nucleotide substitutions in the control region of mitochondrial DNA in humans and chimpanzees. Mol Biol Evol 10:512–526

Taylor DR, Zeyl C, Cooke E (2002) Conflicting levels of selection in the accumulation of mitochondrial defects in *Saccharomyces cerevisiae*. Proc Natl Acad Sci USA 99:3690–3694

Thomas JA, Welch JJ, Lanfear R, Bromham L (2010) A generation time effect on the rate of molecular evolution in invertebrates. Mol Biol Evol 27:1173–1180

Timmis J (2004) Endosymbiotic gene transfer: organelle genomes forge eukaryotic chromosomes. Nat Rev Genet 5:123–135

Tolkovsky AM (2009) Mitophagy. Biochim Biophys Acta 1793:1508–1515

Trifunovic A et al (2004) Premature ageing in mice expressing defective mitochondrial DNA polymerase. Nature 429:417–423

Van Dyck E, Foury F, Stillman B, Brill SJ (1992) A single-stranded DNA binding protein required for mitochondrial DNA replication in *S. cerevisiae* is homologous to *E. coli* SSB. EMBO J 11:3421–3430

Vanderstraeten S, Van den Brule S, Hu J, Foury F (1998) The role of 3′–5′ exonucleolytic proofreading and mismatch repair in yeast mitochondrial DNA error avoidance. J Biol Chem 273:23690–23697

Venditti C, Meade A, Pagel M (2006) Detecting the node-density artifact in phylogeny reconstruction. Syst Biol 55:637–643

Venditti C, Meade A, Pagel M (2008) Phylogenetic mixture models can reduce node-density artifacts. Syst Biol 57:286–293

Wall MK, Mitchenall LA, Maxwell A (2004) Arabidopsis thaliana DNA gyrase is targeted to chloroplasts and mitochondria. Proc Natl Acad Sci USA 101:7821–7826

Wallace DC (2005) A mitochondrial paradigm of metabolic and degenerative diseases, aging, and cancer: a dawn for evolutionary medicine. Annu Rev Genet 39:359–407

Wolfe KH, Li WH, Sharp PM (1987) Rates of nucleotide substitution vary greatly among plant mitochondrial, chloroplast, and nuclear DNAs. Proc Natl Acad Sci USA 84:9054–9058

Yang D, Oyaizu Y, Oyaizu H, Olsen GJ, Woese CR (1985) Mitochondrial origins. Proc Natl Acad Sci USA 82:4443–4447

Zaegel V et al (2006) The plant-specific ssDNA binding protein OSB1 is involved in the stoichiometric transmission of mitochondrial DNA in arabidopsis. Plant Cell 18:3548–3563

Zubko MK, Day A (2002) Differential regulation of genes transcribed by nucleus-encoded plastid RNA polymerase, and DNA amplification, within ribosome-deficient plastids in stable phenocopies of cereal albino mutants. Mol Genet Genomics 267:27–37

Chapter 7
Gene Transfer to the Nucleus

Mathieu Rousseau-Gueutin, Andrew H. Lloyd, Anna E. Sheppard,
and Jeremy N. Timmis

7.1 Introduction

The cytoplasmic organelles of eukaryotes – mitochondria and chloroplasts – were once free-living prokaryotic organisms (Bock and Timmis 2008; Timmis et al. 2004). From these ancestral prokaryotes, eukaryotes acquired the novel biochemistry of oxidative phosphorylation and photosynthesis. The eukaryotes were initiated by the engulfment of an α-purple bacterium with a precursor to the nucleated cell in the region of 1.3 billion years ago and this was followed by a second engulfment, this time of a cyanobacterium, that led to carbon fixing eukaryotes. Both these events are widely considered to be unique though there is some evidence to the contrary.

During the time that these two or three genomes have cohabited, major changes in gene disposition have occurred; the net effect being a gross reduction in the genome sizes of the erstwhile prokaryotic ancestors of the cytoplasmic organelles. Consequently, the proteomes of mitochondria and plastids hardly reflect their small endogenomes. Most of the genes for their biogenesis and function now reside in the

M. Rousseau-Gueutin • A.H. Lloyd • J.N. Timmis (✉)
School of Molecular and Biomedical Science, The University of Adelaide,
South-Australia 5005, Australia
e-mail: mathieu.rousseau@adelaide.edu.au; andrew.lloyd@adelaide.edu.au;
jeremy.timmis@adelaide.edu.au

A.E. Sheppard,
School of Molecular and Biomedical Science, The University of Adelaide,
South-Australia 5005, Australia

Department of Evolutionary Ecology and Genetics, Zoological Institute Christian-Albrechts
University of Kiel, Am Botanischen Garten 1-9, Kiel 24118, Germany
e-mail: asheppard@zoologie.uni-kiel.de

C.E. Bullerwell (ed.), *Organelle Genetics*,
DOI 10.1007/978-3-642-22380-8_7, © Springer-Verlag Berlin Heidelberg 2012

nucleus where they have migrated and become functional while their products retain their original function in the organelles. There is evidence for another class of genes that have similarly transferred to the nucleus and achieved function, but this function is novel and unrelated to their role in the prokaryotic ancestor (Martin et al. 2002). We shall refer to this process (for both classes) as "functional gene transfer" – a process that accounts for the origin of thousands of nuclear genes in eukaryotes. Many of these functional gene relocations probably took place quite soon after the onset of the endosymbioses but there is also evidence of recent events, most notably in the angiosperms. This genetic voyage from the old world of prokaryotes into the new world of the nucleus involves a culture shock. Accommodating sequence changes must occur for activity that are difficult to envisage other than by evoking the very long time period that has been available and acknowledging that thousands of genes have accomplished the trip using individual strategies. These are rare events but many have been fully characterised at the molecular level.

Nuclear genome sequencing and the application of real-time experimentation to the process of endosymbiosis has begun to clarify some of its features. The main emergent observations are that the nuclei of essentially all species contain numerous genomic tracts that are identical or very similar to extant cytoplasmic organellar DNA sequences. These are referred to as *nu*clear integrants of *mito*chondrial DNA (*numts*: Lopez et al. 1994) and *plasti*d DNA (*nupts*: Timmis et al. 2004), which together comprise *norgs* (*n*uclear integrants of cytoplasmic *org*anellar DNA: Leister 2005). Genes in this class of organelle-derived nuclear DNA are not usually expressed. It has been possible to recapitulate the movement of *norgs* to the nucleus in real time in a few systems (Bock and Timmis 2008; Ricchetti et al. 1999). These experiments suggest that, in evolutionary terms, the nucleus is constantly bombarded by a deluge of organellar DNA, which is regularly incorporated into chromosomes. Only a few experiments show what happens to this DNA after nuclear insertion (Sheppard and Timmis 2009).

We will deal now in detail with the processes outlined above and highlight their evolutionary significance.

7.2 Genome Sizes in Putative Ancestral Prokaryotes and the Extant Cytoplasmic Organelles and Nuclei: Gene Numbers and the Times of Coevolution

Fully functional mitochondrial genomes show a large variation in size between species compared with plastids. The majority of multicellular animals have mitochondrial genomes of between 16 and 17 kb in size and encoding 12 or 13 proteins. Only a most basal multicellular animal, *Trichoplax adhaerens,* has so far been found to possess a larger mitochondrial genome of about 43 kb, but it encodes only a few more proteins (Dellaporta et al. 2006). Single celled organisms show more variation as they contain the both the largest and most gene-rich mitochondrial genomes known as well as the smallest (Timmis et al. 2004). The mitochondrial hydrogenosomes of anaerobic protists have entirely lost their genomes along with oxidative phosphorylation (Hackstein et al. 2006). Plants also show considerable variation and some of the biggest, though not necessarily gene-rich mitochondrial genomes known: that of *Arabidopsis thaliana* (Arabidopsis) is ~367 kb in size with 57 identified genes (Unseld et al. 1997). The mtDNA of the *Cucurbitaceae* is considerably larger and more variable and has been estimated to range from 390 to 2,900 kb (Ward et al. 1981). Two of the cucurbit mitochondrial genomes have now been fully sequenced and confirm these earlier estimates (Alverson et al. 2010). *Cucurbita pepo* (squash/zucchini) has a mitochondrial genome of just under 1 Mb. Interestingly, accumulation of intergenic sequence, often short repeats or sequence of chloroplast origin, rather than the retention of more genes is the main factor contributing to the large size of these genomes. The number of sequenced mitochondrial genomes is currently (September 2010) 2,243 and is expanding rapidly (http://www.ncbi.nlm.nih.gov/genomes/genlist.cgi?taxid=2759&type=4&name=Eukaryotae%20Organelles). A recent list of sequenced mitochondria and chloroplast genomes is found on the NCBI website. Any extant bacteria that are candidates for the mitochondrial ancestor have larger genomes sufficient to support their free-living lifestyle. Within the α-proteobacteria, *Caulobacter crescentus* has a 4.017 Mb genome capable of encoding 3,767 different proteins and *Bradyrhizobium japonicum*, with a larger genome of about 9.1 Mb could code for well more than 8,000 proteins (Timmis et al. 2004). It has been suggested that the mitochondrial ancestor may have been an intracellular parasite that evolved initially to rely upon the host cell for some proteins. Today's obligate parasitic bacterial genomes are certainly much smaller than those of their free-living relatives but that of *Rickettsia prowazekii,* for example (Andersson et al. 1998), still codes for about ten times as many as the most bacterium-like mitochondrial genomes currently known (Hazkani-Covo et al. 2010).

The largest chloroplast genome currently known is from *Floydiella terrestris*, an alga of the chlorophycean lineage (Brouard et al. 2010). The 521 kb genome encodes 70 proteins, which is fewer than species such as the red alga, *Porphyra purpurea* that encodes more than 200 proteins from a smaller chloroplast genome (Reith and Munholland 1995). The flowering plants all encode about 80 proteins in their plastid genomes. Potential cyanobacterial ancestors of the plastid, by comparison with currently available species, must have contained genomes sufficient to encode from 1,884 (*Prochlorococcus marinus*) to more than 7,000 proteins (*Nostoc punctiforme*) (Timmis et al. 2004). Like the mitochondrial genomes that degenerated after the requirement of the Krebs cycle was relaxed, plants and protists that have abandoned photosynthesis have often (Krause 2008), though not always (Wickett et al. 2008), quickly discarded much of their plastid genome. The chloroplast genomes of the parasitic plants *Epifagus virginiana* and *Rhizantella gardneri*, are only 70 kb and 59.2 kb and code for 21 and 20 proteins (mainly proteins of the translation machinery), respectively (Delannoy et al. 2011; Wolfe et al. 1992). In comparison with the chloroplast genome of *Phalaenopsis aphrodite*, which can be taken to resemble that of the photosynthetic ancestor of *R. gardneri*, around 70% of the genes, including all the photosynthetic genes and the genes involved in RNA metabolism, were lost or transferred to the nucleus after the switch to a parasitic and non-photosynthetic lifestyle (Delannoy et al. 2011). The apicomplexan parasites have a photosynthetic evolutionary history and most retain a vestigial chloroplast called the apicoplast. This organelle has lost genes relating to photosynthesis but it performs essential biochemical roles in *Plasmodium falciparum* and *Toxoplasma gondii,* which cannot survive in its absence. Not surprisingly, this unique plastid-related biochemistry is subject to intense scrutiny with a view to controlling the associated major human diseases (Kalanon and McFadden 2010).

7.3 Disposition of Genes That Control Cytoplasmic Organelle Biogenesis

An outcome of the long period of cohabitation of these two (animals, fungi and some protists) or three (photosynthetic eukaryotes) different genetic compartments is that the major proportion of the genes controlling function and biogenesis of the cytoplasmic organelles now reside in the nucleus as a consequence of mass organelle-to-nucleus gene relocation. The products of these genes are synthesised in the cytoplasm and imported, usually as precursor proteins, into the organelle. There is minor variation in the genes that have been retained in the organelle but generally large-scale migration to the nucleus has involved thousands of genes.

7.4 Ongoing Movement Of Organellar DNA

7.4.1 *Nupts* and *numts* Ancient and Modern

The events that initiate functional gene transfers (see Parts VI and VII) involve the incorporation of cytoplasmic organellar nucleic acids, either directly as DNA or through an RNA intermediate, into the nuclear chromosomes, and this is a frequent, ongoing process. Early experiments revealed copies of DNA in nuclei that were virtually identical to mitochondrial DNA in animals and fungi (van den Boogaart et al. 1982) and both mitochondrial and chloroplast DNA in plants (Ayliffe and Timmis 1992; Timmis and Scott 1983; van den Boogaart et al. 1982). Genome sequencing confirmed these findings and showed that tracts of organellar DNA are present in essentially all nuclear genomes that have been carefully examined (Hazkani-Covo et al. 2010; Leister 2005; Richly and Leister 2004; Timmis et al. 2004). In situ hybridisation has quite recently been optimised for plant chromosomes to visualise both *numts* (Lough et al. 2008) and *nupts* (Roark et al. 2010) in maize. Figure 7.1. shows hybridisation of maize mtDNA probes to metaphase chromosomes and illustrates how the position and number of *numt* loci vary between different

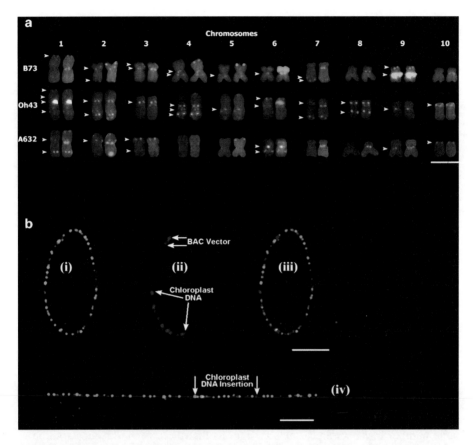

Fig. 7.1 (a) Variation in *numt* loci among karyotypes of maize inbred lines. *Arrowheads* indicate the most consistently observed sites of mitochondrial DNA (mtDNA) probe hybridization (*white*). The mtDNA probe contains a majority of the mitochondrial genome. A karyotyping probe mix (colours) was used to identify the chromosomes. The chromosome to the left shows hybridization of only the mtDNA probe, while both the karyotyping and mtDNA probes are shown on the *right*. This figure is modified from Fig. 2 of Lough et al. 2008 with permission. Bar = 10 μm. (**b**) Fibre-FISH analysis of the chloroplast insertion in BAC OSJNBb0005J14 and in *O. sativa* spp. japonica var. Nipponbare. A–C Fiber-FISH signal derived from a single OSJNBb0005J14 molecule. (i) *Green* signals were derived from labelled OSJNBb0005J14 DNA. (ii) *Red* signals were from a BAC vector (pBeloBAC11) probe and three shotgun clones (OTAWA47, OTAWA35, OTAWC79) that span the inserted chloroplast DNA. (iii) Merged image of (i) and (ii). (iv) A genomic fibre-FISH signal obtained from Nipponbare rice using OSJNBb0005J14 (*green*) and three shotgun clones (*red*) as probes. Bars = 10 μm. This figure is modified from Fig. 4 of Yuan et al. 2002 with permission

inbred lines. This work confirms earlier observations of major *nupt* differences between varieties within species (Ayliffe et al. 1998), and it will be interesting to see the *norg* variation revealed when multiple genomes of human and Arabidopsis become available and, perhaps, relate this to geographic or ecological factors.

Indeed, population studies promise to become fertile ground for determining the biological importance of *norg* integration dynamics (Jacques et al. 2010).

Another example of organelle DNA in the nucleus was beautifully visualised by fibre-FISH (Yuan et al. 2002). While studying a 239-kb BAC cloned nuclear genomic region on the long arm of rice chromosome 10 containing 24 copies of a glutathione S-transferase gene, these authors discovered a 33 kb insertion that was 99.7% similar to two separate regions of the chloroplast genome. They were able to use FISH to confirm the sequence organisation in single molecules of the BAC clone and fibre-FISH to confirm this *nupt* in rice nuclear chromatin (Fig. 7.1b).

The habit of excluding organellar DNA from shotgun assemblies in some cases led to the false conclusion that *norgs* were absent or very rare in some genomes. For example, the nucleus of the honeybee was initially said to be devoid of *numts* (Leister 2005; Pereira and Baker 2004) but subsequently it has turned out to be one of the richest in content (Behura 2007; Pamilo et al. 2007). Shotgun assemblies are still prone to misassembly methods that, in the process of excluding contaminating mtDNA, exclude *norgs,* which must lead to difficulties in closing gaps in the nuclear sequences. Only when a *norg* sequence has decayed significantly is it likely to be included. An analysis of the recent bovine genome assembly (Btau4: Liu et al. 2009a) that was produced with a combination of whole-genome shotgun and BAC sequence (BAC assemblies were partly scaffolded using information from mate pairs, BAC clone vector locations and BAC assembly sequences) shows about 421 *numts* and these are all less than 95% similar to the extant mtDNA except one of only 116 bp, which is identical to mitochondrial DNA (David Adelson, personal communication). This suggests that many more recent transfers have been discarded as contaminating mitochondrial DNA. Interestingly, investigation of a second assembly of the same bovine sequence data (UMD3.1), although about the same number of *numts* are detected, suggests that there is significant variation in their length and sequence similarity to mitochondrial DNA. There is no simple solution to this problem where short sequencing "reads" are the current norm and where BAC contigs are impractical. In contrast, a comparable analysis of the recently available horse genomic data reveals the presence of far more (1,432) *numts*, some of which are both long and highly similar to horse mitochondrial DNA. For example, one numt of 2,762 bp is 98.9% similar to 16.8% of the mitochondrial genome (David Adelson, personal communication). It is likely that different assembly software and procedures will produce markedly different results such that we have very little reliable information about *numts* at this time.

New sequencing technologies that make longer read lengths affordable on a genome-wide scale may help to alleviate this problem, but it is likely that large *numts* and *nupts* will continue to require the use of BACs in order to be characterised fully.

An additional problem is that genome projects have concentrated on representative species, chosen because they have small genomes. In that respect, they are often not truly representative of their phylogenetic clade, which is likely to contain species with much larger genomes. For example, a single average wheat chromosome of its 21 pairs (7 homoeologous groups) is about the same size as the entire rice genome that has been used as a model for grasses (Salamini et al. 2002; Yu et al. 2002). It is likely that species with such large genomes with their massive content of repetitious DNA coupled, in many plants, with homoeologous genes, are able to tolerate very high *norg* contents. This suggests that rice represents cereal genomes in some but no means all respects. At present, only the mammals, because of their homogeneous genome size, escape this anomaly. If the assembly problems referred to above can be solved, we predict a correlation between *norg* content and genome size that far exceeds the current association for *nupts* observed amongst the range of smaller genomes currently available. There will be evolutionary implications if larger genomes turn out to have an increased proportion of their genomes devoted to *norgs*.

7.4.2   Experimental *nupts* and Frequency of DNA Transfer

The comparative rarity of functional gene transfer and the high frequency of *norgs* suggested that DNA transfer was a frequent process but few considered the latter to be approachable by experimentation. However, the ability to transform yeast mitochondria facilitated an elegant experimental demonstration of the process in real time by inserting into mtDNA, selectable marker genes designed for expression in the nucleus (Thorsness and Fox 1990). In plants, it is currently not possible to transform mitochondria but several factors suggested that similar experiments could be aimed at chloroplast-to-nucleus DNA transfer in tobacco. These factors in tobacco were the observation that its nucleus was loaded with *nupts* (some of very large size) and Maliga and colleagues (Maliga 2002; Svab and Maliga 1993) had successfully adapted the chloroplast transformation from Chlamydomonas (Boynton et al. 1988). In addition, the tobacco plastome was the first to be fully sequenced, tissue culture techniques were routine and a single plant could give rise to many thousands of progeny by self fertilisation. Using this system, a selectable marker gene tailored for exclusive nuclear expression was introduced into the inverted repeats of the plastome (Huang et al. 2003). Homoplastomic lines were crossed to female wild type to replace the transplastome by uniparental inheritance and about a quarter of a million progeny screened for expression of the nucleus-specific reporter gene: neomycin – conferring kanamycin resistance. About 1 in 16,000 seedlings screened in this way contained an active *neo* that must have relocated to the nucleus within the male transplastomic parent. It was shown that each event was independent and some of the integration sites were characterised and shown to be linked to kanamycin resistance. Some aspects of these experiments were also carried out using selection of somatic cells, which suggested a lower frequency of transposition of approximately 1 event for every five million cells (Stegemann et al. 2003). These two independent series of experiments together firmly established unexpectedly high movement of plastid DNA to the nucleus that mandated entirely new ways of thinking about endosymbiosis and its role in nuclear evolution (Timmis et al. 2004). The frequencies observed in these screens are almost certainly the tip of the iceberg because the kanamycin-resistant plants must emanate from the relocation of a fairly large tract of DNA (at least the size of the experimental *neo* gene) from its specific location in the transplastome. Inevitably, DNA containing incomplete *neo* genes, either because the transposed fragment is small or because the break point is within the reporter gene, will also be entering the nucleus. Overall plastid DNA ingress must also include all other plastomic regions, which would be expected to transpose at similar rates. None of these events would be picked up in the screens. Given that *nupts* are matched in number by *numts* in sequenced genomes, we expect that similar processes are operating for mitochondrial DNA. The startling conclusion is that a large proportion, perhaps the majority, of tobacco male gametes must contain a newly inserted piece of plastid or mitochondrial DNA – or both!

The apparent discrepancy in the frequencies observed between somatic cells and reproductive cells prompted the notion that the breakdown of pollen plastids and their DNA, which is responsible in part for maternal inheritance may provide many plastome fragments that integrate into the nucleus. In somatic cells the plastids are assumed to be far more stable, accounting for the >2 orders of magnitude lower frequency of nuclear insertion. To test this hypothesis, reciprocal crosses using a transplastomic parent were made and the progenies screened for kanamycin resistance (Sheppard et al. 2008). Even higher transposition frequency was observed in the male parent (1 in 11,000 seedlings) but only one event emerged from the screen when the transplastomic was the female parent – this single plant was atypical and only identified after the screen was modified and prolonged. This large reciprocal difference supports the suggestion that there is a plethora of organelle DNA fragments in cells at some stage during development of the male gametophyte. This finding also suggests that species with biparental inheritance or that mainly reproduce asexually may present less intake of chloroplast DNA. In an attempt to identify the precise stage at which plastid DNA enters the nucleus, a transplastomic with a nuclear-ready GUS gene was prepared (Sheppard et al. 2008). This second nuclear reporter gene was incorporated into the large single copy region of the plastome and it migrated to the nucleus at a frequency comparable to the previous estimates using *neo*. Unfortunately, it was not possible to use histochemistry to search for nuclear expression during pollen development because the doubled CaMV 35 S promoter was not active in this tissue despite previous reports mentioning that it was (Conner et al. 1999). When histochemical staining for GUS was applied to leaf tissue blue spots reflecting nuclear GUS expression were observed at 25- to 300-fold higher frequency than reported in *neo* expression experiments (Sheppard et al. 2008). This is most likely because transient nuclear expression of GUS occurs around 25 times more often than chromosomal integration, although both stable and transient events must involve the migration of plastid DNA to the nucleus. Nevertheless, these experiments demonstrate that within a single plant, there must be large variation with respect to *nupt* content. Presumably, somatic transfer events involving mitochondrial DNA also occur and the processes may be similar in other species. It remains to be seen whether this somatic *norg* variation has functional significance.

7.4.3   The Mutational Fate of *nupts*

The fate of the majority of *norgs*, those that do not achieve functional gene status, is similar to pseudogenes and other non-coding DNA. Analyses have suggested that they are released from the constraints of selection and their sequences begin to decay (Huang et al. 2005; Leister 2005; Noutsos et al. 2005). Like other nuclear sequences, they may be duplicated after insertion and two or more copies may diverge from each other. Some decay rapidly but others persist in the genome for a considerable time. An example of a *numt* that has survived a significant length of time without change is found in primates (Schmitz et al. 2005). Using primers flanking a ~1.6 kb human *numt* that encodes part of the 16 S rRNA in mitochondria, it was possible to amplify the equivalent locus from 18 anthropoid primates that began divergence 40 million years ago. *Numts* in mammalian systems, if they persist in the nucleus, change an order of magnitude more slowly than their counterparts in the mitochondrion such that the consensus sequence of the 1.6 kb *numt* reflects the mitochondrial genome at the time of insertion. During this time, the mitochondrial genome has diverged to give the large variation seen between and within the 18 primate species (Schmitz et al. 2005). Interestingly, these changes involve very significant shifts in the base composition in different species. In contrast, the organelle genomes of plants and fungi evolve much more slowly than their *norgs*, perhaps because of differences between the preserved germ lines in many animal species compared with species that simply differentiate somatic cells into meiocytes. Plants show a C $\rightarrow$ T and G $\rightarrow$ A transition bias in newly integrated *norgs* (Huang et al. 2005). C $\rightarrow$ T and G $\rightarrow$ A transition mutations are attributable to spontaneous deamination of 5-methylcytosines and suggest extensive methylation of these sequences. It appears that mutations in the protein coding DNA of both plant and animal nuclei accumulate at very approximately equal rates. The difference between the two groups is that plant organellar genes are much more stable on average, whereas the mitochondrial genes of mammals are highly heterogeneous between individuals and races.

Inheritance of kanamycin resistance in the progeny of the assumed hemizygotes that result from de novo *nupt* incorporation may be regularly Mendelian or variable. In the most careful analysis so far, about half of the gene transfer lines derived from events that happened in the male parent showed stable inheritance while the other half showed varying degrees of departure from expectation with self-fertilised and backcross progeny showing reduced proportions of kanamycin-resistant plants compared with Mendelian expectations (Sheppard and Timmis 2009). Plants of the majority of the unstable lines were shown to develop somatic heterogeneity that manifested when progeny from individual seed capsules of the same plant were tested for segregation. While some self-fertilised capsules showed the expected 3:1 ratio of resistant:susceptible, others showed significantly fewer resistant progeny, which occasionally fell to 0%. On one branch several adjacent capsules all failed to transmit kanamycin resistance to their progeny suggesting that a single somatic mutation or gene silencing event was responsible. This study went on to show that, in the three lines examined, instability was due to deletion of the *neo* gene and some adjacent DNA that had been incorporated from the transplastome (Sheppard and Timmis 2009). It was not possible to determine the extent of the deletion because the flanking DNA, though considerable in size in these lines, was resent in many places elsewhere in the nucleus from natural *nupts* and of course the same sequences are present in thousands of copies in the plastids of every cell. It is unlikely that somatic deletion is the sole cause of instability as one of the lines examined showed an altered, but relatively consistent, inheritance pattern in all its capsules, suggesting a different, probably meiotic, process.

These experiments show that, while the incorporation of *nupts* is astonishingly frequent, at least half are unstable and very frequently removed within one or two generations. These observations of a dynamic balance between ingress and egress help to explain why the nuclear genome is not continuously expanding in size. The other half of the lines that are genetically stable over a few generations may also decay in the relatively short term but longer term instability is not amenable to experimentation. However, some *nupts* clearly survive for many millions of years suggesting selective advantage (Huang et al. 2005; Leister 2005; Schmitz et al. 2005). It is possible that the allotetraploid status of tobacco accounts for the difference between stable and unstable integration. *Nupts* that insert into an ancestral diploid genome that is differentially targeted for mutational change (Woodhouse et al. 2010) or meiotic reprogramming (Slotkin et al. 2009) may be unstable, whereas those that integrate into the "preserved" genome of the partnership may be more stable. However, it is unwise to speculate further as it must be remembered that allotetraploid tobacco is the only species where these measurements have been possible.

7.5   DNA or RNA?

Organelle-to-nucleus sequence transfer could occur either by direct transfer of DNA or by transfer involving reverse-transcribed RNA. Some studies support a direct DNA mechanism because many *numts* and *nupts* are very long and include non-coding regions. For example, a 620 kb *numt* and a 131 kb *nupt* have been found in Arabidopsis and rice, respectively, which strongly suggest direct DNA transfer (Stupar et al. 2001; Yu et al. 2003). Whole-genome studies provide no evidence for overrepresentation of highly transcribed regions in *numts* and *nupts* (Matsuo et al. 2005; Woischnik and Moraes 2002). However, there are examples of functional mitochondrial gene relocation to the nucleus where splicing and RNA editing appear to have occurred prior to transfer, suggesting the involvement of RNA intermediates (Adams et al. 2000; Grohmann et al. 1992; Nugent and Palmer 1991). Experimental evidence in support of a direct DNA mechanism has been found in yeast, where it has been shown that in a mutant strain with a high rate of mitochondrion-to-nucleus transfer, transfer occurs independently of an RNA intermediate (Shafer et al. 1999). However, it is possible that the mechanism of transfer is variable, so RNA-mediated transfer may also have an important role to play. Recent experiments to investigate the occurrence of RNA-mediated transfer in tobacco have not been able to discover any evidence for RNA intermediates in the process of gene transfer from the chloroplast to the nucleus (Sheppard et al. submitted).

The mode of transfer has implications in the use of transplastomic plants in biotechnological applications. Because plastids are predominantly maternally inherited, transplastomic crop plants offer greatly enhanced transgene containment compared with nuclear transgenics (Maliga 2002). However, there are two mechanisms by which plastid transgenes can escape through pollen at low frequency: occasional paternal transmission of plastids (Ruf et al. 2007; Svab and Maliga 2007) and transfer of transgenes to the nuclear genome (Huang et al. 2003; Sheppard et al. 2008). The latter type of escape would not normally result in transgene expression due to the absence of a nuclear promoter, but fortuitous integration or subsequent rearrangement could bring a transgene into context with an existing nuclear promoter (Lloyd and Timmis 2011). Furthermore, it has been shown that plastid promoters can have weak nuclear activity (Cornelissen and Vandewiele 1989). Therefore, if strict containment of a transgene product is vital, further measures will be required to prevent expression following transfer to the nuclear genome. One possibility is to make the function of plastid transgenes dependent on plastid RNA editing, such that nuclear integrants arising from direct DNA transfer should be non-functional.

7.6 The Changes Required for a Prokaryotic Gene to Become Functional in the Nucleus

Genes that relocated to the nucleus from ancestral prokaryotes were most unlikely to have been expressed immediately (Bock and Timmis 2008). Indeed, it is improbable that they would be transcribed let alone appropriately regulated such that their mRNAs were translated into polypeptides and imported into the organelle to usurp the function of the original organellar gene. Yet, this was achieved by thousands of individual genes. A prokaryotic gene is not always required to go through all these three steps to become functional in the nuclear genome. Indeed, a case has been discovered where a chloroplastic promoter (*psbA*) was immediately transcriptionally active when introduced into the nuclear genome (Cornelissen and Vandewiele 1989; Lloyd and Timmis 2011), and it would be interesting to test whether other prokaryotic promoters possess nuclear activity. Moreover, organellar genes sometimes already contain information for protein targeting into the organelle (Ueda et al. 2007). Finally, the AT richness within the 3′UTR of chloroplastic genes may provide, without sequence change, abundant chance polyadenylation sites for mRNAs that originate in the nucleus, thereby generating stable transcripts (Stegemann and Bock 2006; Lloyd and Timmis 2011). The acquisition of a nuclear promoter, a sequence encoding a transit peptide and a polyadenylation motif could be immediately possible at the time of insertion but the chances are vanishingly small for any individual event. Therefore multiple *norgs* and their corresponding organellar gene are likely to coexist until nuclear mutations and rearrangements activate one of the *norgs*. After that redundant genes must coexist in two separate genetic compartments of the cell until one of them becomes defunct. Results arising from the study of the *cox2* gene in legumes tend to show that, when a gene is present in both mitochondrion and nuclear genomes, there is no selective advantage for a mitochondrial vs. nuclear location of *cox2* in plants. The loss of functionality of a gene in one genome is presumably the result of chance mutations silencing one or other of the *cox2* genes (Adams et al. 1999). This might explain the observation in flowering plants of a frequent loss and transfer to the nucleus of some mitochondrial (Adams et al. 2000, 2001a, b, 2002) or chloroplastic genes (Millen et al. 2001). In summary: in some cases, the nuclear gene will decay, but in this case the possibility remains for the whole process to repeat itself with another *norg* comprising a newly activated copy of the gene. In contrast, once the organellar gene is lost, the situation cannot be reversed and the nuclear gene will be maintained. This helps to explain why so many genes have been transferred when the chances of each individual event are so small.

Many examples of decaying genes may be observed as pseudogenes in organelle genomes, and these are eventually eliminated by deletion (Adams et al. 2000; Millen et al. 2001). Initially, this gene decay process may seem unsurprising in the light of the predictions of Muller's ratchet (Muller 1964), but the polyhaploid status of organellar genomes promotes strong maintenance of the *status quo*. This mechanism must have evolved to counter the hostile oxidative stressed environment of the cytoplasmic organelles. However, there may be an even stronger ability to maintain a new allele in populations of sexually reproducing diploid or amphidiploid organisms.

7.7 Nuclear Genes Deriving from *norgs*: Similar or Different Functions Compared with the Organelle Counterpart

It is probable that most genes that have transferred functionally from the cytoplasmic organelles to the nucleus did so in at least two stages as the likelihood of inserting into a nuclear location that will provide for immediate fortuitous expression appears low (Lloyd and Timmis 2011). It seems more likely that an initially long or fragmented *norg* will be rearranged by nuclear tinkering, that will include all sorts of mutation, which may happen to activate a gene before it degenerates and forfeits its ability to survive in the population. Ingenious phylogenetic comparisons have been able to estimate the total contribution of the cyanobacterial ancestor to the extant Arabidopsis nuclear genome (Martin et al. 2002). Just less than half (9,368) of the total predicted Arabidopsis proteins (those that were sufficiently conserved for primary sequence comparison) were compared with a variety of reference genomes. Of these, 1,700 (18%) showed highest homology to, or branched with, cyanobacterial proteins, suggesting that, if this applied to the entire proteome, 4,500 genes may have been acquired from the ancestral plastid. Most interestingly, more than half of the proteins analysed were not predicted to be targeted to the chloroplast, indicating that many of the initially plastid-derived proteins are involved in unrelated cellular processes and suggesting that endosymbiotic transfer has been an important source of gene innovation in flowering plants. This may be an overestimate however, as TargetP and other chloroplast prediction algorithms regularly used often incorrectly assign non-chloroplast locations to many known chloroplast proteins. This can be up to 40% of the time in some cases (Kleffmann et al. 2004). When similar phylogenetic comparisons were applied to the glaucophyte *Cyanophora paradoxa*, the proportion of nuclear genes of cyanobacterial origin was less (10.7%), and rather fewer of them were predicted to have non-plastid functions (Reyes-Prieto et al. 2006). Other estimates of symbiont-derived nuclear genes in Arabidopsis suggested 4.7% and for the red alga *Cyanidioschyzon merolae* 12.7% (Sato et al. 2005), all of which were assumed to encode plastid proteins. A recent analysis of Arabidopsis, rice, Chlamydomonas and Cyanidioschyzon proteins indicate that ~14% of nuclear genes are of plastid origin (Deusch et al. 2008). A similar study was performed to detect nuclear proteins of alpha-proteobacterial origin. This showed that 630 orthologous groups presented a close evolutionary relationship between alphaproteobacterial and eukaryotic proteins in the 22,525 phylogenies reconstructed (Gabaldón and Huynen 2003). Most of these mitochondrial proteins that derive from the so-called proto-mitochondrion (Gray et al. 1999) are proteins involved in translation, post-translational modification, protein folding and metabolism (Gabaldón and Huynen 2003, 2007; Karlberg et al. 2000). From all these analyses, it must be concluded that plastid and mitochondrial genomes have made a significant contribution to modern nuclear genomes and they are clearly continuing to do so.

A search by Leister and colleagues (Noutsos et al. 2007) revealed a different class of *norg* insertions into nuclear genomes. They showed that nuclear exons encoding novel protein sequences can be generated by insertions of organellar DNA. Many of the insertions arose from non-coding regions of the organelle DNA or from coding regions where the encoded amino acid sequence has been dramatically altered (e.g. due to a frame-shift or a large number of mutations). This work is very significant as it predicts that organelle-derived DNA insertions might have been responsible for many ancient functional exon acquisitions that are not directly detectable because of the short sequences involved and the high level of divergence from the sequence of origin, which is the only sequence available to use in a search.

Recent studies showed that not only exons but also introns can be created by the insertion of DNA of mitochondrial origin. These mitochondrial derived introns were observed in the human genome (74 bp: 100% identity to part of the ATP synthase FO subunit 6), in the crustacean *Daphnia* (64 bp: 96% identity to the 16 s rRNA) and in the unicellular alga *Bigellowiella natans* (74 bp: 86% of identity to *cox1*) and were all of a small size (Curtis and Archibald 2010; Li et al. 2009;

Ricchetti et al. 2004). In the case of *B. natans*, the mitochondrial fragment inserted in the nuclear genome did not possess the 5'-GT...AG-3' splicing elements at the time of the transfer and it is not known how and when it acquired those elements.

Geneticists have previously considered that genes with new functions have arisen after an existing gene duplicated, providing functional redundancy. Under these circumstances, one of the two genes is freed from selection pressure and its sequence can decay. Often this process will lead to genetic oblivion but occasionally to the birth of a new version of the old gene or to one with an entirely new function (neofunctionalisation). Overall, we think the suggestion (Timmis and Scott 1984) that organellar DNA insertion is a major substrate for nuclear tinkering is now well supported by molecular evidence, and it appears that the processes of endosymbiotic evolution are at least as significant as gene duplication in adding to the scope and heterogeneity of the nuclear genome.

7.8 Specific Cases of Gene Relocation Events

Some elegant examples of the apparently unlikely gene relocations have been uncovered for both plastid and mitochondrial genes. Characterisations of nuclear genes that encode cytoplasmic organellar proteins show that they retain, often with little change, their former prokaryotic identities at both the nucleotide and amino acid sequence levels. For example, the iron-sulphur subunit of succinate dehydrogenase (*sdh2*) in *Homo sapiens* (Au et al. 1995), yeast (Lombardo et al. 1990), *Arabidopsis* and maize (Figueroa et al. 1999a, b) is highly similar to the gene that remains in the mitochondrial genome of *Reclinomonas americana* (Burger et al. 1996). There are few species known where *sdh2* has not relocated to the nucleus, and it is assumed to represent a very ancient endosymbiotic transfer. The nuclear genes differ in being preceded by nuclear promoters and sequences encoding transit peptides that direct the precursor protein product, translated on 80 S cytoplasmic ribosomes, into the mitochondria. Likewise, the nuclear small subunit of ribulose bisphosphate carboxylase/oxygenase of *Pisum sativum*, characterised in pioneering experiments (Bedbrook et al. 1980), is prefaced by a strong nuclear promoter and 150 bp of DNA encoding 50 N-terminal amino acids that direct the precursor protein that is translated on 80 S cytoplasmic ribosomes, into the chloroplast.

The majority of endosymbiotic gene transfers are thought to be ancient events and the process appears to have stopped in most animal species. However, functional relocation seems alive and well amongst the angiosperms and several fascinating recent examples have contributed to an understanding of the process. A summary of known cases of recent (based on variation within the flowering plants) functional gene transfers from the mitochondrion or the chloroplast to the nucleus is presented in Table 7.1. One of the first such descriptions was in maize (Figueroa et al. 1999b) where, in contrast with many other plants including *Vicia faba*, the mitochondrial protein *rps14* was not encoded in mitochondrial DNA. A search of maize genomic DNA located an *rps14*-like gene within the long first

Table 7.1 Summary of known cases of functional gene transfers from the mitochondrion or the chloroplast to the nucleus in flowering plants

NCBI taxo.	Order	Family	Genus/Species	Mitochondrion														Chloroplast		
				Cox2	Rpl2	Rpl5	Rps1	Rps2	Rps7	Rps10	Rps11	Rps12	Rps14	Rps16	Rps19	Sdh3	Sdh4	InfA	Rpl22	Rpl32
Liliopsida	Acorales	Acoraceae	*Acorus*																17	
	Poales	Cyperaceae	*Carex*												9					
		Poaceae	*Hordeum vulgare*													4	4			
			Oryza sativa		17	20				16			15		9	4				
			Triticum aestivum		17	20				17			20		9					
			Zea mays							2			11		9					
Stem eudicots	Ranunculales	Papaveraceae	*Papaver*																	
		Ranunculaceae	*Aquilegia formosa*													17	17			
Asterids	Apiales	Apiaceae	*Daucus carota*							2										
	Asterales	Asteraceae	*Lactuca sativa*		17		17		17	2			17			17	17			
	Caryophyllales	Amaranthaceae	*Spinacia oleracea*							2										
		Chenopodiaceae	*Beta vulgaris*		17		17						17			17				
	Ericales	Ericaceae	*Rhododendron*						17											
			Vaccinium corymbosum													17				
	Lamiales	Lamiaceae	*Ocimum basilicum*		17		17			17						17	17			
		Scrophulariaceae	*Triphysaria*		17											17				
	Solanales	Convolvulaceae	*Ipomoea batatas*										17			17				
		Solanaceae	*Solanum lycopersicum*		3		5											18		
Rosids	Brassicales	Brassicaceae	*Arabidopsis thaliana*		3					22			10		5	4	4	18		
	Cucurbitales	Cucurbitaceae	*Cucumis sativus*										17		17					
	Fabales	Fabaceae	*Atylosia*	1																
			Cullen	1																
			Dumasia	1																
			Eriosema	1																
			Erythrina	1																
			Glycine max	7	3										5	4	4	18		
			Lespedeza	1																
			Medicago				5													
			Melilotus		3		14													
			Neonotonia	1																
			Ortholobium	1																
			Phaseolus	1																
			Pisum sativum																12	

(continued)

Table 7.1 (continued)

NCBI taxo. Order	Family	Genus/Species	Mitochondrion Cox2	Rpl2	Rpl5	Rps1	Rps2	Rps7	Rps10	Rps11	Rps12	Rps14	Rps16	Rps19	Sdh3	Sdh4	Chloroplast InfA	Rpl22	Rpl32
		Pseudeminia	1																
		Psoralea	1																
		Ramirezella	1																
		Trigonella				14													
		Vigna	19																
Malpighiales	Euphorbiaceae	Euphorbia cyparissias																	
		Manihot							17			17		17	17	17			
	Salicaceae	Populus		6			6		6	6		6		6	6	6			
Malvales	Malvaceae	Gossypium		3		5			6					5	4	6		8, 21	
Myrtales	Onagraceae	Fuchsia							2										
		Oenothera									13								
Oxalidales	Oxalidaceae	Oxalis							2										
Rosales	Rosaceae	Fragaria x ananassa			17					17	17								
		Malus domestica											17	17	17				
		Prunus persica										17	17						
Sapindales	Rutaceae	Citrus reticulata		17		17			17			17		17					

The species or genus in which a functional gene transfer has been described are mentioned in this table. Only the genes that are still present in the mitochondrion or chloroplast genomes of some flowering plants were taken into account. The numbers refer to the publications where the functional gene transfer was described: 1, Adams et al. (1999). 2, Adams et al. (2000). 3, Adams et al. (2001a). 4, Adams et al. (2001b). 5, Adams et al. (2002). 6, Choi et al. (2006). 7, Covello and Gray (1992). 8, Cusack and Wolfe (2007). 9, Fallahi et al. (2005). 10, Figueroa et al. (1999a). 11, Figueroa et al. (1999b). 12, Gantt et al. (1991). 13, Grohmann et al. (1992). 14, Hazle and Bonen (2007). 15, Kubo et al. (1999). 16, Kubo et al. (2000). 17, Liu et al. (2009b). 18, Millen et al. (2001). 19, Nugent and Palmer (1991). 20, Sandoval et al. (2004). 21, Ueda et al. (2007). 22, Wischmann and Schuster (1995).

intron of the gene encoding the iron-sulphur subunit of succinate dehydrogenase (*sdh2*) referred to earlier. Differential RNA splicing resulted in mRNAs encoding either sdh2 or rps14, both of which used the same transit peptide, hijacked from *sdh2* by the insertion of the former *rps14* mitochondrial gene, for mitochondrial importation. Later studies (Cusack and Wolfe 2007) revealed a similar process for a chloroplast gene – with an additional evolutionary twist. The chloroplast ribosomal protein gene *rpl32*, which is present in the plastid genome in most species including *Medicago spp*, was incorporated into an intron of the nuclear gene for chloroplast superoxide dismutase (*sod-cp*) in an ancestor of mangrove and poplar trees. Expression of both proteins was achieved in mangrove by differential splicing of the precursor mRNA and, mirroring the mechanism for rps10 in maize, both were imported into chloroplasts using the transit peptide derived from *sod-cp*. Cusack and Wolfe were able to track a further progression of the endosymbiotic process: in poplar, the bifunctional gene seen in mangrove was duplicated and the copies specialised in producing one of the two chloroplast proteins to achieve complete subfunctionalization (Cusack and Wolfe 2007).

The birth of these new nuclear genes sounds improbable but such events have been well documented amongst the angiosperms, and there is strong evidence that the mitochondrial or chloroplast genes have been transferred many times in separate phylogenetic clades and employing different molecular strategies (Adams et al. 2000; Millen et al. 2001). These latter publications discovered an even more unexpected aspect of endosymbiotic gene transfer in the angiosperms. Mitochondrial *rps10* (mitochondrial ribosomal small subunit protein 10) (Adams et al. 2000) and *infA* (chloroplast translation initiation factor 1) (Millen et al. 2001) had made the organelle-to-nucleus passage many times. By plotting the instances of gene relocation to a robust phylogenetic tree of the angiosperms, the authors were able to conclude that a functional rps10 had moved from mtDNA to the nucleus in 26 different clades while active *infA* had relocated on 24 separate and, therefore, independent occasions. All this occurred during the relatively short period of the radiation of flowering plants. It seems either that there are new evolutionary forces operating in the angiosperms with respect to gene transfer or that there is something unusual about some of the ribosomal protein genes and *infA*. What these forces could be is currently a matter for speculation but the development of the male gametophyte linked to uniparental inheritance could be responsible for the reinvigoration of gene transfer.

7.9   The Genetic Consequences of Continuous Inflow of Organellar DNA to the Nucleus and Its Significance for Nuclear Genetics

The events described above, particularly the invasion of the nucleus by *norgs*, is happening in the few species investigated, at a rate that exceeds normal mutation by mechanisms that include highly active transposons and retrotransposons. Therefore we expect most of the insertion events into active genes to cause a range of disadvantage up to lethality. This is another reason why estimates of the frequency of *norg* incorporation must be considered very conservative. It may also be argued that this disadvantage is not so serious for species that contain a large amount of non-coding DNA (of which *norgs* are themselves components) into which new *norgs* may incorporate without phenotypic effects. The ultimate consequence of this deluge (Martin 2003) of *norgs* is the continuous donation of novel DNA to the nucleus for experimentation leading, in rare cases, to new genes with new functions or modified genes with old functions with the control of expression based in a new and more sophisticated genetic environment.

7.10   The Genetic Reasons for Moving to the Nucleus and for Remaining in the Cytoplasmic Organelle

Some think that, given time, all the genes in present day cytoplasmic organelles will migrate to the nucleus. However, most consider that, given the length of time that has been available already and given the hostile mutagenic environment in mitochondria and chloroplasts, there must be a very profound reason that genes remain untransferred. The reason for the retention of genes in the highly energetic cytoplasmic organellar compartments has been the subject of much speculation (Allen 1993; Allen et al. 2005; Daley and Whelan 2005). These include simple (and incorrect) ideas that some proteins are simply too hydrophobic for import and more complex scenarios envisaging that the assembly of protein complexes may require a starting peptide inside an organelle for correct configuration. Perhaps, the most intriguing idea is that there are organellar redox sensors and redox response regulators encoded in the nucleus that function together in feedback control of redox potential in photosynthesis and respiration. These would place organellar genes under redox regulatory control (Allen 1993). There is a growing body of experimental evidence that supports these ideas (Allen and Puthiyaveetil 2008; Puthiyaveetil and Allen 2009; Puthiyaveetil et al. 2008).

7.11   Conclusion

It is now well established that many of the genes initially present in ancestral mitochondrial and plastid genomes have relocated to the nucleus and it appears that organellar genomes have also made significant contributions to novel functions encoded in the nuclear genome. As the genomes of more species are examined and methods for inferring gene origin and predicting targeting peptides improve, it will be most interesting to see the extent to which organellar genomes have been involved in shaping the functions encoded in nuclear genomes. A related area that requires further investigation is the contribution of shorter organelle sequences to existing nuclear genes. While it has been shown that organelle sequences can be incorporated as novel nuclear exons (Noutsos et al. 2007), it is currently unclear to what extent this process has contributed to nuclear genome evolution. More sophisticated sequence comparison methods may be required to answer this question, as it is likely that many such sequences diverge rapidly once they are incorporated into the nuclear genome.

Functional transfer of organellar DNA to the nucleus almost certainly involves several steps, including transfer of the DNA to the nucleus and subsequent rearrangement to generate a functional sequence context. While the former process has been relatively well studied, and we can see many examples of the "finished product" in nuclear genomes, the processes that result in the generation of a functional nuclear gene following integration into the nucleus remain relatively poorly characterised. To gain better insight into these processes, it will be advantageous to utilise the experimental systems that have been developed for real-time detection of organelle-to-nucleus DNA transfer. While it is known that in tobacco, many plastid DNA integrants in the nucleus are unstable, with deletion of the marker gene occurring in many cases within a single generation, it has so far not been possible to characterise this instability at the sequence level (Sheppard and Timmis 2009). It is very likely that most of the deletion events do not involve precise excision of the integrated plastid DNA and in this way the process may contribute to novel sequence context acquisition.

Finally, it will be interesting to discover whether the results that have been obtained with existing experimental systems are representative of organelle-to-nucleus transfer in a wider sense. Establishment of similar systems in new species will allow us to ascertain whether tobacco is typical, or whether it has an unusually high frequency of plastid DNA incorporation in the nucleus. It would also be interesting to know how these processes operate in natural environments, for example, the extent to which they are influenced by environmental variation. Taken together, it would then be possible to develop an integrated understanding of the contribution of organelle-to-nucleus gene transfer to eukaryotic evolution.

Acknowledgements We thank the Australian Research Council for financial support (Grants DP0667006 and DP0986973). We are grateful to Ashley Lough and Kathy Newton for Fig. 7.1a and to Jiming Jiang and Robin Buell for Fig. 7.1b. This chapter is not intended to be an exhaustive review of the topic. Therefore we apologise for omitting discussion of some important research contributions in the interests of brevity.

References

Adams KL, Song K, Roessler PG, Nugent JM, Doyle JL, Doyle JJ, Palmer JD (1999) Intracellular gene transfer in action: dual transcription and multiple silencings of nuclear and mitochondrial cox2 genes in legumes. Proc Natl Acad Sci USA 96:13863–13868

Adams KL, Daley DO, Qiu YL, Whelan J, Palmer JD (2000) Repeated, recent and diverse transfers of a mitochondrial gene to the nucleus in flowering plants. Nature 408:354–357

Adams KL, Ong HC, Palmer JD (2001a) Mitochondrial gene transfer in pieces: fission of the ribosomal protein gene rpl2 and partial or complete gene transfer to the nucleus. Mol Biol Evol 18:2289–2297

Adams KL, Rosenblueth M, Qiu Y-L, Palmer JD (2001b) Multiple losses and transfers to the nucleus of two mitochondrial succinate dehydrogenase genes during angiosperm evolution. Genetics 158:1289–1300

Adams KL, Qiu YL, Stoutemyer M, Palmer JD (2002) Punctuated evolution of mitochondrial gene content: high and variable rates of mitochondrial gene loss and transfer to the nucleus during angiosperm evolution. Proc Natl Acad Sci USA 99:9905–9912

Allen JF (1993) Control of gene-expression by redox potential and the requirement for chloroplast and mitochondrial genomes. J Theor Biol 165:609–631

Allen JF, Puthiyaveetil S (2008) Chloroplast sensor kinase – the redox messenger of organelle gene expression. Biochim Biophys Acta – Bioenergetics 1777:S108–S109

Allen JF, Puthiyaveetil S, Strom J, Allen CA (2005) Energy transduction anchors genes in organelles. Bioessays 27:426–435

Alverson AJ, Wei XX, Rice DW, Stern DB, Barry K, Palmer JD (2010) Insights into the evolution of mitochondrial genome size from complete sequences of *Citrullus lanatus* and *Cucurbita pepo* (Cucurbitaceae). Mol Biol Evol 27:1436–1448

Andersson SGE, Zomorodipour A, Andersson JO, Sicheritz-Ponten T, Alsmark UCM, Podowski RM, Naslund AK, Eriksson AS, Winkler HH, Kurland CG (1998) The genome sequence of *Rickettsia prowazekii* and the origin of mitochondria. Nature 396:133–140

Au HC, Reamrobinson D, Bellew LA, Broomfield PLE, Saghbini M, Scheffler IE (1995) Structural organization of the gene encoding the human iron-sulfur subunit of succinate-dehydrogenase. Gene 159:249–253

Ayliffe MA, Timmis JN (1992) Plastid DNA-sequence homologies in the tobacco nuclear genome. Mol Gen Genet 236:105–112

Ayliffe MA, Scott NS, Timmis JN (1998) Analysis of plastid DNA-like sequences within the nuclear genomes of higher plants. Mol Biol Evol 15:738–745

Bedbrook JR, Smith SM, Ellis RJ (1980) Molecular-cloning and sequencing of cDNA-encoding the precursor to the small subunit of chloroplast ribulose-1,5-bisphosphate carboxylase. Nature 287:692–697

Behura SK (2007) Analysis of nuclear copies of mitochondrial sequences in honeybee (*Apis mellifera*) genome. Mol Biol Evol 24:1492–1505

Bock R, Timmis JN (2008) Reconstructing evolution: gene transfer from plastids to the nucleus. Bioessays 30:556–566

Boynton JE, Gillham NW, Harris EH, Hosler JP, Johnson AM, Jones AR, Randolphanderson BL, Robertson D, Klein TM, Shark KB, Sanford JC (1988) Chloroplast transformation in Chlamydomonas with high-velocity microprojectiles. Science 240:1534–1538

Brouard JS, Otis C, Lemieux C, Turmel M (2010) The exceptionally large chloroplast genome of the green alga *Floydiella terrestris* illuminates the evolutionary history of the Chlorophyceae. Genome Biol Evol 2:240–256

Burger G, Lang BF, Reith M, Gray MW (1996) Genes encoding the same three subunits of respiratory complex II are present in the mitochondrial DNA of two phylogenetically distant eukaryotes. Proc Natl Acad Sci USA 93:2328–2332

Choi C, Liu Z, Adams KL (2006) Evolutionary transfers of mitochondrial genes to the nucleus in the Populus lineage and coexpression of nuclear and mitochondrial Sdh4 genes. New Phytol 172:429–439

Conner AJ, Mlynarova L, Stiekema WJ, Nap JP (1999) Gametophytic expression of GUS activity controlled by the potato Lhca3.St.1 promoter in tobacco pollen. J Exp Bot 50:1471–1479

Cornelissen M, Vandewiele M (1989) Nuclear transcriptional activity of the plastid *psbA* promoter. Nucleic Acids Res 17:19–29

Covello PS, Gray MW (1992) Silent mitochondrial and active nuclear genes for subunit 2 of cytochrome *c* oxydase (*cox2*) in soybean: evidence for RNA-mediated gene transfer. EMBO J 11:3815–3820

Curtis BA, Archibald JM (2010) A spliceosomal intron of mitochondrial DNA origin. Curr Biol 20:R919–R920

Cusack BP, Wolfe KH (2007) When gene marriages don't work out: divorce by subfunctionalization. Trends Genet 23:270–272

Daley DO, Whelan J (2005) Why genes persist in organelle genomes. Genome Biol 6:110

Delannoy E, Fujii S, des Francs CC, Brundrett M, Small I (2011) Rampant gene loss in the underground orchid *Rhizanthella gardneri* highlights evolutionary constraints on plastid genomes. Mol Biol Evol 28(7):2077–2086

Dellaporta SL, Xu A, Sagasser S, Jakob W, Moreno MA, Buss LW, Schierwater B (2006) Mitochondrial genome of *Trichoplax adhaerens* supports Placozoa as the basal lower metazoan phylum. Proc Natl Acad Sci USA 103:8751–8756

Deusch O, Landan G, Roettger M, Gruenheit N, Kowallik KV, Allen JF, Martin W, Dagan T (2008) Genes of cyanobacterial origin in plant nuclear genomes point to a heterocyst-forming plastid ancestor. Mol Biol Evol 25:748–761

Fallahi M, Crosthwait J, Calixte S, Bonen L (2005) Fate of mitochondrially located S19 ribosomal protein genes after transfer of a functional copy to the nucleus in cereals. Mol Genet Genome 273:76–83

Figueroa P, Gómez I, Carmona R, Holuigue L, Araya A, Jordana X (1999a) The gene for mitochondrial ribosomal protein S14 has been transferred to the nucleus in *Arabidopsis thaliana*. Mol Genet Genome 262:139–144

Figueroa P, Gomez I, Holuigue L, Araya A, Jordana X (1999b) Transfer of *rps14* from the mitochondrion to the nucleus in maize implied integration within a gene encoding the iron-sulphur subunit of succinate dehydrogenase and expression by alternative splicing. Plant J 18:601–609

Gabaldón T, Huynen MA (2003) Reconstruction of the proto-mitochondrial metabolism. Science 301:609

Gabaldón T, Huynen MA (2007) From endosymbiont to host-controlled organelle: the hijacking of mitochondrial protein synthesis and metabolism. PLoS Comput Biol 3:e219

Gantt JS, Baldauf SL, Calie PJ, Weeden NF, Palmer JD (1991) Transfer of rpl22 to the nucleus greatly preceded its loss from the chloroplast and involved the gain of an intro. EMBO J 10:3073–3078

Gray MW, Burger G, Lang BF (1999) Mitochondrial evolution. Science 283:1476–1481

Grohmann L, Brennicke A, Schuster W (1992) The mitochondrial gene encoding ribosomal-protein S12 has been translocated to the nuclear genome in Oenothera. Nucleic Acids Res 20:5641–5646

Hackstein JHP, Tjaden J, Huynen M (2006) Mitochondria, hydrogenosomes and mitosomes: products of evolutionary tinkering! Curr Genet 50:225–245

Hazkani-Covo E, Zeller RM, Martin W (2010) Molecular poltergeists: mitochondrial DNA Copies (numts) in sequenced nuclear genomes. Plos Genet 6:11

Hazle T, Bonen L (2007) Status of genes encoding the mitochondrial S1 ribosomal protein in closely-related legumes. Gene 405:108–116

Huang CY, Ayliffe MA, Timmis JN (2003) Direct measurement of the transfer rate of chloroplast DNA into the nucleus. Nature 422:72–76

Huang CY, Grunheit N, Ahmadinejad N, Timmis JN, Martin W (2005) Mutational decay and age of chloroplast and mitochondrial genomes transferred recently to angiosperm nuclear chromosomes. Plant Physiol 138:1723–1733

Jacques N, Sacerdot C, Derkaoui M, Dujon B, Ozier-Kalogeropoulos O, Casaregola S (2010) Population polymorphism of nuclear mitochondrial dna insertions reveals widespread diploidy associated with loss of heterozygosity in *Debaryomyces hansenii*. Eukaryot Cell 9:449–459

Kalanon M, McFadden GI (2010) Malaria, *Plasmodium falciparum* and its apicoplast. Biochem Soc Transac 38:775–782

Karlberg O, Canbäck B, Kurland CG, Andersson SGE (2000) The dual origin of the yeast mitochondrial proteome. Yeast 17:170–187

Kleffmann T, Russenberger D, von Zychlinski A, Christopher W, Sjolander K, Gruissem W, Baginsky S (2004) The *Arabidopsis thaliana* chloroplast proteome reveals pathway abundance and novel protein functions. Curr Biol 14:354–362

Krause K (2008) From chloroplasts to "cryptic" plastids: evolution of plastid genomes in parasitic plants. Curr Genet 54:111–121

Kubo N, Harada K, Hirai A, K-i K (1999) A single nuclear transcript encoding mitochondrial RPS14 and SDHB of rice is processed by alternative splicing: common use of the same mitochondrial targeting signal for different proteins. Proc Natl Acad Sci USA 96:9207–9211

Kubo N, Jordana X, Ozawa K, Zanlungo S, Harada K, Sasaki T, Kadowaki K (2000) Transfer of the mitochondrial rps10 gene to the nucleus in rice: acquisition of the 5′ untranslated region followed by gene duplication. Mol Gen Genet 263:733–739

Leister D (2005) Origin, evolution and genetic effects of nuclear insertions of organelle DNA. Trends Genet 21:655–663

Li W, Tucker AE, Sung W, Thomas WK, Lynch M (2009) Extensive, recent intron gains in *Daphnia* populations. Science 326:1260–1262

Liu Y, Qin X, Song X-Z, Jiang H, Shen Y, Durbin KJ, Lien S, Kent M, Sodeland M, Ren Y, Zhang L, Sodergren E, Havlak P, Worley K, Weinstock G, Gibbs R (2009a) Bos taurus genome assembly. BMC Genome 10:180

Liu S-L, Zhuang Y, Zhang P, Adams KL (2009b) Comparative analysis of structural diversity and sequence evolution in plant mitochondrial genes transferred to the nucleus. Mol Biol Evol 26:875–891

Lloyd AH, Timmis JN (2011) The origin and characterization of new nuclear genes originating from a cytoplasmic organellar genome. Mol Biol Evol 28(7):2019–2028

Lombardo A, Carine K, Scheffler IE (1990) Cloning and characterization of the iron-sulfur subunit gene of succinate-dehydrogenase from *Saccharomyces cerevisiae*. J Biol Chem 265:10419–10423

Lopez JV, Yuhki N, Masuda R, Modi W, Obrien SJ (1994) Numt, a recent transfer and tandem amplification of mitochondrial-DNA to the nuclear genome of the domestic cat. J Mol Evol 39:174–190

Lough AN, Roark LM, Kato A, Ream TS, Lamb JC, Birchler JA, Newton KJ (2008) Mitochondrial DNA transfer to the nucleus generates extensive insertion site variation in maize. Genetics 178:47–55

Maliga P (2002) Engineering the plastid genome of higher plants. Curr Opin Plant Biol 5:164–172

Martin W (2003) Gene transfer from organelles to the nucleus: frequent and in big chunks. Proc Natl Acad Sci USA 100:8612–8614

Martin W, Rujan T, Richly E, Hansen A, Cornelsen S, Lins T, Leister D, Stoebe B, Hasegawa M, Penny D (2002) Evolutionary analysis of *Arabidopsis*, cyanobacterial, and chloroplast genomes reveals plastid phylogeny and thousands of cyanobacterial genes in the nucleus. Proc Natl Acad Sci USA 99:12246–12251

Matsuo M, Ito Y, Yamauchi R, Obokata J (2005) The rice nuclear genome continuously integrates, shuffles, and eliminates the chloroplast genome to cause chloroplast-nuclear DNA flux. Plant Cell 17:665–675

Millen RS, Olmstead RG, Adams KL, Palmer JD, Lao NT, Heggie L, Kavanagh TA, Hibberd JM, Giray JC, Morden CW, Calie PJ, Jermiin LS, Wolfe KH (2001) Many parallel losses of *infA* from chloroplast DNA during angiosperm evolution with multiple independent transfers to the nucleus. Plant Cell 13:645–658

Muller HJ (1964) The relation of recombination to mutational advance. Mutat Res 1:2–9

Noutsos C, Richly E, Leister D (2005) Generation and evolutionary fate of insertions of organelle DNA in the nuclear genomes of flowering plants. Genome Res 15:616–628

Noutsos C, Kleine T, Armbruster U, DalCorso G, Leister D (2007) Nuclear insertions of organellar DNA can create novel patches of functional exon sequences. Trends Genet 23:597–601

Nugent JM, Palmer JD (1991) RNA-mediated transfer of the gene coxii from the mitochondrion to the nucleus during flowering plant evolution. Cell 66:473–481

Pamilo P, Viljakainen L, Vihavainen A (2007) Exceptionally high density of NUMTs in the honeybee genome. Mol Biol Evol 24:1340–1346

Pereira SL, Baker AJ (2004) Low number of mitochondrial pseudogenes in the chicken (*Gallus gallus*) nuclear genome: implications for molecular inference of population history and phylogenetics. BMC Evol Biol 4:17

Puthiyaveetil S, Allen JF (2009) Chloroplast two-component systems: evolution of the link between photosynthesis and gene expression. Proc Biol Sci 276:2133–2145

Puthiyaveetil S, Kavanagh TA, Cain P, Sullivan JA, Newell CA, Gray JC, Robinson C, van der Giezen M, Rogers MB, Allen JF (2008) The ancestral symbiont sensor kinase CSK links photosynthesis with gene expression in chloroplasts. Proc Natl Acad Sci USA 105:10061–10066

Reith M, Munholland J (1995) Complete nucleotide sequence of the *Porphyra purpurea* chloroplast genome. Plant Mol Biol Rep 13:333–335

Reyes-Prieto A, Hackett JD, Soares MB, Bonaldo MF, Bhattacharya D (2006) Cyanobacterial contribution to algal nuclear genomes a primarily limited to plastid functions. Curr Biol 16:2320–2325

Ricchetti M, Fairhead C, Dujon B (1999) Mitochondrial DNA repairs double-strand breaks in yeast chromosomes. Nature 402:96–100

Ricchetti M, Tekaia F, Dujon B (2004) Continued colonization of the human genome by mitochondrial DNA. PLoS Biol 2:e273

Richly E, Leister D (2004) NUPTs in sequenced eukaryotes and their genomic organization in relation to NUMTs. Mol Biol Evol 21:1972–1980

Roark LM, Hui Y, Donnelly L, Birchler JA, Newton KJ (2010) Recent and frequent insertions of chloroplast DNA into maize nuclear chromosomes. Cytogenet Genome Res 129:17–23

Ruf S, Karcher D, Bock R (2007) Determining the transgene containment level provided by chloroplast transformation. Proc Natl Acad Sci USA 104:6998–7002

Salamini F, Ozkan H, Brandolini A, Schafer-Pregl R, Martin W (2002) Genetics and geography of wild cereal domestication in the near east. Nat Rev Genet 3:429–441

Sandoval P, Leûn G, Gûmez I, Carmona R, Figueroa P, Holuigue L, Araya A, Jordana X (2004) Transfer of RPS14 and RPL5 from the mitochondrion to the nucleus in grasses. Gene 324:139–147

Sato N, Ishikawa M, Fujiwara M, Sonoike K (2005) Mass identification of chloroplast proteins of endosymbiont origin by phylogenetic profiling based on organism-optimized homologous protein groups. Genome Inform 16:56–68

Schmitz J, Piskurek O, Zischler H (2005) Forty million years of independent evolution: a mitochondrial gene and its corresponding nuclear pseudogene in primates. J Mol Evol 61:1–11

Shafer KS, Hanekamp T, White KH, Thorsness PE (1999) Mechanisms of mitochondrial DNA escape to the nucleus in the yeast *Saccharomyces cerevisiae*. Curr Genet 36:183–194

Sheppard AE, Timmis JN (2009) Instability of plastid DNA in the nuclear genome. Plos Genet 5:8

Sheppard AE, Ayliffe MA, Blatch L, Day A, Delaney SK, Khairul-Fahmy N, Li Y, Madesis P, Pryor AJ, Timmis JN (2008) Transfer of plastid DNA to the nucleus is elevated during male gametogenesis in tobacco. Plant Physiol 148:328–336

Slotkin RK, Vaughn M, Borges F, Tanurdzic M, Becker JD, Feijo JA, Martienssen RA (2009) Epigenetic reprogramming and small RNA silencing of transposable elements in pollen. Cell (Cambridge) 136:461–472

Stegemann S, Hartmann S, Ruf S, Bock R (2003) High-frequency gene transfer from the chloroplast genome to the nucleus. Proc Natl Acad Sci USA 100:8828–8833

Stupar RM, Lilly JW, Town CD, Cheng Z, Kaul S, Buell CR, Jiang JM (2001) Complex mtDNA constitutes an approximate 620-kb insertion on *Arabidopsis thaliana* chromosome 2: implication of potential sequencing errors caused by large-unit repeats. Proc Natl Acad Sci USA 98:5099–5103

Svab Z, Maliga P (1993) High-frequency plastid transformation in tobacco by selection for a chimeric aada gene. Proc Natl Acad Sci USA 90:913–917

Svab Z, Maliga P (2007) Exceptional transmission of plastids and mitochondria from the transplastomic pollen parent and its impact on transgene containment. Proc Natl Acad Sci USA 104:7003–7008

Thorsness PE, Fox TD (1990) Escape of DNA from mitochondria to the nucleus in *Saccharomyces cerevisiae*. Nature 346:376–379

Timmis JN, Scott NS (1983) Sequence homology between spinach nuclear and chloroplast genomes. Nature 305:65–67

Timmis JN, Scott NS (1984) Promiscuous DNA – sequence homologies between DNA of separate organelles. Trends Biochem Sci 9:271–273

Timmis JN, Ayliffe MA, Huang CY, Martin W (2004) Endosymbiotic gene transfer: organelle genomes forge eukaryotic chromosomes. Nat Rev Genet 5:123–135

Ueda M, Fujimoto M, Arimura S, Murata J, Tsutsumi N, Kadowaki K (2007) Loss of the rpl32 gene from the chloroplast genome and subsequent acquisition of a preexisting transit peptide within the nuclear gene in Populus. Gene 402:51–56

Unseld M, Marienfeld JR, Brandt P, Brennicke A (1997) The mitochondrial genome of *Arabidopsis thaliana* contains 57 genes in 366,924 nucleotides. Nat Genet 15:57–61

van den Boogaart P, Samallo J, Agsteribbe E (1982) Similar genes for a mitochondrial ATPase subunit in the nuclear and mitochondrial genomes of *Neurospora crassa*. Nature 298:187–189

Ward BL, Anderson RS, Bendich AJ (1981) The mitochondrial genome is large and variable in a family of plants (Cucurbitaceae). Cell 25:793–803

Wickett NJ, Zhang Y, Hansen SK, Roper JM, Kuehl JV, Plock SA, Wolf PG, dePamphilis CW, Boore JL, Goffinet B (2008) Functional gene losses occur with minimal size reduction in the plastid genome of the parasitic liverwort *Aneura mirabilis*. Mol Biol Evol 25:393–401

Wischmann C, Schuster W (1995) Transfer of rps10 from the mitochondrion to the nucleus in *Arabidopsis thaliana*: evidence for RNA-mediated transfer and exon shuffling at the integration site. FEBS Lett 374:152–156

Woischnik M, Moraes CT (2002) Pattern of organization of human mitochondrial pseudogenes in the nuclear genome. Genome Res 12:885–893

Wolfe KH, Morden CW, Palmer JD (1992) Function and evolution of a minimal plastid genome from a nonphotosynthetic parasitic plant. Proc Natl Acad Sci USA 89:10648–10652

Woodhouse MR, Schnable JC, Pedersen BS, Lyons E, Lisch D, Subramaniam S, Freeling M (2010) Following tetraploidy in maize, a short deletion mechanism removed genes preferentially from one of the two homeologs. PLoS Biol 8:6

Yu J, Hu SN, Wang J, Wong GKS, Li SG, Liu B, Deng YJ, Dai L, Zhou Y, Zhang XQ, Cao ML, Liu J, Sun JD, Tang JB, Chen YJ, Huang XB, Lin W, Ye C, Tong W, Cong LJ, Geng JN, Han YJ, Li L, Li W, Hu GQ, Huang XG, Li WJ, Li J, Liu ZW, Li L, Liu JP, Qi QH, Liu JS, Li L, Li T, Wang XG, Lu H, Wu TT, Zhu M, Ni PX, Han H, Dong W, Ren XY, Feng XL, Cui P, Li XR, Wang H, Xu X, Zhai WX, Xu Z, Zhang JS, He SJ, Zhang JG, Xu JC, Zhang KL, Zheng XW, Dong JH, Zeng WY, Tao L, Ye J, Tan J, Ren XD, Chen XW, He J, Liu DF, Tian W, Tian CG, Xia HG, Bao QY, Li G, Gao H, Cao T, Wang J, Zhao WM, Li P, Chen W, Wang XD, Zhang Y, Hu JF, Wang J, Liu S, Yang J, Zhang GY, Xiong YQ, Li ZJ, Mao L, Zhou CS, Zhu Z, Chen RS, Hao BL, Zheng WM, Chen SY, Guo W, Li GJ, Liu SQ, Tao M, Wang J, Zhu LH, Yuan LP, Yang HM (2002) A draft sequence of the rice genome (*Oryza sativa* L. ssp indica). Science 296:79–92

Yu YS, Rambo T, Currie J, Saski C, Kim HR, Collura K, Thompson S, Simmons J, Yang TJ, Nah G, Patel AJ, Thurmond S, Henry D, Oates R, Palmer M, Pries G, Gibson J, Anderson H, Paradkar M, Crane L, Dale J, Carver MB, Wood T, Frisch D, Engler F, Soderlund C, Palmer LE, Tetylman L, Nascimento L, de la Bastide M, Spiegel L, Ware D, O'Shaughnessy A, Dike S, Dedhia N, Preston R, Huang E, Ferraro K, Kuit K, Miller B, Zutavern T, Katzenberger F, Muller S, Balija V, Martienssen RA, Stein L, Minx P, Johnson D, Cordum H, Mardis E, Cheng ZK, Jiang JM, Wilson R, McCombie WR, Wing RA, Yuan QP, Shu OY, Liu J, Jones KM, Gansberger K, Moffat K, Hill J, Tsitrin T, Overton L, Bera J, Kim M, Jin SH, Tallon L, Ciecko A, Pai G, Van

Aken S, Utterback T, Reidmuller S, Bormann J, Feldblyum T, Hsiao J, Zismann V, Blunt S, de Vazeilles A, Shaffer T, Koo H, Suh B, Yang Q, Haas B, Peterson J, Pertea M, Volfovsky N, Wortman J, White O, Salzberg SL, Fraser CM, Buell CR, Messing J, Song RT, Fuks G, Llaca V, Kovchak S, Young S, Bowers JE, Paterson AH, Johns MA, Mao L, Pan HQ, Dean RA (2003) In-depth view of structure, activity, and evolution of rice chromosome 10. Science 300:1566–1569

Yuan Q, Hill J, Hsiao J, Moffat K, Ouyang S, Cheng Z, Jiang J, Buell CR (2002) Genome sequencing of a 239-kb region of rice chromosome 10 L reveals a high frequency of gene duplication and a large chloroplast DNA insertion. Mol Genet Genomics 267:713–720

Part IV
Origins of Organelle Proteomes

Chapter 8
Recycling and Tinkering: The Evolution of Protein Transport to and into Endosymbiotically Derived Organelles

Oliver Mirus and Enrico Schleiff

8.1 Protein Translocation as a Consequence of Cellular Complexity

The function of cellular systems depends on many different solutes. They can be classified into ions, metabolites and fatty acids, RNA, DNA, and polypeptides. These solutes have to be exchanged between different compartments of the cell and the crosstalk and exchange between these cellular components adds up to the cellular performance. In the simplest systems, e.g., Gram-positive bacteria, they are present in two distinct compartments, in the cytoplasm and in the surrounding membrane. Already in such a system, a regulatory system for the distribution of the solutes has to exist. In the course of cellular evolution and the formation of complex cellular ensembles, the demand for sorting and its regulation was raised. Hence, many complex systems were developed to deliver components to their place of function and to store them for the time of requirement (given are references for some examples: Berridge et al. 2003; Schnell and Hebert 2003; Haydon and Cobbett 2007; De Domenico et al. 2008; Heil and Ton 2008; van Meer et al. 2008; Holt and Bullock 2009).

Most of the regulatory and transport processes are performed by proteins. On the one hand, they have to control distribution, storage, and integration of solutes. Here, the processes range from the integration of metals into biomolecules as seen in the iron–sulfur cluster synthesis (e.g., Lill 2009) to the assembly of large molecular

O. Mirus
Department of Biosciences, Molecular Cell Biology of Plants, Institute for Molecular Biosciences, Goethe University, Max-von-Laue Str. 9, 60438 Frankfurt, Germany

E. Schleiff (✉)
Department of Biosciences, Cluster of Excellence Macromolecular Complexes, Molecular Cell Biology of Plants, Institute for Molecular Biosciences and Centre of Membrane Proteomics, Goethe University, Max-von-Laue Str. 9, 60438 Frankfurt, Germany
e-mail: schleiff@bio.uni-frankfurt.de

C.E. Bullerwell (ed.), *Organelle Genetics*,
DOI 10.1007/978-3-642-22380-8_8, © Springer-Verlag Berlin Heidelberg 2012

complexes such as the ribosome (e.g., Fromont-Racine et al. 2003). These two pathways nicely document the concept of cellular function – both take place in at least two different compartments; both are interlinked, as proteins made by ribosomes are required for the iron–sulfur cluster production and iron–sulfur cluster-containing proteins are involved in ribosome biogenesis; and thereby, both pathways demonstrate how deeply the function of endosymbiotically derived organelles is interlinked with the processes of other cellular compartments.

On the other hand, proteinaceous components control the transfer of solutes across the membrane boundary protecting cells from the outside world or compartmentalizing the cell into different reaction rooms. Here, several mechanisms and machines have evolved over time and are now specialized for the transport of a certain subset of solutes (e.g., Mirus et al. 2009a, 2010). Within this article, a brief overview on and comparison of the systems specialized on transport of polypeptides will be given, followed by a detailed discussion of the evolutionary development of protein translocation systems of the chloroplast and mitochondrial systems.

8.2 The Prokaryotic Transport Routes

Initially, proteins might have integrated spontaneously into the membrane surrounding the "protocell" (Pohlschröder et al. 2005; Bohnsack and Schleiff 2010). Subsequently and step by step, proteins evolved to enhance the specificity and the efficiency of the insertion of membrane proteins (Fig. 8.1). At first, a system of the Oxa1/YidC/Alb type (named after the central component of the mitochondrial/bacterial/chloroplast system, see Sects. 8.5 and 8.6, van der Laan et al. 2005a) might have evolved, which catalyzed the insertion of membrane proteins. In line

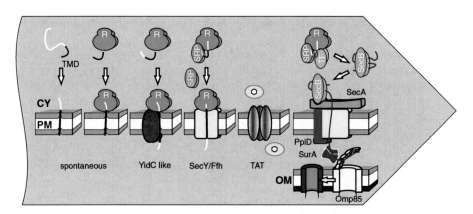

Fig. 8.1 The translocation systems in prokaryotic membranes. The evolutionary development of the translocation components in the bacterial membrane systems is depicted as described in the text. The postulated timeline is indicated by the direction of the *arrow*

with such a proposal, YidC proteins of type 1 (Oxa1, Alb4, YidC1, Funes et al. 2009) are found to interact with 70S ribosomes or ribosome nascent chain complexes due to a C-terminal extension (e.g., Kohler et al. 2009). Further consistent with an ancestral function of YidC stands the observation that in Gram-positive bacteria and eukaryotes YidC isoforms for co- and post-translational function evolved by independent gene duplications (Funes et al. 2009), which might have led to the YidC2-type family including Cox18, Alb4, and YidC2, all not containing the C-terminal ribosome binding extension.

It is hypothesized that all other translocation systems of the plasma membrane including the SEC and the TAT translocon evolved subsequently (Fig. 8.1). The occurrence of the SEC translocon composed of SecYE as the minimal translocon (Tsukazaki et al. 2008) provided an additional capacity for protein translocation. Initially, the action of SEC and YidC might have been independent of each other, because the periplasmic domain of YidC interacting with SecF (Xie et al. 2006) is not present in all proteins of this family (Bohnsack and Schleiff 2010). The TAT pathway is considered to be a subsequent invention with respect to the SEC translocon (Settles and Martienssen 1998), because the insertion of proteins into the plasma membrane is envisioned to precede the requirement of translocation of proteins across the membrane. This TAT translocon is composed of a receptor complex involving TatB and TatC and a pore-forming component TatA (Robinson and Bolhuis 2001; Dabney-Smith and Cline 2009). In eukaryotes, the TAT system can be found in chloroplasts (see Sect. 8.6.3) and the central component TatC is still encoded in mitochondrial genomes of red algae, liverwort, vascular plants (e.g., *Arabidopsis thaliana*), and sponges (e.g., Wu et al. 2000; Wang and Lavrov 2007). However, the function of the mitochondrial TatC is not yet known.

Besides the translocation system, a specific targeting system had to evolve to warrant the proper distribution of the substrate proteins (Fig. 8.1). It is discussed that the bacterial protein Ffh might have constituted the original targeting system. All sequenced genomes contain at least one gene encoding Ffh (termed Srp54 in eukaryotes – signal recognition particle component of 54 kDa) and one gene coding for the RNA molecule, which together form the most *primitive* form of the signal recognition particle (SRP, Grudnik et al. 2009). Ffh is a GTPase, which is recognized by the receptor FtsY, which is a GTPase as well (termed SRα in eukaryotes for SRP receptor α). This system can be envisioned as the first targeting system evolved.

Due to the formation of the outer membrane, the cellular systems became more and more complex and at the same time a higher degree of regulation of protein secretion was required. Two additional factors ensured the transport of secreted proteins to the surface, namely SecB and trigger factor (TF, Fig. 8.1). SecB resembles a chaperone (Ullers et al. 2004) and has evolved after cyanobacteria "branched off" from the tree of life. Trigger factor is a ribosome-associated chaperone, which is involved in the discrimination between the SRP- and the SecB-dependent pathways (Beck et al. 2000) and which might have a general function in initial protein folding today (Merz et al. 2008).

Within the periplasmic space, several steps of protein routing and targeting evolved. Here, proteins at the periplasmic side of the plasma membrane (e.g., PpiD), soluble proteins in the periplasm (e.g., SurA, skp), and proteins in the outer membrane (Omp85) are involved (Fig. 8.1, Bohnsack and Schleiff 2010). Interestingly, the putative substrate binding groove of TF recognizing the signal peptide is structurally related to that of the periplasmic chaperone SurA (Stirling et al. 2006). In addition, overlapping substrate specificity was observed for periplasmic peptidyl prolyl cis-trans isomerases PpiD and SurA (Stymest and Klappa 2008). Thus, the signal-recognizing elements might have either originated from one component or were shaped by convergent evolution (Bohnsack and Schleiff 2010).

Finally, for the insertion of proteins into the outer membrane a protein family termed Omp85 has evolved (Löffelhardt et al. 2007; Schleiff et al. 2010). As for the evolution of components involved in the insertion of proteins into the plasma membrane, it can be envisioned that initially beta-barrel proteins inserted spontaneously into the membrane, because they still have the propensity for self-insertion (e.g., Xu and Colombini 1996). Subsequently, Omp85 might have evolved to catalyze this process when the protein content of the outer membrane became more complex.

Today, all translocation systems are essential as a result (a) of the enhanced complexity in all targeted compartments, (b) of the adaptation of the cellular system to the existence of these machineries and (c) most importantly because many proteins of the bacterial systems known today have to pass at least the plasma membrane.

8.3 Eukaryotic Protein Transport and Translocation Systems

With the development of intracellular structures, the bacterial transport systems became adopted and additional transport systems emerged. In eukaryotes, the translocation systems can be divided into protein-targeting systems (Xu and Massague 2004; Schnell and Hebert 2003) and vesicle-mediated translocation systems (Robinson 2004; Bonifacino and Glick 2004). Even though it is tempting to assume that the latter is a eukaryotic invention, a bacterial origin is discussed as well (Dacks et al. 2008), because the central subunit of the retromer vesicle coat, Vps29, shares similarity with bacterial phosphoesterases (Dacks and Field 2007). However, at stage one cannot distinguish between a true bacterial origin of this process and a "recycling" of a prokaryotic protein for a novel function in vesicle transport. The latter, however, is often found in the evolution of translocation systems (see also Sects. 8.5 and 8.6).

The systems for direct protein transport can be divided according to certain properties, e.g., the folding status of the precursor protein during translocation: The nuclear (Görlich and Kutay 1999), peroxisomal (Erdmann and Schliebs 2005) or TAT translocon (Robinson and Bolhuis 2001) handle folded proteins, whereas mitochondria (Chacinska et al. 2009), chloroplasts (Soll and Schleiff 2004), or the endoplasmic reticulum-localized translocon (Osborne et al. 2005; Rapoport 2007) transport unfolded proteins. Alternatively, the form of the translocon can be used for classification: We find preexisting pores in almost all cases with the exception of the peroxisomes and the TAT translocon (Erdmann and Schliebs 2005; Robinson and Bolhuis 2001) where the translocon is only formed upon substrate recognition. Last but not least, one can divide the machines according to their evolutionary origin into eukaryotic inventions or adaptations of the prokaryotic complexes.

At least three systems are discussed to be of eukaryotic origin: the nuclear pore complex, the ER-associated degradation (ERAD) system of the endoplasmic reticulum and the translocon of the peroxisomes. The nuclear transport system can be divided into the pore itself, proteins of the Importin family and the Ran system. For the nuclear pore complex, a relation to the coatomer II complex is observed (Field and Dacks 2009). The nuclear shuttling system is thought to have evolved from an ancient Importin-β-like progenitor (Malik et al. 1997), because Importin-β contains a sequence of HEAT repeats, which are present in prokaryotic proteins as well (Morimoto et al. 2002). Ran is related to the Ras-GTPase protein family, which has a prokaryotic ancestor (Dong et al. 2007), but Ran itself cannot be linked with a prokaryotic relative. Thus, the nuclear transport system might have evolved by "recycling" a prokaryotic protein as discussed for the origin of the vesicular traffic system.

The translocon of the peroxisomes was discussed as being related to the ERAD machinery (Gabaldon et al. 2006; Schlüter et al. 2006). However, this relation is still under debate (e.g., Duhita et al. 2010), but based on the existing data, one can propose that the peroxisomal system is largely of eukaryotic origin (e.g., Bohnsack and Schleiff 2010 and references therein). The ERAD system itself involves many components (Vembar and Brodsky 2008), but their central players such as the Derlins are of eukaryotic origin. The other complexes are all (at least in parts) of prokaryotic origin. In the next sections, the evolutionary adaptation of the prokaryotic complexes within the endosymbiotically derived organelles will be discussed.

8.4 The Routing to Endosymbiotically Derived Organelles

8.4.1 The Targeting Signals

The evolutionary development of eukaryotic cells after capturing an α-proteobacterium – and subsequently a cyanobacterium – included the transfer of genes into the host nucleus. The transfer frequency from the organelle to the host nucleus was estimated to be 2×10^{-5} per cell per generation for mitochondrial genes based on the *Saccharomyces cerevisiae* system, whereas the opposite transfer was not observed (Thorsness and Fox 1990). A similar value was observed for the transfer of genes from chloroplasts to the nucleus (Stegemann et al. 2003), and the frequency of expression activation was found to be 3×10^{-8} (Stegemann and Bock 2006). In the latter study, it was found that the predominant mode of expression activation is the utilization of the promoter of an upstream nuclear gene. Remarkably, after gene transfer a high frequency of point mutations or deletions occurs in the according gene in the organellar genome, which is discussed as a possible mode for the subsequent gene deletion (Stegemann and Bock 2006).

For retargeting of the proteins from the cytosol to the proper organelle, a signal is required. At present, it is envisioned that this signal is present in the amino acid sequence of the precursor protein, and in many cases it is N-terminally positioned. The initial evolutionary development of the signals has to be seen in the context of the evolution of the translocation machineries, because these signals had to be recognized and discriminated. It is hypothesized that in chloroplasts Toc75 originated from the outer membrane β-barrel protein assembly factor Omp85 (see Sects. 8.2, 8.6, and 8.7), and targeting to bacterial Omp85 was found to be dependent on the presence of a C-terminal phenylalanine or tryptophane residue (Struyve et al. 1991). Indeed, signals for the targeting to the primitive plastid of the glaucocystophyte alga *Cyanophora paradoxa* still contain a phenylalanine at their N terminus, which is essential for recognition and translocation of the precursor proteins into the plastids of this alga (Steiner et al. 2005). The same holds true for signals for transport across the ancestral outer and inner membrane of secondary plastids derived from endosymbiotic red algae (van Dooren et al. 2001; Armbrust et al. 2004; Harb et al. 2004; Ralph et al. 2004; Kilian and Kroth 2005) and to some extent even for targeting signals of red algae (e.g., *Cyanidioschyzon merolae*; Patron and Waller 2007). Remarkably, the phenylalanine of a precursor protein from *C. paradoxa* indeed enables the precursor protein to interact with a cyanobacterial Omp85 and with the chloroplast translocation channel (Wunder et al. 2007). Thus, the initial prerequisite for the translocation across the outer membrane of chloroplasts might have evolved from the specificity of the ancestral Omp85 for phenylalanine or tryptophane residue-containing proteins.

The mitochondrial channel Tom40 most likely originated from the eukaryotic VDAC ancestor by a gene duplication event followed by a functional specialization (see Sect. 8.5). Thus, the origination of the signals cannot be linked to a preexisting motif from bacterial targeting signals but rather had to develop in the new cellular context. Here, two different scenarios can be envisioned. On the one hand, several bacterial proteins are predisposed to be translocated into mitochondria (Baker and Schatz 1987; Lucattini et al. 2004; Walther et al. 2009). In addition, about 25% of randomly generated peptides function as mitochondrial import signals (Lemire et al. 1989). This might lead to the speculation that initially a broad spectrum of signals could be used to drive translocation into mitochondria. On the other hand, some bacterial proteins with a signal sequence are recognized by the mitochondrial translocon (Mukhopadhyay et al. 2005). Hence, it was suggested that the initial translocation path might have originated by initial retransport of inner membrane proteins already containing targeting signals (Cavalier-Smith 2006). Thereby the initial seed for the development of the signal might have been the inherited structure of the signal. Nevertheless, both theories have their limits as the first leaves unanswered how specificity was warranted in the first place, and the second does not explain why the SEC translocon of the endosymbiont was not recycled.

Regardless of the question whether bacterial signals might have been inherited, proteins without a targeting signal present in the cytoplasm of the endosymbiont had to be furnished with a signal after their gene had been transferred into the host genome. It was hypothesized that exon-shuffling might have been the mode of signal addition (Fig. 8.2a, e.g., Bruce 2001). As a base of this notion, earlier observations were considered demonstrating that signal sequences of some chloroplast and mitochondrial precursors are encoded by three distinct exons (Quigley et al. 1988; Liaud et al. 1990; Gregerson et al. 1994; Long et al. 1996). It is argued that over the course of evolutionary adaptation a loss of introns might have occurred. A recent analysis of nuclear genes coding for mitochondrial ribosomal subunits in rice and A. thaliana revealed that 19 of 30 genes with an N-terminal extension not present in the bacterial ancestor have at least one intron positioned as such that the signal occurrence could be explained by exon addition (Bonen and Calixte 2006). However, although analyzing the region encoding the first 100 amino acids for proteins with different intracellular localizations, one does not observe a particular enrichment of introns (Fig. 8.2b). If one analyzes the positioning of introns with respect to the "cleavage site," a slight enrichment is observed (Fig. 8.2c). However, whether this is sufficient to conclude a relation remains questionable. Thus, it remains to be further investigated whether the intron positioning is indeed related to signal sequence occurrence.

Based on the analysis of the mitochondrial ribosomal protein S11 – which contains a signal sequence comparable to the one of the beta-subunit of ATP synthase from plant mitochondria – and of the mitochondrial ribosomal protein S11-2 – which contains a signal sequence comparable to the one of the cytochrome oxidase subunit Vb – a common origin of some classes of signals and their distribution by duplication and recombination was suggested (Fig. 8.2a, Kadowaki et al. 1996).

Alternatively or even in parallel to the exon shuffling hypothesis based on the analysis of the O-acetylserine (thiol)-lyases it was proposed that the 5'UTR of the bacterial gene might have been recycled as a coding region for the transit peptide (Fig. 8.2a, Rolland et al. 1993). Even though both notions are possible, the latter appears only a rare option, because the likeliness that the 5'UTR of the bacterial gene encodes a sequence that can be adapted to a transit peptide is rather low.

At this stage, it is hard to conclude which mechanism might have led to the signal required for retargeting. It cannot be excluded that additional processes have led to the occurrence of signals within proteins. The observation that signals can vary in their length (Zhang and Glaser 2002; Bionda et al. 2010) that the position can be within the mature domain of the precursor protein (e.g., Kutik et al. 2008), and that for short cleavable signals the mature domain participates in the translocation event (e.g., Pfanner et al. 1987; Bionda et al. 2010) opens many possibilities. Thus, not only one mode, but also several alternative routes, might have resulted in the development of signals.

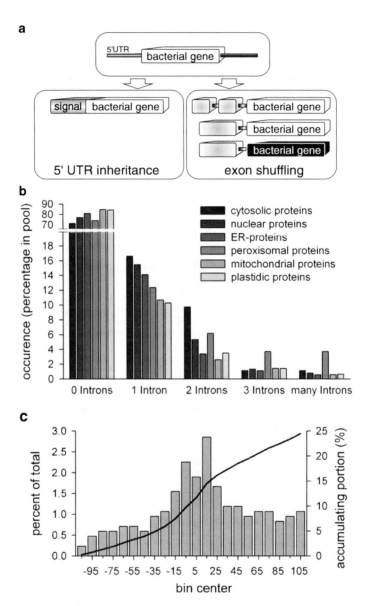

Fig. 8.2 The evolution of the targeting signals. (**a**) The discussed modes of signal evolution are shown. On the *left side*, the bacterial 5′UTR was "recycled" to encode for a signal sequence. On the right side, eukaryotic exons positioned in the 5′-region of the transferred gene encoded for the signal. Subsequent intron loss resulted in a reduced number of exons. Thereby the developed module encoding for a signal was reused for other transferred genes. (**b**) The frequency of introns within the first 100 amino acids encoding region of the genomic DNA sequences coding for 349 confirmed cytosolic proteins, for 355 confirmed ER-proteins, confirmed 347 mitochondrial proteins with N-terminal targeting signal, of 375 confirmed nuclear proteins, of 81 confirmed peroxisomal proteins, and of 914 confirmed chloroplast proteins with N-terminal transit peptides

8.4.2   The Targeting Complexes

The general transport of precursor proteins to chloroplasts and mitochondria requires the aid of molecular chaperones. Here, cytosolic chaperones of the Hsp70-type (e.g., Zhang and Glaser 2002) and of the Hsp90-type (Young et al. 2003; Qbadou et al. 2006) are discussed to be involved. Both chaperones are of prokaryotic origin, but have been adapted to the eukaryotic system (Gupta and Golding 1993; Gupta 1995; Mirus and Schleiff 2009). In the context of protein targeting, the function of Hsp70 has to be seen in relation to the need to maintain an unfolded state of the precursor for the translocation (e.g., Ruprecht et al. 2010) rather than being directly involved in the translocation process. Supporting this notion a specific subclass of chaperones for translocation could not be identified by sequence analysis (Mirus and Schleiff 2009). It might be speculated that Hsp70 is simply being used rather than specifically being involved in the translocation process. For precursor protein degradation, a specific Hsp70 isoform was identified in *Arabidopsis thaliana*, namely Hsc70-IV (Lee et al. 2009). This specific Hsp70 isoform is induced upon biotic and abiotic stress (e.g., Sung et al. 2001; Noël et al. 2007). Thus, the dependence of precursor degradation on Hsc70-IV might rather be related to the chaperone function in global stress response than to specific degradation of precursor proteins in general.

Hsp90 proteins are involved in multiple regulatory pathways within the cell (Young et al. 2004), but an Hsp90 specific for protein translocation has not yet been identified. As our understanding of the routing system is still very limited, it is hard to conclude on its evolutionary origin. However, one can state that it has to be a eukaryotic invention, because all prokaryotic players such as SecA, SecB, TF, or SRP are not involved in the systems for precursor protein transport to mitochondria or chloroplasts known today.

8.5   The Evolutionary Development of the Mitochondrial Translocon

The mitochondrial translocation machinery is composed of at least eight distinct units (Fig. 8.3, Chacinska et al. 2009). (a) Initially, almost all nuclear encoded mitochondrial proteins engage the translocase of the outer mitochondrial membrane (TOM, Sect. 8.5.1). The subsequent distribution of proteins after transfer across the

---

Fig. 8.2 (continued) was analyzed (listed from *left* to *right*). (**c**) The positioning of the intron closest to the predicted cleavage side of 842 plastid proteins was analyzed. Either the distance between the 3′-end of the intron (*minus*) or the 5′-end of the intron (*plus*) and the first base coding for the mature domain was counted and binned in frames of ten bases (the *center* is given). *Bars* give the percentage of the distance according to the bin found and the *line* gives the accumulative percentage of introns from −110 to +110

Fig. 8.3 The evolution of the translocation systems in the mitochondrial membranes. The evolutionary development of the translocation components in the mitochondrial membranes is depicted as described in Sect. 8.5. On *top*, the original bacterial set of proteins is depicted which are subsequently recycled. In the middle, the initial translocon is depicted. Here, components directly recycled are shown in *bright yellow*, proteins evolved by recycling of bacterial proteins adopting a new function are shown in *yellow*, and proteins assumed to be part of the initial translocon but with unknown origin in *orange*. On the *bottom*, the final step of evolution to the translocons known today is shown. *Bright yellow* indicates the components present in the early translocon, *yellow* are the components found in most other species, and *blue* are those components found only in specific branches of the tree of life as described

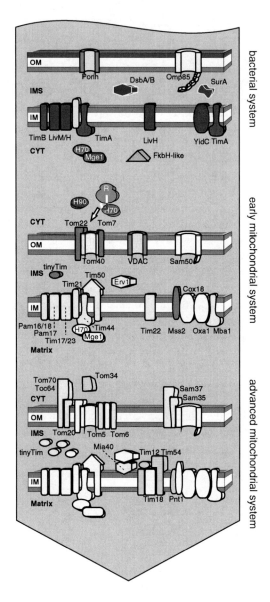

outer membrane involves at least four distinct complexes. (b) The complex annotated as sorting and assembly machinery (SAM) is localized in the outer mitochondrial membrane and inserts β-barrel proteins into the outer membrane from the inner side (Sect. 8.7). (c) Intermembrane space (IMS) chaperones, the so-called tiny Tim proteins (Translocase of the inner mitochondrial membrane of the indicated molecular weight in kDa), are involved in the delivery of proteins to the SAM complex (Sect. 8.5.2). The tiny Tims are also involved in the distribution of precursor proteins in the IMS as they transport inner proteins to (d) the carrier translocase of the inner mitochondrial membrane with Tim22 as the central component (Sect. 8.5.4). (e) Proteins remaining in the IMS engage a system with the Mia40 protein (mitochondrial IMS import and assembly protein of 40 kDa) as the central component (Sect. 8.5.3). (f) Translocation across the inner membrane is driven by the presequence translocase of the inner mitochondrial membrane with the central component Tim23 (Sect. 8.5.4). (g) The translocation process across the inner membrane is further dependent on the energizing Pam machinery (presequence translocase-associated motor; Sect. 8.5.5). Subsequently, protein maturation and folding occurs in the matrix of the organelle. (h) The proteins encoded by the mitochondrial genome are exported into the inner membrane via an Oxa1-dependent system. Thereby, the translocation system subsequent to the Tom machinery is very complex and the same holds true for their evolutionary roots.

8.5.1 The Evolution of the Entrance into Mitochondria

The TOM complex consists of four cytosolically exposed receptor units, Tom5, Tom20, Tom22, and Tom70 (e.g., Chacinska et al. 2009). In addition, a role for Tom34 in precursor protein targeting and translocation in mammals has been proposed (Nuttall et al. 1997). The β-barrel protein Tom40 forms the translocation pore and two additional proteins, Tom6 and Tom7, are discussed to modulate the dynamics of this complex.

The most ancestral complex for translocation of proteins across the outer membrane was composed of Tom40, which originated from a VDAC of bacterial ancestry (Fig. 8.3, Cavalier-Smith 2009). Tom40 can only be detected in species phylogenetically younger than Euglenozoa – e.g., *Giardia intestinalis* (Dagley et al. 2009) or in several apicomplexa (Pusnik et al. 2009) – whereas VDAC (voltage-dependent anion channel, mitochondrial porin) alone is already present in trypanosomatids and euglenids (Pusnik et al. 2009). Based on this observation, it is suggested that the original VDAC might initially have transported both, ions and precursor proteins, and by subsequent gene duplication the two proteins might have evolved independently specializing each in one of the transported substrate types (Cavalier-Smith 2009). Thus, the origination of protein translocation across the mitochondrial outer membrane – which clearly was the bottleneck for mitochondrial protein translocation and mitochondrial development – involved a classical porin-type protein. Thereby, the TOM complex did "newly" evolve in the eukaryotic context although a proteobacterial outer membrane protein was recycled considering VDAC as "intermediate" of the evolutionary development from a bacterial porin to Tom40. The question remaining at stage is why a porin-type protein was able to interact with and perceive a precursor protein. In this respect, it was proposed that the ancestral porin might have been an usher-type porin, which is known to transport proteins as well (Cavalier-Smith 2006). However, this proposal – as appealing as it is – has to be confirmed by future efforts.

Other early evolved Tom components are Tom22 and Tom7. This is in line with the observation that at least Tom40 and Tom22 constitute the minimal functional translocon (Fig. 8.3, Dekker et al. 1998). These proteins can be found in many eukaryotic species including diatoms (Maćasev et al. 2004), but not in trypanosomes (Schneider et al. 2008). It was argued that the addition of Tom22 might have enhanced the specificity of the complex by providing an acidic platform for the interaction with the precursor protein (Cavalier-Smith 2006). In line with this notion, all Tom22 proteins identified have a basic N-terminal domain that is exposed to the cytosol, whereas only in fungi and mammals Tom22 contains an additional acidic domain at the intermembrane space side (Maćasev et al. 2004). However, based on the limited dataset, it is not yet possible to decide whether Tom22 arose in parallel to the gene duplication of the porin leading to VDAC and Tom40, or whether these events were independent.

Tom7 is involved in defining the stability and assembly of the Tom complex. The deletion of fungal Tom7 results in reduced insertion of porins into the outer membrane, whereas the assembly of Tom40 is enhanced (Hönlinger et al. 1996; Sherman et al. 2005; Meisinger et al. 2006). However, the functionality of the protein with respect to the stability of the TOM complex is distinct between *S. cerevisiae* and *Neurospora crassa*. Deletion of Tom7 in *S. cerevisiae* results in a stabilization of the complex composed of Tom40, Tom20, and Tom22 (Hönlinger et al. 1996), whereas the same mutation in *N. crassa* results in a reduced stability (Sherman et al. 2005). Hence, the direct function of this component remains elusive and thereby a proposal why Tom7 might have been required in the ancestral complex cannot be given at this stage.

The TOM complex known from yeast, plants, and humans is assisted by at least four additional receptor components, namely Tom5, Tom20, Tom70, and Tom34. Tom5 is thought to present the last cytosolic receptor domain before translocation across the membrane and can be found in all eukaryotic species (Bohnsack and Schleiff 2010). This receptor is rather small and therefore its evolutionary conservation cannot doubtlessly be determined. However, Tom5 has a helical transmembrane region and thus should be considered a eukaryotic invention.

Tom20, Tom70, and Tom34 are late inventions of the translocation system and may have been added to enhance the specificity or the speed of translocation. Tom20 varies between fungi/metazoa and plants. The plant receptor has an opposite topology and contains two instead of one tetratricopeptide repeat (TPR) motif compared with fungi (Heins and Schmitz 1996). In vertebrates, two Tom20 isoforms are present. One is like the yeast protein (Likić et al. 2005) and the second (termed type II) is characterized by a "glutamine face" specific for the human protein (Schleiff et al. 1997), which is exposed to the outside of the molecule (Schleiff and Sulea 1999; Likić et al. 2005). Hence, parallel evolution after the separation of fungi/metazoa and plants and a subsequent diversification in vertebrates have to be assumed.

Remarkably, analysis of Tom20 and Tom22 from *S. cerevisiae* and *S. castellii* uncovered a case of domain stealing. Here, Tom22 from *S. castellii* has lost an acidic domain, which Tom20 has gained (Hulett et al. 2007). This observation documents that after initial occurrence of the receptor component Tom20 its subsequent evolution has to be seen in concert with the remaining complex components, especially with its co-acting partner Tom22.

The "Tom70-like protein family" is a further example of convergent evolution (Schlegel et al. 2007). The receptor in the yeast (Hase et al. 1984) and mammal systems (Edmonson et al. 2002) is composed of a clamp-type TPR domain and additional eight C-terminally positioned non-clamp type TPRs (e.g., Chan et al. 2006). The clamp-type TPR is designated for the interaction with Hsp90 (Young et al. 2003). However, not all Tom70-dependent precursor proteins require the presence of the clamp-type TPR domain in Tom70 in yeast (Chan et al. 2006). Here, the more C-terminal TPRs are essential for efficient translocation. Whether they form a platform for the interaction with other targeting factors or the precursor protein itself remains elusive. Remarkably, the yeast and mammal receptors do not share any sequence similarity with the protein found in plant mitochondria (Chew et al. 2004). Indeed, the plant receptor belongs to the same family as Toc64, a receptor of the chloroplast outer membrane translocase (Sect. 8.6), which is characterized by a clamp-type TPR domain as well (Qbadou et al. 2006). Thereby, both receptors are functionally related. Nevertheless, this Tom70-like receptor type has to be seen as a late addition to the translocation complex and both, the fungal and the plant representative, are not essential (Yaffe et al. 1989; Qbadou et al. 2006). Thus, a large diversity of this receptor type might exist. In *G. intestinalis,* a protein was categorized as Tom70-like, which contains a C-terminal DnaJ domain and no transmembrane domain at all (Chan et al. 2006). Thus, even though a Tom70 ortholog was not found in lower eukaryotes such as *Dictyostelium discoideum* or *Entamoeba histolytica* (Lithgow and Schneider 2010), a protein with similar function might exist, but it might be composed of at present unexpected domains.

In contrast to other components, Tom34 was only identified in bony vertebrates (Chewawiwat et al. 1999; Schlegel et al. 2007). The protein is a soluble receptor interacting with Hsp90 (Young et al. 1998; Tsaytler et al. 2009), which agrees well with the proposed classification of the TPR domains of this protein (Schlegel et al. 2007). The receptor stimulates protein translocation into mitochondria in fibroblasts (Joseph et al. 2004). Even though the exact function of this receptor remains to be described, it represents an example that the evolutionary development of the translocation complex continued even at the level of components after the split between fungi and metazoa.

8.5.2   The Evolution of the Intermembrane Space Chaperones

The chaperone system within the intermembrane space consists of four tiny Tim proteins with a molecular weight between 8 and 13 kDa, which are annotated as Tim8, Tim9, Tim10, and Tim13 (Chacinska et al. 2009). The proteins are highly conserved in the eukaryotic kingdom (Bauer et al. 1999). In very few species only some or none of the tiny Tims could be identified. For example, in protists such as

Cryptosporidium parvum, only one protein with similarity to the tiny Tims was detected (Gentle et al. 2007). Based on this observation, it was concluded that the family of tiny Tims might have evolved from one common ancestral tiny Tim by subsequent gene duplications (Fig. 8.3, Bauer et al. 1999; Gentle et al. 2007). Even though this conclusion appears to be very likely, it is not yet excluded that the observed existence of none or only one protein is the result of a gene loss or high sequence diversity.

Remarkably, a bacterial ancestor has not yet been discovered (Gentle et al. 2007). However, based on the structural analysis of the oligomeric Tim9-10 complex, a relation to the bacterial chaperone skp is discussed, because both structures revealed "tentacle"-like conformations involved in substrate recognition (Webb et al. 2006). In addition, similar binding properties of Tim10 and SurA with respect to the recognized motifs within target proteins were discovered in vitro (Alcock et al. 2008). However, SurA is not able to complement Tim10, leading to the conclusion that SurA was not able to efficiently transfer the precursor proteins to the evolving TIM and TOM complex and thus was replaced by the tiny Tims in the course of evolution.

Remarkably, the tiny Tim system is one of the few examples where a disease related to mitochondrial dysfunction can directly be linked with protein translocation (Bauer et al. 1999; Jin et al. 1999; Koehler et al. 1999; Rothbauer et al. 2001). The human deafness dystonia peptide (Jin et al. 1996) is an ortholog of the yeast Tim8 and its mutation causes the Mohr–Tranebjaerg syndrome, which is a progressive neurodegenerative disorder (Tranebjaerg et al. 1995). This documents that even though a common ancestor is envisioned for all tiny Tims, diversification during evolution has yielded proteins, which have overlapping but not identical functions.

8.5.3   The System for Intermembrane Space Proteins

Proteins destined for the intermembrane space are recognized by the redox-activated import receptor Mia40, which works in concert with the sulfhydryl oxidase Erv1 (essential for respiration and vegetative growth 1, Fig. 8.3, Chacinska et al. 2009). The system is required for the oxidation of proteins in the IMS and thereby for the formation of disulfide bonds in IMS proteins (Herrmann and Köhl 2007). The Mia40-Erv1 is further discussed as a kind of folding trap, which mediates the unidirectional import of IMS proteins (Herrmann and Köhl 2007). Mia40 itself can be found in fungi, mammals, plants, and red and green algae, but some analyzed unicellular eukaryotes do not encode for this protein (Allen et al. 2008). Most of the latter organisms are parasites, which have adapted to low partial pressures of O_2 and hence Mia40 might not be required, although proteins known to be Mia substrates in yeast are present (Allen et al. 2008). In contrast, Erv1 was identified in most of the species analyzed, and is always present when proteins known to be Mia substrates are encoded (Allen et al. 2008). In view of trypanosomatids representing such system with absence of Mia40 but presence of

Erv1 and Mia substrates, Allen and coworkers suggested that the Mia system evolved from an oxidizing system not involving the Mia40 protein. They suggested a three step scenario, in which Erv1 at first replaced the existing bacterial DsbA and DsbB proteins that catalyze periplasmic disulphide bond formation in proteobacteria (e.g., Heras et al. 2009). However, Erv1 is considered as a eukaryotic invention as it does not share any sequence similarity with bacterial proteins (Herrmann et al. 2009). Subsequently, Erv1 was either co-acting with a disulphide isomerase system based on small molecules, with proteins such as DsbC known from β- and γ-proteobacteria (Kadokura et al. 2003), or independent of such a system, as known for the sulfhydryl oxidase in the endoplasmic reticulum, Ero1 (Frand and Kaiser 1998; Pollard et al. 1998). Finally, Mia40 evolved to provide a receptor for recognition of the incoming reduced precursor proteins. Thus, based on current knowledge one would suggest that Mia40 evolved subsequently to Erv1 and not at the same time, and Mia40 was invented to accelerate the translocation and to enhance the specificity for targeting into the IMS. The latter conclusion is consistent with the observation that not only folding but also import of Mia40-dependent substrates is inhibited in *mia40* yeast mutants (Chacinska et al. 2004; Mesecke et al. 2005).

8.5.4   The Carrier and the Presequence Translocase

The carrier translocase is composed of four subunits, namely the membrane-anchored Tim54, Tim22, Tim18, and the membrane-bound Tim12 (Chacinska et al. 2009). The translocase for presequence-containing precursor proteins involves the components Tim23, Tim17, Tim50, and Tim21 (Chacinska et al. 2009). The two central components, Tim22 and Tim23 belong to the same protein family annotated as PRAT for precursor protein and amino acid transporter (Fig. 8.3, Rassow et al. 1999; Murcha et al. 2007). It was suggested that these proteins originated from the bacterial amino acid permease LivH. The same holds true for Tim17 (Rassow et al. 1999), which is required for the sorting of proteins into the inner membrane and discussed to be involved in the docking of the PAM complex onto the Tim23 machinery (Chacinska et al. 2005). This would suggest that the central component of the translocon for the transfer of precursor proteins across the inner membrane and for the insertion of carrier proteins into the inner membrane evolved by an adaptation of a preexisting bacterial plasma membrane protein.

The carrier translocase component Tim18 is described as a paralog of the mitochondrial succinate dehydrogenase subunit D of Complex II (Sdh4/SdhD) of the electron transport chain (Marcet-Houben et al. 2009), which is related to the bacterial SdhD. Remarkably, SdhD is still encoded by the mitochondrial genome of the red algae *Porphyra purpurea*, *Chondrus crispus* and *C. merolae*, and the jakobid flagellate *Reclinomonas americana* (Burger et al. 1996; Ohta et al. 1998; Unseld et al. 1997). However, it is absent in the liverwort *Marchantia polymorpha*, in green algae such as *Protoheca wickerhamii* and *Clamydomonas reinhardtii*, in fungi such as *Podospora anserina* and *S. cerevisiae,* in humans or in higher plants such as *A. thaliana* (e.g., Anderson et al. 1981; de Zamaroczy and Bernardi 1986; Unseld et al. 1997). Hence, sdhD is a gene, which was transferred into the nuclear genome at a late stage of evolutionary development. This late transfer might have resulted in a gene duplication of sdhD at the very specific branch of *ascomycetes,* as Tim18 could not yet be identified in any other species.

The second component associated with Tim22 is Tim54. Similar to Tim18, Tim54 appears to be a late and fungal-specific addition to the translocon. The protein can be found in yeast, but not in the green lineage (e.g., Figueroa-Martínez et al. 2008) or lower eukaryotes such as *Toxoplasma gondii* or *Plasmodium falciparum* (Sheiner and Soldati-Favre 2008). The third component of the carrier translocase, Tim12, is related to the tiny Tims (Gentle et al. 2007). Hence, its origin is unclear and its occurrence might be the result of gene duplication of Tim10. Furthermore, Tim12 appears to be specific for the fungal branch. At this stage, it is not clear whether one of the existing variants of the tiny Tims in plants or mammals functionally replaces Tim12 or whether Tim12 evolution was enforced by the existence of the Tim54-Tim18 complex.

Besides the two central components, Tim23 and Tim17, the presequence translocase contains two additional proteins in yeast, namely Tim50 and Tim21 (Fig. 8.3). Tim50 was identified in the brown algae *Ectocarpus siliculosus* (CBN76692), in the green algae *C. reinhardtii,* in the moss *Physcomitrella patens,* in plants, e.g., *A. thaliana* (e.g., Figueroa-Martínez et al. 2008), in *human* (Guo et al. 2004) and in the ciliate *Tetrahymena thermophila* (Smith et al. 2007), but not in alveolates such as *T. gondii* or *P. falciparum* (Sheiner and Soldati-Favre 2008). Whether the high sequence diversity found in the latter two species is complicating the detection of components of translocation systems (Bullmann et al. 2010) remains unknown. However, it is tempting to conclude that Tim50 involved in the gating of Tim23 exists in all mitochondrial translocation systems. The latter is consistent with its identification in the reduced translocon of mitosomes (Waller et al. 2009). The protein contains a haloacid dehalogenase domain (HAD), which belongs to the Rossmannoid domains (Burroughs et al. 2006). The HAD region of Tim50 has a "basal 5-stranded core assemblage," which can also be found in FkbH-like bacterial proteins (Burroughs et al. 2006). Thus, it is very likely that Tim50 has evolved by recycling an existing prokaryotic protein.

Tim21 interacts with the Tim23 machinery to replace the presequence translocase-associated motor, which is required to drive insertion of proteins with stop transfer signals into the inner membrane (e.g., Chacinska et al. 2010). Even though the protein was considered to be a eukaryotic invention (Kutik et al. 2009) corresponding sequences can be identified in the entire eukaryotic kingdom and even in some bacteria (Fig. 8.4). The occurrence of Tim21 in all eukaryotic systems is related to its central function in the insertion of presequence-containing inner membrane proteins and parallels the observation made for the presequence translocase-associated motor discussed below. Unfortunately, no functional assignment for the bacterial proteins exist (Fig. 8.4b), and thus it is hard to conclude on their functional relation to Tim21.

a

```
saTim21    M-STMPE----------------------------------------GNMAPRQGWWSRN-
scTim21    MSSSLPRSLLRLGHRKPLFPRYNTFVNSSVITHTSLLRTRLYSNGTGATSGKKDDKTRNK

saTim21    ----WKWVVPVGCLT--PVLMCGCLG--AFVAYFVTSTIKS----TDAYQQAVALVTANP
scTim21    PKPLWPQVKSASTFTFSGILVIGAVGISAIVIYLILSELFSPSGDTQLFNRAVSMVEKNK

saTim21    EVQ--------------EALGTPI-DFGWPRG----SVNTTNGEGRAS--ISVPLEGPK
scTim21    DIRSLLQCDDGITGKERLKAYGELITNDKWTRNRPIVSTKKLDKEGRTHHYMRFHVESKK

saTim21    ASGTMRVEALAEGETWTLDLLQ--VEVPGRPAIDLLDQVGGRQDPELEPLPDEEPPPLEE
scTim21    KIALVHLEAKESKQNYQPDFINMYVDVPGEKRYYLI-------KPKLHPVSN--------

saTim21    DMIPPPEEEVLPPTEEEAPAKKGSEID
scTim21    -----------SKGFLGIRWGPRKD
```

b

Name	Acc. number	species	Phyla
csTim21	Cyan7425_3418	Cyanothece sp. PCC 7425	Cyanobacteria
saTim21	ZP_01459929	Stigmatella aurantiaca	Proteobacteria
rcTim21	ABU59532	Roseiflexus castenholzii	Chloroflexi
mbTim21	XP_001743562	Monosiga brevicollis	Choanoflagellida
scoTim21	XP_003038492	Schizophyllum commune	Fungi
scTim21	P53220	Saccharomyces cerevisiae	Fungi
ceTim21	AAK29831	Caenorhabditis elegans	Metazoa
dmTim21	NP_608929	Drosophila melanogaster	Metazoa
ggTim21	XP_419102	Gallus gallus	Metazoa
hmTim21	XP_002167000	Hydra magnipapillata	Metazoa
hsTim21	Q9BVV7	Homo sapiens	Metazoa
sjTim21	CAX73241	Schistosoma japonicum	Metazoa
taTim21	EDV26031	Trichoplax adhaerens	Metazoa
tnTim21	CAF98004	Tetraodon nigroviridis	Metazoa
bhTim21	CBK24881	Blastocystis hominis	stramenopiles
esTim21	CBN78209	Ectocarpus siliculosus	stramenopiles
piTim21	XP_002997863	Phytophthora infestans	stramenopiles
ptTim21	XP_002176581	Phaeodactylum tricornutum	stramenopiles
atTim21	AT4G00026	Arabidopsis thaliana	Viridiplantae
crTim21	gi|159479166	Chlamydomonas reinhardtii	Viridiplantae
ppTim21	XP_001783295	Physcomitrella patens	Viridiplantae

c

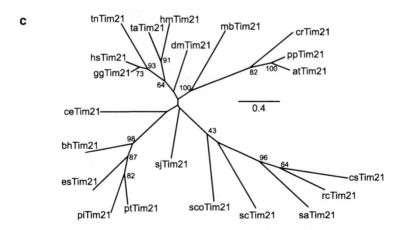

Fig. 8.4 Phylogenetic analysis of sequences related to Tim21 identified in yeast. (**a**) The alignment between Tim21 from *Saccharomyces cerevisiae* and *Stigmatella aurantiaca* is shown, the predicted transmembrane domain is highlighted in *yellow*, identical amino acids in *blue*, and similar amino acids in *green*. (**b**) Based on Tim21, sequences were selected by a reverse blast strategy and (**c**) the phylogenetic relation of these sequences was calculated as previously described (e.g., Mirus et al. 2009a)

8.5.5   The Presequence Translocase-Associated Motor

The presequence translocase-associated motor energizes the translocation process
of presequence-containing precursor proteins across the outer and inner mitochondrial membrane (e.g., Matouschek et al. 2000). The complex is composed of the
mitochondrial chaperone mtHsp70 and its nucleotide exchange factor Mge1, the
membrane anchor for mtHsp70, Tim44, and the three Pam proteins Pam16, Pam17,
and Pam18 (Fig. 8.3). The mitochondrial Hsp70 and its nucleotide exchange factor
are of prokaryotic origin and thereby inherited from the engulfed α-proteobacteria.
The same observation holds true for Tim44, Pam16, and Pam18 (Clements et al.
2009). The latter two contain a so-called DnaJ fold (e.g., Mokranjac et al. 2006)
known to stimulate the ATPase activity of Hsp70 proteins. However, Pam16 does
not contain the typical HPD motif (Kelley 1998) of J-proteins, which was most
likely lost in the course of evolution to a regulatory subunit of the PAM machinery,
because Pam16 is involved in recruiting Pam18 to the presequence translocase
rather than in Hsp70 activation (Kozany et al. 2004; Frazier et al. 2004). Finally,
Pam17 involved in the regulation of the architecture of the presequence translocase
(e.g., van der Laan et al. 2005b) shows a relation to LivM-like proteins based on the
PFAM classification (pfam08566: Pam17; Marchler-Bauer et al. 2009). As mentioned above, Tim23/Tim17 share similarity with LivH, and the latter is proposed to
interact with LivM (Haney and Oxender 1992). This proposal is based on the
overlapping function of both, LivM and LivH (Nazos et al. 1985; Adams et al.
1990). Remarkably, in *Metazoa* Pam17 can only be detected in *Pseudocoelomata*,
Placozoa, and *Cnidaria*, but not in *Coelomata*. The latter holds true for
Viridiplantae as well. However, at present it cannot be excluded that this is due
to an enforced sequence diversification or due to a loss of this component.
Summarizing, one can conclude that the basic machinery for energizing the translocation is built from inherited bacterial proteins, even though they did not necessarily have a functional relation in the mitochondrial ancestor.

8.5.6   Insertion from the Inside of Mitochondria

Some proteins encoded by the mitochondrial genome have to be integrated into the
inner membrane of mitochondria. This translocation is dependent on the Oxa1
protein (see Sect. 8.2) which is of clear α-proteobacterial origin. The process is
further assisted by Mba1, Mdm38, Cox18, Pnt1, and Mss2. Oxa1 contains a
C-terminal α-helical domain that interacts with ribosomes. Mba1 interacts with
the ribosome as well, and it is discussed that this factor is involved in the positioning of the ribosomal exit tunnel to the insertion site of the inner membrane (Ott and
Herrmann 2010). The analysis of Mba1 revealed its inheritance from the α-
proteobacterial ancestor (Smits et al. 2007). Even further, a homology to Tim44
was determined and a common origin of both proteins is proposed (Smits et al.

2007). Mdm38 is conserved throughout the eukaryotic kingdom (Schlickum et al. 2004). The function of Mdm38, however, remains under debate. On the one hand, for the factor in yeast it was reported that Mdm38 interacts with the mitochondrial ribosomes and that its deletion causes a defect in the export of mitochondrial encoded proteins (Frazier et al. 2006; Bauerschmitt et al. 2010). On the other hand, it was discovered that the yeast protein functions as a K^+/H^+ antiporter (Nowikovsky et al. 2004). Its human homolog LETM1, which is the candidate gene for seizures in Wolf–Hirschhorn syndrome (Shanske et al. 2010), acts as Ca^{2+}/H^+ antiporter (Jiang et al. 2009). Thus, it remains questionable whether Mdm38 indeed acts as a protein exporter.

Cox18 (also known as Oxa2) is a paralog of Oxa1 (Bonnefoy et al. 2009) and thereby of prokaryotic origin, but does not contain the C-terminal extension interacting with the ribosome. This Oxa1 paralog is especially required for the insertion of Cox2, which is the only exported protein with a large hydrophilic domain exposed to the intermembrane space. In addition, Pnt1 and Mss2 are also involved in this highly specialized process; however, not much is known about these two factors. Pnt1 appears to be a eukaryotic invention and it might even be fungi specific, because no clear relation to sequences from plants or mammals can be established. Mss2 is a protein with a TPR domain. This domain is related to the once found in components of a putative protein export sorting system in Gram-negative bacteria (Haft et al. 2006). Whether this allows the conclusion that Mss2 is of prokaryotic origin remains questionable. Interestingly, e.g., in *C. reinhardtii*, *A. thaliana* or human, no Mss2 homologs could be identified (Cardol et al. 2005; Gaisne and Bonnefoy 2006). However, in humans different splice variants of Cox18 have been observed, leading to the suggestion that the different forms of Cox18 might act together, explaining the absence of Mss2 and Pnt1 (Gaisne and Bonnefoy 2006). Thus, a fungi-specific invention of Mss2 and Pnt1 can be envisioned, or the genes encoding the two proteins were lost during evolution by inventing a different regulatory mechanism utilizing splicing events.

8.5.7 A Global View on the Evolution of the Mitochondrial Translocases

The evolution of the mitochondrial translocation system can largely be explained by "recycling" of proteobacterial proteins to yield new functions (Fig. 8.3). The only two exceptions known today are Sam50 (see Sect. 8.7), which is still involved in the insertion of β-barrel proteins into the outer mitochondrial membrane, and Oxa1, the member of the YidC/Oxa1/Alb family required for the insertion of mitochondrial encoded proteins into the inner membrane. Remarkably, the direction of the insertion path catalyzed by these two proteins resembles the one in bacteria. This might be one explanation why other systems are the result of functional reassignments rather than the recycling of existing translocation pathways, because all other pathways had to be inverted in their direction. Another explanation might be that on the outer membrane, a translocon for a broad range of proteins does not exist in bacteria, and that a SEC-like translocon, which might have been used in the inner membrane, was already present in the host cell. With respect to the latter fact, it is remarkable that the known protein translocation systems in eukaryotes show little or no overlapping components with exception of chaperones and their regulating proteins (e.g., Bohnsack and Schleiff 2010 and references therein).

Only very few clear eukaryotic inventions exist in the context of the mitochondrial translocons (Fig. 8.3, blue and yellow in the "advanced mitochondrial system"). At this stage, comprehensive phylogenetic studies are missing, and thus it cannot doubtlessly be answered whether the other genes with prokaryotic "origin" were brought in by the endosymbiont or whether they are "gifts" from the host cell. Another drawback of most studies is that investigations of the evolutionary development of the mitochondrial translocon are mainly based on components identified in yeast. However, not all components are conserved in plants or mammals and the question whether adaptation of the fungal system involved gene loss as well cannot be answered. Summarizing, one can conclude that the machinery for protein translocation into mitochondria is built from inherited bacterial proteins, even though they did not necessarily have a functional relation in bacteria. Thereby, the mitochondrial translocon is a large example for "evolutionary recycling."

Remaining is the question why evolution enforced the development of such complex machineries. There might be several answers to this; one might be related to the development of the host cell. Taking the TOM machinery as an example, the origin of the early translocon can be seen as essential for the incorporation of the bacterium as an organelle into the host cell. Here, a rudimentary system was required, (a) initially to provide a pore for protein translocation, (b) subsequently to enhance the specificity and efficiency of the translocation process, and (c) to enable these components to form a dynamic ensemble. This might have led to the first translocation components Tom22, Tom7, and Tom40, the latter even in form of a multifunctional VDAC. At the next level, cells became more complex in their architecture and composition. Therefore, the evolutionary development of "guided" targeting of precursor proteins enforced the development of docking sites for such complexes, e.g., Tom20 and Tom70. In conjunction with the higher complexity of the TOM complex, more components for its assembly evolved (e.g., Tom6, Tom5). At last, in multicellular eukaryotes, mitochondrial functions are regulated with respect to the demand in different cell types and due to a higher complexity of the signaling network, which led to the invention of different isoforms of receptors (e.g., Tom20) or the addition of further components such as Tom34. Thus, the same basic principle as for the evolution of the bacterial translocation systems can be applied, namely enhanced cellular complexity caused adaptation of the translocation system (see Sect. 8.2).

Alternatively, the development of the machineries might be discussed with respect to the ongoing gene transfer from the mitochondrial genome to the host nucleus. Indeed, there is no doubt that this has been the case for the initial

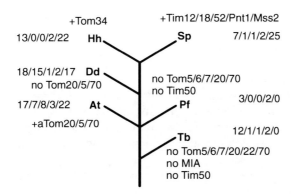

Fig. 8.5 Exploring the link between translocon and mitochondrial genome complexity. Five discussed species were selected and representation of phylogenetic relation is based on Burger et al. (2003). The general complex shown as advanced mitochondrial translocon in Fig. 8.3 is assumed (except components indicated in *blue*) and absence or additional presence of components is indicated according to the discussion in Sect. 8.5. The gene content of mitochondrial genomes is taken from Gray et al. (1998). The numbers of genes coding for components of complex I–V/ ribosomal proteins/other proteins/ribosomal RNA/tRNA are given (Nomenclature: At, *Arabidopsis thaliana*; Dd, *Dictyostelium discoideum*; Hs, *Homo sapiens*; Pf, *Plasmodium falciparum*; Sp, *Schizosaccharomyces pombe*; Tb, *Trypanosoma bruzei*; no, absence of component; a, atypical component; + additional components)

occurrence or the translocons, but appears to be unlikely with respect to diversification. Comparison of the number of genes encoded by the mitochondrial genome and the translocation complex diversity (Fig. 8.5) does not reveal a correlation. In *P. falciparum* or *T. bruzei,* a rather simple translocon is described, but at the same time the mitochondrial genome of both organisms encodes for only a limited number of components when compared to *H. sapiens* or *A. thaliana*. Thus, at this stage it appears more likely that the evolution of translocon complexity is related to the evolution of the host cell and the function which mitochondria have to perform therein.

8.6 The Chloroplast Translocon

The translocation machineries of chloroplast have to catalyze translocation across or insertion into three different membranes, the outer chloroplast envelope, the inner chloroplast envelope, and the thylakoid membranes (e.g., Soll and Schleiff 2004). The knowledge on the distinct targeting pathways into and across the envelope membranes is not yet as advanced as for the mitochondrial system, but basic concepts exist. Translocation across the outer envelope is driven by a TOC complex (translocon on the outer chloroplast envelope membrane (TOC); Sect. 8.6.1). The subsequent translocation across the IMS between outer and inner membrane is not yet well defined, the only soluble IMS component suggested to be involved is Tic22. The transfer into the outer membrane is not yet described, but one putative candidate exists as discussed in Sect. 8.7. The inner envelope translocon is termed TIC and catalyzes the transfer into the stromal compartment (Sect. 8.6.2). Within the stromal compartment, the N-terminal targeting signal is cleaved off (e.g., Soll and Schleiff 2004). This is used either as a signal for the folding of proteins remaining in the stroma or it sets free a signal for thylakoid targeting. Here, different translocation machineries with relation to the bacterial SEC, TAT, and YidC translocon exist for the transfer and integration of precursor proteins (Sect. 8.6.3). Additional machineries besides the mentioned ones exist, however, will not be discussed here, because they are not well described and thus the components involved are not yet fully understood. In addition, evolution of the translocation systems in complex plastids is beyond the scope of this article.

8.6.1   The Translocon in the Outer Envelope Membrane

The chloroplast translocon is composed of the cytosolically exposed Toc64, Toc34, and Toc159 as well as of the pore-forming Toc75 (Oreb et al. 2008, Fig. 8.6, advanced chloroplast system). Toc75 belongs to the Omp85 class (Löffelhardt et al. 2007, see also Sects. 8.2, 8.4.2, and 8.7), while Toc64 contains an amidase as well as a tetratricopeptide repeat domain (e.g., Sohrt and Soll 2000; Mirus and Schleiff 2009, see also Sect. 8.5.1). The other two components, Toc34 and Toc159 are GTPases exposed to the cytosol and are considered as entrance receptors (e.g., Oreb et al. 2008). The complex is further assisted by a DnaJ domain-containing protein (Toc12) exposed to the IMS (Becker et al. 2004), which is involved in the regulation of the IMS-localized Hsp70 (e.g., Marshall et al. 1990).

The central pore-forming component of the TOC translocon, Toc75, clearly belongs to the Omp85 family and is related to the cyanobacterial Omp85 proteins (Alr0075 and Alr2269; Fig. 8.6, e.g., Bredemeier et al. 2007). Consistently, a complex composed of Toc75 and other components was recently determined in cyanelles (Yusa et al. 2008), which are considered as most primitive plastids (Steiner and Löffelhardt 2005). Thus, Toc75 is clearly inherited from the endosymbiotically captured cyanobacteria. Here, one can even go further: it was demonstrated that cyanobacteria of the order *Nostocales* are closely related to the ancestor of chloroplasts (Martin et al. 2002; Deusch et al. 2008). Analysis of three cyanobacteria of this order, namely *Anabaena* sp. PCC 7120, *Nostoc punctiforme* and *Anabaena variabilis* revealed the presence of at least two genes coding for Omp85-like genes (Bredemeier et al. 2007), which are indeed expressed (Nicolaisen et al. 2009). This might explain why, unlike in mitochondria, one of the Omp85 homologs was recycled as the TOC translocation pore. However, the translocation pore-forming Toc75 is not orthologous to Omp85, because Toc75 translocates proteins across the membrane, whereas Omp85 integrates proteins into the membrane, most likely without translocation across the lipid bilayer. Even

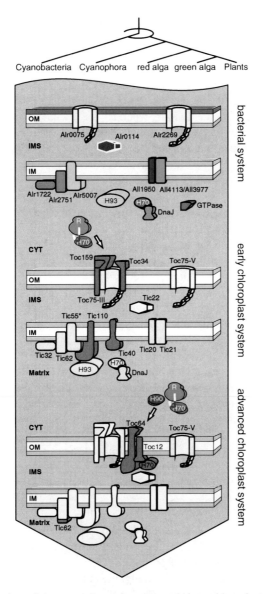

Fig. 8.6 The evolution of the general translocation path into chloroplasts. The evolutionary development of the translocation components in the chloroplast envelope membranes is depicted as described in Sect. 8.6. On *top*, the original bacterial set of proteins is depicted which is subsequently recycled. For these components, the accession numbers for *Anabaena* sp. PCC 7120 are listed. In the *middle*, the initial translocon is depicted. Here, components directly recycled are shown in *bright yellow*, proteins evolved by recycling of bacterial proteins adopting a new function are shown in *yellow*, and proteins assumed to be part of the initial translocon but with unknown origin in *orange*. The exceptions are the components of the cytosolic pathway. Please note Tic55 appears to be lost in *red algae*. On the *bottom*, the final step of evolution to the translocon known in vascular plants is shown. *Bright yellow* indicates the components present in the early translocon and *orange* are the components/domains found in vascular plants only with eukaryotic roots. On *top*, the phylogenetic distribution between algae and plants with respect to chloroplasts is depicted

though the latter proposal still has to be verified, Toc75 can be considered as a recycling product.

The two GTPases Toc159 and Toc34 belong to the TRAFAC class (for translation factor-related, Leipe et al. 2002). This class also contains classic translation factors such as EF-Tu known from prokaryotes (Leipe et al. 2002). However, Toc34 and Toc159 have a different domain structure. Toc34 has a C-terminal extension containing the transmembrane domain, whereas Toc159 has an additional N-terminal acidic domain (A-domain) and a C-terminal 52 kDa transmembrane domain. Nevertheless, the two GTPase domains share a high degree of similarity (Oreb et al. 2008). This leads to the suggestion that (a) the TOC GTPases originated from recycling of a prokaryotic GTPase fold, and (b) a subsequent gene duplication might have occurred to lead to two distinct receptors, which then further evolved to the variety of Toc34 and Toc159 paralogs known today (e.g., Bodyl et al. 2009a). It remains unknown, which of the two receptors initially evolved, because both can be detected in red and green algae (Kalanon and McFadden 2008). Hence, it will be of major importance to obtain sequences from C. paradoxa as it hosts the most ancient plastid.

Toc64 has a rather unique domain structure. It is composed of an IMS-localized amidase domain (Qbadou et al. 2007), where the catalytic function is silenced by a point mutation (Sohrt and Soll 2000), and a C-terminally positioned, cytosolically exposed TPR domain involved in Hsp90 recognition (e.g., Qbadou et al. 2006; Mirus et al. 2009b, Sect. 8.4.1). The protein is a clear invention of plants and can be found as early as in moss (Schlegel et al. 2007). As mentioned earlier, a paralog of the chloroplast Toc64 replaces the mitochondrial Tom70 receptor in plants (Chew et al. 2004). Thus, the Hsp90-recognizing elements appear to be a late invention in evolution. A detailed analysis based on homology models of the TPR domain of the Toc64 paralogs revealed a positioning of the residues involved in functional discrimination of the differently located proteins almost exclusively on the convex surface not involved in chaperone recognition (Mirus et al. 2009b). This observation suggests that the TPR domain has been evolutionarily shaped to be recognized by either the TOM or the TOC components, but not necessarily to discriminate between distinct Hsp90 molecules.

Two components of the TOC translocon face the IMS, namely Hsp70 and Toc12 (e.g., Becker et al. 2004). Whereas the Hsp70 is of eukaryotic type with general prokaryotic roots (Schnell et al. 1994), Toc12 shows a clear relation to bacterial DnaJ proteins. However, even though Toc12 is related to bacterial DnaJ proteins, it has not yet been identified in the green alga Chlamydomonas reinhardtii and even in the moss Physcomitrella patens (Kalanon and McFadden 2008). This might suggest that one of the 26 chloroplast-targeted DnaJ proteins (Chen et al. 2010) has changed its localization from stroma to the intermembrane space or is even dually localized in both compartments.

Thus, the seed of the evolutionary development was the transformation of an Omp85 protein into a translocation pore (Fig. 8.6). The other components were either eukaryotic inventions such as Toc64 and the imsHsp70, or results of the recycling of bacterial proteins as in case of Toc159, Toc34, and Toc12.

8.6.2 The Translocon in the Inner Envelope Membrane

At present, a rather complex translocation system in the inner chloroplast envelope is proposed, and initial results suggest that distinct complexes might exist. One is composed of at least Tic110, Tic62, Tic55, and Tic32 (Küchler et al. 2002), the other of Tic20 and Tic21 (Kikuchi et al. 2009). Here, Tic20, Tic21, and Tic110 are considered as pore-forming proteins, whereas the other components form a redox chain (e.g., Oreb et al. 2008). Additionally, the intermembrane space component Tic22 and the chaperone-interacting, stromally exposed Tic40 act in protein translocation, and association with either Tic110 or Tic20 is suggested. The TIC system is complemented by stromal chaperones of the Hsp70 (e.g., Shi and Theg 2010; Su and Li 2010) and Hsp93 type (e.g., Kovacheva et al. 2007; Su and Li 2010), which act in concert in protein translocation.

Recent studies have elucidated that the TIC translocon is largely of cyanobacterial origin and evolved by massive gene recycling. It is discussed that precursor protein delivery toward the TIC translocon involves Tic22. Tic22 is inherited from cyanobacteria, where a localization of the protein in the thylakoid lumen was reported (Fulda et al. 2002). The function of the protein, however, remains elusive. Tic22 is subsequently recognized by Tic20. For Tic20, a relation to the PRAT family is suggested (e.g., Bodył et al. 2009b). Indeed, in the three *Nostocales* mentioned above, a LivH gene can be identified (Fig. 8.6; NpunF3616; Ava4362). Its complex partner, Tic21, is also of cyanobacterial origin (Lv et al. 2009). In line with a cyanobacterial origin, Tic20 and Tic21 coding sequences are present in red and green algae (Kalanon and McFadden 2008; Lv et al. 2009). Even more remarkably, the gene is still encoded in the genome of the symbiont of the amoeba *Paulinella chromatophora* (Bodył et al. 2009b). In contrast, the other protein discussed to be involved in pore formation, Tic110, is not of cyanobacterial origin and has to be considered as a eukaryotic invention. Nevertheless, a gene coding for Tic110 is found in the red algae *C. merolae* (Kalanon and McFadden 2008), and the protein is present in the glaucocystophyte alga *C. paradoxa* as determined by immunodecoration (Yusa et al. 2008). Thus, Tic110 has to be considered as a very early invention. The translocation process itself is energized by Tic40-associated chaperones. The protein contains a TPR-like domain (Chou et al. 2003), which might have been recycled from existing genes, but in general Tic40 has to be considered as a eukaryotic invention, because it cannot be found in the genomes of sequenced cyanobacteria and even of red algae.

For the three Tic110-associated components Tic32, Tic55, and Tic62, related genes were identified in cyanobacteria (Kalanon and McFadden 2008; Bodył et al. 2009b). However, the Tic62 homolog in cyanobacteria does not contain the domain specific for ferredoxin-NAD(P)-oxido-reductase recognition. This domain can only be found in proteins of vascular plants and thus has to be seen as a late invention (Balsera et al. 2007). Furthermore, Tic55 could not be identified in the sequenced red algae (Kalanon and McFadden 2008) leading to the proposal of a gene loss in the red lineage.

8.6.3 The Translocating Systems in the Thylakoid Membranes

The chloroplasts contain an additional compartment, the thylakoid membrane system. Four distinct target systems exist. They are named according to their main components cpTAT pathway, cpSEC pathway, cpSRP pathway (cp stands for chloroplast), and spontaneous insertion pathway (Fig. 8.7). Thus, it appears that all modes evolved in bacteria (Fig. 8.1) have been preserved. The TAT pathway is composed of three subunits, TatA, TatB, and TatC, which are termed Tha4, Hcf106, and TatC in the plant system (Robinson and Bolhuis 2001). All three plant proteins localized in the thylakoid membrane clearly evolved from the cyanobacterial version (Yen et al. 2002). Recently, it was discussed that the TatA/TatB genes in *Anabaena* sp. PCC 7120 are both closely related to TatA from *E. coli* (Bohnsack and Schleiff 2010), and it might be suggested that the two proteins from *Anabaena* sp. evolved by gene duplication from the ancestor of the proteobacterial TatA. The

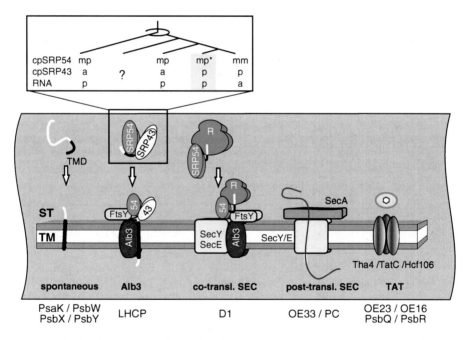

Fig. 8.7 The thylakoid translocation system. Shown are the four modes for protein translocation (please note that co-translational and post-translational translocation by the SEC complex are usually treated as one regime). The only eukaryotically evolved component is shown in *white*. Underneath, first the name of the translocation mode as discussed in Sect. 8.6 and then the names of known substrates are given (OE16/OE23/OE33, 16/23/33-kDa subunit of the oxygen-evolving complex; *PC* plastocyanine). On *top*, the occurrence of the SRP components is indicated and the situation of overlapping occurrence of SRP-RNA, the existence of the RNA-binding motif in cpSRP54 and the presence of SRP43 in *green algae* is highlighted. The phylogenetic distribution is shown as in Fig. 8.6 (*mp* motive present; *mm* motif mutated; *p* present; *a* absent; *Asterisks* please note, not found in *Chlamydomonas*)

two genes present in *Nostocales*, however, are related to the ancestor of the two chloroplast protein-encoding genes.

The second translocon annotated as cpSEC involves a SecA, SecY, and a SecE compound known from bacteria (Sect. 8.2). cpSecA, cpSecY, and cpSecE present in the chloroplasts are of cyanobacterial origin (Vogel et al. 1996; Valentin 1997; Barbrook et al. 1988; Cao and Saier 2003) and it was reported that cpSecE can complement the SecE deletion in bacteria (Fröderberg et al. 2001). However, the absence of an SecG homolog in the thylakoid system suggests that the SEC-translocon was reshaped in the course of evolution to a minimal system as SecG can be found in cyanobacteria (Bohnsack and Schleiff 2010). This reduction on the one hand might reflect the reduced number of substrates recognized by this translocon (Mori and Cline 2001). On the other hand, SecG is discussed to provide the binding site for the SRP receptor (Jiang et al. 2008), and this function is not required in the context of the thylakoid membrane (e.g., Cline and Dabney-Smith 2008). The bacterial SRP receptor cpFtsY with clearly bacterial roots targets the rudimentary SRP to the thylakoid-localized Alb3, which interacts with SecY (Klostermann et al. 2002; Asakura et al. 2008). The latter interaction is required for the co-translational targeting of the plastome-encoded D1 protein (Fig. 8.7) and hence, in this specific case Alb3 might replace SecG in its function as SRP receptor docking site (e.g., Richter et al. 2010). As far as post-translational transport is concerned, the thylakoid SRP is specialized for the transport of LHCP proteins.

The SRP of higher plants represents a drastically reduced form when compared to bacterial SRP complexes and might be the evidence for the transition from the RNA to the protein world via RNA-protein intermediates. In line with this assumption, in vascular plants it has lost its RNA molecule and is only composed of two proteins, cpSRP54 and cpSRP43. The cpSRP54 subunit is related to the bacterial SRP54 components (Franklin and Hoffman 1993), but in the proteins of higher plants the RNA-binding motif is mutated (Richter et al. 2008). The loss of the RNA-binding ability by mutation of the motif correlates with the loss of SRP-RNA from the plastome in higher plants (Rosenblad and Samuelsson 2004). It is discussed that cpSRP43 partially takes over the function of the RNA by exhibiting a similar charge profile at the surface (Grudnik et al. 2009). Accordingly, cpSRP43 is a eukaryotic invention, and a related protein sequence cannot be found in cyanobacteria (Fig. 8.7). Even further, cpSRP43 is present in green algae, mosses and vascular plants, but not in the sequenced genomes of red algae (Tzvetkova-Chevolleau et al. 2007). Thus, it remains to be identified whether the SRP-RNA is indeed functional in green algae and mosses, because in this case RNA-containing and RNA-lacking SRPs would co-exist. By this, it would be an example of an intermediate evolutionary situation: A new functional complex has been established, but the old complex based on the chloroplast-encoded component (in this case SRP-RNA) has not yet been lost.

At stage it is hard to conclude on the evolutionary origin of the components of the SEC pathway, because clear phylogenetic studies have not yet been conducted. Even though it is widely believed that cpSRP54, cpFtsY, cpSecE, cpSecY, and cpSecA are of cyanobacterial origin, on the basis of the existing data, it can only be concluded that they are of bacterial origin. However, it is beyond the scope of this overview to perform an in-depth phylogenetic analysis, an issue which should in future be clarified by the experts in the field.

8.6.4   A Speculation on the Driving Force for Translocon Evolution

As outlined in this section, the chloroplast translocon mostly is of cyanobacterial origin (Figs. 8.6 and 8.7). However, for most of the cyanobacterial proteins related to TIC components, the function has yet to be established. Nevertheless, one can make the same suggestion that as for the mitochondrial translocon the bacterial proteins have been recycled, because most of them are not related to protein translocation in bacteria. However, the translocation complexes found in the thylakoid membranes (Sect. 8.6.3) are clearly related to the bacterial transport systems even though some of the systems have been evolutionarily adapted. Here, the reduction of the substrates transported by the distinct pathways and the alteration of the function has reshaped the translocation components. For example, the SRP molecule has lost most of its components including the SRP-RNA, which reflects the loss of the requirement to enforce a translational arrest for targeting.

In contrast to the mitochondrial system, most of the components are found in red and green algae suggesting a very early occurrence (Fig. 8.6). To this end, the analysis of the translocation complex present in *C. paradoxa* will give further insights. On the one hand, chloroplasts evolved after mitochondria and hence the need for specific receptors on the chloroplast surface might have existed from the beginning, particularly as mitochondrial and chloroplast targeting signals are quite alike. The only known late inventions are the FNR-binding domain of Tic62 found only in vascular plants and Toc12/Toc64 almost exclusively present in land plants. One possible explanation for these late evolutionary modifications might be the parallel occurrence of different plastid types in one organism in land plants and mosses (e.g., Kirk and Tilney-Bassett 1978; Walker and Sack 1990; Kuznetsov et al. 1999). Either the distinct protein demand or the required drastic changes during plastid transitions might have enforced the development of additional receptor components or receptor domains.

8.7   The Insertion of Beta-Barrel Proteins into the Outer Membranes

In general, the proteome of plastids and mitochondria is a complex mixture of proteins (a) inherited from the endosymbiont, (b) inherited from the other endosymbiont, (c) recycled to perform new functions, or (d) newly evolved in the eukaryotic

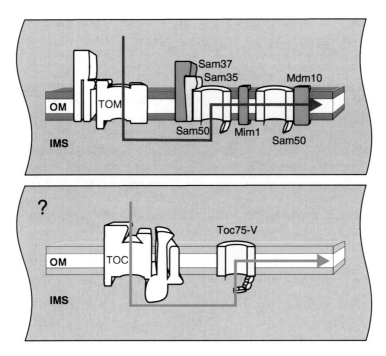

Fig. 8.8 The insertion of β-barrel proteins. The insertion pathway of β-barrel outer membrane proteins of mitochondria (*top*) or plastids (*bottom*) is shown. The mode of insertion for the chloroplast system has not yet been confirmed and the mode of insertion is based on the route taken by Toc75-III (Tranel et al. 1995). The TOM (Fig. 8.3) and TOC translocon (Fig. 8.6) are only depicted as scheme. The additional components found in yeast are shown in *blue*, and Sam35, which shares homology to human metaxin is shown in *yellow*

context (e.g., Leister 2003). One class of proteins clearly belonging to class (a) is the β-barrel proteins of the outer membranes of the endosymbiotic organelles. Remarkably, the insertion path for the assembly of outer membrane proteins appears to be conserved as well (e.g., Paschen et al. 2005; Schleiff and Soll 2005; Löffelhardt et al. 2007; Schleiff et al. 2010) and involves Omp85 orthologs (see Sect. 8.2). These proteins are thought to be translocated across the outer membrane by the TOM/TOC machinery (Fig. 8.8), and subsequently they will be inserted into the outer membrane from the intermembrane space. By this, the final targeting to the outer membrane before insertion is still conserved when compared to bacteria.

The discovery of a second major Omp85-like protein in the chloroplast outer envelope (Eckart et al. 2002) annotated as Toc75-V/Oep80 with closer relation to the ancestral sequences from cyanobacteria than Toc75-III (Bredemeier et al. 2007) clearly suggested an orthologous character of the gene product. The gene is essential (Patel et al. 2008) as expected for a protein involved in the assembly of β-barrel proteins in the outer envelope membrane, which include the protein translocating channel Toc75-III. However, the final experimental proof for this function has not yet been provided.

The notion that Toc75-V is involved in the assembly of β-barrel proteins in the outer envelope membrane is further supported by comparison with the mitochondrial system. Here, an Omp85 homolog was discovered to be involved in β-barrel protein assembly as well. The protein was termed Tob55 (topogenesis of mitochondrial outer membrane β-barrel proteins), Sam50 (sorting and assembly of mitochondria), or mitochondrial Omp85 (Kozjak et al. 2003; Paschen et al. 2003; Gentle et al. 2004; Humphries et al. 2005) and it is of clear proteobacterial origin (e.g., Bredemeier et al. 2007). Meanwhile, in yeast, a machinery for the β-barrel outer membrane assembly including Sam37, Sam35, Mim1, and Mdm10 has been discovered (e.g., Chacinska et al. 2009). However, neither of these components was identified in *Encephalitozoon cuniculi* mitosomes (Walther et al. 2009), and at least for Sam37, a eukaryotic origin is suggested (Cavalier-Smith 2006). This conclusion might be extended even further, stating that these components are specific to a branch of fungi, because no homologous sequences can be detected, e.g., in plants or humans (with the exception of the weak homology between human metaxin and Sam35, Soll and Schleiff 2004). Therefore, the only component which exists in all branches of life is the Omp85-like translocator. The remaining components might have evolved independently in the different branches of life.

8.8   Concluding Remarks

The translocation systems are by and large products of evolutionary recycling of bacterial proteins with a function mostly other than protein transport. At present, one can conclude that the path of insertion of outer membrane β-barrel proteins, the Oxa1 translocation system in mitochondria and the systems hosted in the thylakoid membrane are (in parts) orthologous to the systems of the ancestral bacteria. For most of the other factors, only homologs can be identified, but in many cases the bacterial protein related to the ancestor of the translocation component performs functions distinct from protein translocation. Thus, new strategies have been evolved during evolution based on existing protein folds – for which here the term "evolutionary recycling" is suggested. Recently, the evolutionary development of protein translocation machineries was also described as "tinkering" (Alcock et al. 2010) – in the sense of playing with something and trying to get it to work. In our opinion that nicely describes the evolutionary step after the evolutionary extraction of useful materials and its reuse. Thus, we would describe the evolutionary process leading to the translocation machineries known today as "Recycling and Tinkering": The use of existing structures in a different context and subsequent adaptations of the translocation complexes and their components to meet the requirements for the integration of the organelles into the cellular context, which is also enhanced in its complexity during evolution.

As outlined, for many components their origin can be traced, but especially for the translocation systems in chloroplasts some additional information is required to fully reconstitute the evolutionary development. On the one hand, sequencing of the glaucocystophyte alga *C. paradoxa* will bring new insights into the initial translocon evolved shortly after uptake of the cyanobacteria. On the other hand, a detailed analysis of some of the thylakoid translocation complex components is still required to confirm the common belief that they are all of cyanobacterial origin.

Acknowledgements We are grateful to Sascha Straus, Stefan Simm, Andreas Weber, and Arndt von Haeseler for open and critical discussions concerning this issue. The work was supported by grants from the Deutsche Forschungsgemeinschaft and from the Volkswagenstiftung to ES.

References

Adams MD, Wagner LM, Graddis TJ, Landick R, Antonucci TK, Gibson AL, Oxender DL (1990) Nucleotide sequence and genetic characterization reveal six essential genes for the LIV-I and LS transport systems of *Escherichia coli*. J Biol Chem 265:11436–11443

Alcock FH, Grossmann JG, Gentle IE, Likić VA, Lithgow T, Tokatlidis K (2008) Conserved substrate binding by chaperones in the bacterial periplasm and the mitochondrial intermembrane space. Biochem J 409:377–387

Alcock F, Clements A, Webb C, Lithgow T (2010) Evolution. Tinkering inside the organelle. Science 327:649–650

Allen JW, Ferguson SJ, Ginger ML (2008) Distinctive biochemistry in the trypanosome mitochondrial intermembrane space suggests a model for stepwise evolution of the MIA pathway for import of cysteine-rich proteins. FEBS Lett 582:2817–2825

Anderson S, Bankier AT, Barrell BG, de Bruijn MH, Coulson AR, Drouin J, Eperon IC, Nierlich DP, Roe BA, Sanger F, Schreier PH, Smith AJ, Staden R, Young IG (1981) Sequence and organization of the human mitochondrial genome. Nature 290:457–465

Armbrust EV, Berges JA, Bowler C, Green BR, Martinez D, Putnam NH, Zhou S, Allen AE, Apt KE, Bechner M, Brzezinski MA, Chaal BK, Chiovitti A, Davis AK, Demarest MS, Detter JC, Glavina T, Goodstein D, Hadi MZ, Hellsten U, Hildebrand M, Jenkins BD, Jurka J, Kapitonov VV, Kröger N, Lau WW, Lane TW, Larimer FW, Lippmeier JC, Lucas S, Medina M, Montsant A, Obornik M, Parker MS, Palenik B, Pazour GJ, Richardson PM, Rynearson TA, Saito MA, Schwartz DC, Thamatrakoln K, Valentin K, Vardi A, Wilkerson FP, Rokhsar DS (2004) The genome of the diatom *Thalassiosira pseudonana*: ecology, evolution, and metabolism. Science 306:79–86

Asakura Y, Kikuchi S, Nakai M (2008) Non-identical contributions of two membrane-bound cpSRP components, cpFtsY and Alb3, to thylakoid biogenesis. Plant J 56:1007–1017

Baker A, Schatz G (1987) Sequences from a prokaryotic genome or the mouse dihydrofolate reductase gene can restore the import of a truncated precursor protein into yeast mitochondria. Proc Natl Acad Sci USA 84:3117–3121

Balsera M, Stengel A, Soll J, Bölter B (2007) Tic62: a protein family from metabolism to protein translocation. BMC Evol Biol 7:43

Barbrook AC, Lockhart PJ, Howe CJ (1988) Phylogenetic analysis of plastid origins based on secA sequences. Curr Genet 34:336–341

Bauer MF, Rothbauer U, Mühlenbein N, Smith RJ, Gerbitz K, Neupert W, Brunner M, Hofmann S (1999) The mitochondrial TIM22 preprotein translocase is highly conserved throughout the eukaryotic kingdom. FEBS Lett 464:41–47

Bauerschmitt H, Mick DU, Deckers M, Vollmer C, Funes S, Kehrein K, Ott M, Rehling P, Herrmann JM (2010) Ribosome-binding proteins Mdm38 and Mba1 display overlapping functions for regulation of mitochondrial translation. Mol Biol Cell 21:1937–1944

Beck K, Wu LF, Brunner J, Müller M (2000) Discrimination between SRP- and SecA/SecB-dependent substrates involves selective recognition of nascent chains by SRP and trigger factor. EMBO J 19:134–143

Becker T, Hritz J, Vogel M, Caliebe A, Bukau B, Soll J, Schleiff E (2004) Toc12, a novel subunit of the intermembrane space preprotein translocon of chloroplasts. Mol Biol Cell 15:5130–5144

Berridge MJ, Bootman MD, Roderick HL (2003) Calcium signalling: dynamics, homeostasis and remodelling. Nat Rev Mol Cell Biol 4:517–529

Bionda T, Tillmann B, Simm S, Beilstein K, Ruprecht M, Schleiff E (2010) Chloroplast import signals – the length requirement for translocation in vitro and in vivo. J Mol Biol 402 (3):510–523

Bodył A, Mackiewicz P, Stiller JW (2009a) Early steps in plastid evolution: current ideas and controversies. Bioessays 31:1219–1232

Bodył A, Mackiewicz P, Stiller JW (2009b) Comparative genomic studies suggest that the cyanobacterial endosymbionts of the amoeba *Paulinella chromatophora* possess an import apparatus for nuclear-encoded proteins. Plant Biol (Stuttg) 12:639–649

Bohnsack MT, Schleiff E (2010) The evolution of protein targeting and translocation systems. Biochim Biophys Acta 1803:1115–1130

Bonen L, Calixte S (2006) Comparative analysis of bacterial-origin genes for plant mitochondrial ribosomal proteins. Mol Biol Evol 23:701–712

Bonifacino JS, Glick BS (2004) The mechanisms of vesicle budding and fusion. Cell 116:153–166

Bonnefoy N, Fiumera HL, Dujardin G, Fox TD (2009) Roles of Oxa1-related inner-membrane translocases in assembly of respiratory chain complexes. Biochim Biophys Acta 1793:60–70

Bredemeier R, Schlegel T, Ertel F, Vojta A, Borissenko L, Bohnsack MT, Groll M, von Haeseler A, Schleiff E (2007) Functional and phylogenetic properties of the pore-forming beta-barrel transporters of the Omp85 family. J Biol Chem 282:1882–1890

Bruce BD (2001) The paradox of plastid transit peptides: conservation of function despite divergence in primary structure. Biochim Biophys Acta 1541:2–21

Bullmann L, Haarmann R, Mirus O, Bredemeier R, Hempel F, Maier UG, Schleiff E (2010) Filling the gap, evolutionarily conserved Omp85 in plastids of chromalveolates. J Biol Chem 285:6848–6856

Burger G, Lang BF, Reith M, Gray MW (1996) Genes encoding the same three subunits of respiratory complex II are present in the mitochondrial DNA of two phylogenetically distant eukaryotes. Proc Natl Acad Sci USA 93:2328–2332

Burger G, Gray MW, Lang BF (2003) Mitochondrial genomes: anything goes. Trends Genet 19:709–716

Burroughs AM, Allen KN, Dunaway-Mariano D, Aravind L (2006) Evolutionary genomics of the HAD superfamily: understanding the structural adaptations and catalytic diversity in a superfamily of phosphoesterases and allied enzymes. J Mol Biol 361:1003–1034

Cao TB, Saier MH Jr (2003) The general protein secretory pathway: phylogenetic analyses leading to evolutionary conclusions. Biochim Biophys Acta 1609:115–125

Cardol P, González-Halphen D, Reyes-Prieto A, Baurain D, Matagne RF, Remacle C (2005) The mitochondrial oxidative phosphorylation proteome of *Chlamydomonas reinhardtii* deduced from the Genome Sequencing Project. Plant Physiol 137:447–459

Cavalier-Smith T (2006) Origin of mitochondria by intracellular enslavement of a photosynthetic purple bacterium. Proc Biol Sci 273:1943–1952

Cavalier-Smith T (2009) Kingdoms Protozoa and Chromista and the eozoan root of the eukaryotic tree. Biol Lett 6:342–345

Chacinska A, Pfannschmidt S, Wiedemann N, Kozjak V, Sanjuán Szklarz LK, Schulze-Specking A, Truscott KN, Guiard B, Meisinger C, Pfanner N (2004) Essential role of Mia40 in import and assembly of mitochondrial intermembrane space proteins. EMBO J 23:3735–3746

Chacinska A, Lind M, Frazier AE, Dudek J, Meisinger C, Geissler A, Sickmann A, Meyer HE, Truscott KN, Guiard B, Pfanner N, Rehling P (2005) Mitochondrial presequence translocase:

switching between TOM tethering and motor recruitment involves Tim21 and Tim17. Cell 120:817–829

Chacinska A, Koehler CM, Milenkovic D, Lithgow T, Pfanner N (2009) Importing mitochondrial proteins: machineries and mechanisms. Cell 138:628–644

Chacinska A, van der Laan M, Mehnert CS, Guiard B, Mick DU, Hutu DP, Truscott KN, Wiedemann N, Meisinger C, Pfanner N, Rehling P (2010) Distinct forms of mitochondrial TOM-TIM supercomplexes define signal-dependent states of preprotein sorting. Mol Cell Biol 30:307–318

Chan NC, Likić VA, Waller RF, Mulhern TD, Lithgow T (2006) The C-terminal TPR domain of Tom70 defines a family of mitochondrial protein import receptors found only in animals and fungi. J Mol Biol 358:1010–1022

Chen KM, Holmström M, Raksajit W, Suorsa M, Piippo M, Aro EM (2010) Small chloroplast-targeted DnaJ proteins are involved in optimization of photosynthetic reactions in *Arabidopsis thaliana*. BMC Plant Biol 10:43

Chew O, Lister R, Qbadou S, Heazlewood JL, Soll J, Schleiff E, Millar AH, Whelan J (2004) A plant outer mitochondrial membrane protein with high amino acid sequence identity to a chloroplast protein import receptor. FEBS Lett 557:109–114

Chewawiwat N, Yano M, Terada K, Hoogenraad NJ, Mori M (1999) Characterization of the novel mitochondrial protein import component, Tom34, in mammalian cells. J Biochem 125:721–727

Chou ML, Fitzpatrick LM, Tu SL, Budziszewski G, Potter-Lewis S, Akita M, Levin JZ, Keegstra K, Li HM (2003) Tic40, a membrane-anchored co-chaperone homolog in the chloroplast protein translocon. EMBO J 22:2970–2980

Clements A, Bursac D, Gatsos X, Perry AJ, Civciristov S, Celik N, Likic VA, Poggio S, Jacobs-Wagner C, Strugnell RA, Lithgow T (2009) The reducible complexity of a mitochondrial molecular machine. Proc Natl Acad Sci USA 106:15791–15795

Cline K, Dabney-Smith C (2008) Plastid protein import and sorting: different paths to the same compartments. Curr Opin Plant Biol 11:585–592

Dabney-Smith C, Cline K (2009) Clustering of C-terminal stromal domains of Tha4 homo-oligomers during translocation by the Tat protein transport system. Mol Biol Cell 20:2060–2069

Dacks JB, Field MC (2007) Evolution of the eukaryotic membrane-trafficking system: origin, tempo and mode. J Cell Sci 120:2977–2985

Dacks JB, Poon PP, Field MC (2008) Phylogeny of endocytic components yields insight into the process of nonendosymbiotic organelle evolution. Proc Natl Acad Sci USA 105:588–593

Dagley MJ, Dolezal P, Likic VA, Smid O, Purcell AW, Buchanan SK, Tachezy J, Lithgow T (2009) The protein import channel in the outer mitosomal membrane of *Giardia intestinalis*. Mol Biol Evol 26:1941–1947

De Domenico I, McVey Ward D, Kaplan J (2008) Regulation of iron acquisition and storage: consequences for iron-linked disorders. Nat Rev Mol Cell Biol 9:72–81

de Zamaroczy M, Bernardi G (1986) The primary structure of the mitochondrial genome of *Saccharomyces cerevisiae* – a review. Gene 47:155–177

Dekker PJ, Ryan MT, Brix J, Müller H, Hönlinger A, Pfanner N (1998) Preprotein translocase of the outer mitochondrial membrane: molecular dissection and assembly of the general import pore complex. Mol Cell Biol 18:6515–6524

Deusch O, Landan G, Roettger M, Gruenheit N, Kowallik KV, Allen JF, Martin W, Dagan T (2008) Genes of cyanobacterial origin in plant nuclear genomes point to a heterocyst-forming plastid ancestor. Mol Biol Evol 25:748–761

Dong JH, Wen JF, Tian HF (2007) Homologs of eukaryotic Ras superfamily proteins in prokaryotes and their novel phylogenetic correlation with their eukaryotic analogs. Gene 396:116–124

Duhita N, Le HA, Satoshi S, Kazuo H, Daisuke M, Takao S (2010) The origin of peroxisomes: the possibility of an actinobacterial symbiosis. Gene 450:18–24

Eckart K, Eichacker L, Sohrt K, Schleiff E, Heins L, Soll J (2002) A Toc75-like protein import channel is abundant in chloroplasts. EMBO Rep 3:557–562

Edmonson AM, Mayfield DK, Vervoort V, DuPont BR, Argyropoulos G (2002) Characterization of a human import component of the mitochondrial outer membrane, TOMM70A. Cell Commun Adhes 9:15–27

Erdmann R, Schliebs W (2005) Peroxisomal matrix protein import: the transient pore model. Nat Rev Mol Cell Biol 6:738–742

Field MC, Dacks JB (2009) First and last ancestors: reconstructing evolution of the endomembrane system with ESCRTs, vesicle coat proteins, and nuclear pore complexes. Curr Opin Cell Biol 21:4–13

Figueroa-Martínez F, Funes S, Franzén LG, González-Halphen D (2008) Reconstructing the mitochondrial protein import machinery of *Chlamydomonas reinhardtii*. Genetics 179:149–155

Frand AR, Kaiser CA (1998) The ERO1 gene of yeast is required for oxidation of protein dithiols in the endoplasmic reticulum. Mol Cell 1:161–170

Franklin AE, Hoffman NE (1993) Characterization of a chloroplast homologue of the 54-kDa subunit of the signal recognition particle. J Biol Chem 268:22175–22180

Frazier AE, Dudek J, Guiard B, Voos W, Li Y, Lind M, Meisinger C, Geissler A, Sickmann A, Meyer HE, Bilanchone V, Cumsky MG, Truscott KN, Pfanner N, Rehling P (2004) Pam16 has an essential role in the mitochondrial protein import motor. Nat Struct Mol Biol 11:226–233

Frazier AE, Taylor RD, Mick DU, Warscheid B, Stoepel N, Meyer HE, Ryan MT, Guiard B, Rehling P (2006) Mdm38 interacts with ribosomes and is a component of the mitochondrial protein export machinery. J Cell Biol 172:553–564

Fröderberg L, Röhl T, van Wijk KJ, de Gier JW (2001) Complementation of bacterial SecE by a chloroplastic homologue. FEBS Lett 498:52–56

Fromont-Racine M, Senger B, Saveanu C, Fasiolo F (2003) Ribosome assembly in eukaryotes. Gene 313:17–42

Fulda S, Norling B, Schoor A, Hagemann M (2002) The Slr0924 protein of *Synechocystis* sp. strain PCC 6803 resembles a subunit of the chloroplast protein import complex and is mainly localized in the thylakoid lumen. Plant Mol Biol 49:107–118

Funes S, Hasona A, Bauerschmitt H, Grubbauer C, Kauff F, Collins R, Crowley PJ, Palmer SR, Brady LJ, Herrmann JM (2009) Independent gene duplications of the YidC/Oxa/Alb3 family enabled a specialized cotranslational function. Proc Natl Acad Sci USA 106:6656–6661

Gabaldon T, Snel B, van Zimmeren F, Hemrika W, Tabak H, Huynen MA (2006) Origin and evolution of the peroxisomal proteome. Biol Direct 1:8

Gaisne M, Bonnefoy N (2006) The COX18 gene, involved in mitochondrial biogenesis, is functionally conserved and tightly regulated in humans and fission yeast. FEMS Yeast Res 6:869–882

Gentle I, Gabriel K, Beech P, Waller R, Lithgow T (2004) The Omp85 family of proteins is essential for outer membrane biogenesis in mitochondria and bacteria. J Cell Biol 164:19–24

Gentle IE, Perry AJ, Alcock FH, Likić VA, Dolezal P, Ng ET, Purcell AW, McConnville M, Naderer T, Chanez AL, Charrière F, Aschinger C, Schneider A, Tokatlidis K, Lithgow T (2007) Conserved motifs reveal details of ancestry and structure in the small TIM chaperones of the mitochondrial intermembrane space. Mol Biol Evol 24:1149–1160

Görlich D, Kutay U (1999) Transport between the cell nucleus and the cytoplasm. Annu Rev Cell Dev Biol 15:607–660

Gray MW, Lang BF, Cedergren R, Golding GB, Lemieux C, Sankoff D, Turmel M, Brossard N, Delage E, Littlejohn TG, Plante I, Rioux P, Saint-Louis D, Zhu Y, Burger G (1998) Genome structure and gene content in protist mitochondrial DNAs. Nucleic Acids Res 26:865–878

Gregerson RG, Miller SS, Petrowski M, Gantt JS, Vance CP (1994) Genomic structure, expression and evolution of the alfalfa aspartate aminotransferase genes. Plant Mol Biol 25:387–399

Grudnik P, Bange G, Sinning I (2009) Protein targeting by the signal recognition particle. Biol Chem 390:775–782

Guo Y, Cheong N, Zhang Z, De Rose R, Deng Y, Farber SA, Fernandes-Alnemri T, Alnemri ES (2004) Tim50, a component of the mitochondrial translocator, regulates mitochondrial integrity and cell death. J Biol Chem 279:24813–24825

Gupta RS (1995) Phylogenetic analysis of the 90 kD heat shock family of protein sequences and an examination of the relationship among animals, plants, and fungi species. Mol Biol Evol 12:1063–1073

Gupta RS, Golding GB (1993) Evolution of HSP70 gene and its implications regarding relationships between archaebacteria, eubacteria, and eukaryotes. J Mol Evol 37:573–582

Haft DH, Paulsen IT, Ward N, Selengut JD (2006) Exopolysaccharide-associated protein sorting in environmental organisms: the PEP-CTERM/EpsH system. Application of a novel phylogenetic profiling heuristic. BMC Biol 4:29

Haney SA, Oxender DL (1992) Amino acid transport in bacteria. Int Rev Cytol 137:37–95

Harb OS, Chatterjee B, Fraunholz MJ, Crawford MJ, Nishi M, Roos DS (2004) Multiple functionally redundant signals mediate targeting to the apicoplast in the picomplexan parasite *Toxoplasma gondii*. Eukaryot Cell 3:663–674

Hase T, Müller U, Riezman H, Schatz G (1984) A 70-kd protein of the yeast mitochondrial outer membrane is targeted and anchored via its extreme amino terminus. EMBO J 3:3157–3164

Haydon MJ, Cobbett CS (2007) Transporters of ligands for essential metal ions in plants. New Phytol 174:499–506

Heil M, Ton J (2008) Long-distance signalling in plant defence. Trends Plant Sci 13:264–272

Heins L, Schmitz UK (1996) A receptor for protein import into potato mitochondria. Plant J 9:829–839

Heras B, Shouldice SR, Totsika M, Scanlon MJ, Schembri MA, Martin JL (2009) DSB proteins and bacterial pathogenicity. Nat Rev Microbiol 7:215–225

Herrmann JM, Köhl R (2007) Catch me if you can! Oxidative protein trapping in the intermembrane space of mitochondria. J Cell Biol 176:559–563

Herrmann JM, Kauff F, Neuhaus HE (2009) Thiol oxidation in bacteria, mitochondria and chloroplasts: common principles but three unrelated machineries? Biochim Biophys Acta 1793:71–77

Holt CE, Bullock SL (2009) Subcellular mRNA localization in animal cells and why it matters. Science 326:1212–1216

Hönlinger A, Bömer U, Alconada A, Eckerskorn C, Lottspeich F, Dietmeier K, Pfanner N (1996) Tom7 modulates the dynamics of the mitochondrial outer membrane translocase and plays a pathway-related role in protein import. EMBO J 15:2125–2137

Hulett JM, Walsh P, Lithgow T (2007) Domain stealing by receptors in a protein transport complex. Mol Biol Evol 24:1909–1911

Humphries AD, Streimann IC, Stojanovski D, Johnston AJ, Yano M, Hoogenraad NJ, Ryan MT (2005) Dissection of the mitochondrial import and assembly pathway for human Tom40. J Biol Chem 280:11535–11543

Jiang Y, Cheng Z, Mandon EC, Gilmore R (2008) An interaction between the SRP receptor and the translocon is critical during cotranslational protein translocation. J Cell Biol 180:1149–1161

Jiang D, Zhao L, Clapham DE (2009) Genome-wide RNAi screen identifies Letm1 as a mitochondrial Ca^{2+}/H^+ antiporter. Science 326:144–147

Jin H, May M, Tranebjaerg L, Kendall E, Fontán G, Jackson J, Subramony SH, Arena F, Lubs H, Smith S, Stevenson R, Schwartz C, Vetrie D (1996) A novel X-linked gene, DDP, shows mutations in families with deafness (DFN-1), dystonia, mental deficiency and blindness. Nat Genet 14:177–180

Jin H, Kendall E, Freeman TC, Roberts RG, Vetrie DL (1999) The human family of Deafness/Dystonia peptide (DDP) related mitochondrial import proteins. Genomics 61:259–267

Joseph AM, Rungi AA, Robinson BH, Hood DA (2004) Compensatory responses of protein import and transcription factor expression in mitochondrial DNA defects. Am J Physiol Cell Physiol 286:C867–C875

Kadokura H, Katzen F, Beckwith J (2003) Protein disulfide bond formation in prokaryotes. Annu Rev Biochem 72:111–135

Kadowaki K, Kubo N, Ozawa K, Hirai A (1996) Targeting presequence acquisition after mito-chondrial gene transfer to the nucleus occurs by duplication of existing targeting signals. EMBO J 15:6652–6661

Kalanon M, McFadden GI (2008) The chloroplast protein translocation complexes of *Chlamydomonas reinhardtii*: a bioinformatic comparison of Toc and Tic components in plants, green algae and red algae. Genetics 179:95–112

Kelley WL (1998) The J-domain family and the recruitment of chaperone power. Trends Biochem Sci 23:222–227

Kikuchi S, Oishi M, Hirabayashi Y, Lee DW, Hwang I, Nakai M (2009) A 1-megadalton translocation complex containing Tic20 and Tic21 mediates chloroplast protein import at the inner envelope membrane. Plant Cell 21:1781–1797

Kilian O, Kroth PG (2005) Identification and characterization of a new conserved motif within the presequence of proteins targeted into complex diatom plastids. Plant J 41:175–183

Kirk JTO, Tilney-Bassett RAE (1978) The plastids: their chemistry, structure, growth and inheri-tance. Elsevier/North-Holland Biomedical Press, Amsterdam, New York, Oxford

Klostermann E, Droste Gen Helling I, Carde JP, Schünemann D (2002) The thylakoid membrane protein ALB3 associates with the cpSecY-translocase in *Arabidopsis thaliana*. Biochem J 368:777–781

Koehler CM, Leuenberger D, Merchant S, Renold A, Junne T, Schatz G (1999) Human deafness dystonia syndrome is a mitochondrial disease. Proc Natl Acad Sci USA 96:2141–2146

Kohler R, Boehringer D, Greber B, Bingel-Erlenmeyer R, Collinson I, Schaffitzel C, Ban N (2009) YidC and Oxa1 form dimeric insertion pores on the translating ribosome. Mol Cell 34:344–353

Kovacheva S, Bédard J, Wardle A, Patel R, Jarvis P (2007) Further in vivo studies on the role of the molecular chaperone, Hsp93, in plastid protein import. Plant J 50:364–379

Kozany C, Mokranjac D, Sichting M, Neupert W, Hell K (2004) The J domain-related cochaperone Tim16 is a constituent of the mitochondrial TIM23 preprotein translocase. Nat Struct Mol Biol 11:234–241

Kozjak V, Wiedemann N, Milenkovic D, Lohaus C, Meyer HE, Guiard B, Meisinger C, Pfanner N (2003) An essential role of Sam50 in the protein sorting and assembly machinery of the mitochondrial outer membrane. J Biol Chem 278:48520–48523

Küchler M, Decker S, Hörmann F, Soll J, Heins L (2002) Protein import into chloroplasts involves redox-regulated proteins. EMBO J 21:6136–6145

Kutik S, Stojanovski D, Becker L, Becker T, Meinecke M, Krüger V, Prinz C, Meisinger C, Guiard B, Wagner R, Pfanner N, Wiedemann N (2008) Dissecting membrane insertion of mitochondrial beta-barrel proteins. Cell 132:1011–1024

Kutik S, Stroud DA, Wiedemann N, Pfanner N (2009) Evolution of mitochondrial protein biogenesis. Biochim Biophys Acta 1790:409–415

Kuznetsov OA, Schwuchow J, Sack FD, Hasenstein KH (1999) Curvature induced by amyloplast magnetophoresis in protonemata of the moss *Ceratodon purpureus*. Plant Physiol 119:645–650

Lee S, Lee DW, Lee Y, Mayer U, Stierhof YD, Lee S, Jürgens G, Hwang I (2009) Heat shock protein cognate 70-4 and an E3 ubiquitin ligase, CHIP, mediate plastid-destined precursor degradation through the ubiquitin-26 S proteasome system in *Arabidopsis*. Plant Cell 21:3984–4001

Leipe DD, Wolf YI, Koonin EV, Aravind L (2002) Classification and evolution of P-loop GTPases and related ATPases. J Mol Biol 317:41–72

Leister D (2003) Chloroplast research in the genomic age. Trends Genet 19:47–56

Lemire BD, Fankhauser C, Baker A, Schatz G (1989) The mitochondrial targeting function of randomly generated peptide sequences correlates with predicted helical amphiphilicity. J Biol Chem 264:20206–20215

Liaud MF, Zhang DX, Cerff R (1990) Differential intron loss and endosymbiotic transfer of chloroplast glyceraldehyde-3-phosphate dehydrogenase genes to the nucleus. Proc Natl Acad Sci USA 87:8918–8922

Likić VA, Perry A, Hulett J, Derby M, Traven A, Waller RF, Keeling PJ, Koehler CM, Curran SP, Gooley PR, Lithgow T (2005) Patterns that define the four domains conserved in known and novel isoforms of the protein import receptor Tom20. J Mol Biol 347:81–93

Lill R (2009) Function and biogenesis of iron-sulphur proteins. Nature 460:831–838

Lithgow T, Schneider A (2010) Evolution of macromolecular import pathways in mitochondria, hydrogenosomes and mitosomes. Philos Trans R Soc Lond B Biol Sci 365:799–817

Löffelhardt W, von Haeseler A, Schleiff E (2007) The β-barrel shaped polypeptide transporter, an old concept for precursor protein transfer across membranes. Symbiosis 44:33–42

Long M, de Souza SJ, Rosenberg C, Gilbert W (1996) Exon shuffling and the origin of the mitochondrial targeting function in plant cytochrome c1 precursor. Proc Natl Acad Sci USA 93:7727–7731

Lucattini R, Likic VA, Lithgow T (2004) Bacterial proteins predisposed for targeting to mitochondria. Mol Biol Evol 21:652–658

Lv HX, Guo GQ, Yang ZN (2009) Translocons on the inner and outer envelopes of chloroplasts share similar evolutionary origin in *Arabidopsis thaliana*. J Evol Biol 22:1418–1428

Maćasev D, Whelan J, Newbigin E, Silva-Filho MC, Mulhern TD, Lithgow T (2004) Tom22′, an 8-kDa trans-site receptor in plants and protozoans, is a conserved feature of the TOM complex that appeared early in the evolution of eukaryotes. Mol Biol Evol 21:1557–1564

Malik HS, Eickbush TH, Goldfarb DS (1997) Evolutionary specialization of the nuclear targeting apparatus. Proc Natl Acad Sci USA 94:13738–13742

Marcet-Houben M, Marceddu G, Gabaldón T (2009) Phylogenomics of the oxidative phosphorylation in fungi reveals extensive gene duplication followed by functional divergence. BMC Evol Biol 9:295

Marchler-Bauer A, Anderson JB, Chitsaz F, Derbyshire MK, DeWeese-Scott C, Fong JH, Geer LY, Geer RC, Gonzales NR, Gwadz M, He S, Hurwitz DI, Jackson JD, Ke Z, Lanczycki CJ, Liebert CA, Liu C, Lu F, Lu S, Marchler GH, Mullokandov M, Song JS, Tasneem A, Thanki N, Yamashita RA, Zhang D, Zhang N, Bryant SH (2009) CDD: specific functional annotation with the Conserved Domain Database. Nucleic Acids Res 37:D205–D210

Marshall JS, DeRocher AE, Keegstra K, Vierling E (1990) Identification of heat shock protein hsp70 homologues in chloroplasts. Proc Natl Acad Sci USA 87:374–378

Martin W, Rujan T, Richly E, Hansen A, Cornelsen S, Lins T, Leister D, Stoebe B, Hasegawa M, Penny D (2002) Evolutionary analysis of Arabidopsis, cyanobacterial, and chloroplast genomes reveals plastid phylogeny and thousands of cyanobacterial genes in the nucleus. Proc Natl Acad Sci USA 99:12246–12251

Matouschek A, Pfanner N, Voos W (2000) Protein unfolding by mitochondria. The Hsp70 import motor. EMBO Rep 1:404–410

Meisinger C, Wiedemann N, Rissler M, Strub A, Milenkovic D, Schönfisch B, Müller H, Kozjak V, Pfanner N (2006) Mitochondrial protein sorting: differentiation of beta-barrel assembly by Tom7-mediated segregation of Mdm10. J Biol Chem 281:22819–22826

Merz F, Boehringer D, Schaffitzel C, Preissler S, Hoffmann A, Maier T, Rutkowska A, Lozza J, Ban N, Bukau B, Deuerling E (2008) Molecular mechanism and structure of Trigger Factor bound to the translating ribosome. EMBO J 27:1622–1632

Mesecke N, Terziyska N, Kozany C, Baumann F, Neupert W, Hell K, Herrmann JM (2005) A disulfide relay system in the intermembrane space of mitochondria that mediates protein import. Cell 121:1059–1069

Mirus O, Schleiff E (2009) The evolution of tetratricopeptide repeat domain containing receptors involved in protein translocation. Endocyt Cell Res 19:31–50

Mirus O, Strauss S, Nicolaisen K, von Haeseler A, Schleiff E (2009a) TonB-dependent transporters and their occurrence in cyanobacteria. BMC Biol 7:68

Mirus O, Bionda T, von Haeseler A, Schleiff E (2009b) Evolutionarily evolved discriminators in the 3-TPR domain of the Toc64 family involved in protein translocation at the outer membrane of chloroplasts and mitochondria. J Mol Model 15:971–982

Mirus O, Hahn A, Schleiff E (2010) Outer membrane proteins. In: König H, Claus H, Varma A (eds) Prokaryotic cell wall compounds. Structure and biochemistry. Springer, Berlin, pp 175–231

Mokranjac D, Bourenkov G, Hell K, Neupert W, Groll M (2006) Structure and function of Tim14 and Tim16, the J and J-like components of the mitochondrial protein import motor. EMBO J 25:4675–4685

Mori H, Cline K (2001) Post-translational protein translocation into thylakoids by the Sec and Delta pH-dependent pathways. Biochim Biophys Acta 1541:80–90

Morimoto K, Nishio K, Nakai M (2002) Identification of a novel prokaryotic HEAT-repeats-containing protein which interacts with a cyanobacterial IscA homolog. FEBS Lett 519:123–127

Mukhopadhyay A, Ni L, Yang CS, Weiner H (2005) Bacterial signal peptide recognizes HeLa cell mitochondrial import receptors and functions as a mitochondrial leader sequence. Cell Mol Life Sci 62:1890–1899

Murcha MW, Elhafez D, Lister R, Tonti-Filippini J, Baumgartner M, Philippar K, Carrie C, Mokranjac D, Soll J, Whelan J (2007) Characterization of the preprotein and amino acid transporter gene family in Arabidopsis. Plant Physiol 143:199–212

Nazos PM, Mayo MM, Su TZ, Anderson JJ, Oxender DL (1985) Identification of livG, a membrane-associated component of the branched-chain amino acid transport in Escherichia coli. J Bacteriol 163:1196–1202

Nicolaisen K, Mariscal V, Bredemeier R, Pernil R, Moslavac S, López-Igual R, Maldener I, Herrero A, Schleiff E, Flores E (2009) The outer membrane of a heterocyst-forming cyano-bacterium is a permeability barrier for uptake of metabolites that are exchanged between cells. Mol Microbiol 74:58–70

Noël LD, Cagna G, Stuttmann J, Wirthmüller L, Betsuyaku S, Witte CP, Bhat R, Pochon N, Colby T, Parker JE (2007) Interaction between SGT1 and cytosolic/nuclear HSC70 chaperones regulates Arabidopsis immune responses. Plant Cell 19:4061–4076

Nowikovsky K, Froschauer EM, Zsurka G, Samaj J, Reipert S, Kolisek M, Wiesenberger G, Schweyen RJ (2004) The LETM1/YOL027 gene family encodes a factor of the mitochondrial K^+ homeostasis with a potential role in the Wolf–Hirschhorn syndrome. J Biol Chem 279:30307–30315

Nuttall SD, Hanson BJ, Mori M, Hoogenraad NJ (1997) hTom34: a novel translocase for the import of proteins into human mitochondria. DNA Cell Biol 16:1067–1074

Ohta N, Sato N, Kuroiwa T (1998) Structure and organization of the mitochondrial genome of the unicellular red alga Cyanidioschyzon merolae deduced from the complete nucleotide sequence. Nucleic Acids Res 26:5190–5198

Oreb M, Tews I, Schleiff E (2008) Policing Tic 'n' Toc, the doorway to chloroplasts. Trends Cell Biol 18:19–27

Osborne AR, Rapoport TA, van den Berg B (2005) Protein translocation by the Sec61/SecY channel. Annu Rev Cell Dev Biol 21:529–550

Ott M, Herrmann JM (2010) Co-translational membrane insertion of mitochondrially encoded proteins. Biochim Biophys Acta 1803:767–775

Paschen SA, Waizenegger T, Stan T, Preuss M, Cyrklaff M, Hell K, Rapaport D, Neupert W (2003) Evolutionary conservation of biogenesis of beta-barrel membrane proteins. Nature 426:862–866

Paschen SA, Neupert W, Rapaport D (2005) Biogenesis of beta-barrel membrane proteins of mitochondria. Trends Biochem Sci 30:575–582

Patel R, Hsu SC, Bédard J, Inoue K, Jarvis P (2008) The Omp85-related chloroplast outer envelope protein OEP80 is essential for viability in Arabidopsis. Plant Physiol 148:235–245

Patron NJ, Waller RF (2007) Transit peptide diversity and divergence: a global analysis of plastid targeting signals. Bioessays 29:1048–1058

Pfanner N, Müller HK, Harmey MA, Neupert W (1987) Mitochondrial protein import: involvement of the mature part of a cleavable precursor protein in the binding to receptor sites. EMBO J 6:3449–3454

Pohlschröder M, Hartmann E, Hand NJ, Dilks K, Haddad A (2005) Diversity and evolution of protein translocation. Annu Rev Microbiol 59:91–111

Pollard MG, Travers KJ, Weissman JS (1998) Ero1p: a novel and ubiquitous protein with an essential role in oxidative protein folding in the endoplasmic reticulum. Mol Cell 1:171–182

Pusnik M, Charrière F, Mäser P, Waller RF, Dagley MJ, Lithgow T, Schneider A (2009) The single mitochondrial porin of *Trypanosoma brucei* is the main metabolite transporter in the outer mitochondrial membrane. Mol Biol Evol 26:671–680

Qbadou S, Becker T, Mirus O, Tews I, Soll J, Schleiff E (2006) The molecular chaperone Hsp90 delivers precursor proteins to the chloroplast import receptor Toc64. EMBO J 25:1836–1847

Qbadou S, Becker T, Bionda T, Reger K, Ruprecht M, Soll J, Schleiff E (2007) Toc64 – a preprotein-receptor at the outer membrane with bipartide function. J Mol Biol 367:1330–1346

Quigley F, Martin WF, Cerff R (1988) Intron conservation across the prokaryote-eukaryote boundary: structure of the nuclear gene for chloroplast glyceraldehyde-3-phosphate dehydrogenase from maize. Proc Natl Acad Sci USA 85:2672–2676

Ralph SA, Foth BJ, Hall N, McFadden GI (2004) Evolutionary pressures on apicoplast transit peptides. Mol Biol Evol 21:2183–2194

Rapoport TA (2007) Protein translocation across the eukaryotic endoplasmic reticulum and bacterial plasma membranes. Nature 450:663–669

Rassow J, Dekker PJ, van Wilpe S, Meijer M, Soll J (1999) The preprotein translocase of the mitochondrial inner membrane: function and evolution. J Mol Biol 286:105–120

Richter CV, Träger C, Schünemann D (2008) Evolutionary substitution of two amino acids in chloroplast SRP54 of higher plants cause its inability to bind SRP RNA. FEBS Lett 582:3223–3229

Richter CV, Bals T, Schünemann D (2010) Component interactions, regulation and mechanisms of chloroplast signal recognition particle-dependent protein transport. Eur J Cell Biol 89 (12):965–973

Robinson MS (2004) Adaptable adaptors for coated vesicles. Trends Cell Biol 14:167–174

Robinson C, Bolhuis A (2001) Protein targeting by the twin-arginine translocation pathway. Nat Rev Mol Cell Biol 2:350–356

Rolland N, Job D, Douce R (1993) Common sequence motifs coding for higher-plant and prokaryotic O-acetylserine (thiol)-lyases: bacterial origin of a chloroplast transit peptide? Biochem J 293:829–833

Rosenblad MA, Samuelsson T (2004) Identification of chloroplast signal recognition particle RNA genes. Plant Cell Physiol 45:1633–1639

Rothbauer U, Hofmann S, Mühlenbein N, Paschen SA, Gerbitz KD, Neupert W, Brunner M, Bauer MF (2001) Role of the deafness dystonia peptide 1 (DDP1) in import of human Tim23 into the inner membrane of mitochondria. J Biol Chem 276:37327–37334

Ruprecht M, Bionda T, Sato T, Sommer MS, Endo T, Schleiff E (2010) On the impact of precursor unfolding during protein import into chloroplasts. Mol Plant 3:499–508

Schlegel T, Mirus O, von Haeseler A, Schleiff E (2007) The tetratricopeptide repeats of receptors involved in protein translocation across membranes. Mol Biol Evol 24:2763–2774

Schleiff E, Soll J (2005) Membrane protein insertion: mixing eukaryotic and prokaryotic concepts. EMBO Rep 6:1023–1027

Schleiff E, Sulea T (1999) Structure determination and prediction: a study on human Tom20. J Mol Struct THEOCHEM 468:127–134

Schleiff E, Shore GC, Goping IS (1997) Interactions of the human mitochondrial protein import receptor, hTom20, with precursor proteins in vitro reveal pleiotropic specificities and different receptor domain requirements. J Biol Chem 272:17784–17789

Schleiff E, Maier UG, Becker T (2010) Omp85 in eukaryotic systems – one protein family with distinct functions. Biol Chem 392(1–2):21–27

Schlickum S, Moghekar A, Simpson JC, Steglich C, O'Brien RJ, Winterpacht A, Endele SU (2004) LETM1, a gene deleted in Wolf–Hirschhorn syndrome, encodes an evolutionarily conserved mitochondrial protein. Genomics 83:254–261

Schlüter A, Fourcade S, Ripp R, Mandel JL, Poch O, Pujol A (2006) The evolutionary origin of peroxisomes: an ER-peroxisome connection. Mol Biol Evol 23:838–845

Schneider A, Bursać D, Lithgow T (2008) The direct route: a simplified pathway for protein import into the mitochondrion of trypanosomes. Trends Cell Biol 18:12–18

Schnell DJ, Hebert DN (2003) Protein translocons: multifunctional mediators of protein translocation across membranes. Cell 112:491–505

Schnell DJ, Kessler F, Blobel G (1994) Isolation of components of the chloroplast protein import machinery. Science 266:1007–1012

Settles AM, Martienssen R (1998) Old and new pathways of protein export in chloroplasts and bacteria. Trends Cell Biol 8:494–501

Shanske AL, Yachelevich N, Ala-Kokko L, Leonard J, Levy B (2010) Wolf–Hirschhorn syndrome and ectrodactyly: new findings and a review of the literature. Am J Med Genet A 152A:203–208

Sheiner L, Soldati-Favre D (2008) Protein trafficking inside *Toxoplasma gondii*. Traffic 9:636–646

Sherman EL, Go NE, Nargang FE (2005) Functions of the small proteins in the TOM complex of *Neurospora crasssa*. Mol Biol Cell 16:4172–4182

Shi LX, Theg SM (2010) A stromal heat shock protein 70 system functions in protein import into chloroplasts in the moss *Physcomitrella patens*. Plant Cell 22:205–220

Smith DG, Gawryluk RM, Spencer DF, Pearlman RE, Siu KW, Gray MW (2007) Exploring the mitochondrial proteome of the ciliate protozoon *Tetrahymena thermophila*: direct analysis by tandem mass spectrometry. J Mol Biol 374:837–863

Smits P, Smeitink JA, van den Heuvel LP, Huynen MA, Ettema TJ (2007) Reconstructing the evolution of the mitochondrial ribosomal proteome. Nucleic Acids Res 35:4686–4703

Sohrt K, Soll J (2000) Toc64, a new component of the protein translocon of chloroplasts. J Cell Biol 148:1213–1221

Soll J, Schleiff E (2004) Protein import into chloroplasts. Nat Rev Mol Cell Biol 5:198–208

Stegemann S, Bock R (2006) Experimental reconstruction of functional gene transfer from the tobacco plastid genome to the nucleus. Plant Cell 18:2869–2878

Stegemann S, Hartmann S, Ruf S, Bock R (2003) High-frequency gene transfer from the chloroplast genome to the nucleus. Proc Natl Acad Sci USA 100:8828–8833

Steiner JM, Löffelhardt W (2005) Protein translocation into and within cyanelles. Mol Membr Biol 22:123–132

Steiner JM, Yusa F, Pompe JA, Löffelhardt W (2005) Homologous protein import machineries in chloroplasts and cyanelles. Plant J 44:646–652

Stirling PC, Bakhoum SF, Feigl AB, Leroux MR (2006) Convergent evolution of clamp-like binding sites in diverse chaperones. Nat Struct Mol Biol 13:865–870

Struyve M, Moons M, Tommassen J (1991) Carboxy-terminal phenylalanine is essential for the correct assembly of a bacterial outer membrane protein. J Mol Biol 218:141–148

Stymest KH, Klappa P (2008) The periplasmic peptidyl prolyl cis-trans isomerases PpiD and SurA have partially overlapping substrate specificities. FEBS J 275:3470–3479

Su PH, Li HM (2010) Stromal Hsp70 is important for protein translocation into pea and *Arabidopsis* chloroplasts. Plant Cell 22:1516–1531

Sung DY, Vierling E, Guy CL (2001) Comprehensive expression profile analysis of the Arabidopsis Hsp70 gene family. Plant Physiol 126:789–800

Thorsness PE, Fox TD (1990) Escape of DNA from mitochondria to the nucleus in *Saccharomyces cerevisiae*. Nature 346:376–379

Tranebjaerg L, Schwartz C, Eriksen H, Andreasson S, Ponjavic V, Dahl A, Stevenson RE, May M, Arena F, Barker D, Elverland HH, Lubs H (1995) A new X linked recessive deafness syndrome with blindness, dystonia, fractures, and mental deficiency is linked to Xq22. J Med Genet 32:257–263

Tranel PJ, Froehlich J, Goyal A, Keegstra K (1995) A component of the chloroplastic protein import apparatus is targeted to the outer envelope membrane via a novel pathway. EMBO J 14:2436–2446

Tsaytler PA, Krijgsveld J, Goerdayal SS, Rüdiger S, Egmond MR (2009) Novel Hsp90 partners discovered using complementary proteomic approaches. Cell Stress Chaperones 14:629–638

Tsukazaki T, Mori H, Fukai S, Ishitani R, Mori T, Dohmae N, Perederina A, Sugita Y, Vassylyev DG, Ito K, Nureki O (2008) Conformational transition of Sec machinery inferred from bacterial SecYE structures. Nature 455:988–991

Tzvetkova-Chevolleau T, Hutin C, Noël LD, Goforth R, Carde JP, Caffarri S, Sinning I, Groves M, Teulon JM, Hoffman NE, Henry R, Havaux M, Nussaume L (2007) Canonical signal recognition particle components can be bypassed for posttranslational protein targeting in chloroplasts. Plant Cell 19:1635–1648

Ullers RS, Ang D, Schwager F, Georgopoulos C, Genevaux P (2004) Trigger Factor can antagonize both SecB and DnaK/DnaJ chaperone functions in *Escherichia coli*. Proc Natl Acad Sci USA 104:3101–3106

Unseld M, Marienfeld JR, Brandt P, Brennicke A (1997) The mitochondrial genome of *Arabidopsis thaliana* contains 57 genes in 366,924 nucleotides. Nat Genet 15:57–61

Valentin K (1997) Phylogeny and expression of the secA gene from a chromophytic alga – implications for the evolution of plastids and sec-dependent protein translocation. Curr Genet 32:300–307

van der Laan M, Nouwen NP, Driessen AJ (2005a) YidC – an evolutionary conserved device for the assembly of energy-transducing membrane protein complexes. Curr Opin Microbiol 8:182–187

van der Laan M, Chacinska A, Lind M, Perschil I, Sickmann A, Meyer HE, Guiard B, Meisinger C, Pfanner N, Rehling P (2005b) Pam17 is required for architecture and translocation activity of the mitochondrial protein import motor. Mol Cell Biol 25:7449–7458

van Dooren GG, Schwartzbach SD, Osafune T, McFadden GI (2001) Translocation of proteins across the multiple membranes of complex plastids. Biochim Biophys Acta 1541:34–53

van Meer G, Voelker DR, Feigenson GW (2008) Membrane lipids: where they are and how they behave. Nat Rev Mol Cell Biol 9:112–124

Vembar SS, Brodsky JL (2008) One step at a time: endoplasmic reticulum-associated degradation. Nat Rev Mol Cell Biol 9:944–957

Vogel H, Fischer S, Valentin K (1996) A model for the evolution of the plastid sec apparatus inferred from secY gene phylogeny. Plant Mol Biol 32:685–692

Walker LM, Sack FD (1990) Amyloplasts as possible statoliths in gravitropic protonemata of the moss *Ceratodon purpureus*. Planta 181:71–77

Waller RF, Jabbour C, Chan NC, Celik N, Likic VA, Mulhern TD, Lithgow T (2009) Evidence of a reduced and modified mitochondrial protein import apparatus in microsporidian mitosomes. Eukaryot Cell 8:19–26

Walther DM, Papic D, Bos MP, Tommassen J, Rapaport D (2009) Signals in bacterial beta-barrel proteins are functional in eukaryotic cells for targeting to and assembly in mitochondria. Proc Natl Acad Sci USA 106:2531–2536

Wang X, Lavrov DV (2007) Mitochondrial genome of the homoscleromorph *Oscarella carmela* (Porifera, Demospongiae) reveals unexpected complexity in the common ancestor of sponges and other animals. Mol Biol Evol 24:363–373

Webb CT, Gorman MA, Lazarou M, Ryan MT, Gulbis JM (2006) Crystal structure of the mitochondrial chaperone TIM9.10 reveals a six-bladed alpha-propeller. Mol Cell 21:123–133

Wu LF, Ize B, Chanal A, Quentin Y, Fichant G (2000) Bacterial twin-arginine signal peptide-dependent protein translocation pathway: evolution and mechanism. J Mol Microbiol Biotechnol 2:179–189

Wunder T, Martin R, Löffelhardt W, Schleiff E, Steiner JM (2007) The invariant phenylalanine of precursor proteins discloses the importance of Omp85 for protein translocation into cyanelles. BMC Evol Biol 7:236

Xie K, Kiefer D, Nagler G, Dalbey RE, Kuhn A (2006) Different regions of the nonconserved large periplasmic domain of *Escherichia coli* YidC are involved in the SecF interaction and membrane insertase activity. Biochemistry 45:13401–13408

Xu X, Colombini M (1996) Self-catalyzed insertion of proteins into phospholipid membranes. J Biol Chem 271:23675–23682

Xu L, Massague J (2004) Nucleocytoplasmic shuttling of signal transducers. Nat Rev Mol Cell Biol 5:209–219

Yaffe MP, Jensen RE, Guido EC (1989) The major 45-kDa protein of the yeast mitochondrial outer membrane is not essential for cell growth or mitochondrial function. J Biol Chem 264:21091–21096

Yen MR, Tseng YH, Nguyen EH, Wu LF, Saier MH Jr (2002) Sequence and phylogenetic analyses of the twin-arginine targeting (Tat) protein export system. Arch Microbiol 177:441–450

Young JC, Obermann WMJ, Hartl FU (1998) Specific binding of tetratricopeptide repeat proteins to the C-terminal 12-kDa domain of Hsp90. J Biol Chem 273:18007–18010

Young JC, Hoogenraad NJ, Hartl FU (2003) Molecular chaperones Hsp90 and Hsp70 deliver preproteins to the mitochondrial import receptor Tom70. Cell 112:41–50

Young JC, Agashe VR, Siegers K, Hartl FU (2004) Pathways of chaperone-mediated protein folding in the cytosol. Nat Rev Mol Cell Biol 5:781–791

Yusa F, Steiner JM, Löffelhardt W (2008) Evolutionary conservation of dual Sec translocases in the cyanelles of *Cyanophora paradoxa*. BMC Evol Biol 8:304

Zhang XP, Glaser E (2002) Interaction of plant mitochondrial and chloroplast signal peptides with the Hsp70 molecular chaperone. Trends Plant Sci 7:14–21

Chapter 9
Subcellular and Sub-organellar Proteomics as a Complementary Tool to Study the Evolution of the Plastid Proteome

Marcel Kuntz and Norbert Rolland

9.1 Origin and Main Functions of Plastids

Evolutionary studies have indicated that eukaryotes have arisen from an endosymbiotic association between an α-proteobacterium-like organism (i.e., the ancestor of mitochondria) and a host cell. Plant cells are thought to have acquired an additional cyanobacterium-like endosymbiont (the ancestor of plastids) more than 1.6 billion years ago (Yoon et al. 2004; Reyes-Prieto et al. 2007; Bogorad 2008). Association of these three independent organisms (the host cell, the plastid, and the mitochondria ancestors) in an integrated and highly regulated eukaryotic system is not only a fascinating and esthetic achievement of Evolution but was also vital for the appearance, functioning, and adaptation of most terrestrial ecosystems.

Plastids fulfill a number of essential functions, including photosynthesis, assimilation of nitrogen and sulfur, synthesis of amino acids, fatty acids, and many secondary metabolites (Weber et al. 2005; Block et al. 2007; Joyard et al. 2009, 2010) and, as a byproduct of the photosynthesis process, produce molecular oxygen. Plastids communicate and coordinate their various functions with other cell compartments: many plastid-localized biochemical pathways rely on metabolites from the cytosol, and vice versa many cytoplasmic and nuclear functions depend on the supply of molecules produced in the plastids (carbohydrates, fatty acids, specific lipids, nucleotides, alkaloids, isoprenoids, hormone precursors, chlorophyll degradation products, amino acids, vitamins, etc.) (Lunn 2007; Joyard et al. 2009, 2010; Linka and Weber 2010).

When considering the study of organelle evolution, one of the central questions is the techniques that are required and/or currently used to address this question. This chapter will make clear why proteomic studies are relevant approaches and

M. Kuntz • N. Rolland (✉)
Laboratoire de Physiologie Cellulaire & Végétale, CNRS/Université Joseph Fourier/INRA/CEA
Grenoble, 17 rue des Martyrs, F-38000 Grenoble, France
e-mail: norbert.rolland@cea.fr

C.E. Bullerwell (ed.), *Organelle Genetics*,
DOI 10.1007/978-3-642-22380-8_9, © Springer-Verlag Berlin Heidelberg 2012

why subcellular proteomics is essential to provide an in-depth evaluation of the plastid proteome and will provide an inventory of already performed plastid-targeted proteomic analyses. Due to strong bias in proteomics data available for plastid proteomes (up to now, only targeted to higher plants and green algae), we do not aim, in this review, to characterize the entire plastid proteome, but we highlight some of the current data, generated by plastids proteomics, in terms of functions, compartmentation, and evolution of this organelle.

9.2 Why Do We Need Subcellular Proteomic Techniques to Analyze the Evolution of the Plastid Proteome?

9.2.1 The Predicted Plastid Proteome Varies in Size and Composition

As stated above, plastids derive from a cyanobacterium-like endosymbiont that was engulfed by a host cell. During evolution, these plastids have conserved some traits of their endosymbiotic origin. Contemporary plastids from algae and plant cells still contain their own genome. However, this genome (~100 genes) now only codes for a small percent of the protein content of the free-living cyanobacterial ancestor, indicating that many genes have been lost from the ancient plastid genomes or transferred to the nuclear genome of the host cell. It is thus impossible to infer an accurate picture of the chloroplast proteome simply from the plastid genome (Martin et al. 2002). The vast majority of the plastid proteins from algae or higher plants are now nucleus encoded and must be imported to the plastid after translation in the cytosol (Jarvis 2008). When the total genome sequence of *Arabidopsis* became available (The Arabidopsis Genome Initiative 2000), some groups tried to estimate the size of the plastid proteome using specific algorithms that were designed to predict the plastidial localization of these nuclear-encoded proteins (Emanuelsson et al. 1999, 2000; Emanuelsson and von Heijne 2001). The first estimation was that, of approximately 25,000 nuclear genes, the plastid proteome of *Arabidopsis* was containing approximately 2,000–2,500 proteins. Surprisingly, only a relatively small part (30%) of these proteins revealed an unambiguous cyanobacterial origin (based on sequence similarity), suggesting that the contemporary Arabidopsis plastid has recruited proteins originating from the nuclear genome of the host cell (Abdallah et al. 2000). Another study pointed out that approximately 4,500 *Arabidopsis* proteins (around 20% of the nuclear-encoded *Arabidopsis* proteins) were acquired from the cyanobacterial ancestor of plastids and transferred to the nuclear genome during the course of eukaryotic genome evolution (Martin et al. 2002). However, this study also concluded that most of these proteins were predicted to be targeted to cell compartments other than the plastid. In other words, these two above-cited studies suggested that one should not extrapolate predictions or data obtained for the cyanobacterial ancestor to predict the proteome of the higher plant plastids.

This limit in extrapolating prediction was also exemplified when the size and composition of the *Arabidopsis* (dicot) plastid proteome was compared to that of a monocot species, rice. This analysis, based on the use of several independent plastid localization predictors, indicated that the estimated size of the plastid proteome differs markedly between rice (4,800 proteins) and *Arabidopsis* (2,100 proteins) (Richly and Leister 2004). It was thus questioned whether such a variability only results from the difference in nuclear gene number in rice (64,582) when compared to *Arabidopsis* (26,445). Again, search for proteins of cyanobacterial origin within these two predicted plastid proteomes showed the existence of both conserved nucleus-encoded plastid proteins that are predominantly of prokaryotic origin and a large fraction of species-specific (62% in Arabidopsis and 79% in rice) plastid-targeted proteins. Altogether, these data suggested that the degree of chloroplast specialization differs between the two species and that the chloroplast proteome diversity is generated both by gene evolution leading to novel proteins located in the chloroplast and by relocation of conserved gene products due to altered targeting (Richly and Leister 2004). More recently, the level of cyanobacterial gene recruitment detected in *Arabidopsis* was compared with that predicted for the glaucophyte alga *Cyanophora paradoxa* (Reyes-Prieto et al. 2006). This study showed that while only 11% of the algal nuclear genes are of cyanobacterial origin, only a low percentage (11.1%) of the algal nuclear genes might have nonplastid functions. This study indicates that early diverging algal groups have transferred a smaller number of cyanobacterium-like endosymbiont genes to their nucleus as compared to higher plants, but the vast majority of these genes were further recruited for plastid functions (Reyes-Prieto et al. 2006).

To conclude, the plastid proteome is highly variable in size and content over evolutionary time. One should thus not extrapolate predictions or data obtained for the cyanobacterial ancestor to the plastids of algae or higher plants, from one plant to another and even, as discussed below, from one plant tissue to another one.

9.2.2 The Tools Used to Predict Plastid Localization Vary in Efficiency

Because of its procaryotic type genome and of its low compartmentation, the prediction of the cyanobacterial proteome using genome analysis remains relatively simple. This is not true when prediction of subcellular localization is to be performed in highly compartmentalized eukaryotic cells.

In higher plants, TargetP and ChloroP tools (Emanuelsson et al. 1999, 2007) are expected to localize correctly the majority of the plastid proteins, showing relatively low discrepancies between experimental and prediction tool localizations. However, around 30% of the experimentally localized proteins detected in the

Arabidopsis chloroplast have no ChloroP-predictable transit peptides (Ferro et al. 2010). Some of these proteins are encoded by the chloroplast genome (4% of the total proteome) and can be identified without difficulty. Other proteins reside on the outer envelope of the plastid and most of these proteins do not bear a predictable chloroplast transit peptide (Jarvis 2008). In several other cases, the ChloroP prediction can also be wrong. For instance, ChloroP does not predict any transit peptide for the phosphate/triose-phosphate translocator (Ferro et al. 2010), a major envelope protein known to contain a genuine plastid transit peptide that is lost during import to the inner envelope membrane of the chloroplast. Furthermore, some gene structures are not properly predicted and this also leads to erroneous ChloroP predictions for the actual targeting of the deduced protein sequence. Finally, a number of plastid proteins have been reported not to follow a canonical import pathway (Miras et al. 2002, 2007; Nada and Soll 2004; Villarejo et al. 2005; Nanjo et al. 2006; Kitajima et al. 2009; Armbruster ct al. 2009), and these proteins were demonstrated not to contain a classical and cleavable (and thus predictable, using the existing tools) plastid transit peptide. Based on all these observations, the evaluation of a plastid proteome, on the basis of software-based protein localization prediction, yields partial and unreliable data. These arguments strongly suggest that subcellular proteomics is required to provide experimental information about subcellular localization of plastid proteins to define further the rules for protein import into organelles (Baginsky and Gruissen 2004).

Our current knowledge of the experimentally localized plastid proteins comes almost exclusively from studies of plants. However, more recent organelle-targeted proteomic studies have been performed on green algae; the main model being *Chlamydomonas reinhardtii* (for a review, see Rolland et al. 2009). While relatively efficient for higher plants, prediction tools such as TargetP and ChloroP (Emanuelsson et al. 1999, 2007) are not optimized for the protein sequences of *Chlamydomonas*, leading to incorrect subcellular localization for approximately 50% of the experimentally localized chloroplast (Terashima et al. 2010) or mitochondrial (Atteia et al. 2009) proteins.

The plastid proteome in the higher plants or in *Chlamydomonas* has a relatively simple origin via integration of proteins from a single cyanobacterial primary endosymbiont and the host. In these phototrophic organisms harboring primary plastids, most of these proteins are targeted back to the plastid by a transit peptide that directs the proteins across the double membrane (Jarvis 2008). However, such primary plastids (surrounded by a two-membrane system) were subsequently spread to other protist lineages through a series of eukaryote–eukaryote secondary and tertiary endosymbiotic events, resulting in a large diversity of photosynthetic lineages (Keeling 2004; Archibald 2009). Both red and green algae have been involved in secondary endosymbioses. In these algae containing secondary and tertiary plastids, proteins thus need to have the N-terminal bipartite presequences that code for a signal peptide that directs the protein to the host endomembrane system and a plastid transit peptide to lead the protein to the plastid and to be transported across three to four membrane layers surrounding the organelle (McFadden 1999). This feature makes even more difficult the prediction of the plastid proteome within these organisms.

Analyses performed to identify some plastid-targeted sequences are limited to proteins that are identified according to their phylogenetic relationship to other plastid homologs, participation in plastid-located processes, and/or possession of plastid-targeted sequence comprising a signal peptide and a transit peptide (e.g., Nosenko et al. 2010; Minge et al. 2010; Suzuki and Miyagishima 2010). To date, virtually nothing is known about the proteomics-based plastid proteome in these taxa.

9.2.3 The Size and Content of Plastid Proteomes Are Variable in Various Plant Tissues or in Various Environmental Conditions: One Organelle with Many Functional Variations

In higher plants, plastids are in fact a diverse group of organelles sometimes interconvertible during development (Inaba and Ito-Inaba 2010). Proplastids are undifferentiated plastids, present in developing meristems. Etioplasts are formed in dark-grown seedlings. Chloroplasts carry out the light-driven carbon fixation in the green tissues. Chromoplasts accumulate carotenoids in some flowers and fruits, usually with the concomitant loss of photosynthetic capacity and chlorophyll. Amyloplasts are nonpigmented organelles which contain large amounts of starch and play roles in gravitropism and storage of reduced carbon and are found in roots, seeds, tuber, and some fruits.

The proteome of different plastid types has been described, although not to the same extent as chloroplasts, one reason being that extraction of different plastid types with a relatively good level of purity and integrity is difficult from *Arabidopsis*, or impossible (*Arabidopsis* does not form chromoplasts). Other plant species are more adapted to the purification of other plastid types but sometimes genome sequence information or ESTs are lacking for these species, thus strongly limiting large-scale proteomic analyses. Consequently, recent approaches, which combine large-scale sequencing of the transcriptome of the plant species with proteomics analyses, have emerged.

9.3 Which Types of Plastid Proteomes Have Been Analyzed, and Where Can We Find These Data?

9.3.1 The Chloroplast

Chloroplasts, the most studied plastid type, are distributed throughout the cytoplasm of leaf cells and range from about 4–10 μm in size. Prediction of the chloroplast proteome has been the subject of many debates (van Wijk 2004; Sun et al. 2004; Baginsky and Gruissen 2004). The most recent analyses however converge and, *ca.* 3000 proteins are estimated to be required to build a fully functional chloroplast proteome (Jarvis 2008). One of the first massive proteomics-based studies targeted to the whole *Arabidopsis* chloroplast proteome came from the group of Baginsky and coworkers (Kleffmann et al 2004). This work allowed the identification of almost 700 plastid proteins with near-complete protein coverage for key chloroplast pathways, such as carbon fixation and photosynthesis. These data were completed by huge efforts performed by the same group through the proteome analysis from various plastid types (see below). A specific database (plprot) was created that combines not only proteome information of various plastids but also data issued from plastid proteome analyses from other laboratories. This plprot database is accessible at http://www.plprot.ethz.ch (Kleffmann et al. 2006). A more recent large-scale analysis of the purified chloroplasts from *Arabidopsis* leaves allowed identification of 1,325 proteins (Zybailov et al. 2008). Out of these, more than 900 proteins could be unambiguously assigned to the chloroplast, thanks to manual annotation. With this huge amount of data, the PPDB database (http://ppdb.tc.cornell.edu) was generated in which all mass-spectrometry data are projected on identified gene models (Sun et al. 2009). Finally, another repertoire of more than 1,300 chloroplast proteins (Ferro et al. 2010) was recently performed. These analyses allowed building up the AT_CHLORO database (http://www.grenoble.prabi.fr/at_chloro/). During this work, by focusing, in the same set of experiments, on the proteins from the stroma, the thylakoids, and the envelope membranes, the partitioning of each protein in these three chloroplast compartments could then be assessed by using a semiquantitative proteomics approach (Ferro et al. 2010). This analysis was further used to revisit the sub-plastidial compartmentation of the chloroplast metabolisms and its main functions (Joyard et al. 2009, 2010). While the green algae *Chlamydomonas reinhardtii* has been the subject of a variety of proteomic analyses (Rolland et al. 2009), the composition of the chloroplast proteome from *Chlamydomonas* was only recently analyzed. Of a total of 2,315 identified proteins, quantitative analyses based on spectral counting clearly localized 606 of these proteins to the chloroplast, including several new proteins of unknown function induced under anaerobic conditions (Terashima et al. 2010). It is important to note that the above-cited studies experimentally identified 30–50% of genuine plastid proteins that do not contain classical targeting signals, and thus, proteins that would not have been predicted to be targeted to plastids through bioinformatics-based analyses. Very few data are available for plastids deriving from other taxa. Only very recent data targeted to the composition of photosystem I-associated antenna were obtained from the red alga *Cyanidioschyzon merolae* (Busch et al. 2010).

9.3.2 The Amyloplast

The first amyloplast proteome was obtained from wheat (Andon et al. 2002), with relatively few identified proteins. That less than 50% of these proteins could be assigned a function suggested that the proteomes of heterotrophic and autotrophic plastids differ considerably, which is especially apparent for proteins involved in energy metabolism. The main functions of amyloplasts were found to be mostly (85%) restricted to the protein destination/storage, energy metabolism, and unknown categories. Two other studies, also targeted to amyloplasts from developing wheat endosperm, recently revisited the metabolic properties of the amyloplasts (Balmer et al. 2006; Dupont 2008). Out of approximately 200 identified plastid proteins, one-third could be classified within the destination/storage, energy metabolism, and unknown categories and more than half of these proteins were known or predicted to be involved in metabolism and response to stress. In particular, enzymes of amino acid, nucleic acid, and sulfur metabolism were identified, demonstrating the versatility of amyloplasts. Interestingly, in the context of this review, these data were organized into proposed metabolic and biosynthetic pathways illustrated with names of enzymes and compounds (see, Dupont 2008).

9.3.3 The Undifferentiated Plastid

The tobacco BY2 cell culture was originally used as a model for the purification of undifferentiated heterotrophic plastids (Baginsky et al. 2004). Interestingly, in the present context, many of the 160 identified proteins were absent from chloroplast proteomes, suggesting that undifferentiated plastids also contain specific functions. Comparison of the proteome of these undifferentiated plastids from tobacco with data obtained for the differentiated heterotrophic amyloplasts (Andon et al. 2002; see above) was also performed and provided information on prevalent metabolic activities of different plastid types. More recently, plastids from liquid callus cultures of *Arabidopsis* (Dunkley et al. 2004, 2006) were also analyzed to provide complementary information on the proteome of poorly differentiated heterotrophic plastids. However, these studies identified relatively few plastid proteins that could not be discriminated from mitochondrial proteins (for a review, see Sadowski et al. 2008).

9.3.4 The Embryoplast

Protoplasts from *Brassica napus* were identified as a source to isolate plastids (termed embryoplasts) from developing embryos (Jain et al. 2008). Proteomic analysis of these preparations identified 80 proteins, most of them (70%) being plastid encoded, known plastid proteins, or predicted to be plastid-localized by at least one of the tools predicting plastid localization. Surprisingly, more than 50% of these proteins were related to the light reactions of photosynthesis, strongly suggesting that these plastids are more closely related to chloroplasts than are leucoplasts or amyloplasts.

9.3.5 The Etioplast

Etioplasts are chloroplasts that have not been exposed to light and can be technically considered as leucoplasts. Etioplasts contain prolamellar bodies, which are membrane aggregations of semi-crystalline lattices of branched tubules arranged in geometric patterns. Rice etioplasts purified from dark-grown leaves were first used to analyze the proteome of this type of plastids (von Zychlinski et al. 2005). This study identified 240 unique proteins (including some previously unknown plastid proteins), providing new insights into heterotrophic plastid metabolism. Novel etioplast-specific proteins could also be identified by comparing these data with proteomes of *Arabidopsis* chloroplasts and plastids isolated from BY2 cells. In another study, the light-induced proteome dynamics was analyzed to study the development of rice chloroplasts from etioplasts (Kleffmann et al. 2007). The study revealed that the main proportion of total protein mass in etioplasts corresponded to carbohydrate and amino acid metabolisms or to the control of plastid genome expression. Chaperones, proteins for photosynthetic energy metabolism, and enzymes of the tetrapyrrole pathway were identified among the most abundant etioplast proteins (Kleffmann et al. 2006). Differential accumulation of nuclear-encoded or plastid-encoded proteins was investigated to know the presence of these proteins as a function of time after de-etiolation. Interestingly, the ATPase, Clp, and FtsH protease complexes and proteins responsible for defense against oxidative stress were found to be highly abundant in etioplasts, suggesting their major role in biogenesis and functioning of etioplasts (Kanervo et al. 2008).

9.3.6 The Chromoplast

The first chromoplast proteome was obtained from the bell pepper fruit (Siddique et al. 2006). These species offer the possibility of recovering large amounts of chromoplasts with a good level of purity. A total of 151 proteins, including never-before-reported proteins from previous plastid-targeted proteome studies, were identified. As expected, among the most abundant chromoplast proteins were enzymes involved in the synthesis and storage of carotenoids that strongly accumulate in pepper chromoplasts (Deruere et al. 1994).

A tomato-fruit chromoplast proteome has also been obtained (Barsan et al. 2010), revealing the presence of 988 proteins, among which 209 had not been listed at the time of publication in plastid databanks. Since chromoplasts lack chlorophyll, it was consistent that they also lack enzymes involved in chlorophyll biosynthesis and contained enzymes involved in chlorophyll degradation. It was more surprising, although not fully unexpected, to find a number of proteins involved in the PSI and PSII photosystems, corresponding to 22% and 39% of the PSI and PSII proteins of the *Arabidopsis* chloroplast, respectively. It is probable that these polypeptides are nonfunctional proteins stored (either as full size or segments) after the disintegration of photosynthesis complexes in the chromoplasts. More surprisingly, chromoplasts contain the entire set of Calvin-cycle proteins, including Rubisco. Since this cycle cannot be functional in nonphotosynthetic plastids as it is in chloroplasts, these polypeptides may be remaining forms (not yet proteolized), like the above-mentioned thylakoid polypeptides. Another possibility could be that they serve to adjust the content of various carbohydrates for diverse metabolic branches. This possibility could be corroborated by observations of the activities of some of these enzymes in isolated chromoplasts (see references in Egea et al. 2010). Unlike the Calvin cycle, the oxidative pentose phosphate pathway (OxPPP) is known to be functional in chromoplasts in order to produce reducing power either from imported sugars or, at an early stage of chromoplast differentiation, from starch degradation. It was therefore consistent to find the OxPPP enzymes in chromoplasts. Proteins of lipid metabolism and trafficking were also represented, which is consistent with the fact that new membranes are visible in chromoplasts, possibly involved in the synthesis of carotenoids as well as other hydrophobic compounds. The chromoplast proteome also contains all the enzymes of the lipoxygenase pathway required for the synthesis of lipid-derived aroma volatiles. Proteins involved in starch synthesis coexisted with several starch-degrading proteins and starch excess proteins. This chromoplast proteomic analysis suggests that chromoplasts are not subjected to a massive and rapid proteolytic process. However, synthesis of potentially toxic compounds such as chlorophylls has been suppressed. From an evolutionary point of view, one can consider chromoplasts as the most recently evolved plastid form, as suggested by their absence in a number of species and their presence under various forms in others. Evolution may have selected chromoplast differentiation in some flowers and fruit in order to attract animals that will cross-pollinate flowers and help disseminate seeds. The surprisingly similarity (at least qualitatively) between chloroplast and chromoplast proteomes may reflect the recent selection of chromoplast differentiation from chloroplasts in some species.

9.4   Sub-plastidial Proteomes for In-Depth Analysis of Plastid Functions

Chloroplasts contain several key subcompartments including the chloroplast envelope (a double membrane with an intermembrane space between, that surrounds the organelle and plays central roles in sustaining the communication of the chloroplast with the plant cell), the stroma (mainly composed of soluble proteins), and the thylakoid membrane (which is a highly organized internal membrane network formed of flat compressed vesicles and is the center of oxygenic photosynthesis). The thylakoid vesicles delimit another discrete soluble compartment, the thylakoid lumen. Some plastoglobules (plastid-produced lipid bodies) are suggested to originate as protuberances on the thylakoid membrane or from the inner chloroplast envelope membrane. The two limiting envelope membranes (inner and outer membrane of the plastid envelope) are the only consistent membrane structure of the different types of plastids (Joyard et al. 1998). Organelle purification and subfractionation is essential for cataloging proteomes (van Wijk 2004; Baginsky and Gruissen 2004; Rossignol et al. 2006). Furthermore, due to the limits resulting from dissimilar physicochemical properties of soluble (stroma or thylakoid lumen) or membrane (envelope or thylakoid membranes) proteins (Sun et al. 2004), different compartments of the chloroplast were investigated using a broad range of purification and solubilization techniques. Proteomic studies have also been devoted to the study of the sub-plastidial organization of these functions. The sub-plastidial arrangement in various membrane or soluble compartment is not only a remnant of the bacterial ancestor but is of primary importance for the integrated metabolisms of plastids. This can be exemplified by the plastid envelope which delimitates the plastid compartment but is also the location of a number of vital metabolisms (Block et al. 2007).

9.4.1 The Plastid Envelope Membranes

Due to the low relative abundance of chloroplast envelope proteins (less than 1% of chloroplast proteins) when compared to other plastid compartments, the envelope fraction remained poorly characterized until the availability of *Arabidopsis* genome information (The AGI 2000) and the development of proteomics-based approaches targeted to this membrane system. Transcript levels were also relatively low and corresponding ESTs for many envelope proteins were also missing from databases. A decade ago, very few enzymes involved in specific metabolisms, few transporters or ion channels, and some members of the Toc and Tic translocons involved in the plastid targeting of nuclear-encoded chloroplast proteins were known (Joyard et al. 1998; Block et al. 2007). One of the first efforts to analyze the composition of the chloroplast envelope from spinach chloroplasts was based on the use of organic solvents to obtain a specific enrichment of intrinsic proteins from the hydrophobic core of the membrane (Seigneurin-Berny et al. 1999; Ferro et al. 2000, 2002). These first studies identified 54 proteins within purified envelope fractions. Interestingly, most of these envelope proteins were highly hydrophobic, and previously unknown. Many envelope components were known or predicted ion or metabolite transporters. Multiple approaches toward identification of a more exhaustive list of experimentally determined envelope proteins were used on the chloroplast envelope from *Arabidopsis* (Ferro et al. 2003; Froehlich et al. 2003), which identified more than 100 and 350 proteins, respectively (Table 9.1). A deeper analysis revealed that the vast majority of these proteins were involved in ion and metabolite transport, components of the protein import machinery, involved in chloroplast lipid metabolism, and soluble proteins such as proteases and proteins involved in carbon metabolism or in responses to oxidative stress. Almost one-third of the newly identified proteins had no known function (Rolland et al. 2003). Another study targeted the outer envelope membrane of pea chloroplasts (Schleiff et al. 2003). This study combined the selection of β-barrel proteins from the complete *Arabidopsis* genome (Table 9.1) with protein identification from highly purified outer envelope membranes of pea chloroplasts. In addition to already known envelope components, four new proteins of the outer membrane of the chloroplast envelope were identified (Schleiff et al. 2003). As mentioned above, proteomics analysis of the chloroplast envelope is limited by low amounts of the envelope proteins compared to stroma and thylakoid membranes. Use of the model plant, *Arabidopsis*, introduces additional technical problems that limit yield, particularly compared to pea or spinach. These nonmodel plants are easily available throughout the year and remain models of choice for large-scale preparation of pure, high-quality, intact chloroplasts and consequently, larger amounts of envelope membranes as compared to *Arabidopsis*. Bräutigam and coworkers (2008b) explored the potential of a preliminary cDNA database for the nonmodel plant pea created by a small number of massively parallel pyrosequencing runs, for use in proteomics. A pea chloroplast envelope membrane proteome sample was thus analyzed using the species-specific database generated by pyrosequencing. A total of 255 nonredundant proteins were identified using a combination of pea, *Arabidopsis*, or *Medicago* databases (Bräutigam et al. 2008b). Another study used comparative proteomics of chloroplast envelopes extracted from pea (C3 plant) and chloroplast envelopes extracted from the mesophyll cell of the C4 plant maize (Bräutigam et al. 2008a). The aim of this work was to profile quantitative and/or qualitative changes within the chloroplast envelope during adaptation of the mesophyll cell to the requirements of C4 photosynthesis. Interestingly, the envelope membranes from both types of chloroplasts contain many orthologous proteins, but the levels of some of these proteins are very different between the two systems. In particular, several putative transport proteins that are highly abundant in C4 envelopes, but relatively minor in C3 envelopes are, therefore, candidates for the transport of C4 photosynthetic intermediates such as pyruvate, oxaloacetate, and malate (Bräutigam et al. 2008a). These data are complemented by a study of specific differences in the chloroplast membrane proteomes of maize bundle sheath (BS) and mesophyll cells (Majeran et al. 2008). As well as determining various adaptations of photosynthetic functions or metabolic machineries, the study also determined functional differentiation of envelope transporters (Majeran et al. 2008). More recently, a comparison of proplastid and chloroplast envelope proteomes and the corresponding transcriptomes of leaves and shoot apex was performed, which allowed revealing a clearly distinct composition of the proplastid envelope, especially when considering the small molecule and protein transport across proplastid envelope membranes (Bräutigam and Weber 2009). The identification and accurate localization of chloroplast envelope proteins was also recently revisited in *Arabidopsis*. Using a large-scale and semiquantitative proteomics

Table 9.1 Proteomic studies aiming to identify the proteome of plastids and their sub-plastidial compartments from higher plants and green algae

Tissue or cell	Targeted organelle or compartment	Plant or alga	Number of proteins	References
Whole plastid proteomes				
Plastids from higher plants				
Leaves	Chloroplast	*A. thaliana*	690	Kleffmann et al. (2004)
Leaves	Chloroplast	*A. thaliana*	1325	Zybailov et al. (2008)
Leaves	Chloroplast	*A. thaliana*	1323	Ferro et al. (2010)
Developing endosperm	Amyloplast and purified amyloplast membranes	*Triticum aestivum*	171	Andon et al. (2002)
Developing endosperm	Amyloplast	*T. aestivum*	289	Balmer et al. (2006) and Dupont (2008)
Cell culture	Undifferentiated heterotrophic plastid from BY-2 cells	*Nicotiana tabacum*	140	Baginsky et al. (2004)
Dark-grown leaves	Etioplast	*Oryza sativa*	240	von Zychlinski et al. (2005)
Dark-grown leaves	Etioplast inner membranes	*T. aestivum*	21	Blomqvist et al. (2006)
Dark-grown seedlings	Etioplast to chloroplast transition	*O. sativa*	369	Kleffmann et al. (2007)
Etiolated seedlings	Etioplast to chloroplast transition	*P. sativum*	16	Kanervo et al. (2008)
Bell pepper red fruits	Chromoplast	*Capsicum annuum*	151	Siddique et al. (2006)
Tomato fruit	Chromoplast	*Solanum lycopersicum*	988	Barsan et al. (2010)
Developing embryos	Embryoplast (starting from protoplasts)	*Brassica napus*	80	Jain et al. (2008)
Chloroplast from green algae				
Cells grown in aerobic versus anaerobic conditions	Chloroplast	*Chlamydomonas reinhardtii*	606	Terashima et al. (2010)
Proteome of sub-plastidial fractions				
Envelope membranes				
Leaves	Hydrophobic core of the envelope membranes	*Spinacia oleracea*	54	Ferro et al. (2002)
Leaves	Outer envelope membrane	*Pisum sativum*	16	Schleiff et al. (2003)
Leaves	Hydrophobic core of the envelope membranes	*Arabidopsis thaliana*	106	Ferro et al. (2003)

(continued)

Table 9.1 (continued)

Tissue or cell	Targeted organelle or compartment	Plant or alga	Number of proteins	References
Leaves	Whole envelope membranes	*A. thaliana*	350	Froehlich et al. (2003)
Leaves	Whole envelope membranes	*P. sativum*	255	Bräutigam et al. (2008b)
Leaves	Whole envelope membranes (C3 and C4 plants)	*P. sativum, Zea mays*	420	Bräutigam et al. (2008a)
Leaves	Mixed thylakoid and envelope membranes (C3 and C4 cells)	*Z. mays*	610	Majeran et al. (2008)
Leaves	Whole envelope membranes	*A. thaliana*	460	Ferro et al. (2010)
Stroma				
Leaves	Chloroplast stroma (C3 and C4 cells)	*Z. mays*	400	Majeran et al. (2005)
Leaves	Chloroplast stroma	*A. thaliana*	241	Peltier et al. (2006)
Leaves	Chloroplast stroma	*A. thaliana*	550	Zybailov et al. (2008)
Thylakoid membranes and lumen				
Leaves	Thylakoid lumen	*P. sativum*	60	Peltier et al. (2000)
Leaves	Thylakoid lumen	*A. thaliana*	81	Peltier et al. (2002)
Leaves	Thylakoid lumen	*A. thaliana*	36	Schubert et al. (2002)
Leaves	Thylakoid lumen	*S. oleracea*	22	Schubert et al. (2002)
Leaves	Thylakoid membrane	*A. thaliana*	154	Friso et al. (2004)
Leaves	Mixed thylakoid and envelope membranes (C3 and C4 cells)	*Z. mays*	610	Majeran et al. (2008)
Plastoglobules				
Leaves	Plastoglobules	*A. thaliana*	25	Vidi et al. (2006)
Leaves	Plastoglobules	*A. thaliana*	32	Ytterberg et al. (2006)

approach (spectral count), together with an in-depth investigation of the literature, the envelope localization could be assessed for 300 proteins exclusively detected in the chloroplast envelope and 460 proteins when considering proteins enriched in the envelope fraction and also shared with another chloroplast subcompartment (Ferro et al. 2010). All these data provide evidence that envelope membranes are indeed one of the most complex and dynamic systems within the plant cell.

9.4.2   The Stroma

Relatively few studies were performed on the stroma with the specific aim of characterizing the stromal proteome or identifying proteome dynamics (for reviews, see Baginsky and Gruissen 2004; van Wijk 2004). Most of the available data on stromal components were derived from targeted biochemical and molecular approaches and from a global knowledge of the compartmentation of the cell metabolism, whereas envelope or thylakoid membranes were targeted in various proteomics studies (Lunn 2007). BS and mesophyll cells of maize leaves were chosen to perform a quantitative comparative proteome analysis targeted on the chloroplast stroma (Majeran et al. 2005). The aim of this study was to expend our knowledge on the plastid functions that are affected in the stroma to accommodate C4 photosynthesis. Given the complexity of the stromal proteome, only a small number of stromal protein complexes in *Arabidopsis* had been characterized. Using highly purified chloroplasts extracted from *Arabidopsis* leaves, 241 proteins were identified from the stroma, representing about 99% of the stromal protein mass (Peltier et al. 2006). The analysis covered most known chloroplast functions, ranging from protein biogenesis and protein fate to primary and secondary metabolism, and a number of new components were identified. The stroma proteome of *Arabidopsis* was more recently revisited, resulting in the identification of 550 stromal proteins (Zybailov et al. 2008). A qualitative and quantitative proteomic analysis was also performed to examine changes in the stroma and lumen proteomes of *Arabidopsis* leaves during cold shock as well as short- and long-term cold acclimation (Goulas et al. 2006). This study identified 43 differentially expressed proteins, providing new insights into the cold response and acclimation of *Arabidopsis*.

9.4.3   The Thylakoid Lumen

The initial study on the thylakoid lumen was on spinach and pea (Kieselbach et al. 1998, 2000; Peltier et al. 2000). To fully utilize the benefit of the *Arabidopsis* genome sequence, another proteomics study was then performed on its luminal and peripheral thylakoid proteome (Peltier et al. 2002). A total of 81 proteins were identified using MS/MS. Detailed analysis of known or predicted proteins revealed that the main functions of the thylakoid luminal proteome are to support protein folding and proteolysis of thylakoid proteins and to protect against oxidative stress (Peltier et al. 2002). The very same year, Schubert and coworkers independently reported the thylakoid luminal proteome again in *Arabidopsis* (Schubert et al. 2002). Although only 36 proteins were identified, a comparison was made with the identified 22 spinach thylakoid lumen proteins. Based on these independent experimental and in silico analyses, the entire luminal proteome of *Arabidopsis* was estimated to comprise ~80 proteins. Only one differential proteomics study was used to investigate the thylakoid lumen to reveal the presence of new lumen proteins (Goulas et al. 2006). When combined, the above-cited studies yielded more than 100 proteins (Kieselbach and Schröder 2003; van Wijk 2004). Interestingly, these studies have shown that chloroplast lumen proteins play an important role for the regulation of photosynthesis but are not restricted to the generation of the pH gradient that fuels ATP synthesis. However, many of the predicted luminal proteins were found to be present at concentrations at least 10,000-fold lower than proteins of the photosynthetic apparatus (Peltier et al. 2002). It is thus expected that previously unidentified/undetectable luminal proteins could be recovered during more recent studies targeted to the chloroplast (e.g., Zybailov et al. 2008).

9.4.4   The Thylakoid Membrane

Initial mass spectrometry-based studies of the thylakoid membrane proteins in spinach, pea, and *Arabidopsis* were essentially performed on antennae or reaction-center subunits to identify the composition of the photosynthetic complexes and their post-translational modifications (Whitelegge et al. 1998; Vener et al. 2001; Zolla et al. 2002, 2003; Gomez et al. 2002). Other than identifying the most abundant LHC proteins, these studies were very useful for comparison of the LHC proteins within a single plant or among different plant species. Then, a set of 58 nuclear-encoded thylakoid membrane proteins with experimentally assigned *N*-termini from four plant species was reported (Gomez et al. 2003). Information thus obtained was used to test, on thylakoid membrane proteins, the various existing tools predicting plastid localization and/or cleavage sites in experimentally identified transit peptides. SignalP was demonstrated to efficiently predict the cleavage site of soluble proteins targeted to the thylakoid lumen, thus suggesting that the mechanism for the targeting of thylakoid integral proteins inserted via spontaneous mechanism may be related to the secretory mechanism of Gram-negative bacteria (Gomez et al. 2003). The two first in-depth analyses of the thylakoid membrane were published in 2004, resulting in the identification of 154 and 240 proteins, respectively (Friso et al. 2004; Peltier et al. 2004). A recent study of thylakoid membrane dynamics, especially on environmentally modulated phosphoproteome was targeted to the photosynthetic membranes of *C. reinhardtii* (Turkina et al. 2006). The study revealed that major changes in phosphorylation are clustered at the interface between the PSII core and its LHCII antennae. These data also suggest that the controlling mechanisms for photosynthetic state transitions and LHCII uncoupling from PSII under high light stress allow thermal energy dissipation. Still relying on intact mass measurements of membrane proteins, all PSI and LHC proteins were analyzed in ten different plant species, and PSI proteins present within stroma lamellae of the thylakoid membrane were identified (Zolla et al 2007). Hippler and coworkers also investigated the impact of iron deficiency on protein dynamics linked to functional properties of respiratory and photosynthetic machineries in the green algae *C. reinhardtii* (Naumann et al. 2007). The

study used differential proteomics coupled to physiological measurements of respiration or photosynthesis efficiency to mainly reveal that iron-deprivation induces
a transition from photoheterotrophic to primarily heterotrophic metabolism. This
further suggests that a hierarchy exists within organelles of a single cell for iron
allocations, and that this hierarchy of iron allocation is closely linked to the
metabolic state of the cell. A more recent study providing information on the
thylakoid composition and dynamics was a comparative analysis of chloroplast
membrane proteomes in maize mesophyll and BS cells (Majeran et al. 2008). The
study complements previous data published by the same group on the stromal
compartment of chloroplast (Majeran et al. 2005) with an aim to understand how
plastids accommodate C4 photosynthesis. Hundreds of proteins were demonstrated
to differentially accumulate in membranes extracted from mesophyll and BS
plastids.

9.4.5 The Plastoglobules

Plastoglobules are plastid-produced lipid bodies containing galactolipids,
prenylquinones, and pigments; they are thought surrounded by a lipid monolayer
studded with proteins. Plastoglobules are suggested to originate as protuberances on
the thylakoid membrane or from the inner chloroplast envelope membrane. Recent
data have demonstrated that plastoglobules only form on the thylakoid membrane
and are surrounded by a half-bilayer membrane that is continuous with the thylakoid outer leaflet, and that they remain structurally coupled to thylakoids throughout their life span (Austin et al. 2006). Two recent studies have advanced our
knowledge on the plastoglobule proteins and their functions (Vidi et al. 2006;
Ytterberg et al. 2006). The identified proteins fall into three categories: plastoglobulins/
PAP/fibrillins, chloroplast and chromoplast metabolic proteins, and unclassified
proteins (for a review, see Brehelin et al. 2007).

9.5 Conclusion

To conclude, pure computation-based predictions are limited in predicting plastid
proteomes. The origin of protein sequences might be misleading in suggesting
plastid targeting since cyanobacterial proteins appear to have been recruited by
other cell compartments. Indeed, the size of the plastid proteome is variable over
evolutionary time and one should probably not extrapolate predictions from one
alga to another, one group of plants to another. Tools used to predict plastid
targeting of proteins also appear to be limited since 30–50% of experimentally
based plastid proteomes are not predicted to be targeted to plastids. Proteomics also
proved to be limited when the inventories of plastid proteomes are to be performed.
It is clear that the size and content of plastid proteome is variable in different tissues

(e.g., proplastids and chloroplasts) or in various environmental conditions (e.g., light and dark). It is important to note here that members of the *Arabidopsis* proteomics community involved in developing many of these proteomics resources (Joshi et al. 2011) decided to create a summary aggregation portal that is capable of retrieving proteomics data from a series of online resources including every plastid proteomics data (http://gator.masc-proteomics.org/). New DNA sequencing and mass spectrometry technologies, in combination with increasing amounts of genome sequence data from plants and algae, will open up experimental possibilities to identify a more complete set of plastid proteins as well as their expression levels from various species. Complementary with the prediction of the complete plastid proteomes through analysis of targeting signals, plastid proteomics is expected to provide many new insights into the evolution of plastid metabolism, biogenesis, adaptation, and functions.

References

Abdallah F, Salamini F, Leister D (2000) A prediction of the size and evolutionary origin of the proteome of chloroplasts of *Arabidopsis*. Trends Plant Sci 5:141–142

Andon N-L, Hollingworth S, Koller A, Greenland AJ, Yates J-R 3rd, Haynes P-A (2002) Proteomic characterization of wheat amyloplasts using identification of proteins by tandem mass spectrometry. Proteomics 2:1156–1168

Archibald JM (2009) The puzzle of plastid evolution. Curr Biol 19(2):R81–R88

Armbruster U, Hertle A, Makarenko E, Zühlke J, Pribil M, Dietzmann A, Schliebner I, Aseeva E, Fenino E, Scharfenberg M, Voigt C, Leister D (2009) Chloroplast proteins without cleavable transit peptides: rare exceptions or a major constituent of the chloroplast proteome? Mol Plant 2:1325–1335

Atteia A, Adrait A, Brugière S, Tardif M, van Lis R, Deusch O, Dagan T, Kuhn L, Gontero B, Martin W, Garin J, Joyard J, Rolland N (2009) A proteomic survey of *Chlamydomonas reinhardtii* mitochondria sheds new light on the metabolic plasticity of the organelle and on the nature of the alpha-proteobacterial mitochondrial ancestor. Mol Biol Evol 26:1533–1148

Austin JR II, Frost E, Vidi P-A, Kessler F, Staehelina LA (2006) Plastoglobules are lipoprotein subcompartments of the chloroplast that are permanently coupled to thylakoid membranes and contain biosynthetic enzyme. Plant Cell 18:1693–1703

Baginsky S, Gruissen W (2004) Chloroplast proteomics: potentials and challenges. J Exp Bot 55:1213–1220

Baginsky S, Siddique A, Gruissem W (2004) Proteome analysis of tobacco bright yellow-2 (BY-2) cell culture plastids as a model for undifferentiated heterotrophic plastids. J Proteome Res 3:1128–1137

Balmer Y, Vensel WH, DuPont FM, Buchanan BB, Hurkman WJ (2006) Proteome of amyloplasts isolated from developing wheat endosperm presents evidence of broad metabolic capability. J Exp Bot 57:1591–1602

Barsan C, Sanchez-Bel P, Rombaldi C, Egea I, Rossignol M, Kuntz M, Zouine M, Latché A, Bouzayen M, Pech J-C (2010) Characteristics of the tomato chromoplast revealed by proteomic analysis. J Exp Bot 61:2413–2431

Block MA, Douce R, Joyard J, Rolland N (2007) Chloroplast envelope membranes: a dynamic interface between plastids and the cytosol. Photosynth Res 92:225–244

Bogorad L (2008) Evolution of early eukaryotic cells: genomes, proteomes, and compartments. Photosynth Res 95:11–21

Bräutigam A, Hofmann-Benning S, Weber AP (2008a) Comparative proteomics of chloroplast envelopes from C3 and C4 plants reveals specific adaptations of the plastid envelope to C4 photosynthesis and candidate proteins required for maintaining C4 metabolite fluxes. Plant Physiol 148:568–579

Bräutigam A, Shrestha RP, Whitten D, Wilkerson CG, Carr KM, Froehlich JE, Weber AP (2008b) Low-coverage massively parallel pyrosequencing of cDNAs enables proteomics in non-model species: comparison of a species-specific database generated by pyrosequencing with databases from related species for proteome analysis of pea chloroplast envelopes. J Biotechnol 136:44–53

Bräutigam A, Weber AP (2009) Proteomic analysis of the proplastid envelope membrane provides novel insights into small molecule and protein transport across proplastid membranes. Mol Plant 2:1247–1261

Brehelin C, Kessler F, van Wijk KJ (2007) Plastoglobules: versatile lipoprotein particles in plastids. Trends Plant Sci 12:260–266

Busch A, Nield J, Hippler M (2010) The composition and structure of photosystem I-associated antenna from *Cyanidioschyzon merolae*. Plant J 62:886–897

Deruere J, Romer S, d'Harlingue A, Backhaus RA, Kuntz M, Camara B (1994) Fibril assembly and carotenoid overaccumulation in chromoplasts: a model for supramolecular lipoprotein structures. Plant Cell 6:119–133

Dunkley TP, Hester S, Shadforth IP, Runions J, Weimar T, Hanton SL, Griffin JL, Bessant C, Brandizzi F, Hawes C, Watson RB, Dupree P, Lilley KS (2006) Mapping the *Arabidopsis* organelle proteome. Proc Natl Acad Sci U S A 103:6518–6523

Dunkley TP, Watson R, Griffin JL, Dupree P, Lilley KS (2004) Localization of organelle proteins by isotope tagging (LOPIT). Mol Cell Proteomics 3:1128–1134

Dupont FM (2008) Metabolic pathways of the wheat (*Triticum aestivum*) endosperm amyloplast revealed by proteomics. BMC Plant Biol 8:39

Egea I, Barsan C, Bian W, Purgatto E, Latché A, Chervin C, Bouzayen M, Pech J-C (2010) Chromoplast differentiation: current status and perspectives. Plant Cell Physiol 51: 1601–1611

Emanuelsson O, Brunak S, von Heijne G, Nielsen H (2007) Locating proteins in the cell using TargetP, SignalP and related tools. Nat Protoc 2:953–971

Emanuelsson O, Nielsen H, von Heijne G (1999) ChloroP, a neural network-based method for predicting chloroplast transit peptides and their cleavage sites. Protein Sci 8:978–984

Emanuelsson O, Nielsen H, Brunak S, von Heijne G (2000) Predicting subcellular localization of proteins based on the N-terminal amino acid sequence. J Mol Biol 300:1005–1016

Emanuelsson O, von Heijne G (2001) Prediction of organellar targeting signals. Biochim Biophys Acta 1541:114–119

Ferro M, Brugière S, Salvi D, Seigneurin-Berny D, Court M, Moyet L, Ramus C, Miras S, Mellal M, Le Gall S, Kieffer-Jaquinod S, Bruley C, Garin J, Joyard J, Masselon C, Rolland N (2010) AT_CHLORO: a comprehensive chloroplast proteome database with subplastidial localization and curated information on envelope proteins. Mol Cell Proteomics 9:1063–1084

Ferro M, Salvi D, Brugiere S, Miras S, Kowalski S, Louwagie M, Garin J, Joyard J, Rolland N (2003) Proteomics of the chloroplast envelope membranes from *Arabidopsis thaliana*. Mol Cell Proteomics 2:325–345

Ferro M, Salvi D, Riviere-Rolland H, Vermat T, Seigneurin-Berny D, Grunwald D, Garin J, Joyard J, Rolland N (2002) Integral membrane proteins of the chloroplast envelope: identification and subcellular localization of new transporters. Proc Natl Acad Sci USA 99:11487–11492

Ferro M, Seigneurin-Berny D, Rolland N, Chapel A, Salvi D, Garin J, Joyard J (2000) Organic solvent extraction to identify hydrophobic chloroplast membrane proteins. Electrophoresis 21:3517–3526

Ferro M, Tardif M, Reguer E, Cahuzac R, Bruley C, Vermat T, Nugues E, Vigouroux M, Vandenbrouck Y, Garin J, Viari A (2008) PepLine: a software pipeline for high-throughput direct mapping of tandem mass spectrometry data on genomic sequences. J Proteome Res 7:1873–1883

Friso G, Giacomelli L, Ytterberg AJ, Peltier JB, Rudella A, Sun Q, Wijk KJ (2004) In-depth analysis of the thylakoid membrane proteome of *Arabidopsis thaliana* chloroplasts: new proteins, new functions, and a plastid proteome database. Plant Cell 16:478–499

Froehlich JE, Wilkerson CG, Ray WK, McAndrew RS, Osteryoung KW, Gage DA, Phinney BS (2003) Proteomic study of the *Arabidopsis thaliana* chloroplastic envelope membrane utilizing alternatives to traditional two-dimensional electrophoresis. J Proteome Res 2:413–425

Gomez SM, Nishio JN, Faull KF, Whitelegge JP (2002) The chloroplast grana proteome defined by intact mass measurements from liquid chromatography mass spectrometry. Mol Cell Proteomics 1:46–59

Gomez SM, Bil KY, Aguilera R, Nishio JN, Faull KF, Whitelegge JP (2003) Transit peptide cleavage sites of integral thylakoid membrane proteins. Mol Cell Proteomics 2:1068–1085

Goulas E, Schubert M, Kieselbach T, Kleczkowski LA, Gardestrom P, Schroder W, Hurry V (2006) The chloroplast lumen and stromal proteomes of *Arabidopsis thaliana* show differential sensitivity to short- and long-term exposure to low temperature. Plant J 47:720–734

Inaba T, Ito-Inaba Y (2010) Versatile roles of plastids in plant growth and development. Plant Cell Physiol 51:1847–1853

Jain R, Katavic V, Agrawal GK, Guzov VM, Thelen JJ (2008) Purification and proteomic characterization of plastids from *Brassica napus* developing embryos. Proteomics 8:3397–3405

Jarvis P (2008) Targeting of nucleus-encoded proteins to chloroplasts in plants. New Phytol 179:257–285

Joshi H, Hirsch-Hoffmann M, Baerenfaller K, Gruissem W, Baginsky S, Schmidt R, Shulze WX, Sun Q, van Wijk KJ, Egelhofer V, Wienkoop S, Weckwerth W, Bruley C, Rolland N, Toyoda T, Nakagami H, Jones AME, Briggs SP, Castleden I, Tanz SK, Millar AH, Heazlewood JL (2011) MASCP Gator: an aggregation portal for the visualization of *Arabidopsis* proteomics data. Plant Physiol 155:259–270

Joyard J, Teyssier E, Miege C, Berny-Seigneurin D, Marechal E, Block MA, Dorne AJ, Rolland N, Ajlani G, Douce R (1998) The biochemical machinery of plastid envelope membranes. Plant Physiol 118:715–723

Joyard J, Ferro M, Masselon C, Seigneurin-Berny D, Salvi D, Garin J, Rolland N (2009) Chloroplast proteomics and the compartmentation of plastidial isoprenoid biosynthetic pathways. Mol Plant 2:1154–1180

Joyard J, Ferro M, Masselon C, Seigneurin-Berny D, Salvi D, Garin J, Rolland N (2010) Subplastidial compartmentation of lipid biosynthetic pathways. Prog Lipid Res 49:128–158

Keeling PJ (2004) Diversity and evolutionary history of plastids and their hosts. Am J Bot 91 (10):1481–1493

Kanervo E, Singh M, Suorsa M, Paakkarinen V, Aro E, Battchikova N, Aro E-M (2008) Expression of protein complexes and individual proteins upon transition of etioplasts to chloroplasts in pea (*Pisum sativum*). Plant Cell Physiol 49:396–410

Kieselbach T, Hagman AB, Schröder WP (1998) The thylakoid lumen of chloroplasts. Isolation and characterization. J Biol Chem 273:6710–6716

Kieselbach T, Bystedt M, Hynds P, Robinson C, Schröder WP (2000) A peroxidase homologue and novel plastocyanin located by proteomics to the *Arabidopsis* chloroplast thylakoid lumen. FEBS Lett 480:271–276

Kieselbach T, Schröder WP (2003) The proteome of the chloroplast lumen of higher plants. Photosynth Res 78:249–264

Kitajima A, Asatsuma S, Okada H, Hamada Y, Kaneko K, Nanjo Y, Kawagoe Y, Toyooka K, Matsuoka K, Takeuchi M, Nakano A, Mitsui T (2009) The rice alpha-amylase glycoprotein is targeted from the Golgi apparatus through the secretory pathway to the plastids. Plant Cell 21:2844–2858

Kleffmann T, Russenberger D, von Zychlinski A, Christopher W, Sjolander K, Gruissem W, Baginsky S (2004) The *Arabidopsis thaliana* chloroplast proteome reveals pathway abundance and novel protein functions. Curr Biol 14:354–362

Kleffmann T, Hirsch-Hoffmann M, Gruissem W, Baginsky S (2006) plprot: a comprehensive proteome database for different plastid types. Plant Cell Physiol 47:432–436

Kleffmann T, von Zychlinski A, Russenberger D, Hirsch-Hoffmann M, Gehrig P, Gruissem W, Baginsky S (2007) Proteome dynamics during plastid differentiation in rice. Plant Physiol 143:912–923

Linka N, Weber AP (2010) Intracellular metabolite transporters in plants. Mol Plant 3:21–53

Lunn JE (2007) Compartmentation in plant metabolism. J Exp Bot 58:35–47

McFadden GI (1999) Plastids and protein targeting. J Eukaryot Microbiol 46(4):339–346

Majeran W, Cai Y, Sun Q, van Wijk KJ (2005) Functional differentiation of bundle sheath and mesophyll maize chloroplasts determined by comparative proteomics. Plant Cell 17:3111–4031

Majeran W, Zybailov B, Ytterberg J, Dunsmore J, Sun Q, van Wijk KJ (2008) Consequences of C4 differentiation for chloroplast membrane proteomes in maize mesophyll and bundle sheath cells. Mol Cell Proteomics 7:1609–1638

Martin W, Rujan T, Richly E, Hansen A, Cornelsen S, Lins T, Leister D, Stoebe B, Hasegawa M, Penny D (2002) Evolutionary analysis of *Arabidopsis*, cyanobacterial, and chloroplast genomes reveals plastid phylogeny and thousands of cyanobacterial genes in the nucleus. Proc Natl Acad Sci U S A 99(24):12246–12251

Minge A, Shalchian-Tabrizi K, Tørresen OK, Takishita K, Probert I, Inagaki Y, Klaveness D, Jakobsen KS (2010) A phylogenetic mosaic plastid proteome and unusual plastid-targeting signals in the greencolored dinoflagellate *Lepidodinium chlorophorum*. BMC Evol Biol 10:191

Miras S, Salvi D, Ferro M, Grunwald D, Garin J, Joyard J, Rolland N (2002) Non-canonical transit peptide for import into the chloroplast. J Biol Chem 277:47770–47778

Miras M, Salvi D, Piette L, Seigneurin-Berny D, Grunwald D, Reinbothe C, Joyard J, Reinbothe S, Rolland N (2007) TOC159- and TOC75-independent import of a transit sequence less precursor into the inner envelope of chloroplasts. J Biol Chem 282:29482–29492

Nada A, Soll J (2004) Inner envelope protein 32 is imported into chloroplasts by a novel pathway. J Cell Sci 117:3975–3982

Nanjo Y, Oka H, Ikarashi N, Kaneko K, Kitajima A, Mitsui T, Muñoz FJ, Rodríguez-López M, Baroja-Fernández E, Pozueta-Romero J (2006) Rice plastidial N-glycosylated nucleotide pyrophosphatase/phosphodiesterase is transported from the ER-golgi to the chloroplast through the secretory pathway. Plant Cell 18:2582–2592

Naumann B, Busch A, Allmer J, Ostendorf E, Zeller M, Kirchhoff H, Hippler M (2007) Comparative quantitative proteomics to investigate the remodeling of bioenergetic pathways under iron deficiency in *Chlamydomonas reinhardtii*. Proteomics 7:3964–3979

Nosenko T, Lidie KL, Van Dolah FM, Lindquist E, Cheng JF, US Department of Energy-Joint Genome Institute, Bhattacharya D (2010) Chimeric plastid proteome in the Florida "Red Tide" dinoflagellate *Karenia brevis*. Mol Biol Evol 23:2026–2038

Peltier JB, Cai Y, Sun Q, Zabrouskov V, Giacomelli L, Rudella A, Ytterberg AJ, Rutschow H, van Wijk KJ (2006) The oligomeric stromal proteome of *Arabidopsis thaliana* chloroplasts. Mol Cell Proteomics 5:114–133

Peltier JB, Ytterberg AJ, Sun Q, van Wijk KJ (2004) New functions of the thylakoid membrane proteome of *Arabidopsis thaliana* revealed by a simple, fast, and versatile fractionation strategy. J Biol Chem 279:49367–49383

Peltier JB, Emanuelsson O, Kalume DE, Ytterberg J, Friso G, Rudella A, Liberles DA, Soderberg L, Roepstorff P, von Heijne G, van Wijk KJ (2002) Central functions of the lumenal and peripheral thylakoid proteome of *Arabidopsis* determined by experimentation and genome-wide prediction. Plant Cell 14:211–236

Peltier JB, Friso G, Kalume DE, Roepstorff P, Nilsson F, Adamska I, van Wijk KJ (2000) Proteomics of the chloroplast: systematic identification and targeting analysis of lumenal and peripheral thylakoid proteins. Plant Cell 12:319–341

Reyes-Prieto A, Hackett JD, Soares MB, Bonaldo MF, Bhattacharya D (2006) Cyanobacterial contribution to algal nuclear genomes is primarily limited to plastid functions. Curr Biol 16:2320–2325

Reyes-Prieto A, Weber AP, Bhattacharya D (2007) The origin and establishment of the plastid in algae and plants. Annu Rev Genet 41:147–168

Richly E, Leister D (2004) An improved prediction of chloroplast proteins reveals diversities and commonalities in the chloroplast proteomes of *Arabidopsis* and rice. Gene 329:11–16

Rolland N, Ferro M, Seigneurin-Berny D, Garin J, Douce R, Joyard J (2003) Proteomics of chloroplast envelope membranes. Photosynth Res 78:205–230

Rolland N, Atteia A, Decottignies P, Garin J, Hippler M, Kreimer G, Lemaire SD, Mittag M, Wagner V (2009) Chlamydomonas proteomics. Curr Opin Microbiol 12:285–291

Rossignol M, Peltier JB, Mock HP, Matros A, Maldonado AM, Jorrín JV (2006) Plant proteome analysis: a 2004–2006 update. Proteomics 6:5529–5548

Sadowski PG, Groen AJ, Paul Dupree P, Lilley KS (2008) Sub-cellular localization of membrane proteins. Proteomics 8:3991–4011

Schleiff E, Eichacker LA, Eckart K, Becker T, Mirus O, Stahl T, Soll J (2003) Prediction of the plant beta-barrel proteome: a case study of the chloroplast outer envelope. Protein Sci 12: 748–759

Schubert M, Petersson UA, Haas BJ, Funk C, Schroder WP, Kieselbach T (2002) Proteome map of the chloroplast lumen of *Arabidopsis thaliana*. J Biol Chem 277:8354–8365

Seigneurin-Berny D, Rolland N, Garin J, Joyard J (1999) Technical advance: differential extraction of hydrophobic proteins from chloroplast envelope membranes: a subcellular-specific proteomic approach to identify rare intrinsic membrane proteins. Plant J 19:217–228

Siddique MA, Grossmann J, Gruissem W, Baginsky S (2006) Proteome analysis of bell pepper (*Capsicum annuum* L.) chromoplasts. Plant Cell Physiol 47:1663–1673

Sun Q, Emanuelsson O, van Wijk KJ (2004) Analysis of curated and predicted plastid subproteomes of *Arabidopsis*. Subcellular compartmentalization leads to distinctive proteome properties. Plant Physiol 135:723–734

Sun Q, Zybailov B, Majeran W, Friso G, Olinares PD, van Wijk KJ (2009) PPDB, the plant proteomics database at Cornell. Nucleic Acids Res 37:D969–D974

Suzuki K, Miyagishima SY (2010) Eukaryotic and eubacterial contributions to the establishment of plastid proteome estimated by large-scale phylogenetic analyses. Mol Biol Evol 27:581–590

Terashima M, Specht M, Naumann B, Hippler M (2010) Characterizing the anaerobic response of *Chlamydomonas reinhardtii* by quantitative proteomics. Mol Cell Proteomics 9:1514–1532

The Arabidopsis Genome Initiative (2000) Analysis of the genome sequence of the flowering plant *Arabidopsis thaliana*. Nature 408:796–815

Turkina MV, Kargul J, Blanco-Rivero A, Villarejo A, Barber J, Vener AV (2006) Environmentally modulated phosphoproteome of photosynthetic membranes in the green alga *Chlamydomonas reinhardtii*. Mol Cell Proteomics 5:1412–1425

van Wijk KJ (2004) Plastid proteomics. Plant Physiol Biochem 42:963–977

Vener AV, Harms A, Sussman MR, Vierstra RD (2001) Mass spectrometric resolution of reversible protein phosphorylation in photosynthetic membranes of *Arabidopsis thaliana*. J Biol Chem 276:6959–6966

Vidi PA, Kanwischer M, Baginsky S, Austin JR, Csucs G, Dormann P, Kessler F, Brehelin C (2006) Tocopherol cyclase (VTE1) localization and vitamin E accumulation in chloroplast plastoglobule lipoprotein particles. J Biol Chem 281:11225–11234

Villarejo A, Burén S, Larsson S, Déjardin A, Monné M, Rudhe C, Karlsson J, Jansson S, Lerouge P, Rolland N, von Heijne G, Grebe M, Bako L, Samuelsson G (2005) Evidence for a protein transported through the secretory pathway en route to the higher plant chloroplast. Nat Cell Biol 7:1224–1231

von Zychlinski A, Kleffmann T, Krishnamurthy N, Sjolander K, Baginsky S, Gruissem W (2005) Proteome analysis of the rice etioplast – metabolic and regulatory networks and novel protein functions. Mol Cell Proteomics 4:1072–1084

Weber AP, Schwacke R, Flugge UI (2005) Solute transporters of the plastid envelope membrane. Annu Rev Plant Biol 56:133–164

Whitelegge JP, Gundersen CB, Faull KF (1998) Electrospray-ionization mass spectrometry of intact intrinsic membrane proteins. Protein Sci 7:1423–1430

Yoon HS, Hackett JD, Ciniglia C, Pinto G, Bhattacharya D (2004) A molecular timeline for the origin of photosynthetic eukaryotes. Mol Biol Evol 21:809–818

Ytterberg AJ, Peltier JB, van Wijk KJ (2006) Protein profiling of plastoglobules in chloroplasts and chromoplasts. A surprising site for differential accumulation of metabolic enzymes. Plant Physiol 140:984–997

Zolla L, Rinalducci S, Timperio AM, Huber CG (2002) Proteomics of light-harvesting proteins in different plant species. Analysis and comparison by liquid chromatography-electrospray ionization mass spectrometry. Photosystem I. Plant Physiol 130:1938–1950

Zolla L, Rinalducci S, Timperio AM (2007) Proteomic analysis of photosystem I components from different plant species. Proteomics 7:1866–1876

Zolla L, Timperio AM, Walcher W, Huber CG (2003) Proteomics of light-harvesting proteins in different plant species. Analysis and comparison by liquid chromatography-electrospray ionization mass spectrometry. Photosystem II. Plant Physiol 131:198–214

Zybailov B, Rutschow H, Friso G, Rudella A, Emanuelsson O, Sun Q, van Wijk KJ (2008) Sorting signals, N-terminal modifications and abundance of the chloroplast proteome. PLoS One 3(4): e1994

Part V
Evolution of Organelle Transcription

Chapter 10
Mitochondrial Gene Expression and Dysfunction in Model Protozoa

Christian Barth, Luke A. Kennedy, and Paul R. Fisher

10.1 Introduction to Protozoan Mitochondrial Genomes

Protozoan mitochondrial genomes are extraordinarily diverse in size, structure and gene organisation. In almost all cases, however, the protozoan mitochondrial DNA (mtDNA) carries a core set of genes, encoding proteins involved in oxidative phosphorylation and energy production, some ribosomal proteins, two or three ribosomal RNAs and in most cases transfer RNAs (Gray et al. 2004). When compared to their mammalian counterparts, protozoa tend to have the larger mitochondrial genome. While the typical mammalian (and vertebrate) mitochondrial genome ranges in size from 16 to 19 kb, the mtDNA in most protozoa is at least three to four times larger. However, exceptions do exist at both ends of the scale; the apicomplexan endoparasites *Plasmodium falciparum, P. vivax* or *P. simium*, for example, harbour linear mtDNA molecules smaller than 6 kb, carrying only three protein-coding and two ribosomal RNA genes (Wilson and Williamson 1997). The mitochondrial genomes of their close relatives, the dinoflagellates, share the same gene complement, but are substantially bigger due to gene duplications and a large number of short inverted repeat sequences (Nash et al. 2008). In sharp contrast to this, the mitochondrial genome of the jakobid flagellate *Reclinomonas americana* is by far the gene richest and least derived mtDNA characterised to date (Lang et al. 1997). It carries 97 protein and RNA encoding genes on a 69 kb circular DNA molecule, with at least 18 unique genes of known function not found in the mtDNA of any other species. Amongst these are also the genes encoding the α_2, β and β' and sigma subunits of the prokaryotic RNA polymerase, which is particularly noteworthy, as the mitochondrial genomes of all other species are transcribed by a nuclear-encoded T3/T7 bacteriophage-type single subunit RNA polymerase (Cermakian et al. 1997; Li et al. 2001). By comparing the gene-rich *R. americana*

C. Barth (✉) • L.A. Kennedy • P.R. Fisher
Department of Microbiology, La Trobe University, Melbourne, Australia
e-mail: c.barth@latrobe.edu.au

C.E. Bullerwell (ed.), *Organelle Genetics*,
DOI 10.1007/978-3-642-22380-8_10, © Springer-Verlag Berlin Heidelberg 2012

mitochondrial genome with that of other species, it also becomes apparent that the observed genome sizes do not necessarily correlate with the number of genes they encode. This is particularly obvious when comparing the mtDNA of *R. americana* with that of the cress *Arabidopsis thaliana*, for example. With 367 kb in size, the plant mitochondrial genome is much larger than the *R. americana* mitochondrial DNA, but codes for only 57 genes (Unseld et al. 1997).

Considering the large number of representatives within the group of protozoa, only a relatively small selection of mitochondrial sequence data is available from these organisms. One of the first amoebozoan mitochondrial genomes to be completely sequenced was that of the cellular slime mould *Dictyostelium discoideum* (Ogawa et al. 2000).

10.1.1 The Mitochondrial Genome of Dictyostelium discoideum

The cellular slime mould *D. discoideum* has long been regarded as a valuable and attractive tool for the study of eukaryotic cell biology. The organism combines typical eukaryotic cellular and molecular biology with the experimental tractability of a microorganism in which biochemical, classical and molecular genetic as well as cell biological approaches are readily adopted. Its developmental lifecycle (Fig. 10.1) is unique amongst protozoa and at the different stages of development

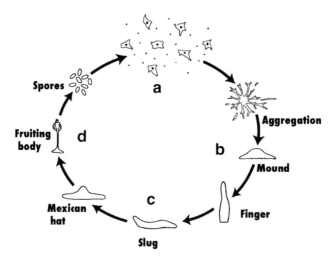

Fig. 10.1 Developmental lifecycle of *Dictyostelium discoideum*. The unicellular and independent *D. discoideum* cells, the amoebae (**a**), feed on bacteria and decaying matter. When the food source is exhausted, starvation triggers the amoebae to aggregate, resulting in the formation of a mound (**b**). The mound enters a developmental cycle involving complex morphological changes, producing a multicellular slug (**c**) that migrates towards the light and warmth. The slug subsequently gives rise to a fruiting body (**d**) composed of a spore head supported by a stalk (modified from Brown and Strassmann, CC 3.0 Strassmann)

D. discoideum features both plant- and animal-like characteristics (Otto and Kessin 2001). Based on these properties, *D. discoideum* has been used as a model system to gain insight into the complex nature of cell differentiation processes, as well as signal transduction and cell motility (Wilczynska et al. 1997; Coates and Harwood 2001; Maeda 2005; MacWilliams et al. 2006; Annesley and Fisher 2009). More recently, *D. discoideum* has also been used as a model system to study mitochondrial disorders (Kotsifas et al. 2002; Barth et al. 2007; Bokko et al. 2007). Contributing to this, a detailed transcription map of the D. *discoideum* mitochondrial genome has been established and the mode of transcription has been studied in detail (Barth et al. 1999, 2001; Le et al. 2009).

The *D. discoideum* mitochondrial genome is a circular DNA molecule and is 55,564 bp in size (Ogawa et al. 2000). It codes for 33 proteins, six open reading frames (ORFs) of unknown function, two ribosomal RNA genes and 18 transfer RNA genes (Fig. 10.2). Most genes are tightly packed and some even overlap, however, intergenic spacers ranging in size from a few nucleotides to more than 2 kb do also exist. All genes are involved in biological processes that are typically localised to the mitochondria, namely respiration and translation. In addition to the standard set of genes, the *D. discoideum* mtDNA also contains the genes of some NADH dehydrogenase and ATP synthase subunits (such as *nad7, nad9, nad11* and *atp1*), which are nuclear-encoded in many other organisms and are post-transcriptionally imported into the mitochondria of these organisms (Cole and Williams 1994; Cole et al. 1995). Some other notable features of the *D. discoideum* mtDNA include the fusion of the adjacent cytochrome oxidase subunit 1 and 2 genes (*cox1/2*), which form a single ORF with no apparent stop codon present at the 3′ end of *cox1* (Ogawa et al. 1997; Pellizzari et al. 1997). The two genes, however, are expressed

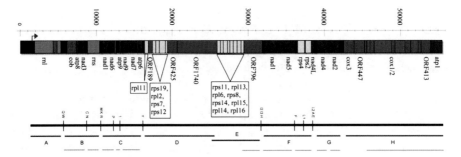

Fig. 10.2 Schematic representation of the gene content, organisation and transcription of the mitochondrial genome of *Dictyostelium discoideum. Top*: The mtDNA is a 55,564 bp circular DNA molecule encoding 17 components of the electron transport chain and oxidative phosphorylation apparatus (*coloured* boxes), two ribosomal RNAs (*blue*), 15 ribosomal proteins (*yellow*), 18 tRNAs (not shown), four group I introns (*grey*) and five ORFs of unknown function (*purple*). All genes are located on the same strand in the same orientation. Non-coding sequences are shown in *black. Bottom*: The genome is transcribed from a single transcription start site (*red arrow*) into a large polycistronic transcript (*thick black line*). The excision of tRNA molecules (indicated by *purple* letters) leads to the generation of secondary transcripts (a–h), which are further processed to form mature mono-, di- or tricistronic transcripts

as two individual proteins that migrate separately on SDS-PAGE (Bisson et al. 1985). In addition to its peculiar gene organisation, which is shared by *Acanthamoeba castellanii* (Lonergan and Gray 1996), the *D. discoideum cox1/2* gene is interrupted by four introns, which possess the potential to form conserved secondary RNA structures characteristic of group-I introns (Ogawa et al. 1997). Another group-I intron in the *rnl* gene coding for the large subunit ribosomal RNA (Angata et al. 1995) is located at the same site as the introns found in the mitochondrial *rnl* gene from the green alga *Scenedesmus obliquus* (Kück et al. 1990) and from the colourless alga *Prototheca wickerhamii* (Wolff et al. 1994). In contrast to animals, where all mitochondrial ribosomal proteins are encoded by nuclear DNA (Bonen 1991), a number of the *D. discoideum* mitochondrial ribosomal proteins are mitochondrially encoded. *D. discoideum* shares this feature with plants and many lower eukaryotes, including other protozoa (Pritchard et al. 1990; Wolff et al. 1994). The number of mitochondrially encoded ribosomal protein genes varies greatly amongst these species, from only one in *Saccharomyces cerevisiae* and other fungi (Burke and RajBhandary 1982; Zamaroczy and Bernardi 1986) to 17 in the oomycete *Phytophthora infestans* (Lang and Forget 1993). Although the single ribosomal protein genes are frequently found within intronic sequences of large ribosomal subunit RNA genes (Burke and RajBhandary 1982; Cummings et al. 1989), multiple genes are usually arranged in gene clusters. The *D. discoideum* mtDNA contains two such clusters of ribosomal protein genes (Fig. 10.2) and the gene arrangement within the clusters is similar to that found in *Escherichia coli* (Zurawski and Zurawski 1985), and *Marchantia polymorpha* (Oda et al. 1992) and is almost identical to that of *A. castellanii* (Burger et al. 1995; Iwamoto et al. 1998).

10.2 Transcription of Protozoan Mitochondrial Genomes

The mechanisms mediating mitochondrial gene expression in different organisms are as diverse as the size and gene organisation of their mtDNAs. The basic mechanism has been most thoroughly investigated in animals, particularly *Homo sapiens*, *Mus musculus* and *Xenopus laevis*, and in fungi (predominantly in *S. cerevisiae*) (Shadel and Clayton 1993; Taanman 1999; Foury and Kucej 2001; Asin-Cayuela and Gustafsson 2007), revealing significant species-specific differences in the number and structure of mitochondrial promoters (Parisi and Clayton 1991; Antoshechkin and Bogenhagen 1995). The study of mitochondrial gene expression in these organisms has also led to the development of in vitro transcription/translation systems, which greatly facilitated the identification and characterisation of most of the key components (Docherty 1996; Hatzack et al. 1998). Transcription requires a relatively specific core enzyme, the mitochondrial RNA polymerase, and in many cases one or more mitochondrial transcription factors, which may be essential for promoter recognition or stabilisation of the polymerase-promoter complex (Jan et al. 1999).

10.2.1 The Mitochondrial RNA Polymerase in Dictyostelium discoideum

In contrast to the complex multi-subunit RNA polymerases that are responsible for the transcription of nuclear genomes, transcription of mitochondrial genes has been found to be catalysed by a much simpler, single-subunit RNA polymerase (McAllister 1993; Cermakian et al. 1997; Cheetham and Steitz 2000). Given that the alleged eubacterial ancestor of mitochondria presumably possessed a multi-subunit ($\alpha_2\beta\beta'$ σ) α-proteobacterial RNA polymerase, this finding was unexpected and it has been postulated that during evolution, a bacteriophage-type polymerase replaced the ancestral polymerase, thereby transmitting the replication and transcription mechanisms of bacteriophages to mitochondria (Gray 1992; Rousvoal et al. 1998). As indicated earlier in this chapter, the replacement did not occur in *R. americana*, where fully functional copies of the ancestral $\alpha_2\beta\beta'\sigma$ proteobacterial RNA polymerase genes have been retained in the mitochondrial genome (Lang et al. 1997; Gray et al. 1998). Similarly, the mitochondrial genome of the brown alga *Pylaiella littoralis* still not only harbours non-functional traces of the ancestral α-proteobacterial transcription system, but it also codes for a single-subunit RNA polymerase (Rousvoal et al. 1998; Oudot-Le Secq et al. 2001). In all other organisms, the gene sequences coding for the single-subunit RNA polymerase have been subsequently transferred from the mitochondrial genome to the nucleus. Although the function of the mitochondrial RNA polymerase has been well characterised in yeast and in mammals, very little is known about the mechanisms and components involved in the control of mitochondrial transcription in protozoa.

Based on the widespread similarities amongst mitochondrial RNA Polymerase protein and gene sequences, the gene for the *D. discoideum* mitochondrial RNA Polymerase, *rpmA*, has been identified, cloned and sequenced (AY040092; Le et al. 2009). The *rpmA* gene consists of a continuous open reading frame of 2,850 nucleotides in length and codes for a protein of 950 amino acid residues with a molecular mass of approximately 109 kDa. The mitochondrial localisation of the enzyme has been confirmed, and the evolutionary relationship between the *D. discoideum* mitochondrial RNA polymerase and others has been determined in sequence alignments and phylogenetic trees. The *D. discoideum* protein was also the first protozoan mitochondrial RNA polymerase used in functional studies to demonstrate the protein's ability to specifically initiate transcription from mitochondrial transcription initiation sequences (Le et al. 2009).

10.2.2 Transcription of the Mitochondrial Genome in Dictyostelium discoideum

The mode of transcription in the mitochondria of *D. discoideum* has been studied extensively using northern hybridization and primer extension analyses to detect mitochondrial transcripts and to identify their 5′ ends, Reverse Transcription-Polymerase Chain Reaction (RT-PCR) to identify RNA transcripts not detectable by hybridization, in vitro capping to label the 5′ ends of transcripts that have been generated by transcription initiation, electrophoretic mobility shift assays (EMSA) for the identification of protein-binding DNA sequences and in bacterio transcription experiments to examine the role of these sequences in transcription.

Northern hybridization experiments provided the first indication that the genes in the *D. discoideum* mitochondrial genome are co-transcribed in clusters and that the resulting polycistronic transcripts are subject to co- or post-transcriptional processing. Using gene-specific probes directed against all genes present in the mitochondrial genome, eight large, polycistronic transcripts were detected, some of which were found to be further processed to form smaller mono-, di- or tricistronic RNA molecules (Barth et al. 1999, 2001). Larger transcripts were not detected in any of the hybridisation studies, and based on the assumption that the *D. discoideum* mitochondrial genome as a rather large genome (56 kb) would most probably be transcribed from multiple promoters, it was concluded at the time that the *D. discoideum* mitochondrial genes are located in eight major transcription units. The assumption leading to this conclusion was based on earlier findings in yeast mitochondria, where, depending on the genome size (19.4 kb in *Schizosaccharomyces pombe* to more than 80 kb in *S. cerevisiae*; Bullerwell et al. 2003; Schäfer et al. 2005), the number of promoters ranges from two to at least 20 individual promoters (Costanzo and Fox 1990).

In an attempt to identify a similar consensus in the *D. discoideum* mitochondrial genome, the 5′ ends of the eight major transcripts were mapped in primer extension analyses and the sequences upstream of the 5′ ends were aligned to identify possible consensus sequences. Although a short consensus had been identified, database searches with the identified sequence demonstrated that this sequence was not confined to the sequences upstream of the eight putative transcription start sites. These findings pointed towards the possibility that some of the 5′ ends determined by primer extension analysis represented sites of RNA processing rather than transcription initiation, indicating the possibility that even larger RNA transcripts are produced in the mitochondria of *D. discoideum*. Northern hybridization may have been not sensitive enough to detect any larger transcripts if their levels had been to low due to rapid processing.

In order to establish whether transcripts larger than those reported exists in the mitochondria of *D. discoideum*, RT-PCR was employed, a method more sensitive than northern hybridization and therefore more suitable for the detection of RNA transcripts at low levels. Depending on the sensitivity of the probe, detection by northern hybridization requires multiple copies of an RNA transcript to be present, while RT requires only very little RNA template to synthesise a cDNA, which is subsequently amplified in a PCR step to observable levels (Freeman et al. 1999). RT-PCR involves two steps, the first is known as the reverse transcription step and the second is the amplification step. Reverse transcription is mediated by Moloney Murine Leukaemia Virus Reverse Transcriptase (M-MLV RT), an enzyme that transcribes RNA sequences into a complementary DNA (cDNA) sequence by

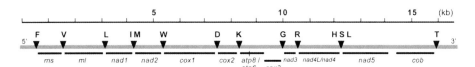

Fig. 10.3 Processing of the primary H-strand transcript by tRNA excision in mammalian mitochondria. Schematic representation of the primary polycistronic transcript (*grey line*) derived from the Heavy strand of the human mitochondrial DNA, and the monocistronic processing products (*black lines*) resulting from co-transcriptional excision of tRNA molecules (*triangles*). The relevant amino acids are indicated in the three letter code, sizes are shown in kilobases (kb)

extending a gene-specific 3′ primer, forming an RNA:DNA hybrid. This RNA:DNA hybrid is then denatured and the cDNA is amplified exponentially in the PCR by DNA-dependent *Taq* polymerase I (Dale and Schantz 2002).

By performing RT-PCR reactions over the intervening regions between the eight major transcripts in the *D. discoideum* mitochondria, products were obtained for all intervening regions, except for the region between transcript H and A upstream of the *rnl* sequence (Fig. 10.3). The successful amplification of the intervening regions in these experiments demonstrated the existence of transcribed sequences spanning the gaps between the eight major RNA transcripts observed in the northern hybridization studies, suggesting that the eight major transcripts were themselves derived from processing of one or more even larger primary transcripts. The fact that all but one of the intervening regions were amplified by RT-PCR further suggested that the 5′ end of transcript A was not derived from processing but was generated by transcription initiation.

The finding that the 5′ end of transcript A has been generated by transcription initiation was confirmed in in vitro capping experiments, a method that has been used successfully for the identification of transcriptional start sites in the mitochondrial genomes of many organisms, including humans (Yoza and Bogenhagen 1984) and yeast (Christianson and Rabinowitz 1983). Generally, mitochondrial transcripts are not capped like nuclear mRNA precursors (Montoya et al. 1981) and mitochondrial transcripts, whose 5′ ends were formed by transcription initiation, can therefore be capped in vitro. Transcripts that have been generated by transcription initiation have a triphosphate ribonucleotide at their 5′ end, whereas transcripts derived from processing have a monophosphate ribonucleotide at their 5′ end. Only in the presence of a di- or triphosphate can the capping enzyme guanylyl transferase add a "cap" in form of a GTP to the 5′ end of the transcript (Keith et al. 1982). In vitro capping by guanylyl transferase with [a-^{32}P] GTP therefore allows the specific labelling of only those 5′ termini that originate from the initiation of transcription and can be used to discriminate primary transcripts from those transcripts that are derived from co- or post-transcriptional processing (Auchincloss and Brown 1989; Binder and Brennicke 1993).

To investigate whether the 5′ end of transcript A was the only 5′ end that had been created by transcription initiation, *D. discoideum* mitochondrial RNA was capped in vitro in the presence of radioactively labelled [a-^{32}P] GTP and then used as probe in Southern hybridization experiments. For the Southern blots, DNA fragments representing the intervening regions between the eight major transcripts were amplified by PCR, separated by agarose gel electrophoresis, transferred onto nylon membrane and were then allowed to hybridise with the in vitro capped RNA. Only one of the DNA fragments, the fragment spanning the region between transcript H and A was detected by the capped mitochondrial RNA, indicating that only transcript A had been capped by guanylyl transferase. This result clearly identified the 5′ end of transcript A to be generated by transcription initiation, and since none of the other DNA fragments were found to hybridise to capped RNA, the 5′ end of transcript A must represent the only transcription start site. These findings complemented the results obtained in the RT-PCR experiments, suggesting that the *D. discoideum* mitochondrial genome is indeed transcribed from a single initiation site.

The exact location of the identified transcription initiation site in the *D. discoideum* mitochondrial genome has been determined by primer extension analysis, and the non-coding sequences upstream of the start site have been examined further, as they may contain the promoter and other regulatory sites. Since these sites would be binding sites for the RNA polymerase and possible transcription factors, a series of Electrophoretic Mobility Shift Assays (EMSA) was performed to examine the protein-binding capacity of the sequences contained in the non-coding region. The EMSA is based on the fact that the migration of DNA fragments is retarded during electrophoresis if protein is bound to them (Ausubel et al. 1994).

A number of DNA fragments, each 200–300 bp in length, together representing the entire length of the non-coding sequences upstream of the transcription initiation site (~ 2 kb; Fig. 10.2) were amplified by PCR and radioactively end labelled with [α-^{32}P] ATP. Each DNA fragment was then individually incubated with concentrated mitochondrial protein extract and the DNA-protein samples were analysed on non-denaturing polyacrylamide gels and their migration pattern was compared to those of protein-free DNA samples. The presence of band shifts for some but not all samples, and the absence of band shifts in appropriate control reactions, suggested that some DNA fragments had been specifically bound by DNA-binding proteins. Since most of the binding sites were located towards the 3′ end of the non-coding sequences just upstream of the identified start of transcription, these sequences were of particular interest as they may provide the binding sites for proteins directly involved in the transcription process, which would support the conclusions drawn from the RT-PCR and in vitro capping experiments.

The ultimate proof that a given DNA sequence has the potential to serve as a binding site for components of the transcription apparatus and to function as a transcription initiation site is to demonstrate its function in appropriate transcription experiments. One of the most useful tools to investigate the mechanism of transcription is an in vitro transcription system. Edwards and colleagues presented the first in vitro transcription system for the study of transcription in yeast mitochondria in 1982. This was followed by the development of mitochondrial in vitro transcription systems for other organisms, allowing detailed studies of promoter and regulatory sequences in the mitochondrial genome of many organisms, such as humans (Walberg and Clayton 1983), *Xenopus leavis* (Bogenhagen and Yoza 1986),

Neurospora crassa (Kennell and Lambowitz 1989) and *Triticum aestivum* (Hanic-Joyce and Gray 1991). The in vitro transcription systems have also been invaluable tools for the identification of the protein components involved in mitochondrial transcription. In many cases, the provision of a linear DNA template containing the promoter and control sequences in a reaction supplemented with mitochondrial protein extract is sufficient for RNA synthesis to occur, and the generation of any transcripts can be monitored either by providing radiolabelled ribonucleotides in the transcription reaction, or by blotting the transcription products onto a nylon membrane and subsequent detection using radiolabelled transcript-specific probes. In many cases, however, and depending on the organism and the reaction conditions, in vitro transcription is not as straightforward, resulting in no transcription products or unsatisfactory results. This is frequently due to the presence of nucleases in the mitochondrial lysates, as observed in the mitochondria of fungi (*N. crassa* and *S. cerevisiae*; Chow and Fraser 1983; Dake et al. 1988) and animals (Ikeda et al. 1997). Many of these potent Mg^{2+} or Mn^{2+}-dependent enzymes exhibit both RNase and DNase activities, and several studies have suggested that the mitochondrial nucleases are involved in apoptosis by demonstrating that mammalian cells induced to undergo apoptosis released nucleases from their mitochondria (Meng et al. 1998; van Loo et al. 2001). Walberg and Clayton (1983) were able to remove or at least reduce the nuclease activity present in mitochondrial lysates by fractionation and used the partially purified mitochondrial RNA polymerase in run-off transcription assays to identify specific light strand transcription in the displacement loop region of human mitochondrial DNA. Others have successfully removed the nucleases from mitochondrial lysates by immunoprecipitation, using antibodies raised against endo- and exonucleases in *N. crassa* and *S. cerevisiae* (Chow and Fraser 1983; Dake et al. 1988). The problems arising from the presence of nucleases can also be overcome by isolating the mitochondrial RNA polymerase directly from mitochondrial protein extracts or by expressing the RNA polymerase gene in *E. coli* cells and using the purified enzyme for in vitro transcription. However, since the mitochondrial RNA polymerase is a rather large enzyme, this approach may also be limited due to low expression levels, insolubility, or misfolding of the expressed protein leading to low enzymatic activities. As some mitochondrial RNA polymerases are known to require additional co-factors or transcription factors for efficient promoter recognition and binding (Virbasius and Scarpulla 1994; Shadel and Clayton 1995; Falkenberg et al. 2007), the lack of essential components of the transcription machinery poses another limitation to in vitro transcription experiments.

The presence of nuclease activities in mitochondrial protein extracts, poor solubility of the mitochondrial RNA polymerase expressed in and purified from *E. coli* cells, and possibly inappropriate reactions conditions also hampered the in vitro transcription experiments conducted to demonstrate the function of the identified transcription initiation site in the *D. discoideum* mitochondrial genome. However, in anticipation that a small amount of the *D. discoideum* mitochondrial RNA polymerase expressed in *E. coli* was folded correctly and active, the above-mentioned problems were overcome by performing the transcription reaction directly in the bacterial host rather than purifying the recombinant protein from the bacteria and using it in vitro. In order to perform an in bacterio transcription reaction, *E. coli* cells expressing the *D. discoideum* mitochondrial RNA polymerase gene (*rpmA*) were transformed with a construct harbouring a fragment of the *D. discoideum* mitochondrial DNA containing the identified transcription initiation site including some upstream non-coding sequences that may contain the promoter and other regulatory sites. After induction of the *rpmA* gene, mitochondrial DNA-specific transcripts were detected in the *E. coli* cells, not only demonstrating that transcription from the provided template had been initiated specifically by the *D. discoideum* mitochondrial RNA polymerase, but also confirming the position and function of the identified single transcription initiation site in the *D. discoideum* mitochondrial genome (Le et al. 2009).

Based on the above observations, conducting the transcription experiments directly in a bacterial host rather than in vitro offers several advantages; in bacterio transcription can solve the problems arising from low in vitro activity of the expressed RNA polymerase, which may be due to poor expression and purification efficiencies or misfolding of the enzyme, or due to inappropriate reaction conditions. In the latter case, it is possible that the *E. coli* host even provides essential but unknown resources otherwise not present in an in vitro reaction.

10.3   Mitochondrial Transcript Processing

A distinct feature of many mitochondrial transcriptomes is the generation of polycistronic transcripts, and it is likely that, as in bacterial systems, the co-transcription of genes is used for the coordinate regulation of genes. In contrast to the bacterial transcripts, however, mitochondrial polycistronic transcripts undergo varying degrees of processing to mature messenger RNAs (mRNA), ribosomal RNAs (rRNAs) and transfer RNA (tRNAs). The processing mechanisms and the signals that dictate these events are currently under investigation by research groups from various disciplines, including those examining the role of mitochondria in apoptosis, ageing and disease.

Processing of mitochondrial RNA molecules can be achieved using conserved sequence elements (Osinga et al. 1984; Schuster and Brennicke 1989), by excision of tRNAs via endonucleolytic cleavage (Ojala et al. 1981; Montoya et al. 1981; Burger et al. 1985), by removal of non-coding nucleotides via exonucleolytic cleavage (Dziembowski et al. 2003; Stewart and Beckenbach 2009) and through various secondary structures formed by the nascent transcripts (Agsteribbe and Hartog 1987; Montoya et al. 2006). Current knowledge suggests that the processing, modification and translation of RNA transcripts is coupled in mitochondria and that some of these processes, once thought to be post-transcriptional, actually occur co-transcriptionally (Shadel 2004).

Researchers investigating the mechanism of mitochondrial transcription and RNA processing have been able to exploit particular features of mitochondrial RNA metabolism. In vitro capping of mitochondrial RNAs to identify transcription initiation sites has been utilised for more than 30 years, as has primer extension and northern hybridization analysis of the patterns of mature and precursor forms of the transcripts, and advances in molecular biology have greatly increased the sensitivity of these methods. In addition, circularization of RNA transcripts by ligation, followed by RT-PCR, cloning and sequence analysis allows a direct examination of the 5' and 3' termini of the RNA molecules.

10.3.1 Transcript Processing in Mammalian Mitochondria

The genetic content, structure and organisation of the mitochondrial genome in mammals are highly conserved. Mammalian mtDNA is a double stranded, closed circular molecule of approximately 16.5 kb in length, encoding two ribosomal RNAs, 13 polypeptides involved in OXPHOS complexes and 22 tRNAs (Anderson et al. 1981). The two strands of the molecule differ in G–C content and can be separated by density gradient, accordingly, they have been designated the Heavy (H) and Light (L) strand (Wolstenholme 1992; Taanman 1999). Almost all the genetic information is encoded on the H strand, with only one polypeptide and eight tRNAs encoded by the L-strand. The gene organisation is extremely compact, there are no introns and the intergenic regions, when present, are limited to a few nucleotides in length. A non-coding region, the displacement loop (D-loop), contains all the major regulatory elements required for the expression of both strands of the genome. There are three sites of initiation within the D-loop, two (IT_{H1} and IT_{H2}) responsible for the generation of large polycistronic precursors from the H strand, and the third (IT_L) responsible for transcription of large polycistronic precursors from the L-strand (Montoya et al. 1982; Bogenhagen et al. 1984).

The mechanism by which the polycistronic transcripts synthesised from the three initiation sites are processed to mature RNA molecules was first elucidated nearly 3 decades ago in seminal studies by Attardi and colleagues (Ojala et al. 1980, 1981; Montoya et al. 1981). A peculiar feature of the H-mtDNA is that, with few exceptions, each of the rRNA and polypeptide encoding sequences is flanked on both sides by one or more tRNA genes. According to the "tRNA punctuation" model of mitochondrial RNA processing, polycistronic precursor RNAs are processed, giving rise to mature rRNAs, mRNAs and tRNAs after precise endonucleolytic cleavage on both sides of the pre-tRNA molecules (Ojala et al. 1981; Montoya et al. 2006; Fig. 10.3).

In humans, 5' endonucleolytic cleavage occurs first and is accomplished by a mitochondrial RNase P. The human enzyme is distinct from orthologs found in prokaryotes and yeast in that it does not require any catalytic RNA component (Rossmanith et al. 1995). It is probable that the subsequent 3' endonucleolytic cleavage in human mitochondrial pre-tRNAs is performed by an RNase Z, encoded by the ELAC2 gene. Interestingly, this gene has been implicated in susceptibility to prostate cancer (Takaku et al. 2003; Montoya et al. 2006). In some cases, tRNA excision from the primary transcripts leads to mature mRNAs and rRNAs with mature ends, others are flanked by small (1–2 nucleotides) spacers and several overlap by one residue with the adjacent RNA molecule (Montoya et al. 1981; 2006). As some of the polypeptide genes do not encode a translational stop codon, but rather end in a single U or UA (Ojala et al. 1981), the completion of the stop codon (UAA) is reliant upon polyadenylation; hence, inaccurate cleavage by RNase P could result in non-functional gene products (Montoya et al. 1981; Clayton 1984; Shadel and Clayton 1993).

The excision of tRNAs from the primary transcript accounts for the release of most mature RNA molecules and the generation of their termini; however, the termini of some transcripts cannot be generated by tRNA excision alone. These termini are the 3′ ends of ATPase subunits 8 and 6 (*atp8*, *atp6*) and of NADH dehydrogenase subunits 4L and 5 (*nad4L*, *nad5*), as well as the 5′ ends of cytochrome c oxidase subunit 3 (*cox3*) and of cytochrome b (*cob*), respectively (Ojala et al. 1981; Montoya et al. 1981; Temperley et al. 2010; Fig. 10.3). Rossmanith and colleagues (1995) proposed the generation of these termini to be catalysed by the human RNase P in a manner analogous to the cleavage of 4.5S pre-rRNA by the *E. coli* RNase P. Alternatively, the 5′ ends of the *cox3* and *cob* transcripts are immediately contiguous to tRNA genes encoded on the L strand, and considering that the antisense sequence of a tRNA gene would still be capable of forming a cloverleaf-like structure, it is possible that the 5′ termini arise from a variation of the strict punctuation model (Montoya et al. 2006).

In other cases where tRNA excision fails to provide the processing signal, it is possible that the RNA molecule is capable of forming secondary structures similar to the cloverleaf structure of tRNAs, or it folds in a way that is also recognised by the tRNA processing enzymes. Indeed, several studies have demonstrated that the 5′ termini of mitochondrial RNA molecules are capable of forming extensive secondary structures, and it has been suggested that the secondary structures also play pivotal roles in post-transcriptional events, for example, by possessing functions analogous to the 5′ cap structure found in mature nuclear mRNAs (Denslow et al. 1989; Liao and Spremulli 1990; Montoya et al. 2006). The cap structure is known to aid not only in the protection from 5′to 3′ exoribonuclease activity, but also in the interaction with translation initiation factors and the main components of the translation machinery (Banerjee 1980, Proudfoot et al. 2002).

Liao and Spremulli (1989, 1990) demonstrated the effect of sequence length and secondary structure formation on the interaction of bovine mitochondrial mRNA with ribosomal subunits and revealed that a minimum transcript length of ~400 nucleotides was required for efficient binding. This study is of interest with regard to the translation of mature dicistronic mRNAs. The juxtaposition of the nad4L/nad4 and atp8/6 genes and their transcription and subsequent processing as dicistronic messengers possibly provides insight into the mechanisms of ribosomal recognition, binding and attachment. Both 5′ genes in these dicistronic transcripts (*nad4L* and *atp8*) are much shorter than those shown to be required for efficient ribosomal binding. In the absence of the required recognition sites associated with 5′ untranslated regions or 5′ cap structures, their persistence as a larger transcript may be a mechanism to increase the efficiency of translation, suggesting a different role for the prevalence and persistence of co-transcription of genes into polycis-tronic molecules (Taanman 1999).

The human mitochondrial genetic system exhibits apparent simplicity; however, the generic mechanism of RNA cleavage and maturation may only hold true at a mechanistic level (Temperley et al. 2010). Not only in various mammalian but also in other metazoan mtDNAs, modification of the classical tRNA punctuation model, as proposed by Attardi and colleagues (Ojala et al. 1980, 1981; Montoya et al. 1981), is necessary in order to elucidate species-specific variations in mitochondrial RNA processing. With respect to this, it is noteworthy that in some metazoans, such as *Drosophila melanogaster*, the classic punctuation model has been shown to sufficiently explain the generation of all 3′ ends of RNAs. However, additional endonucleolytic cleavage is required in order to process 5′ non-coding nucleotides present on at least one RNA, and occurs only after the mature 3′ termini has been formed (Stewart and Beckenbach 2009). The same study has also revealed that the polyadenylation of newly generated 3′ ends of mitochondrial RNAs is coupled to tRNA cleavage events. In a transcript containing the *nad3* gene sequence and a downstream cluster of five tRNAs, the tRNAs were found to be removed sequen-tially in a 3′–5′ direction, but each processing intermediate was polyadenylated to a level similar to that of mature mRNA transcripts. This coupling of transcription, transcript processing and RNA modification implies a highly coordinated and complex system of mitochondrial gene expression.

It is clear that despite characterization of the basic mechanisms of transcription and transcript processing in mammals, there remains a multitude of activities involving cis- and trans-acting elements to be elucidated. The identification and characterisation of these elements requires a coordinated approach utilising classical methodologies in combination with evolving fields such as comparative genomics.

10.3.2 Transcript Processing in Yeast Mitochondria

The study of mitochondrial transcription and RNA processing in yeast has contributed greatly to our understanding of the mechanisms by which the mtDNA is expressed, replicated and maintained. The size of the *S. cerevisiae* mitochondrial genome (~80 kb) can vary between strains due to recombination and subsequent polymorphism involving long intergenic A–T rich spacers and G–C clusters. The mitochondrial DNA encodes seven polypeptides involved in oxidative phosphory-lation, one ribosomal protein (*var* 1), two rRNAs, 24 tRNAs and the 9S RNA component of RNase P (Foury et al. 1998).

Transcription of the mtDNA is initiated at no less than 20 individual sites dispersed around the genome, all containing a highly conserved nonanucleotide motif (NTATAAGTA). Deletion mutagenesis studies have demonstrated that this consensus sequence alone is capable of initiating transcription in vitro, but the relative strength of the promoter is dependent upon the composition of the flanking sequences (Levens et al. 1981; Christianson and Rabinowitz 1983; Tracy and Stern 1995; Biswas 1999). Transcription from these promoters results in the creation of polycistronic precursors that are composed of two or more coding sequences that require maturation in order to form functional RNA molecules. As in animal mtDNA, tRNA genes serve as important processing signals in budding yeasts. The excision of the tRNAs is accomplished via the actions of two distinct endonucleolytic cleavages involving proteins functionally analogous to the human RNase P and RNase Z. In a similar fashion to some of the transcripts generated by tRNA punctuation in human mtDNA, tRNA excision in yeast does not always generate precise 5′ and 3′ termini, and therefore nearly all transcripts undergo further processing via additional endonucleolytic cleavages (Tzagaloff and Myers 1986; Schäfer et al. 2005). The 3′ termini of mitochondrial mRNAs in S. cerevisiae are specified by a highly conserved dodecamer sequence found at variable distances downstream of the coding sequence in the 3′ UTR (Osinga et al. 1984). The 12-mer motif is proposed to act as a recognition signal that recruits the degradosome complex (mtEXO), comprising the RNA helicase Suv3p and the 3′–5′ exoribonuclease Dss1p (Dziembowski et al. 2003). It has been proposed that the dodecamer binding protein (DBP), a site-specific RNA binding protein, prevents the degradation of newly processed transcripts by binding to the recognition site (Li and Zassenhaus 2000).The 5′ end processing of the transcript containing the COB mRNA, on the other hand, seems to be mediated by the Cbp1p protein, which has also been shown to be linked to the stability of the mature mRNA (Chen and Dieckmann 1994, Shadel 2004).

Unlike the transcripts of animal mtDNAs, transcripts of yeasts and other fungi are not polyadenylated (Tzagaloff and Myers 1986). While polyadenylation of processed transcripts is required for the stability of the nascent RNA in animal mitochondria (Slomovic et al. 2005), in budding yeasts newly processed transcripts seem to be protected from exonucleolytic degradation via protein interactions, perhaps with the processing machinery itself. Alternatively, it has been suggested that the newly processed RNA molecules have the ability to form nuclease-resistant secondary structures (Chen and Dieckmann 1994).

Investigations into the processing events in yeast mitochondria also provided support for the hypothesis that transcription is directly coupled to processing and translation. The N-terminal domain of the S. cerevisiae mitochondrial RNAP (sc-mtRNAP) contains a binding site for Nam1p, a protein implicated in post-transcriptional mitochondrial RNA metabolism (Rodeheffer et al. 2001). The association of the sc-mtRNAP with Nam1p is thought to result in multiple protein interactions involving several RNA processing factors but also translation initiation factors, thereby delivering the transcription complex to the mitochondrial translation machinery. The N-terminal domain of the sc-mtRNAP therefore seems to provide the nucleation point for the coupling of transcription and translation (Rodeheffer et al. 2001; Shadel 2004). A similar link between transcription, processing and translation may also exist in human mitochondria, as the human mitochondrial RNA polymerase possesses an N-terminal extension that contains so-called tandem pentatricopeptide repeat (PPR) domains (Shadel 2004). PPR motifs are predominantly found in proteins involved in various functions in mitochondrial RNA metabolism, including RNA processing, editing, transcription and translation (Delannoy et al. 2007).

Overall, the processing of mitochondrial transcripts seems to be more elaborate in *S. cerevisiae* than in metazoans. However, there are similarities that suggest a common heritage for the evolution of the processing systems, primarily (a) the generation of polycistronic transcripts from a few or multiple promoters, and (b) tRNA punctuation, whereby precise endonucleolytic excision, most likely co-transcriptional, occurs and liberates coding sequences (Schäfer 2005). It seems that the primary determinant for this mode of transcript maturation evidenced in many species is dependent not only on phylogenetic lineage, but also on the particular arrangements within the respective mtDNAs.

The mitochondrial DNA of the fission yeast *Schizosaccharomyces pombe* is similar, in both size and arrangement, to that of humans. All the coding sequences are arranged on one strand with a high gene density (Bullerwell et al. 2003; Schäfer 2005). The transcription and subsequent maturation of *S. pombe* mitochondrial RNAs combines features of both the metazoan and fungal systems. The generation of two large polycistronic transcripts is initiated from two major promoter sites that share almost identical sequence homology to the conserved nonanucleotide motifs of the budding yeasts. With the exception of two genes (*cox3, rnl*), the 3′ ends of the tRNA genes are immediately contiguous to the 5′ termini of mature transcripts that contain 5′ untranslated sequences of varying length (Schäfer 2005). The processing via endonucleolytic cleavage is therefore the primary mechanism for producing the 5′ termini, as is the case in metazoan mitochondria. The 3′ ends of mature transcripts nearly all terminate slightly downstream (-2) of a conserved C-rich element termed the C-core motif, with the accuracy of processing determined by the G–C content of the motif (Schäfer et al. 2005; Hoffmann et al. 2008). This 3′ processing motif is proposed to be functionally analogous to the dodecamer sequence of the budding yeasts and is highly conserved in other species of fission yeasts (Bullerwell et al. 2003; Schäfer 2005; Hoffmann et al. 2008). In *S. pombe*, knockout strains lacking a functional degradosome are impaired in their ability to perform downstream processing of transcripts, while the steady-state levels of mitochondrial RNAs remain unaffected. This implies that, in this organism, homologs of the *S. cerevisiae* mtEXO components (Suv3p, Dss1p) play only a minor role in mitochondrial RNA degradation but are involved primarily in other cellular pathways such as RNA processing (Hoffmann et al. 2008).

The prevalence of a conserved mechanism of transcript processing in a wide distribution of species with similarities in transcription initiation and genome organisation could possibly reflect a relatively ancestral state of the mitochondrial transcription initiation and transcript processing machinery (Schäfer 2005). That is, high gene density coupled with low promoter numbers, tRNA punctuation and alternative modes of transcript processing. In contrast, the high promoter numbers and low gene density found in mitochondrial DNAs of plants and the budding yeasts could represent an adaptation brought about by the increased genome sizes that arose through duplication or recombination events (Foury et al. 1998; Schäfer

2005). Investigation into the mechanisms of mitochondrial gene expression in those organisms that retain ancestral features and exhibit similarities with more highly derived species will evidently contribute greatly to the elucidation of questions regarding the evolution of the mitochondrial genome. The mitochondrial genome of *D. discoideum* shares many features with metazoan, fungal and plant mitochondrial genomes (Gray et al. 1998; Ogawa et al. 2000; Barth et al. 2007).

10.3.3   *Transcript Processing in* D. discoideum *Mitochondria*

Investigations into the transcription of the *D. discoideum* mitochondrial genome led to the identification of eight major polycistronic transcripts that are generated from processing of a single, large polycistronic precursor (Fig. 10.2). The confirmation of the existence and function of a single transcription initiation site was described earlier in this chapter. Northern hybridization studies performed at the time also demonstrated that most of the eight transcripts were further processed into mono-, di- and tricistronic RNAs (Barth et al. 1999, 2001; Le et al. 2009). Given the location and distribution of the tRNA genes in the *D. discoideum* mitochondrial genome, the smaller mature RNA molecules are presumably generated from the larger transcriptional units by endonucleolytic cleavage. The tRNA genes are dispersed around the genome and their location often coincides with the location of processing sites at the RNA level (Fig. 10.2), suggesting that transcript processing in the *D. discoideum* mitochondrial genome also involves the excision of tRNAs from larger transcripts. Amongst others, *D. discoideum* shares this form of mitochondrial RNA maturation with humans, where it has been demonstrated that the single, polycistronic transcript derived from transcription of the entire heavy strand does not exist in its entirety at any given time due to the co-transcriptional release of the tRNA molecules from the primary transcript (Ojala et al. 1980, 1981). This may also be the case in *D. discoideum* mitochondria, where the failure to detect any transcripts larger than the eight major transcripts observed in the northern hybridization studies indicates that the processing of the primary transcript must also occur very efficiently (Barth et al. 2001; Le et al. 2009).

Apart from liberating most of the secondary and tertiary transcripts by tRNA excision, the absence of punctuating tRNAs in other parts of the *D. discoideum* mitochondrial genome necessitates the presence of other processing signals. Some termini that cannot be generated by tRNA excision are the 3' terminus of the secondary transcript G, and the 3' termini of the tertiary transcripts *nad4, nad2,* and *cox3* (Fig. 10.2). As discussed earlier in this chapter, in other organisms, where tRNA punctuation is not the sole mechanism of generating mature transcripts, putative processing signals have been shown to include conserved sequence motifs (Osinga et al. 1984; Schuster and Brennicke 1989; Schäfer et al. 2005) antisense tRNA sequences (Montoya et al. 2006) or secondary structures formed by the nascent RNA transcript (Ojala et al. 1981; Clayton 1984). The secondary processing signals are proposed to either mimic the tRNA processing signals and utilise the same processing enzymes (Rossmanith et al. 1995; Montoya et al. 2006) or recruit an alternative processing machinery (Dziembowski et al. 2003; Shadel 2004; Hoffmann et al. 2008).

Any conserved sequence motifs that could serve as processing signals have not been identified in *D. discoideum* mitochondria, suggesting that non-tRNA punctuated processing is most likely achieved via mechanisms involving an alternative processing machinery. In recent studies in humans and fungi, the roles of DExH/D RNA helicases and PPR proteins have been implicated in various aspects of organelle RNA maturation (Delannoy et al. 2007; Hoffmann et al. 2008; Szczesny et al. 2010). Through comparative genomics and in silico analysis, a number of genes encoding members of these families have been identified in *D. discoideum* and cloned, and the mitochondrial localization of some of these putative processing enzymes has been confirmed in vivo (Kennedy and Barth, unpublished). The characterization of these proteins and their exact function in RNA metabolism in the mitochondria of *D. discoideum* is the subject of current investigations.

10.4 Mitochondrial Dysfunction

Mitochondrial dysfunction in humans leads to a great variety of clinical outcomes that can adversely affect any organs or tissues, but most commonly includes neurological and neurodegenerative disease (Zeviani and Carelli 2007; Francione and Fisher 2010). Collectively, neurodegenerative diseases are forecast by the UN to eclipse cancer as the second major cause of death worldwide by 2040. Mitochondrial dysfunction is a central feature of the disease process in most of them, even in cases where the disease has a primary non-mitochondrial cause. These include Alzheimer's, Parkinson's, Huntington's and Motor Neuron diseases, Multiple Systems Atrophy and Rett Syndrome. Parkinson's Disease (PD), for example, is one of the most common neurodegenerative disorders and, like other such diseases, is most often of unknown aetiology (Vila et al. 2008; Hatano et al. 2009). However a minority of PD cases are familial and monogenic. Their study has allowed the identification so far of 11 nuclear-encoded PD-associated proteins, at least 6 of which are partly or entirely localised in the mitochondria.

Overtly mitochondrial diseases (those known to result from genetic defects directly affecting the mitochondria) were previously considered to affect about 1 in 5,000 individuals (Zeviani and Carelli 2007; Di Donato 2009), making them collectively one of the more common human genetic diseases. However, more recent estimates are that as many as 1 in 250 individuals may be affected (Schäfer et al. 2004, 2008; Elliott et al. 2008). Part of the reason for the uncertainty is the complexity and variability of the human disease phenotypes arising from mitochondrial dysfunction. The unpredictability of mitochondrial disease outcomes in humans arises from complexities of human mitochondrial biology, development and ageing which are overlaid upon the underlying cellular mechanisms. This has hindered our understanding of the fundamental disease processes. In contrast, the protozoan *D. discoideum* combines the experimental tractability of a well-established model system with a unique lifecycle (Fig. 10.1) that offers a wide range of consistent, readily assayed disease phenotypes (Table 10.1).

The *D. discoideum* mitochondrial disease model originated with isolation of a phototaxis-deficient mutant in which a subset of the mitochondrial genomes had been disrupted by plasmid insertion into the large ribosomal RNA subunit gene (*rnl*) (Wilczynska et al. 1997). This situation, in which only a subset of the mitochondrial genomes is mutant, is known as heteroplasmy and it occurs in maternally inherited human mitochondrial diseases. Subsequent targeted, heteroplasmic disruption of *rnl* and eight other mitochondrial genes in *D. discoideum* revealed a consistent pattern of phenotypic outcomes (Wilczynska et al. 1997; Francione 2008). These included impaired growth and increased ability of cells infected with *Legionella pneumophila* to support intracellular proliferation of the bacterial pathogen, defects in photosensory and thermosensory transduction and aberrant morphogenesis in the multicellular stages (Table 10.1). The same pattern of aberrant phenotypes was observed in strains in which expression of chaperonin 60, an essential mitochondrial protein, had been genetically inhibited ("knocked down" by a technique called antisense inhibition). Chida et al. (2004) reported similar phenotypes when ethidium bromide treatment was used to selectively interfere with replication of the mitochondrial genome, thereby producing cells with depleted levels of mitochondrial DNA (Table 10.1).

In each of these cases, the nature of the genetic defect is such that it would cause a generalised respiratory deficiency affecting multiple respiratory complexes. Chaperonin 60 knockdown is expected to cause a generalised respiratory defect, because of the essential role of this protein in folding both mitochondrially and nuclear-encoded proteins in the mitochondria. Depletion of the entire mitochondrial genome with ethidium bromide would reduce the levels of expression of all mitochondrial genes including the mitochondrially encoded subunits of Complexes I, III, IV and V. Francione (2008) showed that expression of the entire mitochondrial genome was depressed in mutants in which any of nine different mitochondrial genes had been targeted for disruption. This is consistent with the fact that the entire mitochondrial genome is transcribed unidirectionally from a single promoter and the mature mitochondrial RNAs are all derived from the resulting transcript by endonucleolytic processing (see earlier sections of this article).

In humans, as in *D. discoideum*, different genetic defects affecting the mitochondria can lead to similar signs and symptoms, but the same genetic defect can also lead to markedly different and unpredictable clinical outcomes. This also was true in the *D. discoideum* model. Compared to the effects of generalise respiratory defects described above, quite different and more limited phenotypic outcomes were observed for mutations that affected division (*fszA* and *fszB*) or subcellular localization (*cluA*) of the mitochondria (Zhu et al. 1997; Gilson et al. 2003). These defects would not be expected to impair oxidative phosphorylation. Furthermore, not every strain with a generalised respiratory defect exhibited all of the aberrant phenotypes listed in Table 10.1. The differences turned out to be

Table 10.1 Phenotypes associated with mitochondrial dysfunction in *D. discoideum*

Method of generating mitochondrial dysfunction	Phenotype									*Legionella* susceptibility	References
	Growth on bacteria	Growth in broth	Phagocytosis	Pinocytosis	Phototaxis	Thermotaxis	Morphogenesis	Aggregation	Chemotaxis		
Pharmacological, expected to affect respiration											
Ethidium bromide inhibition of mtDNA replication	−	−				−	− (Stalky)	−			Chida et al. (2004)
Genetic, expected to affect respiration											
Heteroplasmic mitochondrial gene disruption (*rnl*, *nad5*, *cob*, *nad2*, *atp6*, *atp1*, *cox3*, *ORF1740*, *ORF796*)	−		+	+		−	− (Stalky)	−			Wilczynska et al. (1997), Francione (2008) and Francione et al. (2009)
Heteroplasmic *rps4* disruption	+/−	+/−									Inazu et al. (1999) and Fisher (unpublished)
Chaperonin 60 antisense inhibition	−	−	+	+		−	− (Stalky)	−			Kotsifas et al. (2002); Bokko et al. (2007); Francione et al. (2009)
Genetic, respiratory complex-specific defect in respiration											
MidA knockout producing specific Complex I deficiency	−	−	−	−	−	−	+/− (Stalky)	+			Torija et al. (2006) and Carilla-Latorre et al. (2010)
Genetic, not known to affect respiration											
Nuclear *fszA*, *fszB* disruption	+ (fszA⁻) − (fszB⁻)	+ (fszA⁻) − (fszB⁻)			+	+	+	+			Gilson et al. (2003) and Fisher (unpublished)
Nuclear *cluA* disruption	Defective cytokinesis	Defective cytokinesis	+			+	+	+			Zhu et al. (1997)
Nuclear *torA* disruption	−		+					+	−		van Es et al. (2001)
Nuclear Dd-TRAP1 RNAi inhibition		−						−			Morita et al. (2004)

+ wild-type phenotype; − aberrant phenotype; +/− mildly aberrant phenotype; *Shaded cells* phenotype not reported caused primarily by differences in the severity of the underlying genetic defect. This was most clearly shown by genetic dose–response curves relating the severity of the mutant phenotype to the number of copies of the chaperonin 60 antisense-inhibition construct that were integrated stably into the genome in the individual knockdown strains. Thus, Kotsifas et al. (2002) reported that although phototaxis was impaired significantly in all of their chaperonin 60 knockdown strains, growth in liquid medium was significantly slower only when the copy number of the antisense construct was greater than about 60. Bokko et al. (2007) and Francione et al. (2009) also reported striking copy-number dependence of the various aberrant phenotypes in chaperonin 60 knockdown strains (Table 10.1). On the basis of their phenotypes and those of a large collection of heteroplasmic mitochondrial gene disruptants, Francione (2008) suggested a hierarchy of abnormalities that can be ranked in order of appearance as the mitochondrial dysfunction increases in severity. The ranking indicates the sensitivity of the phenotype to the degree of mitochondrial dysfunction and can be written as phototaxis/thermotaxis > *Legionella* susceptibility > growth > multicellular development > aggregation >> phagocytosis and macropinocytosis.

In human mitochondrial disease, the outcomes also depend partly on the severity of the underlying genetic disorder (e.g. the tissue-dependent, age-dependent proportion of mutant mitochondrial genomes in heteroplasmic mitochondrial disease). This phenomenon in humans has been referred to as the threshold effect. It arises from nonlinearities (e.g. protein levels are not a simple linear function of the cognate mRNA levels) and homeostatic feedbacks in the effects that primary genetic defects ultimately exert on mitochondrial ATP generation (Rossignol et al. 2003). These nonlinearities and regulatory processes influence the steady-state levels of mitochondrial mRNA, protein, protein activity, electron transport rate, mitochondrial membrane potential and ATP generation. As a result, steady-state ATP levels are diminished significantly only when the cell's homeostatic mechanisms are overwhelmed by the severity of the underlying genetic disorder.

Can the threshold effect on ATP levels provide sufficient explanation for similar threshold effects on the cytopathological outcomes of mitochondrial disease? If so, then disease phenotypes would result from an ATP insufficiency that occurs only when cellular homeostatic mechanisms fail. The *D. discoideum* model has made clear that this is not the case. Bokko et al. (2007) and Francione et al. (2009) showed instead that diverse phenotypes of mitochondrially diseased *D. discoideum* are caused by chronic activity of AMP-activated Protein Kinase (AMPK), an energy-sensing protein kinase that is itself a central component of the cellular mechanisms for ATP homeostasis. Overexpression of a constitutively active form of AMPK caused the same phenotypic outcomes as mitochondrial dysfunction, while antisense inhibition of AMPK expression in mitochondrially diseased cells suppressed all of the aberrant phenotypes.

AMPK is a heterotrimeric protein kinase that is activated with exquisite sensitivity by increases in the AMP/ATP ratio. Its normal role in healthy cells is to inhibit a variety of energy consuming processes (e.g. cell cycle progression and growth) and to stimulate energy production (e.g. by fatty acid oxidation) and mitochondrial biogenesis. Bokko et al. (2007) confirmed that in *D. discoideum* AMPK stimulates mitochondrial biogenesis and ATP production, as it does in human cells. In otherwise healthy cells facing a temporary energy shortage, the result is restoration of cellular ATP generation that enables a return to normality. In a mitochondrially diseased cell however, mitochondrial energy producing capacity is permanently compromised genetically, so that ATP levels remain in the normal range only as long as and because AMPK is chronically activated (Fig. 10.4). The result is an abnormal steady state in the diseased cell in which the downstream consequences of AMPK activity are permanent cytopathological features.

Not all cellular energy consuming functions are regulated by AMPK. Of the various phenotypes studied in the *Dictyostelium* mitochondrial disease model, phagocytosis and macropinocytosis were impervious to AMPK signalling (Bokko et al. 2007; Francione et al. 2009). These two nutrient uptake mechanisms used by *D. discoideum* were accordingly also unaffected by heteroplasmic mitochondrial gene disruption or by chaperonin 60 knockdown. The specificity of AMPK signalling pathways thus explains why some cellular energy-requiring functions are affected in mitochondrial disease and others are not.

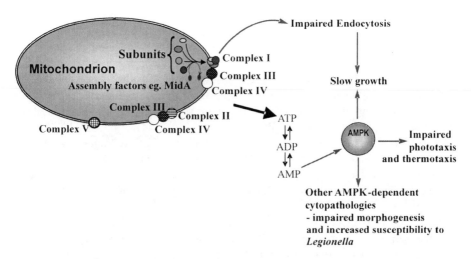

Fig. 10.4 Cytopathological pathways in *D. discoideum* mitochondrial dysfunction. Two cytopathological pathways are shown. 1. Generalised mitochondrial respiratory dysfunction affecting multiple oxidative phosphorylation complexes compromises mitochondrial ATP generation, leading to chronic AMPK activation. The resulting dysregulation of intracellular signalling produces multiple cytopathological outcomes. Not shown is the homeostatic feedback by which AMPK stimulates mitochondrial biogenesis and ATP production. In mitochondrially diseased cells, chronic AMPK activation can thereby maintain ATP at normal levels, while at the same time causing chronic downstream cytopathologies. 2. In addition to AMPK-dependent pathways, Complex I-specific dysfunction specifically impairs endocytic pathways (phagocytosis and macropinocytosis) in an AMPK-independent manner. Different, more limited cytopathologies may be caused by mutations that affect other aspects of mitochondrial biology without impairing ATP production

As noted above, heteroplasmic mitochondrial gene disruption and chaperonin 60 antisense inhibition are both expected to impair the levels and activities of multiple respiratory complexes. However, in many cases of human mitochondrial disease there are specific defects in particular respiratory complexes. In particular, about 40% of human mitochondrial disease cases involve a specific Complex I deficiency. There is no a priori reason why a specific Complex I deficiency should cause different cytopathology from the generalised respiratory defects described above. However, recent discoveries using the *D. discoideum* model have revealed that Complex I-specific deficiency produces additional adverse phenotypic consequences that are independent of AMPK and superimposed upon those that result from chronic AMPK hyperactivity (Table 10.1, Fig. 10.4). During a search for genes shared by *D. discoideum* and humans, but not the yeast *S. cerevisiae*, Ricardo Escalante's group in Spain discovered a protein called MidA (Torija et al. 2006). This protein proved to be a methyl transferase and mitochondrial Complex I assembly factor whose absence in a knockout mutant caused reduced levels and activities of Complex I accompanied by increased levels and activities of other respiratory complexes (Carilla-Latorre et al. 2010). The increase in these other respiratory complexes suggest a compensatory feedback mechanism, possibly via AMPK, that stimulates mitochondrial biogenesis but, because of the absence of MidA, is unable to restore Complex I activity. Like other mitochondrially diseased strains, the mutant exhibited defects in phototaxis and thermotaxis that were AMPK dependent. Unlike them, it also exhibited severe AMPK-independent defects in phagocytosis and macropinocytosis and these deficiencies in nutrient uptake resulted in secondary growth defects that could not be suppressed by AMPK antisense inhibition (Carilla-Latorre et al. 2010). The results suggest the existence of additional cytopathological pathways that are specifically elicited by Complex I deficiency and do not involve AMPK signalling (Fig. 10.4).

Apart from the obvious reduction in mitochondrial capacity to generate ATP, a clear and immediate consequence of OXPHOS defects is an increase in the leakage of electrons from the electron transport chain to generate reactive oxygen species (ROS) (Zeviani and Carelli 2007; Di Donato 2009; Francione and Fisher 2010). The resulting mitochondrial damage produces a vicious feedback cycle that contributes to the degenerative nature of many mitochondrial diseases and ultimately activates various forms of programmed cell death. Most of the work on downstream pathways in mitochondrial and neurodegenerative diseases has therefore focused on the well understood mitochondrial cell death mechanisms – the processes of apoptosis that are initiated by collapse of the mitochondrial membrane potential in the mitochondrial permeability transition and release into the cytoplasm of proapoptotic molecules like cytochrome c and Apoptosis Inducing Factor (AIF). However, death is but the most obvious endpoint of cytopathological processes resulting from mitochondrial dysfunction. Normal cellular functions (e.g. dopamine synthesis and secretion in dopaminergic neurons) may be disturbed by mitochondrial defects that are sublethal at the cellular level but can cause pathological outcomes at the whole organism level. From the study of these sublethal cellular outcomes in the model protozoan *D. discoideum*, an overall picture of mitochondrial disease is emerging of a pathological disturbance in signalling networks that is more complex and nuanced than simple ATP insufficiency and cell death.

References

Agsteribbe E, Hartog M (1987) Processing of precursor RNAs from mitochondria of *Neurospora crassa*. Nucleic Acid Res 15:7249–7263

Anderson S, Bankier AT, Barrell BG, de Bruijn MH, Coulson AR, Drouin J, Eperon IC, Nierlich DP, Roe BA, Sanger F, Schreier PH, Smith AJ, Staden R, Young IG (1981) Sequence and organisation of the human mitochondrial genome. Nature 290:457–465

Angata K, Ogawa S, Yanagisawa K, Tanaka Y (1995) A group-I intron in the mitochondrial large-subunit ribosomal RNA-encoding gene of *Dictyostelium discoideum*: same site localisation in alga and *in vitro* self-splicing. Gene 153:49–55

Annesley SJ, Fisher PRF (2009) *Dictyostelium discoideum* – a model for many reasons. Mol Cell Biochem 329:73–91

Antoshechkin I, Bogenhagen DF (1995) Distinct roles for two purified factors in transcription of *Xenopus* mitochondrial DNA. Mol Cell Biol 15:7032–7042

Asin-Cayuela J, Gustafsson CM (2007) Mitochondrial transcription and its regulation in mammalian cells. Trends Biochem Sci 32:111–117

Auchincloss AH, Brown GG (1989) Soybean mitochondrial transcripts capped in vitro with guanylyltransferase. Biochem Cell Biol 67:315–319

Ausubel FM, Brent R, Kingston RE, Moore DD, Seidman JG, Smith JA, Struhl K (1994) Current protocols in molecular biology. Green Publishing Associates and Wiley, Toronto

Banerjee AK (1980) 5′-Terminal cap structure in eukaryotic messenger ribonucleic acids. Microbiol Rev 44:175–205

Barth C, Greferath U, Kotsifas M, Fisher PR (1999) Polycistronic transcription and editing of the mitochondrial small subunit (SSU) ribosomal RNA in *Dictyostelium discoideum*. Curr Genet 36:55–61

Barth C, Greferath U, Kotsifas M, Tanaka Y, Alexander S, Alexander H, Fisher PR (2001) Transcript mapping and processing of mitochondrial RNA in *Dictyostelium discoideum*. Curr Genet 39:355–364

Barth C, Le P, Fisher PR (2007) Mitochondrial biology and disease in *Dictyostelium*. Int Rev Cytol 263:207–252

Binder S, Brennicke A (1993) Transcription initiations sites in mitochondria of *Oenothera berteriana*. J Biol Chem 268:7849–7855

Bisson R, Schiavo G, Papini E (1985) Cytochrome c oxidase from the slime mould *Dictyostelium discoideum*: purification and characterisation. Biochemistry 24:7845–7852

Biswas TK (1999) Nucleotide sequences surrounding the nonanucleotide promoter motif influence the activity of the yeast mitochondrial promoter. Biochemistry 38:9693–9703

Bogenhagen DF, Yoza BK (1986) Accurate in vitro transcription of *Xenopus laevis* mitochondrial DNA from two bidirectional promoters. Mol Cell Biol 6:2543–2550

Bogenhagen DF, Applegate EF, Yoza BK (1984) Identification of a promoter for transcription of the heavy strand of human mtDNA: in vitro transcription and deletion mutagenesis. Cell 36:1105–1113

Bokko PB, Francioni L, Ahmed AU, Bandala-Sanchez E, Annesley SJ, Huang X, Khurana T, Kimmel AR, Fisher PR (2007) Diverse mitochondrial cytopathologies are caused by AMPK signalling. Mol Biol Cell 18:1874–1886

Bonen L (1991) The mitochondrial genome: so simple yet so complex. Curr Opin Genet Dev 1:515–522

Bullerwell CE, Leigh J, Forget L, Lang BF (2003) A comparison of three fission yeast mitochondrial genomes. Nucleic Acids Res 31:759–768

Burger G, Helmer-Cittrich MH, Nelson MA, Werner S, Macino G (1985) RNA processing in *Neurospora crassa* mitochondria: transfer RNAs punctuate a large precursor transcript. EMBO J 4:197–204

Burger G, Plante I, Lonergan KM, Gray MW (1995) The mitochondrial DNA of the amoeboid protozoon, *Acanthamoeba castellanii*: complete sequence, gene content and genome organization. J Mol Biol 245:522–537

Burke JM, RajBhandary UL (1982) Intron within the large rRNA gene of *Neurospora crassa* mitochondria: a long open reading frame and a consensus sequence possibly important in splicing. Cell 31:509–520

Carilla-Latorre S, Gallardo ME, Annesley SJ, Calvo-Garrido J, Graña O, Accari SL, Smith PK, Valencia A, Garesse R, Fisher PR, Escalante R (2010) MidA is a putative mitochondrial methyltransferase required for mitochondrial complex I function. J Cell Sci 123:1674–1683

Cermakian N, Ikeda TM, Miramontes P, Lang BF, Gray MW, Cedergren R (1997) On the evolution of the single-subunit RNA polymerase. J Mol Evol 45:671–681

Cheetham GM, Steitz TA (2000) Insight into transcription: structure and function of single DNA-dependent RNA polymerase. Curr Opin Struct Biol 10:117–123

Chen W, Dieckmann C (1994) Cbp1p is required for message stability following 5′ processing of COB mRNA. J Biol Chem 269:16574–16578

Chida J, Yamaguchi H, Amagai A, Maeda Y (2004) The necessity of mitochondrial genome DNA for normal development of *Dictyostelium* cells. J Cell Sci 117:3141–3152

Chow TY, Fraser MJ (1983) Purification and properties of single strand DNA-binding endo-exonuclease of *Neurospora crassa*. J Biol Chem 258:12010–12018

Christianson T, Rabinowitz M (1983) Identification of multiple transcriptional initiation sites on the yeast mitochondrial genome by in vitro capping with guanylyltransferase. J Biol Chem 258:14025–14033

Clayton DA (1984) Transcription of the mammalian mitochondrial genome. Annu Rev Biochem 53:573–594

Coates JC, Harwood AJ (2001) Cell-cell adhesion and signal transduction during *Dictyostelium* development. J Cell Sci 114:4349–4358

Cole RA, Williams KL (1994) The *Dictyostelium discoideum* mitochondrial genome: a primordial system using the universal code and encoding hydrophilic proteins atypical of metazoan mitochondrial DNA. J Mol Evol 39:579–588

Cole RA, Slade M, Williams K (1995) *Dictyostelium discoideum* mitochondrial DNA encodes a NADH: ubiquinone oxido-reductase subunit, which is nuclear encoded in other eukaryotes. J Mol Evol 40:616–621

Costanzo MC, Fox TD (1990) Control of mitochondrial gene expression in *Saccharomyces cerevisiae*. Annu Rev Genet 24:91–113

Cummings DJ, Domenico JM, Nelson J (1989) DNA sequence and secondary structures of the large subunit rRNA coding regions and its two class-I introns of mitochondrial DNA from *Podospora anserina*. J Mol Evol 28:242–255

Dake E, Hofmann TJ, McIntire S, Hudson A, Zassenhaus HP (1988) Purification and properties of the major nuclease from mitochondria of *Saccharomyces cerevisiae*. J Biol Chem 263:7691–7702

Dale JW, Schantz MV (2002) From genes to genomes: concepts and applications of DNA technology. Wiley, West Sussex

Delannoy E, Stanley WA, Bond CS, Small ID (2007) Pentatricopeptide repeat (PPR) proteins as sequence-specificity factors in post-transcriptional processes in organelles. Biochem Soc Trans 35:1643–1647

Denslow ND, Michaels GS, Montoya J, Attardi G, O'Brian TW (1989) Mechanism of mRNA binding to Bovine mitochondrial ribosomes. J Biol Chem 264:8328–8338

Di Donato S (2009) Multisystem manifestations of mitochondrial disorders. J Neurol 256:693–710

Docherty K (1996) Gene transcription: DNA binding proteins essential techniques. Wiley, Chichester

Dziembowski A, Piwowarski J, Hoser R, Minczuk M, Dmochowska A, Siep M, van der Spek H, Grivell L, Stepien PP (2003) The yeast mitochondrial degradosome; its composition, interplay between helicase and RNase activities and the role in mitochondrial metabolism. J Biol Chem 278:1603–1611

Edwards JC, Levens D, Rabinowitz M (1982) Analysis of transcriptional initiation of yeast mitochondrial DNA in a homologous in vitro transcription system. Cell 31:337–346

Elliott HR, Samuels DC, Eden JA, Relton CL, Chinnery PF (2008) Pathogenic mitochondrial DNA mutations are common in the general population. Am J Hum Genet 83:254–260

Falkenberg M, Larsson NG, Gustafsson CM (2007) DNA replication and transcription in mammalian mitochondria. Annu Rev Biochem 76:679–699

Foury F, Kucej M (2001) Yeast mitochondrial biogenesis: a model system for humans? Curr Opin Chem Biol 6:106–111

Foury F, Roganti T, Lecrenier N, Purnelle B (1998) The complete sequence of the mitochondrial genome of *Saccharomyces cerevisiae*. FEBS Lett 440:325–331

Francione L (2008) Mitochondrial disease in *Dictyostelium discoideum*. PhD Thesis. La Trobe University

Francione L, Fisher PR (2010) Cytopathological mechanisms in mitochondrial disease. Curr Chem Biol 4:32–48

Francione L, Smith PK, Accari SL, Taylor PE, Bokko PB, Bozzarro S, Beech PL, Fisher PR (2009) *Legionella pneumophila* multiplication is enhanced by chronic AMPK signalling in mitochondrially diseased *Dictyostelium* cells. Dis Model Mech 2:479–489

Freeman WM, Walker SJ, Vrana KE (1999) Quantitative RT-PCR: pitfalls and potential. Biotechniques 26:112–122

Gilson PR, Yu X-C, Hereld D, Barth C, Savage A, Kiefel BR, Lay S, Fisher PR, Margolin W, Beech PL (2003) Two *Dictyostelium* orthologs of the prokaryotic cell division protein FtsZ localize to mitochondria and are required for the maintenance of normal mitochondrial morphology. Eukaryot Cell 2:1315–1326

Gray MW (1992) The endosymbiont hypothesis revised. Int Rev Cytol 14:233–357

Gray MW, Lang BF, Cedergren R, Golding GB, Lemieux C, Sankoff D, Turmel M, Brossard N, Delage E, Littlejohn TG, Plante I, Rioux P, Saint-Louis D, Zhu Y, Burger G (1998) Genome structure and gene content in protist mitochondrial DNAs. Nucleic Acids Res 26:865–878

Gray MW, Lang BF, Burger G (2004) Mitochondria of protists. Annu Rev Genet 38(477):524

Hanic-Joyce PJ, Gray MW (1991) Accurate transcription of a plant mitochondrial gene in vitro. Mol Cell Biol 11:2035–2039

Hatano T, Kubo S-I, Sato S, Hattori N (2009) Pathogenesis of familial Parkinson's disease: new insights based on monogenic forms of Parkinson's disease. J Neurochem 111:1075–1093

Hatzack F, Dombrowski S, Brennicke A, Binder S (1998) Characterisation of DNA-binding proteins from pea mitochondria. Plant Physiol 116:519–527

Hoffmann B, Nickel J, Speer F, Schäfer B (2008) The 3' ends of mature transcripts are generated by a proccessosome complex in fission yeast mitochondria. J Mol Biol 377:1024–1037

Ikeda S, Hasegawa H, Kaminaka S (1997) A 55-kDa endonuclease of mammalian mitochondria: comparison of its subcellular localisation and endonucleolytic properties with those of endonuclease G. Acta Med Okayama 51:55–62

Inazu Y, Chae SC, Maeda Y (1999) Transient expression of a mitochondrial gene cluster including *rps4* is essential for the phase-shift of *Dictyostelium* cells from growth to differentiation. Dev Genet 25:339–352

Iwamoto M, Pi M, Kurihara M, Morio T, Tanaka Y (1998) A ribosomal protein gene cluster is encoded in the mitochondrial DNA of *Dictyostelium discoideum*: UGA termination codons and similarity of gene order to *Acanthamoeba castellanii*. Curr Genet 33:304–310

Jan P, Stein T, Hehl S, Lisowsky T (1999) Expression studies and promoter analysis of the nuclear gene for mitochondrial transcription factor 1 (MTF1) in yeast. Curr Genet 36:37–48

Keith JM, Venkatesan S, Gershowitz A, Moss B (1982) Purification and characterisation of the mRNA capping enzyme GTP:RNA guanylyltransferase from wheat germ. Biochemistry 21:327–333

Kennell JC, Lambowitz AM (1989) Development of an in vitro transcription system for *Neurospora crassa* mitochondrial DNA and identification of transcription initiation sites. Mol Cell Biol 9:3603–3613

Kotsifas M, Barth C, Lay ST, de Lozanne A, Fisher PR (2002) Chaperonin 60 and mitochondrial disease in *Dictyostelium*. J Muscle Res Cell Motil 23:839–852

Kück U, Godehardt I, Schmidt U (1990) A self-splicing group II intron in the mitochondrial large subunit rRNA (LSUrRNA) gene of the eukaryotic alga *Scenedesmus obliquus*. Nucleic Acids Res 18:2691–2697

Lang BF, Forget L (1993) The mitochondrial genome of *Phytophthora infestans*. In: O'Brien SJ (ed) Genetic maps. Cold Spring Harbor Laboratory, New York, pp 3.133–3.135

Lang BF, Burger G, O'Kelly CJ, Cedergren R, Golding GB, Lemieux C, Sanko D, Turmel M, Gray MW (1997) An ancestral mitochondrial DNA resembling a eubacterial genome in miniature. Nature 387:493–497

Le P, Fisher PR, Barth C (2009) Transcription of the *Dictyostelium discoideum* mitochondrial genome occurs from a single initiation site. RNA 15:2321–2330

Levens D, Ticho B, Ackerman E, Rabinowitz M (1981) Transcriptional initiation and 5' termini of yeast mitochondrial RNA. J Biochem 25:5226–5232

Li H, Zassenhaus HP (2000) Phosphorylation is required for high affinity binding of DBP, a yeast mitochondrial site-specific RNA binding protein. Curr Genet 37:356–363

Li J, Maga JA, Cermakian N, Cedergren R, Feagin JE (2001) Identification and characterization of a *Plasmodium falciparum* RNA polymerase gene with similarity to mitochondrial RNA polymerases. Mol Biochem Parasitol 113:261–269

Liao HX, Spremulli L (1989) Interaction of bovine mitochondrial ribosomes with messenger RNA. J Biol Chem 264:7518–7522

Liao HX, Spremulli L (1990) Effects of length and secondary structure on the interaction of bovine mitochondrial ribosomes with messenger RNA. J Biol Chem 265:11761–11765

Lonergan KM, Gray MW (1996) Expression of a continuous open reading frame encoding subunits 1 and 2 of cytochrome c oxidase in the mitochondrial DNA of *Acanthamoeba castellanii*. J Mol Biol 257:1019–1030

MacWilliams H, Doquang K, Pedrola R, Dollamn G, Grassi D, Peis T, Tsang A, Ceccarelli A (2006) A retinoblastoma ortholog controls stalk/spore preference in *Dictyostelium*. Development 133:1287–1297

Maeda Y (2005) Regulation of growth and differentiation in *Dictyostelium*. Int Rev Cytol 244:287–332

McAllister WT (1993) Structure and function of the bacteriophage T7 RNA polymerase (or, the virtues of simplicity). Cell Mol Biol Res 39:385–391

Meng XW, Fraser MJ, Ireland CM, Feller JM, Ziegler JB (1998) An investigation of a possible role for mitochondrial nuclease in apoptosis. Apoptosis 3:395–405

Montoya J, Ojala D, Attardi G (1981) Distinctive features of the 5'-terminal sequences of the human mitochondrial mRNAs. Nature 290:465–470

Montoya J, Christianson T, Levens D, Rabinowitz M, Attardi G (1982) Identification of initiation sites for heavy-strand and light-strand transcription in human mitochondrial DNA. Proc Natl Acad Sci U S A 79:7195–7199

Montoya J, Lopez-Perez M, Ruiz-Pesini E (2006) Mitochondrial DNA transcription and diseases: past, present and future. Biochim Biophys Acta 1757:1179–1189

Morita T, Amagai A, Maeda Y (2004) Translocation of the *Dictyostelium* TRAP1 homologue to mitochondria induces a novel prestarvation response. J Cell Sci 117:5759–5770

Nash EA, Nisbet RER, Barbrook AC, Howe CJ (2008) Dinoflagellates: a mitochondrial genome all at sea. Trends Genet 24:328–355

Oda K, Yamato K, Ohta E, Nakamura Y, Takemura M, Nozato N, Kohchi T, Ogura Y, Kanegae T, Akashi K, Ohyama K (1992) Gene organisation deduced from the complete sequence of liverwort *Marchantia polymorpha* mitochondrial DNA. A primitive form of plant mitochondrial genome. J Mol Biol 223:1–7

Ogawa S, Matsuo K, Angata K, Yanagisawa K, Tanaka Y (1997) Group-I introns in the cytochrome c oxidase genes of *Dictyostelium discoideum*: two related ORFs in one loop of a group-I intron, a cox1/2 hybrid gene and an unusually large cox3 gene. Curr Genet 31:80–88

Ogawa S, Yoshino R, Angata K, Iwamoto M, Pi M, Kuroe K, Matsuo K, Morio T, Urushihara H, Yanagisawa K, Tanaka Y (2000) The mitochondrial DNA of *Dictyostelium discoideum*: complete sequence, gene content and genome organisation. Mol Gen Genet 263:514–519

Ojala D, Merkel C, Gelfand R, Attardi G (1980) The tRNA punctuate the reading of genetic information in human mitochondrial DNA. Cell 22:393–403

Ojala D, Montoya J, Attardi G (1981) tRNA punctuation model of RNA processing in human mitochondria. Nature 290:470–474

Osinga KA, De Vries E, Van der Horst G, Tabak HF (1984) Processing of yeast mitochondrial messenger RNAs at a conserved dodecamer sequence. EMBO J 3:829–834

Otto GP, Kessin RH (2001) The intriguing biology of *Dictyostelium discoideum* meeting report: International *Dictyostelium* conference 2001. Protist 152:243–248

Oudot-Le Secq MP, Fontaine JM, Rousvoal S, Kloareg B, Loiseaux-de-Goër S (2001) The complete sequence of a brown algal mitochondrial genome, the ectocarpale *Pylaiella littoralis* (L.) Kjellm. J Mol Evol 53:80–88

Parisi MA, Clayton DA (1991) Similarity of human mitochondrial transcription factor 1 to high mobility group proteins. Science 252:965–970

Pellizzari R, Anjard C, Bisson R (1997) Subunits I and II of *Dictyostelium* cytochrome c oxidase are specified by a single open reading frame transcribed into a large polycistronic RNA. Biochim Biophys Acta 1320:1–7

Pritchard AE, Seilhamer JJ, Mahalingam R, Sable CL, Venuiti SE, Cummings DJ (1990) Nucleotide sequence of the mitochondrial genome of *Paramecium*. Nucleic Acids Res 18:173–180

Proudfoot NJ, Furger A, Dye MJ (2002) Integrating mRNA processing with transcription. Cell 108:501–512

Rodeheffer MS, Boone BE, Bryan AC, Shadel GS (2001) Nam1p, a protein involved in RNA processing and translation, is coupled to transcription through an interaction with yeast mitochondrial RNA polymerase. J Biol Chem 276:8616–8622

Rossignol R, Faustin B, Rocher C, Malgat M, Mazat J-P, Letellier T (2003) Mitochondrial threshold effects. Biochem J 370:751–762

Rossmanith W, Tullo A, Potuschak T, Karwan R, Sbisa E (1995) Human mitochondrial tRNA processing. J Biol Chem 270:12885–12891

Rousvoal S, Oudot MP, Fontaine JM, Kloareg B, Goer S (1998) Witnessing the evolution of transcription in mitochondria: the mitochondrial genome of the primitive brown alga *Pylaiella littoralis* (L.) Kjellm. encodes a T7-like RNA polymerase. J Mol Biol 277:1047–1058

Schäfer B (2005) RNA maturation in mitochondria of *S. cerevisiae* and *S. pombe*. Gene 354:80–85

Schäfer AM, Taylor RW, Turnbull DM, Chinnery PF (2004) The epidemiology of mitochondrial disorders – past, present and future. Biochim Biophys Acta 1659:115–120

Schäfer B, Hansen M, Lang BF (2005) Transcription and RNA-processing in fission yeast mitochondria. RNA 11:785–795

Schäfer AM, McFarland R, Blakely EL, He L, Whittaker RG, Taylor RW, Chinnery PF, Turnbull DM (2008) Prevalence of mitochondrial DNA disease in adults. Ann Neurol 63:35–39

Schuster W, Brennicke A (1989) Conserved sequence elements at putative processing sites in plant mitochondria. Curr Genet 15:187–192

Shadel GS (2004) Coupling the mitochondrial transcription machinery to human disease. Trends Genet 20:513–519

Shadel GS, Clayton DA (1993) Mitochondrial transcription initiation, variation, and conservation. J Biol Chem 268:16083–16086

Shadel GS, Clayton DA (1995) A *Saccharomyces cerevisiae* mitochondrial transcription factor sc-mtTFB, shares features with sigma factors but is functionally distinct. Mol Cell Biol 15:2101–2108

Slomovic S, Laufer D, Geiger D, Schuster G (2005) Polyadenylation and degradation of human mitochondrial RNA: the prokaryotic past leaves its mark. Mol Cell Biol 25:6427–6435

Stewart JB, Beckenbach A (2009) Characterization of mature mitochondrial transcripts in *Drosophila*, and the implications for the tRNA punctuation model in arthropods. Gene 445:49–57

Szczesny R, Borowksi LS, Brzezniak LK, Dmochowska A, Gewartowski K, Bartnik E, Stepien PP (2010) Human mitochondrial turnover caught in flagranti: involvement of hSuv3p helicase in RNA surveillance. Nucleic Acids Res 38:279–298

Taanman J (1999) The mitochondrial genome: structure, transcription, translation, and replication. Biochim Biophys Acta 1410:103–123

Takaku H, Minagawa A, Takagi M, Nashimoto M (2003) A candidate prostate cancer susceptibility gene encodes tRNA 3′ processing endoribonuclease. Nucleic Acids Res 31:2272–2728

Temperley RJ, Wydro M, Lightowlers RN, Chrzanowska-Lightowlers ZM (2010) Human mitochondrial mRNAs – like members of all families, similar but different. Biochim Biophys Acta 1797:1081–1085

Torija P, Vicente JJ, Rodrigues TB, Robles A, Cerdán S, Sastre L, Calvo RM, Escalante R (2006) Functional genomics in *Dictyostelium*: MidA, a novel conserved protein is required for mitochondrial function and development. J Cell Sci 119:1154–1164

Tracy RL, Stern DB (1995) Mitochondrial transcription initiation: promoter structures and RNA polymerases. Curr Genet 28:205–216

Tzagaloff A, Myers AM (1986) Genetics of mitochondrial biogenesis. Annu Rev Biochem 55:249–285

Unseld M, Marienfeld JR, Brandt P, Brennicke A (1997) The mitochondrial genome of *Arabidopsis thaliana* contains 57 genes in 366,924 nucleotides. Nat Genet 15:57–61

van Es S, Wessels D, Soll DR, Borleis J, Devreotes PN (2001) Tortoise, a novel mitochondrial protein, is required for directional responses of *Dictyostelium* in chemotactic gradients. J Cell Biol 152:621–632

van Loo G, Schotte P, van Gurp M, Demol H, Hoorelbeke B, Gevaert K, Rodriguez I, Ruiz-Carrillo A, Vandekerckhove J, Declercq W, Beyaert R, Vandenabeele P (2001) Endonuclease G: a mitochondrial protein released in apoptosis and involved in caspase-independent DNA degradation. Cell Death Differ 8:1136–1142

Vila M, Ramonet D, Perier C (2008) Mitochondrial alterations in Parkinson's disease: new clues. J Neurochem 107:317–328

Virbasius JV, Scarpulla RC (1994) Activation of the human mitochondrial transcription factor A gene by nuclear respiratory factors: a potential regulatory link between nuclear and mitochondrial gene expression in organelle biogenesis. Proc Natl Acad Sci U S A 91:1309–1313

Walberg MW, Clayton DA (1983) In vitro transcription of human mitochondrial DNA: identification of specific light strand transcripts from the displacement loop region. J Biol Chem 258:1268–1275

Wilczynska Z, Barth C, Fisher PR (1997) Mitochondrial mutations impair signal transduction in *Dictyostelium discoideum* slugs. Biochem Biophys Res Commun 234:39–43

Wilson RJM, Williamson DH (1997) Extrachromosomal DNA in the Apicomplexa. Microbial Mol Biol Rev 61:1–16

Wolff G, Plante I, Lang BF, Kück U, Burger G (1994) Complete sequence of the mitochondrial DNA of the chlorophyte alga *Prototheca wickerhamii*. Gene content and genome organization. J Mol Biol 237:75–86

Wolstenholme DR (1992) Animal mitochondrial DNA: structure and evolution. Int Rev Cytol 141:173–216

Yoza BK, Bogenhagen DF (1984) Identification and in vitro capping of a primary transcript of human mitochondrial DNA. J Biol Chem 259:3909–3915

Zamaroczy M, Bernardi G (1986) The primary structure of the mitochondrial genome of *Saccharomyces cerevisiae*. Gene 47:155–177

Zeviani M, Carelli V (2007) Mitochondrial disorders. Curr Opin Neurol 20:564–571

Zhu Q, Hulen D, Liu T, Clarke M (1997) The cluA⁻ mutant of *Dictyostelium* identifies a novel class of proteins required for dispersion of mitochondria. Proc Natl Acad Sci USA 9: 7308–7313

Zurawski G, Zurawski SM (1985) Structure of the *Escherichia coli* S10-ribosomal protein operon. Nucleic Acids Res 13:4521–4526

Chapter 11
Mechanism and Regulation of Mitochondrial Transcription in Animal Cells

Paola Loguercio Polosa, Marina Roberti, and Palmiro Cantatore

11.1 Introduction

All eukaryotic cells contain at least two separate genetic systems localized in the nucleus and in mitochondria, respectively. According to the endosymbiotic theory, mitochondria derive from an α-proteobacterium, which enabled anaerobic cells to utilize oxygen (Gray et al. 1999). During evolution, the genome of this bacterium lost many genes that moved to the nuclear DNA; nevertheless, the endosymbiont still retained its own genome and separate machinery for mtDNA replication, transcription and translation. The number of the conserved genes, as well as the size of mtDNA, varies greatly between different organisms. In plants, the mitochondrial genome has a large size but a comparatively small coding capacity; for example, the mtDNA of *A. thaliana* is 376 kbp long, but codes only for 32 proteins (Meinke et al. 1998, 2009). Mitochondrial genomes from lower eukaryotes present a wide variation in terms of size and number of genes. The mtDNA of the protozoan *Reclinomonas americana* is 69 kbp long and codes for 97 genes (Lang et al. 1997), whereas the genome of yeast (68–85 kbp) possesses fewer genes (six protein-coding genes, two rRNA genes, and 24 tRNA genes) (Lecrenier and Foury 2000). In metazoans, the mtDNA is a circular molecule of 15–17 kbp; it lacks introns and, with few exceptions, codes for 13 polypeptides, two rRNAs, and 22 tRNAs, which

P. Loguercio Polosa • M. Roberti
Dipartimento di Biochimica e Biologia Molecolare "Ernesto Quagliariello", Università degli Studi di Bari, Via Orabona, 4, 70125 Bari, Italia

P. Cantatore (✉)
Dipartimento di Biochimica e Biologia Molecolare "Ernesto Quagliariello", Università degli Studi di Bari, Via Orabona, 4, 70125 Bari, Italia

Istituto di Biomembrane e Bioenergetica, CNR, Via Orabona, 4, 70125 Bari, Italia
e-mail: p.cantatore@biologia.uniba.it

C.E. Bullerwell (ed.), *Organelle Genetics*,
DOI 10.1007/978-3-642-22380-8_11, © Springer-Verlag Berlin Heidelberg 2012

are required for translation (Falkenberg et al. 2007). The remaining proteins, which represent about 95% of the mitochondrial polypeptides and include enzymes and factors required for mtDNA maintenance and expression, are nuclear coded, translated in the cytoplasm, and imported into the mitochondrion. This situation implies a reduced autonomy of the organelle that, to perform its functions, must rely on the coordinate expression of two separate genetic systems (Scarpulla 2008). However, mtDNA-coded polypeptides have a fundamental role in cell metabolism since they are subunits of the respiratory chain complexes. The question why mitochondria retained their genetic material and need a dedicated and complex machinery to replicate and express their genome is still largely open. Different explanations have been proposed; the prevailing view suggests that the separate localization of genes involved in oxidative phosphorylation is required for an optimal metabolic regulation of respiratory complexes biogenesis (Allen 2003).

The coordinated expression of nuclear and mitochondrial genes in response to cell requirements and changes in physiological conditions is regulated by a series of complex signaling pathways, known as nucleo-mitochondrial interactions. Signals may originate either from the extracellular environment, affecting the gene expression of nuclear and/or mitochondrial genes, or from mitochondria, influencing the expression of nuclear-encoded mitochondrial proteins (retrograde response) (for review see Liu and Butow 2006; McBride et al. 2006; Hock and Kralli 2009). The correct course of these events is of key importance for cell life; dysfunctions affecting mitochondrial metabolism have been associated with aging (Passos et al. 2007; Larsson 2010) and a growing number of complex diseases, which affect a vast number of organs, especially those that rely mostly on aerobic metabolism (for review see Du and Yan 2010; Wallace et al. 2010). These diseases are often associated with mutations either in mtDNA-coded genes or in nuclear genes coding for proteins involved in mitochondrial metabolism. These include genes controlling mtDNA synthesis (mtDNA polymerase and TWINKLE helicase) (for review see Copeland 2008, 2010), mitochondrial translation (mitochondrial ribosomal proteins and translation factors), and coupling between transcription and translation (Jacobs and Turnbull 2005; Bonawitz et al. 2006; Wang et al. 2007).

A full understanding of the mechanisms governing mtDNA expressionis of crucial relevance to understand the complex role of mtDNA in human diseases and aging, also in the light of the compact mtDNA gene organization which underlies different modes of expression and regulation with respect to nDNA-encoded genes.

In this review, we will focus on mtDNA transcription and its regulation in mammalian cells, describing the most recent achievements, which allow proposing a detailed mechanism, although not yet exhaustive. In addition, we will report information obtained in other metazoans, highlighting the relationship between the different gene organization and the modes of mitochondrial transcription.

11.2   The Basic Mechanism of Transcription in Mammals

Mammalian mtDNA represents an exceptional example of genetic economy. It codes only for 37 genes (two rRNAs, 13 polypeptides, and 22 tRNAs), lacks introns, and possesses a major noncoding sequence (D-loop region), which contains most of the regulatory signals for replication and transcription. The basic features of gene organization include the almost complete absence of intergenic sequences, the proximity of rRNA genes 16 S and 12 S, and the presence of one or more tRNA genes or tRNA-like sequences placed between most of rRNA- or mRNA coding-genes.

Based on the work mostly done in Hela and mouse cells, using in vivo or in vitro approaches with partially purified components, the basic mechanism of transcription of mammalian mtDNA was determined since the mid-1980s (Fig. 11.1). According to these studies, mtDNA is transcribed by three transcription units. One of them is responsible for the production of L-strand-coded transcripts, whereas the other two direct the synthesis of H-strand-coded RNAs. L-strand transcription depends on the LSP promoter and starts from the I_L initiation point, contained in the LSP and placed about 100 bp upstream of the major H-strand replication origin. Transcription from I_L produces a polycistronic RNA that is processed to give rise to the replication primer 7 S RNA, ND6 mRNA, and eight tRNAs. H-strand transcription is controlled by the HSP promoter and generates two units. One starts from I_{H1} placed 16 nt upstream of the tRNA^{Phe} gene and produces a primary transcript that gives rise to rRNAs 16 S and 12 S, tRNA^{Phe}, and tRNA^{Val}. The other unit starts from $I_{H2,}$ placed two nucleotides upstream of the 12 S rRNA gene and is responsible for the synthesis of a polycistronic transcript covering almost the entire mtDNA. Once processed, it originates 10 mRNAs and 14 tRNAs (Montoya et al. 1982, 1983; Falkenberg et al. 2007). The peculiar arrangement of the tRNA genes and the tRNA-like sequences, which flank rRNA or mRNA genes, led to propose the so-called punctuation mechanism, in which the cloverleaf structure of the tRNA or the tRNA-like sequences would function as a recognition signal for endonucleases involved in the processing of polycistronic transcripts (Ojala et al. 1981). Two enzymes have been recently characterized, RNase Z (Dubrovsky et al. 2004) and RNase P, and are responsible for the processing of the 3′ and 5′ end of tRNAs, respectively. The structure of RNase P has been investigated for a long time; initial evidence suggested that human mitochondrial RNase P, as its yeast counterpart, was an RNA-containing enzyme (Puranam and Attardi 2001). A recent paper (Holzmann et al. 2008) has instead demonstrated that human mitochondrial RNase P is a protein-only enzyme composed of three subunits. The first (MRPP-1) is a likely tRNA methylase; the second (MRPP-2) is a member of the family of short-chain dehydrogenases/reductases; and the third (MRPP-3) contains a putative metallonuclease domain and two pentatricopeptide (PPR) motifs. The latter subunit is weakly associated with the other two and probably represents the catalytic component of the complex.

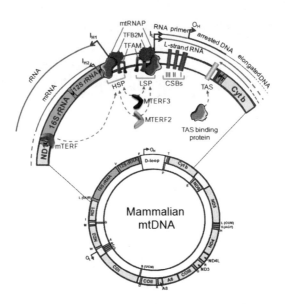

Fig. 11.1 Schematic representation of gene organization and regulatory regions of mammalian mtDNA. The gene organization of mtDNA, reporting the H- and L-strand coded genes, is shown in the circular map. The inset shows the portion of mtDNA and the factors involved in the regulation of transcription. O_H and O_L are H- and L-strand replication origins according to the asymmetric model (Brown and Clayton 2006). I_L is the L-strand transcription initiation site; I_{H1} and I_{H2} are the two H-strand transcription initiation sites. The three-component transcription machinery, formed by mtRNAP, TFAM, and TFB2M, is shown. Three characterized proteins of the MTERF family, involved in the regulation of transcription, are also shown. mTERF (MTERF1) binds inside the tRNA$^{Leu(UUR)}$ gene and mediates transcription termination. The simultaneous binding of mTERF to the HSP promoter and termination site would cause mtRNAP recycling mediated by rDNA looping, thus determining a high rate of synthesis of two rRNAs and two tRNAs. MTERF2 and MTERF 3 are two regulatory factors that act, respectively, as positive and negative modulators of mitochondrial transcription by interacting with the components of the transcription apparatus. CSBI, II, and III are three conserved sequences known to be the site of transition from RNA primer to H-DNA. TAS is a conserved region where the nascent H-DNA is thought to terminate in the presence of the TAS-binding protein (Madsen et al. 1993). At this site, termination of the H-strand polycistronic transcript initiating from I_{H2} takes place

The final step in the processing of mitochondrial RNAs is polyadenylation. Animal mitochondrial mRNAs possess a poly(A) tail of variable length, of around 50 nt on average, whereas rRNAs show a much shorter tail of only few nucleotides. The length of the tail seems to be regulated by a combination of two competing activities: the polyadenylation activity of the poly(A) polymerase (PAP) and the deadenylation activity of the polynucleotide phosphorylase (PNPase). The polyadenylation process has a key role in completing the UAA stop codons, as many mitochondrial mRNAs terminate with either U or UA (Temperley et al. 2010). Moreover, it has been demonstrated that the poly(A) tail serves to stabilize the RNAs since the partial inactivation of human mitochondrial PAP causes a shorter tail and a lower stability of some mitochondrial mRNAs (Nagaike et al. 2005).

11.3   Mitochondrial Transcription in Invertebrates

Mitochondrial genomes in metazoans show a basic invariance in the gene content but remarkable differences in the gene organization. The most studied examples have been sea urchin and *Drosophila*.

Sea urchins (*phylum* Echinoderms) are among the most developed invertebrates and have been used as model system in developmental studies. The relevance of these studies has been enhanced by the completion of the genomic sequence of *Strongylocentrotus purpuratus* (Sea Urchin Genome Sequencing Consortium 2006), which has revealed a closer similarity to humans than was expected from morphological analyses. The gene organization of the sea urchin mtDNA displays several important differences with respect to vertebrates (Cantatore et al. 1989) (Fig. 11.2a). The main variations concern separation of the two ribosomal genes, the clustering of 15 tRNA genes, and the reduced dimension (about 130 bp) of the main noncoding region, which is located in the tRNA gene cluster, downstream of the 12 S rRNA gene. Mapping of mature and precursor mitochondrial transcripts suggested that, in this organism, transcription proceeds via multiple and partially overlapping transcription units, which might start in correspondence of six small conserved AT-containing sequences scattered along the genome (Cantatore et al. 1990).

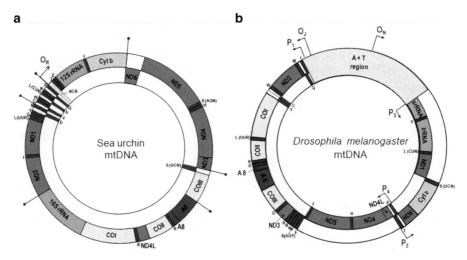

Fig. 11.2   Gene organization of sea urchin and *Drosophila* mtDNA. (**a**) *P. lividus* mtDNA map. O$_R$ indicates the leading strand replication origin (Jacobs et al. 1989). Lagging strand origin is thought to take place at multiple sites. *Dots* mark the six, small AT-rich sequences possibly acting as transcriptional start sites (Cantatore et al. 1990). NCR, main noncoding region. (**b**) *D. melanogaster* mtDNA map. O$_J$ and O$_N$ are the major coding and minor coding strand replication origin, respectively (Saito et al. 2005). P1, P2, P3, and P4 are hypothetical transcription initiation sites placed in proximity of the transcribed genes (Roberti et al. 2009)

The *Drosophila melanogaster* mtDNA is about 19.5 kb long and contains a large noncoding region of 4.6 kbp, which accounts for the different size with respect to vertebrates (Fig. 11.2b). Unlike human and sea urchin genomes, the *Drosophila* mitochondrial genes are almost equally distributed between the two strands and form four clusters located alternatively on the two strands. One of these clusters contains the two ribosomal genes that are placed in adjacent positions (Lewis et al. 1995). The transcription mechanism of the *Drosophila* mtDNA is based on early studies of RNA mapping, which suggested the existence of multiple transcription units starting at the beginning of the gene clusters and ending at their 3' ends (Berthier et al. 1986). An alternative mechanism, based on the existence of two promoters, one for each strand, located in the AT-rich region, has been proposed by Roberti et al. (2006) on the basis of transcript-mapping studies in cells depleted of the termination factor DmTTF (see below).

11.4 The Components of the Mitochondrial Transcription Initiation Machinery

To obtain a full understanding of the mitochondrial transcription mechanism and its regulation, a large effort has been dedicated to clarify the structure and function of the components of the mitochondrial transcription apparatus. Work performed mostly in mammals, particularly in humans, and also in invertebrates, provided an extensive knowledge of the basic structure of the transcription apparatus. However, information is far from exhaustive, since the role of some components is not clear and one cannot rule out the possibility that other not-yet-characterized factors might be involved. Figure 11.3 summarizes the basic functions of the characterized components of the mitochondrial transcription apparatus.

11.4.1 Mitochondrial RNA Polymerase

The mitochondrial RNA polymerase (mtRNAP) is an enzyme consisting of a single subunit having high similarity with the RNA polymerase of the T-odd series bacteriophages. This similarity and the observation that also components of the mtDNA replication apparatus (DNA-polymerase gamma and TWINKLE helicase) are similar to their phage counterparts suggested that genes participating in mtDNA replication and transcription were imported from a T-odd phage during mitochondrial endosymbiosis (Shutt and Gray 2006). Human mtRNAP (1,230 aa residues) contains nine sequence motifs, located in the C-terminal portion, which display a high similarity with the bacteriophage enzyme (Tiranti et al. 1997). The catalytic domain and NTP binding site are contained in this region. The less conserved N-terminal domain contains two PPR repeats, which are thought to be involved in

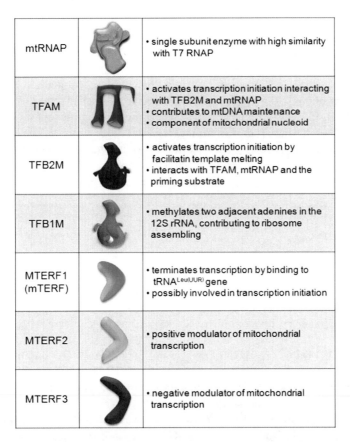

mtRNAP		• single subunit enzyme with high similarity with T7 RNAP
TFAM		• activates transcription initiation interacting with TFB2M and mtRNAP • contributes to mtDNA maintenance • component of mitochondrial nucleoid
TFB2M		• activates transcription initiation by facilitatin template melting • interacts with TFAM, mtRNAP and the priming substrate
TFB1M		• methylates two adjacent adenines in the 12S rRNA, contributing to ribosome assembling
MTERF1 (mTERF)		• terminates transcription by binding to tRNA$^{Leu(UUR)}$ gene • possibly involved in transcription initiation
MTERF2		• positive modulator of mitochondrial transcription
MTERF3		• negative modulator of mitochondrial transcription

Fig. 11.3 Summary of the basic properties of the proteins involved in mammalian mitochondrial transcription. The list also includes rRNA methyl transferase TFB1M because it was initially considered a transcription factor. The form of mtRNAP is based on the structure of T7 RNA polymerase; the form of TFAM is based on the L-shaped structure of an HMG-box domain; the forms of TFB2M and TFB1M are based on the known structures of *S. cerevisiae* mtTFB and bacterial methyltransferase (Bonawitz et al. 2006). The representation of MTERF1, MTERF2 and MTERF3 is based on the tridimensional structures reported by Yakubovskaya et al. (2010) and Spåhr et al. (2010)

interaction with RNA. PPR motifs have been described also in other vertebrate and invertebrate mtRNAPs.

In yeast, there is evidence that mtRNAP may also play a role in coupling transcription and translation. Interaction of Nam1 and Sls1 proteins with the N-terminal domain of mtRNAP appears to stimulate the delivery of synthesized RNA to the mitochondrial translational apparatus (Rodeheffer and Shadel 2003). A similar situation may occur in mammals. Shadel and colleagues have reported that the mitochondrial ribosomal protein MLRP12 interacts with human mtRNAP and activates mitochondrial transcription in vitro (Wang et al. 2007). However, a recent report by the group of Gustafsson (Litonin et al. 2010) does not confirm this evidence since the authors were unable to demonstrate any stimulating effect of the MLRP12 protein on mitochondrial transcription. This is clearly an issue that needs further investigation for a full understanding of the mechanisms regulating the expression of mitochondrial genes.

Sea urchin mtRNAP has been recently characterized (Loguercio Polosa et al. 2007). It is amongst the longest organelle enzymes characterized to date (1,439 amino acids). In its C-terminal part are present the same conserved motifs found in human mtRNAP, whereas the N-terminal region is more variable. The larger size of the protein is mainly due to an N-terminal extension, which contains two PPR motifs and a polyserine segment of unknown function.

11.4.2 TFAM

TFAM (Transcription Factor A Mitochondrial) was initially identified as a transcription-activating factor on the basis of its ability to specifically bind two mtDNA regions located upstream of LSP and HSP and to activate transcription catalyzed by a mitochondrial extract containing mtRNAP activity (Fisher et al. 1987; Fisher and Clayton 1988).

TFAM (25 kDa) is a member of high-mobility group (HMG) proteins; it contains two HMG boxes, involved in nonsequence-specific DNA binding, and a C-terminal tail, which is required for transcription activation and specific DNA binding. The protein is also able to bend and unwind mtDNA (Parisi and Clayton 1991, Fisher et al. 1992, Dairaghi et al. 1995). Recent work showed that TFAM binds mtDNA cooperatively as a homodimer using HMG box A and is responsible for compacting the mtDNA molecule into nucleoids (Kaufman et al. 2007; Gangelhoff et al. 2009). The role of TFAM in mtDNA maintenance had been underlined by early TFAM knockout studies in various organs, showing an association between protein content and mtDNA copy number (Poulton et al. 1994; Larsson et al. 1998; Wang et al. 1999; Silva et al. 2000; Rantanen et al. 2001; Wredenberg et al. 2002; Hansson et al. 2004). This role, which was somehow expected given the participation of TFAM in the synthesis of the 7 S RNA replication primer, seems instead to occur independently of the role of TFAM in transcription. This conclusion emerges from studies demonstrating that overexpression, in TFAM-depleted cells, of a TFAM mutant lacking the C-terminal activation domain caused only increase of mtDNA copy number but not of mitochondrial transcription (Kanki et al. 2004). Similarly, in mouse cells lacking TFAM, overexpression of human TFAM, which is not able to activate mitochondrial transcription in mouse, caused an increase in mtDNA level only (Ekstrand et al. 2004). These data overall suggested that TFAM is the main organizer of the mitochondrial chromatin; in this context the positive correlation between TFAM and mtDNA content is thought to be due to the increased mtDNA stability dependent on the packaging function of the factor.

11.4.3   TFBMs

For a long time, TFAM was considered the only mitochondrial transcription activating factor, in spite of the observation that in yeast mitochondrial transcription is driven by another factor, named mtf1, having no similarity with TFAM (Schinkel et al. 1987). The possibility that at least another factor, besides TFAM, could activate transcription, was reinforced by the observation that, while TFAM stimulated transcription in the presence of a partially purified mtRNAP fraction, it was unable to activate transcription in the presence of purified recombinant mtRNAP (Falkenberg et al. 2002). On the basis of these premises, the authors searched the protein database for a transcription activating factor that could have some similarity to yeast mtf1. Such search produced two homologous proteins, which displayed similarity with the putative mtf1 homolog of *Schizosaccaromyces pombe* (Falkenberg et al. 2002). Those factors, named TFB1M and TFB2M, when added to a transcription system constituted by purified recombinant TFAM and mtRNAP, were able to activate transcription from both LSP and HSP promoters. The activating function required the presence of TFAM; moreover, the stimulating effect of TFB2M was stronger than that of TFB1M. The observation that the two factors displayed similarity with bacterial rRNA methyl transferases prompted a series of studies to reveal the actual function. These studies demonstrated that TFB1M, and to a much lower extent TFB2M, were able to methylate in vitro two adjacent adenines in the 3′ terminal conserved stem-loop region of bacterial 16 S rRNA (Seidel-Rogol et al. 2003; Cotney and Shadel 2006). Since the methyl transferase activity was independent of the transcription stimulating activity (McCulloch and Shadel 2003), it was suggested that the two factors, probably originating from a gene duplication event, might serve different functions. This suggestion was confirmed by studying the effect of TFB1/2 M depletion and overexpression in *D. melanogaster*. TFB2M depletion decreased mitochondrial replication and transcription, whereas TFB1M depletion had no effect on these processes. On the contrary, depletion of TFB1M depressed mitochondrial protein synthesis. Accordingly, TFB2M overexpression in human and *Drosophila* cells increases mtDNA copy number and transcription (Matsushima et al. 2004, 2005; Adán et al. 2008; Cotney et al. 2007). In a recent study, Litonin et al. (2010) have shown that human TFB1M cannot activate transcription in vitro and that only TFAM and TFB2M are required as transcription activators.

Further investigation clarified that the role of TFB1M in mammalian mitochondria is to methylate two adjacent adenines in 12 S rRNA, an event required for ribosome assembly. In particular, Metodiev et al. (2009) showed that disruption of TFB1M gene led to the loss of 12 S rRNA methylation, preventing the assembly of ribosomes and inhibiting mitochondrial protein synthesis. However, also hypermethylation seems to produce negative effects. Cotney et al. (2009) observed that cells containing the pathogenic A1555G mtDNA mutation, which was associated with maternally inherited deafness, display 12 S rRNA hypermethylation, causing aberrant mitochondrial biogenesis that predisposes cells to stress-induced death. Linkage analysis (Bykhovskaya et al. 2004) suggested that TFB1M acts as a nuclear modifier of the disease, that is, a mutation that reduces TFB1M activity may protect cells with the A1555G mutation by restoring the 12 S rRNA methylation level close to normal.

11.4.4 The Basic Structure of the Core Transcription Machinery

The core mitochondrial transcription machinery is a three-member complex constituted by TFAM, mtRNAP, and TFB2M (Falkenberg et al. 2002; Lodeiro et al. 2010). The role of each component in transcription initiation has been investigated by means of several experimental approaches. Early footprinting experiments (Gaspari et al. 2004) demonstrated that mtRNAP contributes to specific promoter recognition by interacting directly with the promoter at several nucleotides located near the transcription start site. Recently, it was demonstrated that the transcription initiation complex (TIC) covers a DNA region comprised between −35 and +10 bp, with the region between −35 and −15 bp being occupied by TFAM and that between −15 and +10 bp contacted by the TFB2M–mtRNAP complex (Sologub et al. 2009). Protein–DNA crosslinking studies and catalytic autolabeling experiments revealed that TFB2M displays multiple roles in transcription initiation. Beside interacting with TFAM and mtRNAP, the factor contributes to promoter melting, is involved in positioning the templating base (+1) in the mtRNAP active site, and interacts directly with the priming substrate. These interactions may serve to stabilize the open promoter complex, avoiding the reannealing of the nontemplate strand. During elongation, when the transcription bubble is stabilized by the interactions between the 3′ end of the RNA and the template strand, TFB2M would be no longer needed and may dissociate from the complex, as occurs for yeast mttf1. TFB2M displacement may be the key event of transcription that marks the transition from initiation to elongation.

Recently, some authors have reported the possibility that transcription could occur in the presence of a more simplified transcription apparatus, in which the role of TFAM or TFB2M appears to be dispensable (Shutt et al. 2010; Litonin et al. 2010). The finding that primer generation at oriL requires mtRNAP only (Fusté et al. 2010) supports these observations.

11.5 The Termination Factor

11.5.1 Human mTERF

The existence of two partially overlapping H-strand transcription units committed to the synthesis of rRNAs and mRNAs raised a number of questions about their reciprocal regulation. Early kinetic and biochemical experiments showed that the two precursor transcripts were synthesized at different rates (Montoya et al. 1983). Subsequent studies identified a 28-bp DNA sequence placed downstream of the 3′ end of 16 S rRNA and recognized by a 39-kDa protein (Kruse et al. 1989; Fernandez-Silva et al. 1997). This protein, named mTERF, is able to promote the in vitro termination of the ribosomal transcription unit in the presence of a mitochondrial extract providing mtRNAP activity. Transcription termination experiments with recombinant mTERF were also performed. Termination was bidirectional and more effective when the mTERF binding site was placed in the reverse orientation with respect to the H-strand promoter. This observation suggested a more efficient termination of the L-strand than H-strand-coded transcripts (Asin-Cayuela et al. 2005), consistent with the absence of L-strand-coded genes downstream of the mTERF target site. Further investigations into the role of human mTERF and its homologs in other organisms have indicated that its role is more complex than initially thought. By using a construct containing the I_{H1} initiation site and the mTERF binding site, Martin et al. (2005) showed that native mTERF not only terminates transcription but also causes transcription activation. This effect was ascribed to the ability of one single mTERF molecule to simultaneously bind the termination site and a sequence placed near I_{H1}. The double interaction causes the formation of a loop structure that might promote the recycling of the transcription machinery through the direct passage of mtRNAP from the termination site to the I_{H1} initiation site. This would determine a higher rate of transcription of rRNAs with respect to mRNAs. Interestingly, the ability of mTERF to interact with the I_{H1} region was much more evident for the native protein than for the recombinant version, a result indicating that either post-translational modifications or co-purifying cofactors might be involved in increasing mTERF binding affinity for the I_{H1}-containing region.

Despite this wealth of information obtained by in vitro approaches, the in vivo role of mTERF is still not clear. The mitochondrial mutation A3243G, which was associated with the MELAS disease (Mitochondrial Encephalomyopathy Lactic Acidosis and Stroke-like episodes) and caused in vitro reduced binding of mTERF (Hess et al. 1991), did not affect either the in vivo occupancy of the protein, nor consistently altered the pattern of mitochondrial transcripts (Chomyn et al. 1992). Manipulations of mTERF levels in vivo by overexpression or RNA interference had complex effects on the steady-state level of both H-strand and L-strand-coded mitochondrial RNAs. It had also some influence on mitochondrial replication affecting replication pausing at the canonical binding sequence and at newly identified binding sites in the D-loop region (Hyvärinen et al. 2007, 2010a).

Two recently published papers throw light on the structure of mTERF and its binding mode to mtDNA (Jiménez-Menéndez et al. 2010; Yakubovskaya et al. 2010). The protein displays a modular architecture consisting of eight to nine repeated motifs, named mterf motifs. Each motif, of about 35 residues, consists of two α-helices and a short 3/10 helix, forming a left-handed triangular superhelix folded around a central hydrophobic core. The motifs form a solenoid-like structure, which twists to the right acquiring a convex and a concave surface.

The resulting structure has a curved shape resembling a half doughnut (or an entire croissant!). Binding of mTERF to DNA causes partial melting within the DNA duplex, with flipping of three nucleotides. Electrostatic interactions occur between the DNA backbone (phosphate groups) and the positively charged residues on the protein's concave surface. These interactions do not confer any sequence specificity and explain why the protein is able to bind a double-stranded DNA with an arbitrary sequence, albeit with low affinity. Sequence-specific interactions take also place, due to hydrogen bonds between five arginines to guanines and adenines. A model for mTERF–DNA binding has been proposed. The protein initially contacts DNA unspecifically and with low affinity; then, when the five arginines establish the correct interactions with DNA bases, a protein conformational change takes place that bends and unwinds DNA and in turn promotes the eversion of three nucleotides. Base flipping, which is stabilized by stacking interactions and hydrogen bonds to the bases and phosphates, is critical for stable protein–DNA binding, which in turn is necessary for promoting transcription termination. Pathogenic mutations that interfere with arginine–guanine interactions negatively affect the termination capacity of mTERF. The protein appears to preferentially interact with L-strand DNA, in agreement with the observation that the termination activity was more efficient in arresting L-strand transcripts, as a means to prevent L-strand transcripts from interfering with rRNA synthesis (Nam and Kang 2005). Finally, the extensive surface of DNA–protein interactions would exclude a single mTERF molecule being able to contact two DNA duplexes, in contrast with the hypothesis of mtDNA looping as mediated by a single mTERF molecule. Conversely, mTERF, alone or in combination with other factors (see below), could positively contribute to transcription initiation through its ability to partially melt DNA.

11.5.2   Transcription Termination Factors in Invertebrates

Functional studies on mTERF homologs in invertebrates have provided important insights on the role of these factors. In sea urchin, the 348-aa protein mtDBP has been characterized, that binds specifically two homologous sequences located in *P. lividus* mtDNA (Fig. 11.2a). One site is placed in the noncoding region (NCR), at the 3′ end of the short D-loop; the other contains the 3′ ends of ND5 and ND6 genes, which are transcribed on opposite strands (Loguercio Polosa et al. 1999). The protein, bound to the target site in the NCR, was able to arrest bidirectional transcription catalyzed by not-purified human mtRNAP (Fernandez-Silva et al. 2001). In the presence of recombinant sea urchin mtRNAP, mtDBP arrested transcription unidirectionally (Loguercio Polosa et al. 2007), that is, only when the enzyme approached the protein binding site in the direction of L-strand transcription. On the contrary, when mtRNAP moves in the opposite direction, mtDBP seems to be not necessary to transcription termination, which occurs in a sequence-dependent manner. Similarly in humans, a factor-independent, sequence-dependent transcription termination event has been observed for the synthesis of the H-strand replication primer (Pham et al. 2006; Wanrooij et al. 2010).

Beside functioning as a transcription terminator, mtDBP also displays a contrahelicase activity (Loguercio Polosa et al. 2005). Since one of the protein binding sites is placed at the 3′ end of the short sea urchin D-loop, the contrahelicase function may be involved in regulating D-loop expansion and, therefore, mtDNA replication. In support of this hypothesis, it was shown that, when transcription takes place in the direction opposite to H-DNA replication, mtDBP dissociates from DNA; this may abrogate the helicase block and allow resumption of mtDNA replication. Therefore, mtDBP with its dual role may be the molecular tool for regulating D-loop expansion during sea urchin mtDNA replication.

An mTERF homolog has been characterized also in *Drosophila* (Roberti et al. 2003). It has been named DmTTF and binds two sites placed at the 3' ends of clusters of genes transcribed in opposite directions, with one site located between tRNA^{Glu} and tRNA^{Phe} and the other between tRNA$^{Ser(UCN)}$ and cyt b (Fig. 11.2b). In vitro transcription experiments showed that DNA-bound DmTTF was able to terminate transcription catalyzed by human mtRNAP contained in a mitochondrial extract (Roberti et al. 2005). In addition, in vivo evidence for the role of this protein has been produced. Roberti et al. (2006) showed that perturbation of the DmTTF level by RNA interference had remarkable effects on the level of mitochondrial transcripts. In particular, DmTTF depletion increased the level of those transcripts mapping on both strands downstream of the two DmTTF binding sites. On the contrary, DmTTF overexpression caused a decrease in the level of RNAs mapping downstream of the two protein binding sites on both strands (Roberti et al. 2009). These results indicate that the protein acts as transcription termination factor also in vivo.

Unexpectedly, it was also found that the level of those transcripts mapping between the AT-rich region and the protein-binding sites was decreased in DmTTF-depleted cells. This may be due to the possibility that, as reported for mTERF, the protein may function as a transcriptional activator and therefore its depletion would cause a lower transcription of those genes placed immediately downstream of the promoters located in the AT-rich region. Alternatively, the decrease of the transcripts mapping upstream of DmTTF-binding sites might be the result of the reduced availability of the polymerase engaged in aberrant transcription beyond the DmTTF-binding sites.

11.6   The MTERF Protein Family

mTERF homologs, initially described in mammals and in some invertebrates, have been found in all metazoans and plants; they constitute a wide protein family named MTERF family. As first reported by Linder et al. (2005), in vertebrates there are four MTERF paralogous genes that define four subfamilies termed MTERF1, 2, 3, and 4, which have been probably generated by gene duplication events.

The MTERF1 subfamily includes the transcription termination factor mTERF; proteins of this subfamily, as well as those belonging to the MTERF2 subfamily, are unique to vertebrates. The MTERF3 and MTERF4 subfamilies include proteins from vertebrates and also from insects and worms and probably represent the most ancestral MTERF genes in metazoans. In the evolutionary tree depicted by Linder, sea urchin mtDBP and *Drosophila* DmTTF do not belong to any of the MTERF subfamilies, even if they are more closely related to the MTERF1 and MTERF2 rather than the MTERF3 and MTERF4 subfamilies. A further protein that does not belong to any subfamily is the uncharacterized *Drosophila* protein CG7175, which, together with DmTTF, might have been generated by a further gene duplication event that occurred in insect lineage (Roberti et al. 2009).

Functional studies on MTERF family members have produced interesting data; at the moment, we have several indications about the roles of mammalian MTERF3 and MTERF2, while the function of MTERF4 is still unknown.

11.6.1 MTERF3

The MTERF3 subfamily contains proteins from vertebrates and invertebrates, including worms and insects. Proteins of this subfamily are the most conserved among MTERF proteins, suggesting a crucial function in the cell. MTERF3 knock out experiments (Park et al. 2007) confirmed this view, as mouse embryos devoid of this protein died before birth. Tissue-specific knockout caused, in the heart only, abnormal mitochondrial biogenesis consisting of mitochondrial proliferation and reduced activity of respiratory complexes I, III, and IV. The lack of the protein caused also an increased steady-state level of mitochondrial transcripts probably due to a more efficient initiation at both H- and L-strand promoters. In vitro DNA-binding experiments indicated that the protein had the capacity to bind mtDNA, though not specifically. ChIP experiments showed interaction of MTERF3 with a large mtDNA region comprising the two transcription promoters. These data suggest that the protein may function as a transcriptional repressor; however, attempts to demonstrate its role in vitro in the presence of a reconstituted transcription system have been unsuccessful, suggesting that additional factors are required for MTERF3 function.

Recently, the crystal structure of human MTERF3 has been published (Spåhr et al. 2010). The structure, which is very similar to that of mTERF, contains seven mterf motifs, each forming a three α-helical structure. Similarly to mTERF, mterf motif repeats form a slightly twisted half-doughnut structure, containing a path of positively charged amino acids exposed on the protein surface, in the correct position to interact with DNA. It is interesting to observe that MTERF3 possesses only two out of the five arginines involved in the sequence-specific DNA binding and contains only one out of the three amino acids needed to stabilize the base flipping; this may indicate a different DNA-binding mode of this protein.

The conservation of mterf motifs suggests that also the other members of the MTERF family share a structure similar to that of mTERF and MTERF3.

The function of MTERF3 has been also investigated in *Drosophila* by analyzing the effect on mitochondrial nucleic acid metabolism of overexpression and depletion of the protein. *Drosophila* cells overexpressing D-MTERF3 had a decreased level of those transcripts, such as 12 S rRNA, COI, and ND2, mapping on either strands immediately downstream of the AT-rich region, where mitochondrial promoters should be located. Therefore, also *in Drosophila*, the available data point toward a role of D-MTERF3 as a negative regulator of mitochondrial transcription. Interestingly, D-MTERF3 knock down did not influence the level of mitochondrial transcripts but rather produced an overall decrease of the mitochondrial protein synthesis. This effect is probably dependent on the concomitant downregulation in D-MTERF3-depleted cells of TFB1M, a factor involved in the mitochondrial ribosome biogenesis (Roberti et al. 2009).

11.6.2 MTERF2

MTERF2 is a likely product of a recent gene duplication event in the vertebrate ancestor lineage, which also originated MTERF1. MTERF2 is the most abundant protein among the members of the MTERF family and is preferentially localized in the mitochondrial nucleoids (Pellegrini et al. 2009). MTERF2 knock out produced viable mice with defects in muscle and brain performance, only when animals were subjected to a ketogenic diet (Wenz et al. 2009). Such animals exhibited a decline in the content and activity of muscle respiratory complexes; they also showed concomitant enhanced mitochondrial proliferation and increased level of other MTERF proteins, as well as of proteins of the basal transcription machinery. A generalized decrease of transcripts coded by both strands and an unbalanced tRNA pool were also observed. On the basis of these evidences, the authors proposed that the protein might behave as a transcriptional activator, possibly interacting with other MTERF family proteins and/or the TIC components.

Conversely, there are still some points that need to be clarified. One concerns the DNA-binding capacity of the protein. While ChIP experiments by Wenz et al. (2009) indicated a preferential binding of the protein to the promoter region, other reports tend to exclude a sequence-specific binding (Pellegrini et al. 2009; Hyvärinen et al. 2010b). In addition, the observation that protein depletion causes an increase of mitochondrial biogenesis and that its overexpression results in a modest mtDNA copy number decrease may instead suggest a role of this protein in mtDNA replication or in both replication and transcription (Hyvärinen et al. 2010b).

11.7 Regulation of Transcription

Transcription regulation is a key step in the control of mitochondrial genes' expression. Considering that the steady-state level of a transcript is determined by the ratio between its synthesis and degradation rate, the content of mitochondrial RNA can be regulated at several levels: (a) components of the core transcription apparatus and auxiliary factors; (b) energetic state of the cell; (c) selection of different transcription units; (d) transcription termination; and (e) stabilization of mitochondrial transcripts.

As regards the first point, nuclear transcription factors such as NRF-1 and NRF-2 and the PGC-1 family coactivators control the content of many proteins involved in mitochondrial transcription. In turn, coactivators may be controlled by cellular or extracellular signals, thus constituting a complex regulatory system, able to adapt mitochondrial gene expression to cell requirements. This vast and complex topic is treated by recent reviews and articles (Diaz and Moraes 2008; Scarpulla 2008; Bruni et al. 2010) and will not be discussed here.

Transcription may be modulated also by members of MTERF family. Functional data on MTERF3 and MTERF2 suggest that they play an opposite role in mitochondrial transcription (Park et al. 2007; Wenz et al. 2009). These factors may interact simultaneously or alternatively with the three-member core transcription apparatus and finely regulate transcription. More work is necessary to test this hypothesis; in addition, it cannot be ruled out that also the transcription termination factor mTERF (MTERF1), which is able to melt DNA (Yakubovskaya et al. 2010), the still uncharacterized MTERF4, or other proteins may participate in transcription regulation.

Transcription rate may be controlled by the energetic state of the cell. Early in vitro studies showed that a high concentration of ATP is needed to form the initiation complex (Gaines et al. 1987; Narasimhan and Attardi 1987). A plausible reason is that ATP is the priming substrate and, therefore, a decrease of its level may attenuate ATP interaction with TFB2M. Therefore, TFB2M may work as molecular sensor for ATP level, thus relating transcript abundance with respiration-dependent ATP synthesis. This kind of regulation mechanism has been proposed in yeast, where a correlation between in vivo changes of transcript content and in vitro sensitivity of mitochondrial promoters to ATP concentration was found (Amiott and Jaehning 2006).

A further level at which mitochondrial RNA synthesis rate can be regulated is the utilization of promoters LSP and HSP, which direct initiation of transcription at I_L and I_{H1}/I_{H2} sites, respectively. Early in vivo pulse-labeling transcription experiments showed that L-strand-coded RNAs were labeled at a higher rate than H-strand-coded transcripts (Cantatore and Attardi 1980). This result has been confirmed by in vitro experiments (Chang and Clayton 1984; Falkenberg et al. 2002), showing that LSP promoter is more active than HSP. The reason for this difference is not clear, also considering the limited coding capacity of the L-strand. One possible explanation could be the need of an efficient synthesis of the RNA primer for H-strand replication given its high turnover (Gelfand and Attardi 1981). Recently, Shutt et al. (2010) hypothesized that transcription initiation at the two promoters could be regulated by the amount of TFAM associated with nucleoids.

Another important point concerns the regulation of the two H-strand-dependent transcription units. The existence of two units initiating at I_{H1} and I_{H2} and producing precursors of rRNAs and mRNAs, respectively, relies on 5'-end mapping experiments of in vivo synthesized primary transcripts labeled with guanylyl transferase (Montoya et al. 1982; Martin et al. 2005). In addition, Montoya et al. (1983) demonstrated that the precursors emanating from I_{H1} and I_{H2} are synthesized in vivo with distinct kinetic properties. However, in vitro experiments with purified recombinant transcription components have failed to prove initiation at I_{H2} (Litonin et al. 2010; Shutt et al. 2010); a possible explanation is that some factors, crucial for initiation at I_{H2}, are missing in the reconstituted system.

The relative content of ribosomal and messenger RNAs may be also regulated at the level of termination, given the existence of a transcription termination event at the end of the rRNA gene unit mediated by mTERF. Moreover, this protein might be responsible for the different transcription rate of ribosomal and messenger RNAs. According to the report of Martin et al. (2005), the rRNA/mRNA ratio may be regulated by the formation of an mTERF-dependent DNA loop, which allows the recycling of the transcription machinery, thus determining a higher content of rRNAs with respect to mRNAs. However, structural studies on mTERF tend to rule out the same molecule interacting contemporaneously with two DNA target sites (Yakubovskaya et al. 2010). This point remains therefore still open and awaits further investigations; it is possible that DNA looping could require the participation of additional factors or post-translational modifications of mTERF. In addition, it remains to be clarified whether transcription termination at the 3' end of the 16 S rRNA gene occurs for both H-strand transcription units or only for that initiating at I_{H1}. The answer to this question might derive from an in vitro system able to initiate transcription from I_{H2}.

Another site where transcription is thought to stop is at the 3' end of the H-strand mRNA transcription unit emanating from I_{H2}. Here, termination should occur in correspondence with conserved sequences (TAS), previously associated with the termination of the D-loop (Madsen et al. 1993; Camasamudram et al. 2003; Freyer et al. 2010).

Finally, early data on mitochondrial RNAs half-life demonstrated the existence of regulation at the level of stability. In Hela cells, it has been reported that both mature ribosomal and messenger RNAs are metabolically unstable, with a half-life of 25–90 min for the mRNAs and 2.5–3.5 h for the rRNAs (Gelfand and Attardi 1981). However, the rate of mitochondrial RNA decay may vary: early work done in the laboratory of G. Attardi ascribed the permanence of mitochondrial translation in anucleated African green monkey cells treated with an inhibitor of mitochondrial transcription to a stabilization of mitochondrial transcripts (England et al. 1978). The existence of such mechanisms has been invoked to explain changes in the level of mitochondrial RNA species that are not directly ascribed to changes in the transcription rate (see for example Ostronoff et al. 1995; Cantatore et al. 1990, 1998).

A recent observation from Freyer et al. (2010) further supports the existence of regulation of gene expression at the level of mitochondrial RNA stability. The authors expressed human TFAM in the heart of TFAM-lacking mice and, in agreement with previous reports, found that human TFAM was not able to stimulate mouse mtDNA transcription. Instead, rather surprisingly, the steady-state level of the transcripts was near to normal. Also, this finding can be explained by the existence of a mitochondrial RNA stabilization mechanism. It might depend both on the action of not-yet-characterized nucleases and on the regulation of enzymes controlling the polyadenylation state of mitochondrial mRNAs, given the relationship between polyadenylation and mitochondrial mRNA stability (Nagaike et al. 2005; Temperley et al. 2010).

11.8 Conclusions and Perspectives

After the discovery that mitochondria possess their own genome, the topics of DNA replication, transcription, and translation in the organelle became the objects of intensive research in many laboratories. As regards mitochondrial transcription in mammals, a basic and simplified mechanism, based on the existence of three transcription units and the punctuation model for the mitochondrial RNAs processing, was proposed in the mid-80s. Further investigations delineated the structure of the basal transcription apparatus comprising a phage-like mtRNAP and the two transcription factors TFAM and TFB2M. Studies in sea urchin and *Drosophila* showed that in invertebrates, despite the conservation of the basal transcription apparatus, the transcription mechanism is different from that of mammals. This suggests that during evolution the differences in the mitochondrial gene organization probably elicited changes in the transcription mechanisms.

It is interesting to compare the features of mammalian mtRNAP with those of bacteriophage, prokaryotic, and nuclear RNA polymerases. The mitochondrial enzyme resembles the T-odd phage counterpart as regards its single-polypeptide composition, for the conservation of nine motifs located in the C-terminal part and for its ability to contribute to promoter recognition. However, it appears that, different from the phage polymerase and similar to the multisubunit enzyme, transcription initiation by mtRNAP requires the additional factors TFAM and TFB2M. Then, promoter clearance would require dissociation of the TFB2M factor, as occurs with the release of sigma subunit in bacteria and of TFBII factor in eukaryotes. These similarities suggest that, to perform a very compelling task such as transcription, mtRNAP has evolved toward a unique system that contains features from both phage, prokaryotic, and eukaryotic enzymes.

Recent studies have been focused on the role in mitochondrial transcription of additional components, which are the members of a large protein family named

MTERF family. The first member of this family is the transcription termination factor mTERF. It may act to avoid transcription pausing caused by head-on collision of the transcription apparatuses, thus safeguarding the integrity of the genome. mTERF seems to display multiple functions as it probably stimulates transcription and has some effects on mtDNA replication. In sea urchin, the transcription termination factor mtDBP controls also mtDNA replication via an antihelicase activity. In *Drosophila*, the mTERF homolog DmTTF regulates transcription at the level of termination and possibly initiation, but the actual evidence does not show an apparent effect on mtDNA replication. Interestingly, *C. elegans* and *C. briggsae*, which possess MTERF3 and MTERF4, do not contain a putative homolog of the transcription termination factor. This may be due to the dispensability of a termination factor in worms, determined by the peculiar organization of worm mitochondrial genes, all of which map on the same mtDNA strand. This might involve a simplified transcription mode that does not require a protein that regulates the traffic of the transcription machineries.

The characterization of two components of the MTERF family, MTERF2 and MTERF3, has extended the complexity of the transcription systems. The two proteins appear to differently modulate mitochondrial RNA synthesis possibly interacting with the components of the basal transcription apparatus. It has been suggested that these interactions are needed to finely adapt mitochondrial transcription efficiency to the metabolic requirements of the cell. This topic introduces the complex scenario of the regulation of transcription; still unknown aspects include the mechanisms responsible for (a) promoter selection; (b) mitochondrial RNA stabilization; and (c) the interplay between transcription and translation. Investigation on these issues will need the full characterization of the participants to the transcription machinery and the study of the enzymes responsible for the turnover of mitochondrial RNAs.

Notes Added in Proof

After submission of our review, a study appeared on the characterization of MTERF4, which demonstrates that the protein directly controls mitochondrial ribosomal biogenesis and translation (Cámara Y, Asin-Cayuela J, Park CB, Metodiev MD, Shi Y, Ruzzenente B, Kukat C, Habermann B, Wibom R, Hultenby K, Franz T, Erdjument-Bromage H, Tempst P, Hallberg BM, Gustafsson CM, Larsson NG. (2011) MTERF4 regulates translation by targeting the methyltransferase NSUN4 to the mammalian mitochondrial ribosome. Cell Metab. 13:527–539).

Acknowledgments This work was supported by grants from University of Bari, Progetto di Ricerca di Ateneo, COFIN-PRIN 2008, and Telethon-Italy (Grant GGP06233). The authors would like to thank F. Fracasso for his precious assistance in preparing the figures.

References

Adán C, Matsushima Y, Hernández-Sierra R, Marco-Ferreres R, Fernández-Moreno MA, González-Vioque E, Calleja M, Aragón JJ, Kaguni LS, Garesse R (2008) Mitochondrial transcription factor B2 is essential for metabolic function in *Drosophila melanogaster* development. J Biol Chem 283:12333–12342

Allen JF (2003) Why chloroplasts and mitochondria contain genomes. Comp Funct Genom 4:31–36

Amiott EA, Jaehning JA (2006) Mitochondrial transcription is regulated via an ATP "sensing" mechanism that couples RNA abundance to respiration. Mol Cell 22:329–338

Asin-Cayuela J, Schwend T, Farge G, Gustafsson CM (2005) The human mitochondrial transcription termination factor (mTERF) is fully active *in vitro* in the non-phosphorylated form. J Biol Chem 280:25499–25505

Berthier F, Renaud M, Alziari S, Durand R (1986) RNA mapping on Drosophila mitochondrial DNA: precursors and template strands. Nucleic Acids Res 14:4519–4533

Bonawitz ND, Clayton DA, Shadel GS (2006) Initiation and beyond: multiple functions of the human mitochondrial transcription machinery. Mol Cell 24:813–825

Brown TA, Clayton DA (2006) Genesis and wanderings: origins and migrations in asymmetrically replicating mitochondrial DNA. Cell Cycle 5:917–921

Bruni F, Loguercio Polosa P, Gadaleta MN, Cantatore P, Roberti M (2010) Nuclear respiratory factor 2 induces the expression of many but not all human proteins acting in mitochondrial DNA transcription and replication. J Biol Chem 285:3939–3948

Bykhovskaya Y, Mengesha E, Wang D, Yang H, Estivill X, Shohat M, Fischel-Ghodsian N (2004) Human mitochondrial transcription factor B1 as a modifier gene for hearing loss associated with the mitochondrial A1555G mutation. Mol Genet Metab 82:27–32

Camasamudram V, Fang JK, Avadhani NG (2003) Transcription termination at the mouse mitochondrial H-strand promoter distal site requires an A/T rich sequence motif and sequence specific DNA binding proteins. Eur J Biochem 270:1128–1140

Cantatore P, Attardi G (1980) Mapping of nascent light and heavy strand transcripts on the physical map of HeLa cell mitochondrial DNA. Nucleic Acids Res 8:2605–2625

Cantatore P, Roberti M, Rainaldi G, Gadaleta MN, Saccone C (1989) The complete nucleotide sequence, gene organization, and genetic code of the mitochondrial genome of *Paracentrotus lividus*. J Biol Chem 264:10965–10975

Cantatore P, Roberti M, Loguercio Polosa P, Mustich A, Gadaleta MN (1990) Mapping and characterization of *Paracentrotus lividus* mitochondrial transcripts: multiple and overlapping transcription units. Curr Genet 17:235–245

Cantatore P, Petruzzella V, Nicoletti C, Papadia F, Fracasso F, Rustin P, Gadaleta MN (1998) Alteration of mitochondrial DNA and RNA level in human fibroblasts with impaired vitamin B12 coenzyme synthesis. FEBS Lett 432:173–178

Chang DD, Clayton DA (1984) Precise identification of individual promoters for transcription of each strand of human mitochondrial DNA. Cell 36:635–644

Chomyn A, Martinuzzi A, Yoneda M, Daga A, Hurko O, Johns D, Lai ST, Nonaka I, Angelini C, Attardi G et al (1992) MELAS mutation in mtDNA binding site for transcription termination factor causes defects in protein synthesis and in respiration but no change in levels of upstream and downstream mature transcripts. Proc Natl Acad Sci USA 89:4221–4225

Copeland WC (2008) Inherited mitochondrial diseases of DNA replication. Annu Rev Med 59:131–146

Copeland WC (2010) The mitochondrial DNA polymerase in health and disease. Subcell Biochem 50:211–222

Cotney J, Shadel GS (2006) Evidence for an early gene duplication event in the evolution of the mitochondrial transcription factor B family and maintenance of rRNA methyltransferase activity in human mtTFB1 and mtTFB2. J Mol Evol 63:707–717

Cotney J, Wang Z, Shadel GS (2007) Relative abundance of the human mitochondrial transcription system and distinct roles for h-mtTFB1 and h-mtTFB2 in mitochondrial biogenesis and gene expression. Nucleic Acids Res 35:4042–4054

Cotney J, McKay SE, Shadel GS (2009) Elucidation of separate, but collaborative functions of the rRNA methyltransferase-related human mitochondrial transcription factors B1 and B2 in mitochondrial biogenesis reveals new insight into maternally inherited deafness. Hum Mol Genet 18:2670–2682

Dairaghi DJ, Shadel GS, Clayton DA (1995) Addition of a 29 residue carboxyl-terminal tail converts a simple HMG box-containing protein into a transcriptional activator. J Mol Biol 249:11–28

Diaz F, Moraes CT (2008) Mitochondrial biogenesis and turnover. Cell Calcium 44:24–35

Du H, Yan SS (2010) Mitochondrial medicine for neurodegenerative diseases. Int J Biochem Cell Biol 42:560–572

Dubrovsky EB, Dubrovskaya VA, Levinger L, Schiffer S, Marchfelder A (2004) *Drosophila* RNase Z processes mitochondrial and nuclear pre-tRNA 3′ ends *in vivo*. Nucleic Acids Res 32:255–262

Ekstrand MI, Falkenberg M, Rantanen A, Park CB, Gaspari M, Hultenby K, Rustin P, Gustafsson CM, Larsson NG (2004) Mitochondrial transcription factor A regulates mtDNA copy number in mammals. Hum Mol Genet 13:935–944

England JM, Costantino P, Attardi G (1978) Mitochondrial RNA and protein synthesis in enucleated African green monkey cells. J Mol Biol 119:455–462

Falkenberg M, Gaspari M, Rantanen A, Trifunovic A, Larsson N-G, Gustafsson CM (2002) Mitochondrial transcription factors B1 and B2 activate transcription of human mtDNA. Nat Genet 31:289–294

Falkenberg M, Larsson NG, Gustafsson CM (2007) DNA replication and transcription in mammalian mitochondria. Annu Rev Biochem 76:679–699

Fernandez-Silva P, Martinez-Azorin F, Micol V, Attardi G (1997) The human mitochondrial transcription termination factor (mTERF) is a multizipper protein but binds to DNA as a monomer, with evidence pointing to intramolecular leucine zipper interactions. EMBO J 16:1066–1079

Fernandez-Silva P, Loguercio Polosa P, Roberti M, Di Ponzio B, Gadaleta MN, Montoya J, Cantatore P (2001) Sea urchin mtDBP is a two-faced transcription termination factor with a biased polarity depending on the RNA polymerase. Nucleic Acids Res 29:4736–4743

Fisher RP, Clayton DA (1988) Purification and characterization of human mitochondrial transcription factor 1. Mol Cell Biol 8:3496–3509

Fisher RP, Topper JN, Clayton DA (1987) Promoter selection in human mitochondria involves binding of a transcription factor to orientation-independent upstream regulatory elements. Cell 50:247–258

Fisher RP, Lisowsky T, Parisi MA, Clayton DA (1992) DNA wrapping and bending by a mitochondrial high mobility group-like activator protein. J Biol Chem 267:3358–3367

Freyer C, Park CB, Ekstrand MI, Shi Y, Khvorostova J, Wibom R, Falkenberg M, Gustafsson CM, Larsson NG (2010) Maintenance of respiratory chain function in mouse hearts with severely impaired mtDNA transcription. Nucleic Acids Res 38(19):6577–6588

Fusté JM, Wanrooij S, Jemt E, Granycome CE, Cluett TJ, Shi Y, Atanassova N, Holt IJ, Gustafsson CM, Falkenberg M (2010) Mitochondrial RNA polymerase is needed for activation of the origin of light-strand DNA replication. Mol Cell 37:67–78

Gaines G, Rossi C, Attardi G (1987) Markedly different ATP requirements for rRNA synthesis and mtDNA light strand transcription versus mRNA synthesis in isolated human mitochondria. J Biol Chem 262:1907–1915

Gangelhoff TA, Mungalachetty PS, Nix JC, Churchill ME (2009) Structural analysis and DNA binding of the HMG domains of the human mitochondrial transcription factor A. Nucleic Acids Res 37:3153–3164

Gaspari M, Falkenberg M, Larsson NG, Gustafsson CM (2004) The mitochondrial RNA polymerase contributes critically to promoter specificity in mammalian cells. EMBO J 23:4606–4614

Gelfand R, Attardi G (1981) Synthesis and turnover of mitochondrial ribonucleic acid in HeLa cells: the mature ribosomal and messenger ribonucleic acid species are metabolically unstable. Mol Cell Biol 1:497–511

Gray MW, Burger G, Lang BF (1999) Mitochondrial evolution. Science 283:1476–1481

Hansson A, Hance N, Dufour E, Rantanen A, Hultenby K, Clayton DA, Wibom R, Larsson NG (2004) A switch in metabolism precedes increased mitochondrial biogenesis in respiratory chain-deficient mouse hearts. Proc Natl Acad Sci USA 101:3136–3141

Hess JF, Parisi MA, Bennett JL, Clayton DA (1991) Impairment of mitochondrial transcription termination by a point mutation associated with the MELAS subgroup of mitochondrial encephalomyopathies. Nature 351:236–239

Hock MB, Kralli A (2009) Transcriptional control of mitochondrial biogenesis and function. Annu Rev Physiol 71:177–203

Holzmann J, Frank P, Löffler E, Bennett KL, Gerner C, Rossmanith W (2008) RNase P without RNA: identification and functional reconstitution of the human mitochondrial tRNA processing enzyme. Cell 135:462–474

Hyvärinen AK, Pohjoismäki JL, Reyes A, Wanrooij S, Yasukawa T, Karhunen PJ, Spelbrink JN, Holt IJ, Jacobs HT (2007) The mitochondrial transcription termination factor mTERF modulates replication pausing in human mitochondrial DNA. Nucleic Acids Res 35:6458–6474

Hyvärinen AK, Kumanto MK, Marjavaara SK, Jacobs HT (2010a) Effects on mitochondrial transcription of manipulating mTERF protein levels in cultured human HEK293 cells. BMC Mol Biol 11:72

Hyvärinen AK, Pohjoismäki JL, Holt IJ, Jacobs HT (2010b) Overexpression of MTERFD1 or MTERFD3 impairs the completion of mitochondrial DNA replication. Mol Biol Rep 38 (2):1321–1328

Jacobs HT, Turnbull DM (2005) Nuclear genes and mitochondrial translation: a new class of genetic disease. Trends Genet 21:312–314

Jacobs HT, Herbert ER, Rankine J (1989) Sea urchin egg mitochondrial DNA contains a short displacement loop (D-loop) in the replication origin region. Nucleic Acids Res 17:8949–8965

Jiménez-Menéndez N, Fernández-Millán P, Rubio-Cosials A, Arnan C, Montoya J, Jacobs HT, Bernadó P, Coll M, Usón I, Solà M (2010) Human mitochondrial mTERF wraps around DNA through a left-handed superhelical tandem repeat. Nat Struct Mol Biol 17:891–893

Kanki T, Ohgaki K, Gaspari M, Gustafsson CM, Fukuoh A, Sasaki N, Hamasaki N, Kang D (2004) Architectural role of mitochondrial transcription factor A in maintenance of human mitochondrial DNA. Mol Cell Biol 22:9823–9834

Kaufman BA, Durisic N, Mativetsky JM, Costantino S, Hancock MA, Grutter P, Shoubridge EA (2007) The mitochondrial transcription factor TFAM coordinates the assembly of multiple DNA molecules into nucleoid-like structures. Mol Biol Cell 18:3225–3236

Kruse B, Narasimhan N, Attardi G (1989) Termination of transcription in human mitochondria: identification and purification of a DNA binding protein factor that promotes termination. Cell 58:391–397

Lang BF, Burger G, O'Kelly CJ, Cedergren R, Golding GB, Lemieux C, Sankoff D, Turmel M, Gray MW (1997) An ancestral mitochondrial DNA resembling a eubacterial genome in miniature. Nature 387:493–497

Larsson NG (2010) Somatic mitochondrial DNA mutations in mammalian aging. Annu Rev Biochem 79:683–706

Larsson NG, Wang J, Wilhelmsson H, Oldfors A, Rustin P, Lewandoski M, Barsh GS, Clayton DA (1998) Mitochondrial transcription factor A is necessary for mtDNA maintenance and embryogenesis in mice. Nat Genet 18:231–236

Lecrenier N, Foury F (2000) New features of mitochondrial DNA replication system in yeast and man. Gene 246:37–48

Lewis DL, Farr CL, Kaguni LS (1995) *Drosophila melanogaster* mitochondrial DNA: completion of the nucleotide sequence and evolutionary comparisons. Insect Mol Biol 4:263–278

Linder T, Park CB, Asin-Cayuela J, Pellegrini M, Larsson NG, Falkenberg M, Samuelsson T, Gustafsson CM (2005) A family of putative transcription termination factors shared amongst metazoans and plants. Curr Genet 48:265–269

Litonin D, Sologub M, Shi Y, Savkina M, Anikin M, Falkenberg M, Gustafsson CM, Temiakov D (2010) Human mitochondrial transcription revisited: only TFAM and TFB2M are required for transcription of the mitochondrial genes in vitro. J Biol Chem 285:18129–18133

Liu Z, Butow RA (2006) Mitochondrial retrograde signaling. Annu Rev Genet 40:159–185

Lodeiro MF, Uchida AU, Arnold JJ, Reynolds SL, Moustafa IM, Cameron CE (2010) Identification of multiple rate-limiting steps during the human mitochondrial transcription cycle in vitro. J Biol Chem 285:16387–16402

Loguercio Polosa P, Roberti M, Musicco C, Gadaleta MN, Quagliariello E, Cantatore P (1999) Cloning and characterisation of mtDBP, a DNA-binding protein which binds two distinct regions of sea urchin mitochondrial DNA. Nucleic Acids Res 27:1890–1899

Loguercio Polosa P, Deceglie S, Roberti M, Gadaleta MN, Cantatore P (2005) Contrahelicase activity of the mitochondrial transcription termination factor mtDBP. Nucleic Acids Res 33:3812–3820

Loguercio Polosa P, Deceglie S, Falkenberg M, Roberti M, Di Ponzio B, Gadaleta MN, Cantatore P (2007) Cloning of the sea urchin mitochondrial RNA polymerase and reconstitution of the transcription termination system. Nucleic Acids Res 35:2413–2427

Madsen CS, Ghivizzani SC, Hauswirth WW (1993) Protein binding to a single termination-associated sequence in the mitochondrial DNA D-loop region. Mol Cell Biol 13:2162–2171

Martin M, Cho J, Cesare AJ, Griffith JD, Attardi G (2005) Termination factor-mediated DNA loop between termination and initiation sites drives mitochondrial rRNA synthesis. Cell 123:1227–1240

Matsushima Y, Garesse R, Kaguni LS (2004) *Drosophila* mitochondrial transcription factor B2 regulates mitochondrial DNA copy number and transcription in schneider cells. J Biol Chem 279:26900–26905

Matsushima Y, Adán C, Garesse R, Kaguni LS (2005) *Drosophila* mitochondrial transcription factor B1 modulates mitochondrial translation but not transcription or DNA copy number in Schneider cells. J Biol Chem 280:16815–16820

McBride HM, Neuspiel M, Wasiak S (2006) Mitochondria: more than just a powerhouse. Curr Biol 16:R551–R560

McCulloch V, Shadel GS (2003) Human mitochondrial transcription factor B1 interacts with the C-terminal activation region of h-mtTFA and stimulates transcription independently of its RNA methyltransferase activity. Mol Cell Biol 23:5816–5824

Meinke DW, Cherry JM, Dean C, Rounsley SD, Koornneef M (1998) *Arabidopsis thaliana*: a model plant for genome analysis. Science 282:679–682

Meinke D, Sweeney C, Muralla R (2009) Integrating the genetic and physical maps of *Arabidopsis thaliana*: identification of mapped alleles of cloned essential (EMB) genes. PLoS One 4:e7386

Metodiev MD, Lesko N, Park CB, Cámara Y, Shi Y, Wibom R, Hultenby K, Gustafsson CM, Larsson NG (2009) Methylation of 12 S rRNA is necessary for in vivo stability of the small subunit of the mammalian mitochondrial ribosome. Cell Metab 9:386–397

Montoya J, Christianson T, Levens D, Rabinowitz M, Attardi G (1982) Identification of initiation sites for heavy-strand and light-strand transcription in human mitochondrial DNA. Proc Natl Acad Sci USA 79:7195–7199

Montoya J, Gaines GL, Attardi G (1983) The pattern of transcription of the human mitochondrial rRNA genes reveals two overlapping transcription units. Cell 34:151–159

Nagaike T, Suzuki T, Katoh T, Ueda T (2005) Human mitochondrial mRNAs are stabilized with polyadenylation regulated by mitochondria-specific poly(A) polymerase and polynucleotide phosphorylase. J Biol Chem 280:19721–19727

Nam SC, Kang C (2005) DNA light-strand preferential recognition of human mitochondria transcription termination factor mTERF. J Biochem Mol Biol 38:690–694

Narasimhan N, Attardi G (1987) Specific requirement for ATP at an early step of in vitro transcription of human mitochondrial DNA. Proc Natl Acad Sci USA 84:4078–4082

Ojala D, Montoya J, Attardi G (1981) tRNA punctuation model of RNA processing in human mitochondria. Nature 290:470–474

Ostronoff LK, Izquierdo JM, Cuezva JM (1995) mt-mRNA stability regulates the expression of the mitochondrial genome during liver development. Biochem Biophys Res Commun 217:1094–1098

Parisi MA, Clayton DA (1991) Similarity of human mitochondrial transcription factor 1 to high mobility group proteins. Science 252:965–969

Park CB, Asin-Cayuela J, Cámara Y, Shi Y, Pellegrini M, Gaspari M, Wibom R, Hultenby K, Erdjument-Bromage H, Tempst P, Falkenberg M, Gustafsson CM, Larsson N (2007) MTERF3 is a negative regulator of mammalian mtDNA transcription. Cell 130:273–285

Passos JF, von Zglinicki T, Kirkwood TB (2007) Mitochondria and ageing: winning and losing in the numbers game. Bioessays 29:908–917

Pellegrini M, Asin-Cayuela J, Erdjument-Bromage H, Tempst P, Larsson NG, Gustafsson CM (2009) MTERF2 is a nucleoid component in mammalian mitochondria. Biochim Biophys Acta 1787:296–302

Pham XH, Farge G, Shi Y, Gaspari M, Gustafsson CM, Falkenberg M (2006) Conserved sequence box II directs transcription termination and primer formation in mitochondria. J Biol Chem 28:24647–24652

Poulton J, Morten K, Freeman-Emmerson C, Potter C, Sewry C, Dubowitz V, Kidd H, Stephenson J, Whitehouse W, Hansen FJ, Parisi M, Brown G (1994) Deficiency of the human mitochondrial transcription factor h-mtTFA in infantile mitochondrial myopathy is associated with mtDNA depletion. Hum Mol Genet 3:1763–1769

Puranam RS, Attardi G (2001) The RNase P associated with HeLa cell mitochondria contains an essential RNA component identical in sequence to that of the nuclear RNase P. Mol Cell Biol 21:548–561

Rantanen A, Jansson M, Oldfors A, Larsson NG (2001) Downregulation of Tfam and mtDNA copy number during mammalian spermatogenesis. Mamm Genome 12:787–792

Roberti M, Loguercio Polosa P, Bruni F, Musicco C, Gadaleta MN, Cantatore P (2003) DmTTF, a novel mitochondrial transcription termination factor that recognises two sequences of Drosophila melanogaster mitochondrial DNA. Nucleic Acids Res 31:1597–1604

Roberti M, Fernandez-Silva P, Loguercio Polosa P, Fernandez-Vizarra E, Bruni F, Deceglie S, Montoya J, Gadaleta MN, Cantatore P (2005) In vitro transcription termination activity of the Drosophila mitochondrial DNA-binding protein DmTTF. Biochem Biophys Res Commun 331:357–362

Roberti M, Bruni F, Loguercio Polosa P, Gadaleta MN, Cantatore P (2006) The Drosophila termination factor DmTTF regulates in vivo mitochondrial transcription. Nucleic Acids Res 34:2109–2116

Roberti M, Loguercio Polosa P, Bruni F, Manzari C, Deceglie S, Gadaleta MN, Cantatore P (2009) The MTERF family proteins: mitochondrial transcription regulators and beyond. Biochim Biophys Acta 1787:303–311

Rodeheffer MS, Shadel GS (2003) Multiple interactions involving the amino-terminal domain of yeast mtRNA polymerase determine the efficiency of mitochondrial protein synthesis. J Biol Chem 278:18695–18701

Saito S, Tamura K, Aotsuka T (2005) Replication origin of mitochondrial DNA in insects. Genetics 171:1695–1705

Scarpulla RC (2008) Transcriptional paradigms in mammalian mitochondrial biogenesis and function. Physiol Rev 88:611–638

Schinkel AH, Koerkamp MJ, Touw EP, Tabak HF (1987) Specificity factor of yeast mitochondrial RNA polymerase. Purification and interaction with core RNA polymerase. J Biol Chem 262:12785–12791

Sea Urchin Genome Sequencing Consortium (2006) The genome of the sea urchin *Strongylocentrotus purpuratus*. Science 314:941–952

Seidel-Rogol BL, McCulloch V, Shadel GS (2003) Human mitochondrial transcription factor B1 methylates ribosomal RNA at a conserved stem-loop. Nat Genet 33:23–24

Shutt TE, Gray MW (2006) Bacteriophage origins of mitochondrial replication and transcription proteins. Trends Genet 22:90–95

Shutt TE, Lodeiro MF, Cotney J, Cameron CE, Shadel G (2010) Core human mitochondrial transcription apparatus is a regulated two-component system *in vitro*. Proc Natl Acad Sci USA 107:12133–12138

Silva JP, Köhler M, Graff C, Oldfors C, Magnusson MA, Berggreen PO, Larsson NG (2000) Impaired insulin secretion and beta-cell loss in tissue-specific knockout mice with mitochondrial diabetes. Nat Genet 26:336–340

Sologub M, Litonin D, Anikin M, Mustaev A, Temiakov D (2009) TFB2 is a transient component of the catalytic site of the human mitochondrial RNA polymerase. Cell 139:934–944

Spåhr H, Samuelsson T, Hällberg BM, Gustafsson CM (2010) Structure of mitochondrial transcription termination factor 3 reveals a novel nucleic acid-binding domain. Biochem Biophys Res Commun 397:386–390

Temperley RJ, Wydro M, Lightowlers RN, Chrzanowska-Lightowlers ZM (2010) Human mitochondrial mRNAs-like members of all families, similar but different. Biochim Biophys Acta 1797:1081–1085

Tiranti V, Savoia A, Forti F, D'Apolito MF, Centra M, Rocchi M, Zeviani M (1997) Identification of the gene encoding the human mitochondrial RNA polymerase (h-mtRPOL) by cyberscreening of the Expressed Sequence Tags database. Hum Mol Genet 6:615–625

Wallace DC, Fan W, Procaccio V (2010) Mitochondrial energetics and therapeutics. Annu Rev Pathol 5:297–348

Wang J, Wilhelmsson H, Graff C, Li H, Oldfors A, Rustin P, Brüning JC, Kahn CR, Clayton DA, Barsh GS, Thorén P, Larsson N (1999) Dilated cardiomyopathy and atrioventricular conduction blocks induced by heart-specific inactivation of mitochondrial DNA gene expression. Nat Genet 21:133–137

Wang Z, Cotney J, Shadel GS (2007) Human mitochondrial ribosomal protein MRPL12 interacts directly with mitochondrial RNA polymerase to modulate mitochondrial gene expression. J Biol Chem 282:12610–12618

Wanrooij PH, Uhler JP, Simonsson T, Falkenberg M, Gustafsson CM (2010) G-quadruplex structures in RNA stimulate mitochondrial transcription termination and primer formation. Proc Natl Acad Sci USA 107:16072–16077

Wenz T, Luca C, Torraco A, Moraes CT (2009) mTERF2 regulates oxidative phosphorylation by modulating mtDNA transcription. Cell Metab 9:499–511

Wredenberg A, Wibom R, Wilhelmsson H, Graff C, Wiener HH, Burden SJ, Oldfors A, Westerblad H, Larsson NG (2002) Increased mitochondrial mass in mitochondrial myopathy mice. Proc Natl Acad Sci USA 99:15066–15071

Yakubovskaya E, Mejia E, Byrnes J, Hambardjieva E, Garcia-Diaz M (2010) Helix unwinding and base flipping enable human MTERF1 to terminate mitochondrial transcription. Cell 141:982–993

Chapter 12
Transcription and Transcription Regulation in Chloroplasts and Mitochondria of Higher Plants

Andreas Weihe, Karsten Liere, and Thomas Börner

12.1 Introduction

Mitochondria and plastids are considered being descendants of an alpha-proteobacterium and a cyanobacterium, respectively, taken up as endosymbionts by the ancestral cells of eukaryotes and of the green lineage of eukaryotes, respectively (see Chaps. 1 and 2). During continuing co-evolution of the endosymbionts and host cells, a massive loss of genes from the endosymbiont/organellar genomes has occurred. Many of those genes have been transferred to the nucleus, which still is a relatively frequent and ongoing process (see Chap. 7). By acquiring plastid and/or mitochondrial targeting sequences, a considerable amount of these gene products were rerouted back into the organelles. Similarly, novel nuclear-encoded proteins also became part of the proteomes of mitochondria and plastids (eukaryotization; Hengeveld and Fedonkin 2004; see Chap. 9). Surprisingly, the transcriptional machinery in the plastids of higher plants turned out to be more complex as known from their cyanobacterial ancestors. Cyanobacteria, as other bacteria, use one RNA polymerase (RNAP) to transcribe all genes. The core enzyme consists of several subunits encoded by the *rpoA, B, C1,* and *C2* genes and is complemented by a σ-factor (constituting the holoenzyme) for promoter recognition and initiation of transcription (Kaneko et al. 1996). Cyanobacteria possess several σ-factors to transcribe different sets of genes (Imamura et al. 2003). Although having a smaller genome, plastids of angiosperms apparently need different RNAPs and different σ-factors for transcription (Hess and Börner 1999; Liere and Maliga 2001). Also mitochondria have completely changed the transcriptional apparatus as compared to their bacterial ancestors, mainly due to the fact that they are using phage-type

A. Weihe • K. Liere • T. Börner (✉)
Institut für Biologie/Genetik, Humboldt-Universität zu Berlin, Chausseestr. 117, 10115 Berlin, Germany
e-mail: andreas.weihe@rz.hu-berlin.de; karsten.liere@rz.hu-berlin.de; thomas.boerner@rz.hu-berlin.de

C.E. Bullerwell (ed.), *Organelle Genetics*,
DOI 10.1007/978-3-642-22380-8_12, © Springer-Verlag Berlin Heidelberg 2012

RNA polymerase(s) for transcription. This chapter describes the components of the transcriptional apparatus in mitochondria and plastids of higher plants and their evolution and their roles in transcription and its regulation.

12.2 The Transcriptional Apparatus of Chloroplasts

12.2.1 Plastid RNA Polymerases

12.2.1.1 PEP: The Plastid-Encoded Plastid RNA Polymerase

The plastid chromosomes of algae and higher plants possess genes for core subunits of a cyanobacterial-type RNAP, which is commonly abbreviated as PEP (for plastid-encoded plastid RNA polymerase). Although PEP might be responsible for transcription of all plastid genes in algae, it is complemented by a second nuclear-encoded plastid RNAP activity in higher plants (NEP; see Sect. 12.2.1.2). In higher plants, the genes encoding PEP core subunits are encoded on the plastome (reviewed in Lysenko and Kuznetsov 2005). The *rpoBC* operon is under control of a NEP promoter in monocotyledonous and dicotyledonous plants (see Sect. 12.2.2). Transcript levels of *rpo* genes are low compared with genes for proteins involved in photosynthesis (e.g., Hess et al. 1993; Legen et al. 2002). The PEP holoenzyme consists of the core subunits complemented by a σ-factor for promoter recognition. These are encoded by nuclear genes (see Sect. 12.2.4.2) in all embryophytes ensuring together with NEP a control of plastid transcription by the nucleus.

PEP can be isolated from plastids as both a soluble enzyme and an insoluble form together with DNA as the so-called "transcriptionally active chromosome" (TAC; e.g., Krause and Krupinska 2000; Pfalz et al. 2006). Proteomic data of different PEP and TAC preparations demonstrated that the basic eubacterial-type holoenzyme of higher plants is assembled within various accessory protein factors, which vary depending on the isolation procedure (Schweer et al. 2010b). The TAC complex consists of more than 50 accessory proteins; phage-type polymerases have not been detected yet within the TAC complex (e.g., Krause and Krupinska 2000; Pfannschmidt et al. 2000; Loschelder et al. 2004; Suzuki et al. 2004; Schröter et al. 2010). The PEP therefore seems to be a good example of the eukaryotization of the plastid, since not only the addition of several accessory components to the eubacterial core enzyme but also the nuclear origin of these factors assign further eukaryotic characteristics.

12.2.1.2 NEP: The Nuclear-Encoded Plastid RNA Polymerase

The existence of one or more *n*uclear-*e*ncoded *p*lastid RNA *p*olymerase(s) (NEP) was suggested by comparing the effects of inhibitors of translation on cytoplasmic and plastidial ribosomes, respectively (Ellis and Hartley 1971). The detection of RNAP activities in plastids lacking PEP due to impaired protein synthesis provided evidence for a nuclear location of gene(s) encoding this activity (Bünger and Feierabend 1980; Siemenroth et al. 1981; Falk et al. 1993; Han et al. 1993; Hess et al. 1993). Additional evidence for a NEP activity in plastids came from the detection of RNA synthesis in nonphotosynthetic plastids of parasitic plants with functionally reduced plastids lacking the plastid genes for PEP core subunits (reviewed in Krause 2008, see Chap. 4). In addition, transplastomic tobacco plants with inactivated PEP genes still transcribed plastid genes. The albino phenotype of these plants indicated that the NEP activity alone is not sufficient for the development of photosynthetically active chloroplasts (Allison et al. 1996; Hajdukiewicz et al. 1997; Krause et al. 2000; Legen et al. 2002).

Angiosperms harbor small *RpoT* gene families encoding enzymes related to the RNAPs of T3/7-type bacteriophages. In addition to the common mitochondrial RpoT polymerase (RpoTm), a plastid-targeted RpoT polymerase (RpoTp) and an RpoT polymerase targeted to both mitochondria and plastids (RpoTmp) may be found (Fig. 12.1; reviewed in Shiina et al. 2005; Liere and Börner 2007a, b; see Sects. 12.3 and 12.4). Heterologously expressed RpoTp, RpoTmp, and RpoTm enzymes of *Arabidopsis* are active RNAPs that prefer circular over linear DNA templates. RpoTm and RpoTp (not RpoTmp) exhibit an intrinsic ability to recognize several mitochondrial and at least one NEP promoter in vitro (Kühn et al. 2007). The organellar localization and expression patterns supported the assumption of these RpoT enzymes to represent the mitochondrial transcriptase (RpoTm) and NEP (RpoTp; Chang et al. 1999; Emanuel et al. 2004, 2006; Kusumi et al. 2004). First confirmation of NEP being represented by RpoTp (in part together with RpoTmp; see below) was provided by studies on transgenic *Nicotiana* and *Arabidopsis* plants that overexpressed RpoTp and exhibited an increased usage of certain NEP promoters (Liere et al. 2004). Furthermore, mutation of the *Arabidopsis RpoTp* gene led to impaired chloroplast biogenesis and altered accumulation of plastid transcripts (Hricová et al. 2006). Similar observations were

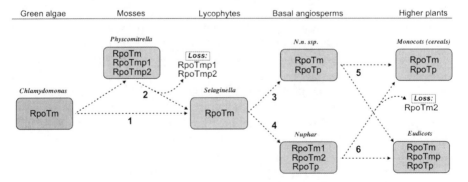

Fig. 12.1 Evolution of plant phage-type RNA polymerases: a hypothetical scenario. Gene duplications and/or loss of genes occurred several times during evolution, giving rise to small *RpoT* gene families. 1 vs. 2, 3 vs. 4, and 5 vs.6 designate alternative routes of evolution

made in *Arabidopsis* plants with reduced RpoTp transcript levels due to the expression of antisense RNA (C. Emanuel et al., unpublished data). Apart from RpoTp, RpoTmp was supposed to play a role not only in mitochondrial but also in plastid gene expression (Baba et al. 2004; Hricová et al. 2006; Courtois et al. 2007), i.e., both polymerases represent the NEP activity. *Arabidopsis* lines with impaired *RpoTmp* function were delayed in chloroplast biogenesis and showed altered plastid transcript levels (Baba et al. 2004, cf. Kühn et al. 2009). *RpoTp/RpoTmp* double mutants exhibited a more severe phenotype than both of the single mutants and were extremely retarded in growth (Hricová et al. 2006).

12.2.2   *Plastid Promoters*

Due to their cyanobacterial origin, many plastid promoters contain a variant of the -35 (TTGaca) and -10 (TAtaaT) consensus sequences of typical σ^{70}-type *E. coli* promoters. Moreover, plastid promoters of the eubacterial σ^{70}-type are accurately recognized by the *E. coli* RNAP (for reviews, see Liere and Maliga 2001; Weihe 2004; Liere and Börner 2007a, b). Because PEP recognizes such plastid σ^{70}-type promoters, they are also often termed PEP promoters. Further regulatory sequences in addition to the promoter core were also identified in the proximity of PEP promoters (see Sect. 12.2.4).

NEP transcripts are, with a few exceptions, rarely detectable in chloroplasts. Therefore, plants with reduced or eliminated transcriptional activity by PEP laid the foundation for the unambiguous identification of transcription initiation sites for a nuclear-encoded transcription activity (i.e., NEP; (Maliga 1998; Hess and Börner 1999; Liere and Maliga 2001). Given the similarity of their respective RNAP activities, NEP promoters analyzed thus far obviously resemble mitochondrial and phage promoters in their structural organization (see Sect. 12.3.3). Based on their sequence properties, they can be grouped into three types (Weihe and Börner 1999; Liere and Maliga 2001). A conserved YRTa-motif is typical for type-I promoters and critical for promoter recognition embedded in a small DNA fragment (-15 to $+5$) upstream of the transcription initiation site ($+1$) (P*atpB*-289; Kapoor and Sugiura 1999, P*accD*-129; Liere and Maliga 1999b, P*rpoB*-345; Liere and Maliga 1999a; Xie and Allison 2002). A subset of type-I NEP promoters (type-Ib) possesses a second, conserved sequence motif (ATAN$_{0-1}$GAA) ~18–20 bp upstream of the YRTa-motif, designated box II or GAA-box (Silhavy and Maliga 1998; Kapoor and Sugiura 1999). A functional role of this element in promoter recognition has been shown in mutational analyses of the tobacco P*atpB*-289 promoter in in vitro and in vivo transcription experiments (Kapoor and Sugiura 1999; Xie and Allison 2002).

A second group of NEP promoters (type-II) lack the YRTa-motif and so far comprises the so-called "nonconsensus" NEP promoters. The best-characterized tobacco P*clpP*-53 was dissected using a transplastomic in vivo approach demonstrating that critical promoter sequences are located mainly downstream of the transcription initiation site (-5 to $+25$; Sriraman et al. 1998). Interestingly, the *clpP*-53 promoter motif and transcription initiation site are conserved among monocots, eudicots, conifers, and liverworts. The lack of transcription in rice from the conserved P*clpP*-53 homolog has been attributed to the lack of a distinct NEP enzyme not present in monocots (e.g., RpoTmp; Sriraman et al. 1998; Liere et al. 2004; Courtois et al. 2007; Swiatecka-Hagenbruch et al. 2008). A further non-YRTa-type NEP promoter recognized by RpoTmp is the *rrn* operon Pc promoter in spinach and *Arabidopsis* (Baeza et al. 1991; Iratni et al. 1994, 1997; Sriraman et al. 1998; Swiatecka-Hagenbruch et al. 2007). The Pc promoter solely drives *rrn* operon transcription in spinach. Although it contains typical σ^{70}-elements which are active as the *rrn* operon promoter in other species, transcription initiates from a site between the conserved $-10/-35$ PEP promoter elements. However, sequences relevant for transcription initiation from Pc have yet to be identified.

12.2.3   Division of Labor Between PEP and NEP

First insights into the division of labor between PEP and NEP were obtained from investigations on the use of PEP vs. NEP promoters in different tissues and under the influence of different endogenous and exogenous factors. Although genes exist that are transcribed from a single promoter, transcription of plastid genes and operons by multiple promoters seems to be a common feature. There is a high diversity in promoter usage for some genes in different species (reviewed in Liere and Börner 2007a, 2007b). Both NEP and PEP promoters together are found upstream of many genes. Consequently, both promoter types are believed to differentially express their cognate gene during plant development (reviewed in Maliga 1998; Liere and Maliga 2001). While NEP promoters are generally active in youngest and nongreen tissues in early leaf development, PEP activity increases during maturation of proplastids to photosynthetically active chloroplasts (e.g., Baumgartner et al. 1993; Hajdukiewicz et al. 1997; Kapoor et al. 1997; Emanuel et al. 2004; Courtois et al. 2007; Swiatecka-Hagenbruch et al. 2008). For example, RpoTmp was shown to function specifically in the transcription of the *Arabidopsis* *rrn* operon during seed germination and early plant development (Courtois et al. 2007; Swiatecka-Hagenbruch et al. 2008) and may have a more general function in chloroplasts during later developmental stages. The identification of thylakoid membrane proteins (NIPs; *N*EP *i*nteracting *p*roteins) interacting with RpoTmp led to a model to explain the developmental switch from NEP to PEP transcription (Azevedo et al. 2008). However, in the light that the plastid transcriptional apparatus together with the plastid DNA (organized in nucleoids) is membrane associated (Sato 2001; Sato et al. 2003; Karcher et al. 2009; Schweer et al. 2010b), it seems unlikely that binding to the thylakoid membrane indicates inactivation of the enzyme. Furthermore, since genes exclusively transcribed by NEP encode housekeeping functions like the *rpoB* gene/operon and *rps15*, NEP should be still necessary in mature chloroplasts. Indeed, NEP and PEP are active throughout leaf development in maize and *Arabidopsis* (Cahoon et al. 2004; Zoschke et al. 2007).

12.2.4 Regulation of Plastid Transcription and Transcription Factors

12.2.4.1 Exogenous and Endogenous Factors Affecting Transcription of Plastid Genes

Exogenous and endogenous factors such as light and plastid type were shown to differentially modulate the transcriptional activity of plastid genes (Mullet 1993; Allison and Maliga 1995; Mayfield et al. 1995; Link 1996; Liere and Börner 2007a; Liere et al. 2011) involving the interaction of the core RNAP with specific σ- and/or other regulatory factors. In silico analyses revealed about 48–78 nuclear *Arabidopsis* genes of putative plastid-localized transcription factors (Wagner and Pfannschmidt 2006; Schwacke et al. 2007), which may expand the regulatory capacity of the plastid transcription machinery.

Transcriptional activities of most plastid-encoded genes show an increase early in light-induced plastid development to rapidly build up the photosynthesis apparatus in higher plants. Moreover, light-dependent plastid transcription also occurs in leaves during greening as well as in mature leaves. Well-known examples are genes such as *psbA*, *psbD-psbC*, *petG*, *rbcL*, and *atpB* (reviewed in Shiina et al. 2005; Liere and Börner 2007a, 2007b; Liere et al. 2011).

Specific photoreceptors are responsible for the perception of particular wavelengths, e.g., cryptochromes and phototropins for blue and phytochromes for red light detection (Chory 2010), and involved in transcriptional activation of photosynthesis-related genes in chloroplasts (Chun et al. 2001; Thum et al. 2001). While red light only partially increased plastid transcription, blue light further enhanced overall plastid transcription activity in dark-adapted mature leaves. Therefore, global activation of plastid transcription after dark adaptation is likely to be mediated by cryptochromes. When exposed to blue/UV-A light, an *Arabidopsis* *phyA*-mutant displayed lower *psbA* and *rrn16* transcript activities than the wild type suggesting a further role for PhyA in light reception (Chun et al. 2001). By downregulating plastid transcription of genes normally induced by light, green light might play a balancing/antagonistic role to blue light in controlling gene expression during early photomorphogenic development (Dhingra et al. 2006). Interestingly, cryptochromes are discussed to be sensors of blue/green light ratios under natural radiation (Sellaro et al. 2010).

One of the *psbD* operon promoters is activated by the blue light. Although this particular promoter comprises two conserved upstream elements (AAG-box, PGT-box) and cognate binding factors (AGF, PGTF; see reviews by Shiina et al. 2005; Liere and Börner 2007a, b), it has been shown that AthSig5 acts as the mediator of blue-light signaling in activating *psbD* BLRP transcription in blue light (Tsunoyama et al. 2002, 2004; Nagashima et al. 2004; Onda et al. 2008). It is assumed that the signal transduction pathway involves reception of blue light by cryptochromes and PhyA (Thum et al. 2001; Mochizuki et al. 2004), further mediation by a protein phosphatase PP7 (Moller et al. 2003), and subsequent induction of *Sig5* expression (Mochizuki et al. 2004). Sig5 associates with AGF (PTF1) and initiates *psbD* transcription in plastids. Furthermore, *psbD* BLRP activity seems also to be regulated in a developmental and tissue-specific manner (Christopher and Hoffer 1998).

Environmental control of plastid gene expression is most intense in differentiation from proplastids to either etioplasts (dark) or chloroplasts (light). Feedback from the plastids depending on its developmental status, metabolism, and/or gene expression is controlling nuclear gene expression by generating the so-called "plastid signals" or "plastid factors" (Fig. 12.2; Bradbeer et al. 1979; Rodermel 2001; Gray 2003; Beck 2005; Brown et al. 2005; Pogson et al. 2008; Woodson and Chory 2008). Certainly, several plastid factors/signals exist. They may neither be plastid gene products (Oelmüller et al. 1986; Lukens et al. 1987) nor include the previously suggested Mg-protoporphyrin IX (Mochizuki et al. 2008; Moulin et al. 2008). The proposed cytosolic signaling pathways and putative organellar signaling molecules remain elusive (Kleine et al. 2009). Various models of plastid signaling in the context of the metabolic network within the cell are discussed (Pfannschmidt 2010). One of the targets of plastid signal(s) is the *RpoTp* gene (encoding NEP; Emanuel et al. 2004), i.e., retrograde signaling likely coordinates expression of PEP and NEP as a prereq-uisite of concerted gene expression in both plastids and nucleus (Fig. 12.2).

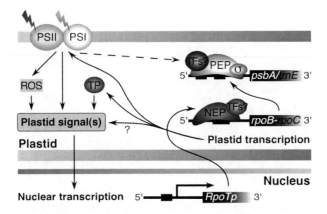

Fig. 12.2 Model on the role of nuclear-encoded phage-type RNA polymerases in regulating plastid gene expression. NEP transcription activity is in part encoded by the nuclear *RpoTp* gene. Expression of the plastid *rpoB* operon encoding subunits of the plastid-encoded RNA-polymerase (PEP) is regulated by NEP transcription. PEP, however, transcribes genes encoding components of the photosynthetic complexes (PSI, PSII) that modulate nuclear transcription by generating various "plastid signals" (e.g., ROS, reactive oxygen species). Redox signals from the PQ pool may be mediated by yet unknown pathways (*dashed line*) to regulate plastid transcription by modifying plastid transcription factors. The trnA^{Glu} encoding *trnE* gene is also transcribed by PEP. The tRNA is required for the synthesis of δ-aminolevulinic acid (ALA), a precursor of the tetrapyrrole (TP) biosynthesis (chlorophyll and heme), which may also provide "plastid signals." Thus, the regulatory network of the nuclear and plastid transcription machineries may be a key element for adjusting the expression of genes located within different compartments of the plant cell in response to exogenous and endogenous factors (Liere et al. 2011)

Effects of the redox state on plastid gene transcription (and on retrograde signaling) were demonstrated by growing plants under light conditions generating an imbalance in excitation energy distribution between the photosystems PSI and PSII photosystems (PSII- and PSI-light, 680 and 700 nm, respectively; Pfannschmidt et al. 1999a, b; Fey et al. 2005; Bräutigam et al. 2009). The change in photosystem stoichiometry correlated with respective changes in the transcriptional rates and transcript amounts of the plastid genes for the reaction center proteins of PSII and PSI, *psbA* and *psaAB*. The redox state signal of the plastochinone pool (PQ) is mediated toward the level of plastid gene expression via plastid kinases STN7 and CSK (Puthiyaveetil and Allen 2008; Pesaresi et al. 2009; Steiner et al. 2009). Furthermore, the combination of redox changes in the PQ and thioredoxin pools might act as cooperative signals that coordinate not only gene expression but also metabolism (Bräutigam et al. 2009) and protein import (Balsera et al. 2010). A recently characterized novel thioredoxin (TRX *z*) was shown to interact with two fructokinase-like proteins (FLNs), both of which appear to be necessary for PEP-dependent gene expression in chloroplasts. TRX *z* was therefore discussed to define together with the two FLNs a thus far unknown protein interaction module essential for chloroplast development (Arsova et al. 2010).

A putative DNA-binding protein of PS II, TSP9, is partially released from PSII upon PQ reduction in spinach and may represent such a signal transducer toward transcription (Carlberg et al. 2003; Zer and Ohad 2003). Identification of an additional protein binding to the light-dependent *psaAB* PEP promoter region (Chen et al. 1993; Cheng et al. 1997) suggested the existence of yet unidentified transcription factors that transmit redox signals. Furthermore, the *Arabidopsis* high chlorophyll fluorescence-mutant *hcf145* shows decreased mRNA stability and transcription of *psaA* (Lezhneva and Meurer 2004). Thus, HCF145 might be involved in transcriptional regulation of the *psaA* operon. Further analysis of this promoter has yet to be reported.

PEP is not only responsible for the redox-regulated transcription at the *psbA* and *psaAB* promoters, but apparently genes for components/subunits of the PEP complex such as *rpoB* (plastid-encoded β-subunit), *AthSig5* (nuclear-encoded σ-factor), and *SibI* (nuclear-encoded Sig1-binding protein; Morikawa et al. 2002) are also regulated via redox control (Fey et al. 2005). Interestingly, the *rpoB* operon is transcribed by RpoTp, suggesting a redox regulation of this enzyme. It seems likely that several distinct redox control pathways control plastid transcription, which depends on environmental conditions such as responses to low or high light (Link 2003; Pfannschmidt and Liere 2005).

A regulatory role, which links chlorophyll synthesis and the developmental switch from NEP to PEP, has been proposed for the plastid-encoded tRNA^{Glu} in *Arabidopsis* (Hanaoka et al. 2005), also required for synthesis of δ-aminolevulinic acid (Schön et al. 1986). However, reinvestigation of the postulated inhibitory activity of tRNA^{Glu} on RpoTp activity demonstrated a rather unspecific effect suggesting that tRNA^{Glu} does not play a role in specifically regulating NEP activity (Bohne et al. 2009).

The so-called "stringent control" enables bacteria to adapt to nutrient-limiting stress conditions (Cashel et al. 1996). The effector molecule is guanosine 5′diphosphate 3′-diphosphate (ppGpp), which binds to the core RNAP modifying its promoter specificity (Toulokhonov et al. 2001; Jishage et al. 2002). Homologs of the bacterial ppGpp synthetases, RelA and SpoT, were found in *Chlamydomonas reinhardtii* (Kasai et al. 2002), *Arabidopsis* (van der Biezen et al. 2000), and tobacco (Givens et al. 2004). Plastid targeting has been demonstrated for some of these RSH termed proteins, suggesting an implication in ppGpp signaling in plastids. RSH expression and plastid ppGpp levels are clearly elevated by light and various abiotic and biotic stress conditions. Furthermore, PEP activity is inhibited by ppGpp in vitro by directly binding to the β′-subunit (Givens et al. 2004; Takahashi et al. 2004; Sato et al. 2009). Thus, it is conceivable that under stress conditions, PEP might be under control of a bacterial-like stringent response mediated by ppGpp, which is also discussed for cyanobacteria (Imamura and Asayama 2009). Interestingly, stress signals specifically induce transcription initiation from the *psbD* BRLP conferred by a special σ-factor, AthSig5 (see Sect. 12.2.4.2; Nagashima et al. 2004; Tsunoyama et al. 2004). However, target genes that are regulated by a plastid stringent control have yet to be identified, which will elucidate the molecular mechanisms of transcriptional responses to plant hormones and environmental stress situations (Zubo et al. 2008, 2011).

12.2.4.2   Nuclear-Encoded Plastid σ-Factors

Similar to bacteria, the eubacterial-type RNAP complex in plastids of higher plants also contains σ-factors that are responsible for promoter recognition and contributing to DNA melting around the initiation site (for recent reviews see Lysenko 2007; Shiina et al. 2009; Schweer 2010; Schweer et al. 2010b; Lerns-Mache 2011). Initially named "sigma-like" factors (SLFs), σ-like activities were biochemically characterized early on in a number of plastids. Since then, genes for σ-factors were cloned from a number of organisms and a unified nomenclature of plant σ-factors has been proposed (http:// sfns.u-shizuoka-ken.ac.jp/pctech/sigma/proposal/index.html; Shiina et al. 2005, 2009; Liere and Börner 2007a, b; Lysenko 2007; Schweer 2010; Schweer et al. 2010b).

Bacterial and plastid σ-factors share three conserved domains involved in binding the core RNA-polymerase (domains 2.1 and 2.3), hydrophobic core formation (2.2), DNA melting (2.3), recognition of the −10 promoter motif (2.4), and recognition of the −35 promoter motif (4.1 and 4.2; Paget and Helmann 2003; Shiina et al. 2009; Schweer et al. 2010b). To specify regulatory determinants, Schweer et al. (2009) transformed chimeric σ-factors (Sig1/6, Sig6/1, Sig3/6, and Sig6/3) into an *Arabidopsis sig6* knockout line. The observed phenotypes and plastid RNA patterns in the resulting plants did point to an important, however not exclusive, role for the highly variable N-terminal segment of plant σ-factors.

Consistent with a prominent role of PEP in leaves, most plastidial σ-factor genes of higher plants are expressed in light-dependent manner in green tissue but are silent in non-photosynthetic roots. The expression of plastid σ-factors seems to be differentially regulated during early *Arabidopsis* development and to be regulated by circadian rhythms (reviewed in Shiina et al. 2005; Liere and Börner 2007a, b; Lysenko 2007; Shiina et al. 2009; Schweer 2010; Schweer et al. 2010b).

Several approaches were employed to address the question of a specific role of σ-factor diversity in transcriptional regulation (Shiina et al. 2005; Liere and Börner 2007b). However, the most conclusive clues for the specificity of each σ-factor have been obtained by extensive characterizations of *Arabidopsis* σ-factor T-DNA insertion mutants, overexpression, or antisense lines (recently reviewed in Shiina et al. 2009; Schweer 2010).

Although about six genes in *Arabidopsis* seem to be controlled by a distinct σ-factor within a certain time frame during plant development (*psaJ* by AthSig2, *psbN* by AthSig3, *ndhF* by AthSig4, *psbD* (BLRP) by AthSig5, and *atpB* by AthSig6), most other genes appear to be controlled by several σ-factors, thereby possessing overlapping functions (Lysenko 2007; Shiina et al. 2009; Schweer 2010; Schweer et al. 2010b; and references therein). Thus far, regulation of plastid gene expression by AthSig2 and AthSig6 seems to be important early in seedling development with a more restricted role in recognition of certain promoters later on. Similarly, in addition to their specific function, AthSig1, AthSig3, and AthSig4 may have overlapping functions in the transcription of photosynthesis genes in mature leaves maybe in response to the developmental and environmental signals. AthSig5 is a highly inducible plastid transcription factor regulated by different signaling pathways in response to environmental stresses. Based on its structural features, a PEP holoenzyme with AthSig5 might be less prone to abortive transcription (Lysenko 2007), which may provide the reason why AthSig5 gained an important role as a stress-inducible transcription factor. Hence, plants, similar to bacteria, use a set of σ-factors to differentially regulate plastid gene expression.

Plant σ-factors are regulated not only at the level of expression but also at the post-translational level in promoter recognition and their binding to the core enzyme. Early on, phosphorylation of σ-factors and the PEP enzyme has been shown to be responsible for changes in chloroplast transcription. A PEP-associated Ser/Thr protein kinase, termed plastid transcription kinase (PTK), has been shown to be involved in plastid σ-factor phosphorylation. Further characterization revealed the catalytic component of PTK to be closely related to the α-subunit of casein kinase 2 (CK2) and was subsequently named cpCK2. Based on the observation that cpCK2 itself is antagonistically regulated by phosphorylation and redox state, cpCK2 was proposed to be part of a signaling pathway controlling PEP activity (for reviews, see Pfannschmidt and Liere 2005; Liere and Börner 2007a, b; Baginsky and Gruissem 2009). In a recent work by Schweer et al. (2009), it has been shown that the regulatory phosphoacceptor sites reside within the highly variable, unconserved regions (UCRs) of plastid σ-factors. In addition, cpCK2 might be assisted by other kinase(s) by pre-phosphorylation ("pathfinder" kinase;

Schweer et al. 2010a). It remains unknown whether cpCK2 is also regulated via extraplastidic signal chains mediated by phyto- and/or chryptochromes.

Bacterial σ-factor activity is controlled by anti-σ factors (Ishihama 2000). Although proteins associated with AthSig1 associated were identified in *Arabidopsis* (SIB1 and T3K9.5; Morikawa et al. 2002), they are not related to any proteins of known function. Since their expression is regulated by light and is developmental and tissue-specific, they thus may be involved in regulation of AthSig1 activity.

12.2.4.3   NEP Transcription Factors

Studies of the mitochondrial RNAP from yeast (*Saccharomyces cerevisiae*), as well as the *Arabidopsis* AthRpoTm and AthRpoTp enzymes revealed promoter specificity to be conferred by the core RNAP alone without auxiliary factors, therefore retaining a characteristic feature of the T7 RNAP (Matsunaga and Jaehning 2004; Kühn et al. 2007). However, with AthRpoTm and AthRpoTp recognizing only part of the investigated promoters and AthRpoTmp displaying no significant promoter specificity but high nonspecific transcription activity in vitro, it is evident that the *Arabidopsis* enzymes need auxiliary factors for transcription *in organello* (Kühn et al. 2007).

Thus far, identification of factors involved in specific promoter recognition and transcription initiation by NEP has failed. Based on information on such factors (mtTFA and mtTFB) interacting with the related mitochondrial phage-type RNAPs from humans, mice, *Xenopus laevis*, and yeast, it has been intriguing to speculate on similar factors being present in plant organelles (see Sect. 12.3.2). To date, however, no functional mtTFA or mtTFB homologs have been characterized in plant mitochondria or plastids. BLAST searches of the *Arabidopsis* genome revealed a TFB-like dimethyladenosine transferase gene formerly characterized as PFC1 (Tokuhisa et al. 1998), which possesses an N-terminal transit peptide mediating protein import into plastids of isolated tobacco protoplasts (B. Kuhla, K. Liere, T. Börner; unpublished data). However, neither the phenotype of *PFC1*-knockout mutants nor in vitro transcription studies with recombinant PFC1 and AthRpoTp did support the idea that this TFB-like dimethyladenosine transferase may act as a primary transcription factor for the phage-type RNAPs (M. Swiatecka-Hagenbruch, K. Liere, T. Börner; unpublished data).

Although not a "principal" transcription factor, the spinach CDF2 is involved in NEP transcription by stimulating transcription of the *rrn* operon Pc promoter (Bligny et al. 2000). CDF2 is suggested to exist in two distinct forms. While CDF2-A might repress transcription initiation by PEP at the *rrn16* P1 promoter (termed P2 in spinach), CDF2-B possibly binds a NEP enzyme and initiates specific transcription from the *rrn16* Pc promoter. In addition, a role of RPL4 (plastid ribosomal protein L4, encoded by the nuclear *Rpl4* gene) has been discussed to be involved in NEP transcription, since it co-purifies with the T7-like transcription complex in spinach (Trifa et al. 1998). A function for this protein in the transcription by NEP or PEP, however, has yet to be demonstrated.

12.3 The Transcriptional Apparatus of Plant Mitochondria

12.3.1 RNA Polymerases

In several protist, fungal, animal, and plant species, mitochondrial transcription has been shown to be performed by nuclear-encoded phage-type RNAP core enzymes (reviewed in Hess and Börner 1999; Weihe 2004; see Chap. 11). Further *RpoT* genes encoding, presumably, mitochondrial RNAPs were detected in a number of plants and green algae, but the encoded polymerases are not yet characterized: *Sorghum bicolor, Brassica oleracea, Vitis vinifera, Ricinus communis, Micromonas pusilla* (http://genome.jgi-psf.org/MicpuN2/MicpuN2.home.html), *Ostreococcus tauri* (http://genome.jgi-psf.org/Ostta4/Ostta4.home.html), and *Chlamydomonas reinhardtii* (http://genome.jgi-psf.org/Chlre3/Chlre3.home.html). Recently, a single *RpoT* gene was detected and its gene product characterized as a mitochondrial RNAP in the lycophyte *Selaginella moellendorffii* (Yin et al. 2009). Two mitochondrial phage-type RNAPs were identified in the waterlily *Nuphar advena*, a basal angiosperm, which contains three *RpoT* genes (Yin et al. 2010).

In the following, we refer to RpoTm and RpoTmp as mitochondrial RNAPs in plants. Mitochondria of eudicots harbor two different catalytic subunits of phage-type RNAP (RpoTm, RpoTmp), whereas in monocots only one mitochondrial RNAP (RpoTm) is found (Fig. 12.1). The *RpoTm* and *RpoTmp* genes in *Arabidopsis* display overlapping expression patterns, and thus, the two polymerases may transcribe different mitochondrial genes (Emanuel et al. 2006). RpoTm, but not RpoTmp, was shown to recognize mitochondrial promoters in vitro (Kühn et al. 2007). According to a study of *Arabidopsis* T-DNA insertion mutants lacking RpoTmp (Kühn et al. 2009), RpoTm has to be considered as the basic RNAP in mitochondria of eudicots required for the transcription of most, if not all, mitochondrial genes, whereas RpoTmp has been suggested to permit mitochondria to independently control the abundances of complexes I and IV for fine-tuning the capacity of these complexes in response to developmental or metabolic requirements of the organelle (Kühn et al. 2009).

12.3.2 Co-factors of the Mitochondrial RNA Polymerase

The yeast mitochondrial RNAP has recently been shown to utilize, like the T7 protein, a C-terminal loop for promoter recognition (Nayak et al. 2009). In contrast to the RNAPs of bacteriophages, the plant, animal, and fungal RpoT polymerases require auxiliary factors for accurate and efficient transcription initiation in vivo. Such factors conferring promoter recognition have been identified in yeast and mammalian mitochondria and are referred to in the following as mtTFA and mtTFB (see Chap. 11). mtTFA, a small protein of 19 kDa which binds DNA upstream of the transcription initiation site via two HMG boxes, stimulates, but is not necessary for transcription initiation (Fisher et al. 1992; Parisi et al. 1993). Mitochondrial mtTFBs, which are related to a family of rRNA dimethyladenosine transferases, have been shown to be essential for initiation of transcription in yeast and animal mitochondria representing the principal specificity factor (reviewed in Shadel and Clayton 1993; Scarpulla 2008, see also Chap. 11).

Homologs of mtTFA and mtTFB would be good candidates for mitochondrial specificity factors in plants. Indeed, in the completely sequenced genome of *Arabidopsis*, several dimethyladenosine transferase-like open reading frames can be found. The product of one of them (at5g66360) was shown to be targeted to mitochondria and to methylate the conserved adenosines in the mitochondrial rRNA. However, neither in vitro transcription assays nor analysis of a respective mutant line did support function of this protein in mitochondrial transcription (Richter et al. 2010). Computational predictions of the subcellular localization of *Arabidopsis* HMG-box proteins did not reveal an mtTFA-homolog potentially targeted to mitochondria (K. Kühn, A. Weihe, K. Liere, unpublished data).

Potential co-factors of the mitochondrial RNAP were found analyzing proteins binding to promoters in mitochondrial lysates used for in vitro transcription assays. In wheat, a 69-kDa protein was purified that stimulates in vitro transcriptional activity from a *cox2* promoter (Ikeda and Gray 1999a). The encoded protein shows not only similarities to regions 2 and 3 of bacterial σ-factors and the yeast mtTFB, but is also a member of the family of PPR proteins which function in RNA metabolism (Schmitz-Linneweber and Small 2008). Three homologous PPR proteins from *Arabidopsis*, shown to be mitochondrially targeted, bound in an in vitro assay to mitochondrial promoter fragments in an unspecific manner and interacted neither with RpoTm nor with RpoTmp to allow for correct initiation of transcription (K. Kühn, K. Liere, A. Weihe, and T. Börner, unpublished data; Kühn et al. 2007). In pea, two proteins of 43 and 32 kDa binding to the *atp9* promoter were isolated from mitochondria, based on their promoter-binding properties. While the 43-kDa protein showed high similarity to isovaleryl CoA dehydrogenases involved in leucine catabolism (Däschner et al. 2001), identity and function of the 32-kDa protein remain to be investigated. Another candidate factor involved in transcription initiation and regulation is MCT, a maize nuclear gene. Its target is the mitochondrial *cox2* promoter, which is active only when the dominant MCT allele is present (Newton et al. 1995). Whether σ-factors could play a role as components of the phage-type transcription machineries in mitochondria and plastids requires further investigation (see Sect. 12.2.4.2).

12.3.3 Mitochondrial Promoters

Individual transcription units have been found and numerous transcription initiation sites have been mapped in several plant species. One class of plant mitochondrial promoters has been identified which is characterized by the consensus sequence motif CRTA, similar to the YRTA motif of plastid NEP promoters (see Sect. 12.2.2; reviews in Binder and Brennicke 2003; Weihe 2004). In eudicotyledoneous plants, the CRTA motif is part of a nonanucleotide sequence overlapping the initiation site (Binder et al. 1996). Whereas limited data are available for "nonconsensus" promoters lacking common structures and sequence motifs, mitochondrial consensus promoters have been intensively studied in eudicot and monocot plants. Both consensus- and non-consensus-type promoters can be found at transcriptional start sites of all types of RNA: mRNA, rRNA, and tRNA. Among the mitochondrial consensus promoters of eudicots, the conserved nona-nucleotide (CRTAaGaGA, transcription initiation site underlined) shows considerable sequence identity between different genes as well as between different species. From a comparison of 11 unambiguously identified promoters from several plant species including pea, soybean, potato, and *Oenothera* (Dombrowski et al. 1999), an extended consensus sequence was deduced (AAAATATCATAAGAGAAG, 100% conserved positions in bold, transcription initiation site underlined) that is composed of three parts: the conserved nona-nucleotide motif from −7 to +2, containing the transcription initiation site; the less well-conserved AT-box, consisting of predominantly adenosine and thymidine bases located through positions −14 to −8; and the positions +3 and +4, where mainly purines are found. A recent study on the mapping of mitochondrial promoters in *Arabidopsis* revealed alternative promoter motifs (Kühn et al. 2005). Apart from CRTA-type consensus motifs, tetranucleotide core motifs such as ATTA and RGTA were identified, and several promoters showed no consensus sequence at all (Kühn et al. 2005, 2007).

The core sequence of monocot promoters was identified as a CRTA tetranucleotide motif just upstream of the first transcribed nucleotide (Rapp et al. 1993; Caoile and Stern 1997). Determination of transcript termini in *Sorghum* revealed also, as a variant of the CRTA motif, degenerated YRTA, AATA, and CTTA sequences (Yan and Pring 1997). Most consensus-type promoters in monocots share a small sequence element which resides about ten nucleotides further upstream and contains an AT-rich region of six nucleotides (Rapp et al. 1993; Tracy and Stern 1995).

The presence of more than one promoter and multiple transcription initiation sites has been described for both monocots and eudicots (Mulligan et al. 1988; Tracy and Stern 1995; Lupold et al. 1999b; Kühn et al. 2005). Kühn et al. (2005) suggested that multiple promoters could be a way to ensure transcription despite possible mitochondrial genome rearrangements. Thus, the activity of maize *cox2* promoters was shown to respond to their genomic context, thereby indicating consequences of intra- and intergenomic recombination for plant mitochondrial gene expression (Lupold et al. 1999a).

12.3.4   *Regulation of Mitochondrial Transcription*

The present knowledge about regulation of mitochondrial genes is still scarce. *In organello* run-on analyses showed that transcription rates of rRNA genes are 2- to 14-fold higher than those of protein coding genes (Finnegan and Brown 1990; Mulligan et al. 1991). A more recent study comprising all mitochondrial-encoded genes in *Arabidopsis* found no enhanced transcription of rRNA genes, but detected even distinct transcription rates of genes encoding components of the same multisubunit complex (Giegé et al. 2000). These differences are, at least partially, counterbalanced in the steady-state RNA pool, most likely by post-transcriptional processes and different RNA stabilities. Transcription in *Arabidopsis* mitochondria cycles in a diurnal rhythm, while steady-state transcript levels do not vary between light and dark phases (Okada and Brennicke 2006). Comparison of mitochondrial transcriptional rates in *Arabidopsis* and in a cytoplasmic male sterile vs. a fertility restored line of *Brassica napus* identified species-specific transcription rates for several genes (*cox1*, *nad4L*, *nad9*, *ccmB*, *rps7*, and *rrn5*), which are most likely determined by different promoter strength in the mitochondrial DNA (Leino et al. 2005). An observed influence of the nuclear background on both transcription rates and post-transcriptional mechanisms indicates that both processes depend not only on mitochondrial *cis*-elements but also on nuclear *trans*-factors (Edqvist and Bergman 2002).

Post-transcriptional processes play a major role in determining mitochondrial RNA levels (Mulligan et al. 1988; Tracy and Stern 1995; Lupold et al. 1999a; Giegé et al. 2000; Leino et al. 2005; Holec et al. 2008). Other data demonstrate the importance of post-translational processes for the control of protein quantities: the response of mitochondrial gene expression to sugar starvation remained more or less unaffected at the transcriptional, post-transcriptional, and translational levels (Giegé et al. 2005). The observed reduction of ATPase complexes could be attributed to nuclear-encoded components of the ATPase being downregulated. Becoming the rate-liming factor in the assembly of new complexes, correct stoichiometric proportions seemed to be achieved post-translationally.

It has been suggested that post-transcriptional processes are the dominant mechanism for tissue-specific differences in steady-state levels of mitochondrial transcripts (Monéger et al. 1994; Smart et al. 1994; Gagliardi and Leaver 1999). Tissue- or cell-specific differences in mitochondrial gene expression have been correlated with transcript levels in a few reports (Topping and Leaver 1990; Li et al. 1996). A comprehensive analysis of mitochondrial transcription initiation sites in *Arabidopsis* revealed no qualitative differences in promoter usage between leaves and flowers (Kühn et al. 2005).

Recently, a study on *Arabidopsis* mutant lines lacking RpoTmp activity revealed effects on transcription that were not counterbalanced post-transcriptionally, but led to changes at the protein level (Kühn et al. 2009). Moreover, this report suggests that RpoTmp-dependent transcription may be considered as a transcriptional mechanism for controlling the formation of complexes I and IV of the respiration chain *via* the expression of certain mitochondrial genes (Kühn et al. 2009).

It has been shown that the copy numbers of mitochondrial genes in *Arabidopsis* may differ considerably not only from each other but vary also between organs and different developmental stages (Preuten et al. 2010). The control of copy numbers of mitochondrial genes in different tissues and developmental stages might be of functional importance and connected with altered transcript levels and rates of respiration. Muise and Hauswirth (1995) compared mitochondrial gene quantities and transcription in maize and *Brassica hirta* and concluded that a direct relationship exists between gene copy number and transcriptional rate. This is in conflict with data of a study on *Phaseolus vulgaris* (Woloszynska and Trojanowski 2009). Muise and Hauswirth (1995) analyzed run-on transcription, i.e., determined transcription rates, whereas Woloszynska and Trojanowski (2009) studied transcript levels. Transcription and transcript levels are not necessarily positively correlated, and the transcriptional activity of genes may be of only little importance for the final level of their products in plant mitochondria. There are also dynamic changes in mitochondrial gene copy numbers. Thus, the formation of anthers and pollen is associated with increased transcript levels and number of mitochondria and possibly with enhanced mtDNA in certain flower tissues (Warmke and Lee 1978; Geddy et al. 2005 and references therein). Both transcript levels and gene copy numbers were found to be enhanced in photosynthetically inactive white compared to green leaves in the *albostrians* mutant of barley (Hedtke et al. 1999a). Increased mitochondrial gene copies were also positively correlated with transcript and protein levels in an *RpoTmp* mutant of *Arabidopsis* (Kühn et al. 2009). On the other hand, in a recent study, no correlation between altering gene copy numbers and transcript levels were found during leaf development in *Arabidopsis* (Preuten et al. 2010).

12.4 Evolution of the Organellar Transcription Machineries

12.4.1 RNA Polymerases

The mitochondrial genome of baker's yeast, *S. cerevisiae*, was the first nonphage genome shown to be transcribed by a phage-type RNAP (Masters et al. 1987; Schinkel et al. 1988). It is now evident that in nearly all eukaryotes related phage-type polymerases are responsible for mitochondrial transcription (Cermakian et al. 1996). The only exception so far is *Reclinomonas americana*, a freshwater protozoon belonging to the jakobids. This lower eukaryote possesses genes for a eubacterial-type RNAP in its mitochondrial genome, which in other eukaryotic lineages may have been lost during the evolution of the organelle (Lang et al. 1997). This contrasts with the situation in plastids where all algae and plants still use the bacterial-type RNAP (PEP; see Sect. 12.2.1.1) inherited from the cyanobacterial ancestor. An exceptional situation occurs in several parasitic angiosperms which lost, together with chloroplast genes encoding functions in photosynthesis, also the *rpo* genes, thus relying solely on NEP for transcription of their plastid genes (Wolfe et al. 1992a, b; Krause et al. 2003; Berg et al. 2004). In the moss *Physcomitrella* and in angiosperms, duplication(s) of the nuclear gene encoding the mitochondrial RNAP resulted in small families of *RpoT* genes encoding mitochondrial and plastid (see Sect. 12.2.1.2) phage-type RNAPs.

As known so far, the nonplant eukaryotes possess only one type of mitochondrial RNAP encoded by a single nuclear gene. Whether algae need a phage-type RNA polymerase (NEP) in addition to PEP to transcribe their plastid genes is not known yet (see review by Smith and Purton 2002).

Land plants carry small families of *RpoT* genes in their nuclear genomes. The only known exception from this rule is the lycophyte *Selaginella moellendorffii* with a single *RpoT* gene encoding the mitochondrial RNAP (Fig. 12.1; Yin et al. 2009). In eudicotyledonous plant species such as *Nicotiana sylvestris* and *A. thaliana* (Hedtke et al. 1997, 1999b, 2000, 2002; Kobayashi et al. 2001, 2002), the *RpoT* gene family consists of three genes, encoding products that are imported into mitochondria (*RpoTm*), plastids (*RpoTp*), and dually into both organelles (*RpoTmp*). The amphidiploid genome of *N. tabacum* contains six *RpoT* genes with two sets of three genes from the two diploid parental species (Hedtke et al. 2002). All *RpoT* genes of monocots, analyzed so far, encode RNAPs (RpoTm, RpoTp) exclusively targeted to either mitochondria or plastids (Chang et al. 1999; Ikeda and Gray 1999b; Emanuel et al. 2004; Kusumi et al. 2004). Similar to the situation in eudicots, the nuclear genome of the moss *Physcomitrella patens* harbors three *RpoT* genes. Two of them seem to encode gene products being capable of dual targeting as a result of translation initiation at two different in-frame AUG start codons (PpRpoTmp1, PpRpoTmp2; Richter et al. 2002). Interestingly, a third *RpoT* gene found in the *Physcomitrella* genome project database (http://genomeportal. jgi-psf.org/Phypa1_1/Phypa1_1.home.html) encodes an enzyme, which is exclusively targeted to mitochondria (PpRpoTm; U. Richter, personal communication). Thus, *Physcomitrella* seems to use three phage-type RNAPs (RpoTm, RpoTmp1, and RpoTmp2) in mitochondrial transcription and two phage-type RNAPs (RpoTmp1, RpoTmp2) in chloroplast transcription. However, *Physcomitrella* RpoTmp localization is still a matter of debate (Kabeya and Sato 2005). On an interesting side note, *Physcomitrella* possesses two copies of the gene encoding the PEP α-subunit in the nucleus (PpRpoA1, PpRpoA2; see Sect. 12.2.1.1; Sugiura et al. 2003; Kabeya et al. 2007). Analyses of respective knockout mutants suggested that plastid genes are differentially transcribed by distinct PEP enzymes with either PpRpoA1 or PpRpoA2. Therefore, PEP complexes harboring either a single or mixed type of the two α subunits may be involved in modulating PEP transcription in *Physcomitrella* plastids.

Phylogenetic analyses indicate that the *RpoT* gene families of *Physcomitrella* and higher plants have arisen by independent gene duplication events dating after the separation of bryophytes and tracheophytes (Fig. 12.3; Richter et al. 2002). Since the lycophyte *Selaginella* possesses only RpoTm for mitochondrial transcription and only PEP for transcribing the plastid genes (Yin et al. 2009), the NEP

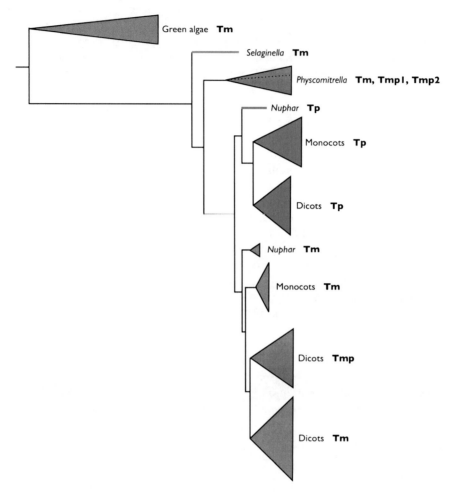

Fig. 12.3 Schematic phylogenetic tree of plant RpoT polymerases based on a Bayesian recon-struction of phylogeny (Yin et al. 2010). Mitochondrial (RpoTm), plastid (RpoTp), and dual targeted (RpoTmp) proteins are shown in *orange*, *green*, and *mixed color*, respectively

polymerases of angiosperms may have arisen only late during the evolution of plants.

Most recently, three *RpoT* genes have been detected in the genome of the basal angiosperm *Nuphar advena*. Two of them encode mitochondrial RNAPs (*RpoTm1*, *RpoTm2*) and one encodes a plastid RNAP (*RpoTp*). The *Nuphar* protein sequences cluster in phylogenetic trees together with the RpoTm and RpoTp enzymes of higher angiosperms, i.e., the *Nuphar* RpoTp is the earliest angiosperm NEP so far described (Figs. 12.1 and 12.3; Yin et al. 2010). The complexity of plastid tran-scription may perhaps have evolved to compensate for degenerating chloroplast promoters (Maier et al. 2008). The intrinsic activity of RpoT polymerases to act as single-polypeptide RNAP (Kühn et al. 2007) might have been sufficient after organellar targeting to support transcription from promoters with simple structures (see Sects. 12.2.2 and 12.3.3), thereby counteracting effects of point mutations in PEP promoters.

12.4.2   Transcription Factors

Most protein factors associated with the transcriptional apparatus, e.g., with the transcriptionally active chromosome (TAC) complex, as well as putative transcription factors of individual plastid or mitochondrial genes are not of obvious bacterial origin, i.e., they have been acquired during the evolution of the eukaryotic lineage of life (see Sect. 12.2.4.3; Sato 2001; Wagner and Pfannschmidt 2006; Schwacke et al. 2007).

The σ-factors, cooperating with the PEP in promoter recognition and transcription initiation in plastids, originate from the cyanobacterial ancestor. Phylogenetic analysis suggests that σ-factors of red and green algal plastids and the group 1 σ-factors of cyanobacteria form a monophyletic group (Minoda et al. 2005). The nuclear genome of the unicellular green algae *Chlamydomonas reinhardtii* harbors only a single gene encoding a σ-factor (CreRpoD; Carter et al. 2004; Bohne et al. 2006). The presence of more than one σ-factor among algae and plants, however, seems to be more common. Independent gene duplications of σ-factor genes occurred several times during the evolution of algae and land plants. The red algae *Cyanidioschizon merolae* possess four σ-factors (Minoda et al. 2005). Higher plant σ-factors also represent a monophyletic group usually encoded by multigene families with at least five subgroups: Sig1, Sig2, Sig3, Sig5, and Sig6. As shown for the algal factors, plant σ-factors are related to cyanobacterial primary (group 1), perhaps also to nonessential primary (group 2) σ^{70}-factors (Lysenko 2006, 2007). Only the Sig5 group might be phylogenetically related to the bacterial alternative σ-factors (Tsunoyama et al. 2002; Shiina et al. 2005; Lysenko 2006). The low expressed *AthSig4* is the only *Arabidopsis Sig* gene without a known ortholog in other plants (Tsunoyama et al. 2002) and is suggested to have originated from partly processed transcripts of *AthSig2*, *AthSig3*, or *AthSig6* (Lysenko 2006, 2007).

12.5   Conclusion

Most eukaryotes harbor only a single nuclear gene encoding a phage-type RNA polymerase which transcribes all mitochondrial genes. During the evolution of plants, however, several independent gene duplications led to the evolvement of small gene families (*RpoT* genes) of phage-type RNA polymerases in the moss *P. patens* and in angiosperms. The *RpoT* gene products act as RNA polymerases in mitochondria and in plastids. Consequently, the transcription machineries of mitochondria and plastids display remarkable similarities. Yet plastids/chloroplasts of all plants (including algae, but excluding certain nonphotosynthetic, parasitic plants) have retained the RNA polymerase of their cyanobacterial ancestor. Therefore, transcription in plastids of *Physcomitrella* and angiosperms is based on a highly complex apparatus with phage-type and bacterial-type polymerases, several sigma factors, and different types of promoters. Moreover, additional eukaryotic protein factors have been added to the genetic apparatus during the evolution of plants and their organelles. Part of these factors is expected to play a role in the regulation of gene expression. While post-transcriptional regulation of gene expression is of major importance in both organelles, transcription of chloroplast genes, too, is affected by exo- and endogenous factors such as light, temperature, circadian clock, and hormones. In contrast, there are only limited data available that would indicate a regulation of gene expression at the level of transcription in plant mitochondria.

Acknowledgments The work of the authors is supported by Deutsche Forschungsgemeinschaft (SFB 429). We are thankful to Kirsten Krause for critically reading the manuscript.

References

Allison LA, Maliga P (1995) Light-responsive and transcription-enhancing elements regulate the plastid *psbD* core promoter. EMBO J 14:3721–3730

Allison LA, Simon LD, Maliga P (1996) Deletion of *rpoB* reveals a second distinct transcription system in plastids of higher plants. EMBO J 15:2802–2809

Arsova B, Hoja U, Wimmelbacher M, Greiner E, Üstün S, Melzer M, Petersen K, Lein W, Börnke F (2010) Plastidial Thioredoxin z interacts with two fructokinase-like proteins in a thiol-dependent manner: evidence for an essential role in chloroplast development in *Arabidopsis* and *Nicotiana benthamiana*. Plant Cell 22:1498–1515

Azevedo J, Courtois F, Hakimi M-A, Demarsy E, Lagrange T, Alcaraz J-P, Jaiswal P, Maréchal-Drouard L, Lerbs-Mache L (2008) Intraplastidial trafficking of a phage-type RNA polymerase is mediated by a thylakoid RING-H2 protein. Proc Natl Acad Sci USA 105:9123–9128

Baba K, Schmidt J, Espinosa-Ruiz A, Villarejo A, Shiina T, Gardestrom P, Sane AP, Bhalerao RP (2004) Organellar gene transcription and early seedling development are affected in the *RpoT;2* mutant of *Arabidopsis*. Plant J 38:38–48

Baeza L, Bertrand A, Mache R, Lerbs-Mache S (1991) Characterization of a protein binding sequence in the promoter region of the 16S rRNA gene of the spinach chloroplast genome. Nucleic Acids Res 19:3577–3581

Baginsky S, Gruissem W (2009) The chloroplast kinase network: new insights from large-scale phosphoproteome profiling. Mol Plant 2:1141–1153

Balsera M, Soll J, Buchanan BB (2010) Redox extends its regulatory reach to chloroplast protein import. Trends Plant Sci 5:515–521

Baumgartner BJ, Rapp JC, Mullet JE (1993) Plastid genes encoding the transcription/translation apparatus are differentially transcribed early in barley (*Hordeum vulgare*) chloroplast development: evidence for selective stabilization of *psbA* mRNA. Plant Physiol 101:781–791

Beck CF (2005) Signaling pathways from the chloroplast to the nucleus. Planta 222:743–756

Berg S, Krause K, Krupinska K (2004) The *rbcL* genes of two *Cuscuta* species, *C. gronovii* and *C. subinclusa*, are transcribed by the nuclear-encoded plastid RNA polymerase (NEP). Planta 219:541–546

Binder S, Brennicke A (2003) Gene expression in plant mitochondria: transcriptional and post-transcriptional control. Philos Trans R Soc Lond B Biol Sci 358:181–188; discussion 188–189

Binder S, Marchfelder A, Brennicke A (1996) Regulation of gene expression in plant mitochondria. Plant Mol Biol 32:303–314

Bligny M, Courtois F, Thaminy S, Chang CC, Lagrange T, Baruah-Wolff J, Stern D, Lerbs-Mache S (2000) Regulation of plastid rDNA transcription by interaction of CDF2 with two different RNA polymerases. EMBO J 19:1851–1860

Bohne A-V, Irihimovitch V, Weihe A, Stern D (2006) *Chlamydomonas reinhardii* encodes a single sigma^{70}-like factor which likely functions in chloroplast transcription. Curr Genet 49:333–340

Bohne A-V, Weihe A, Börner T (2009) Interaction of Arabidopsis phage-type RNA polymerases with transfer RNAs. Endocytobiosis Cell Res 19:63–69

Bradbeer JW, Atkinson YE, Börner T, Hagemann R (1979) Cytoplasmic synthesis of plastid polypeptides may be controlled by plastid-synthesized RNA. Nature 279:816–817

Bräutigam K, Dietzel L, Kleine T, Ströher E, Wormuth D, Dietz K, Radke D, Wirtz M, Hell R, Dörmann P, Nunes-Nesi A, Schauer N, Fernie A, Oliver S, Geigenberger P, Leister D, Pfannschmidt T (2009) Dynamic plastid redox signals integrate gene expression and metabolism to induce distinct metabolic states in photosynthetic acclimation in *Arabidopsis*. Plant Cell 21:2715–2732

Brown NJ, Sullivan JA, Gray JC (2005) Light and plastid signals regulate the expression of the pea plastocyanin gene through a common region at the 5' end of the coding region. Plant J 43:541–552

Bünger W, Feierabend J (1980) Capacity for RNA synthesis in 70S ribosome-deficient plastids of heat-bleached rye leaves. Planta 149:163–169

Cahoon AB, Harris FM, Stern DB (2004) Analysis of developing maize plastids reveals two mRNA stability classes correlating with RNA polymerase type. EMBO Rep 5:801–806

Caoile AGFS, Stern DB (1997) A conserved core element is functionally important for maize mitochondrial promoter activity *in vitro*. Nucleic Acids Res 25:4055–4060

Carlberg I, Hansson M, Kieselbach T, Schroder WP, Andersson B, Vener AV (2003) A novel plant protein undergoing light-induced phosphorylation and release from the photosynthetic thylakoid membranes. Proc Natl Acad Sci USA 100:757–762

Carter ML, Smith AC, Kobayashi H, Purton S, Herrin DL (2004) Structure, circadian regulation and bioinformatic analysis of the unique sigma factor gene in *Chlamydomonas reinhardtii*. Photosynth Res 82:339–349

Cashel M, Gentry DM, Hernandez VJ, Vinella D (1996) The stringent response. In: Neidhardt FC (ed) *Escherichia coli* and *Salmonella typhimurium:* cellular and molecular biology, vol 1. ASM Press, Washington DC, pp 1458–1496

Cermakian N, Ikeda TM, Cedergren R, Gray MW (1996) Sequences homologous to yeast mitochondrial and bacteriophage T3 and T7 RNA polymerases are widespread throughout the eukaryotic lineage. Nucleic Acids Res 24:648–654

Chang C-C, Sheen J, Bligny M, Niwa Y, Lerbs-Mache S, Stern DB (1999) Functional analysis of two maize cDNAs encoding T7-like RNA polymerases. Plant Cell 11:911–926

Chen MC, Cheng MC, Chen SC (1993) Characterization of the promoter of rice plastid *psaA-psaB-rps14* operon and the DNA-specific binding proteins. Plant Cell Physiol 34:577–584

Cheng MC, Wu SP, Chen LFO, Chen SCG (1997) Identification and purification of a spinach chloroplast DNA-binding protein that interacts specifically with the plastid *psaA-psaB-rps14* promoter region. Planta 203:373–380

Chory J (2010) Light signal transduction: an infinite spectrum of possibilities. Plant J 61:982–991

Christopher DA, Hoffer PH (1998) DET1 represses a chloroplast blue light-responsive promoter in a developmental and tissue-specific manner in *Arabidopsis thaliana*. Plant J 14:1–11

Chun L, Kawakami A, Christopher DA (2001) Phytochrome A mediates blue light and UV-A-dependent chloroplast gene transcription in green leaves. Plant Physiol 125:1957–1966

Courtois F, Merendino L, Demarsy E, Mache R, Lerbs-Mache S (2007) Phage-type RNA polymerase RPOTmp transcribes the *rrn* operon from the PC promoter at early developmental stages in *Arabidopsis*. Plant Physiol 145:712–721

Däschner K, Couée I, Binder S (2001) The mitochondrial isovaleryl-coenzyme A dehydrogenase of *Arabidopsis* oxidizes intermediates of leucine and valine catabolism. Plant Physiol 126:601–612

Dhingra A, Bies DH, Lehner KR, Folta KM (2006) Green light adjusts the plastid transcriptome during early photomorphogenic development. Plant Physiol 142:1256–1266

Dombrowski S, Hoffmann M, Guha C, Binder S (1999) Continuous primary sequence requirements in the 18-nucleotide promoter of dicot plant mitochondria. J Biol Chem 274:10094–10099

Edqvist J, Bergman P (2002) Nuclear identity specifies transcriptional initiation in plant mitochondria. Plant Mol Biol 49:59–68

Ellis RJ, Hartley MR (1971) Sites of synthesis of chloroplast proteins. Nature 233:193–196

Emanuel C, Weihe A, Graner A, Hess WR, Börner T (2004) Chloroplast development affects expression of phage-type RNA polymerases in barley leaves. Plant J 38:460–472

Emanuel C, von Groll U, Müller M, Börner T, Weihe A (2006) Development- and tissue-specific expression of the *RpoT* gene family of *Arabidopsis* encoding mitochondrial and plastid RNA polymerases. Planta 223:998–1009

Falk J, Schmidt A, Krupinska K (1993) Characterization of plastid DNA transcription in ribosome deficient plastids of heat-bleached barley leaves. J Plant Physiol 141:176–181

Fey V, Wagner R, Brautigam K, Wirtz M, Hell R, Dietzmann A, Leister D, Oelmuller R, Pfannschmidt T (2005) Retrograde plastid redox signals in the expression of nuclear genes for chloroplast proteins of *Arabidopsis thaliana*. J Biol Chem 280:5318–5328

Finnegan PM, Brown GG (1990) Transcriptional and post-transcriptional regulation of RNA levels in maize mitochondria. Plant Cell 2:71–83

Fisher RP, Lisowsky T, Parisi MA, Clayton DA (1992) DNA wrapping and bending by a mitochondrial high mobility group-like transcriptional activator protein. J Biol Chem 267:3358–3367

Gagliardi D, Leaver CJ (1999) Polyadenylation accelerates the degradation of the mitochondrial mRNA associated with cytoplasmic male sterility in sunflower. EMBO J 18:3757–3766

Geddy R, Mahé L, Brown GG (2005) Cell-specific regulation of a *Brassica napus* CMS-associated gene by a nuclear restorer with related effects on a floral homeotic gene promoter. Plant J 41:333–345

Giegé P, Hoffmann M, Binder S, Brennicke A (2000) RNA degradation buffers asymmetries of transcription in *Arabidopsis* mitochondria. EMBO Rep 1:164–170

Giegé P, Sweetlove LJ, Cognat V, Leaver CJ (2005) Coordination of nuclear and mitochondrial genome expression during mitochondrial biogenesis in *Arabidopsis*. Plant Cell 17:1497–1512

Givens RM, Lin MH, Taylor DJ, Mechold U, Berry JO, Hernandez VJ (2004) Inducible expression, enzymatic activity, and origin of higher plant homologues of bacterial RelA/SpoT stress proteins in *Nicotiana tabacum*. J Biol Chem 279:7495–7504

Gray JC (2003) Chloroplast-to-nucleus signalling: a role for Mg-protoporphyrin. Trends Genet 19:526–529

Hajdukiewicz PTJ, Allison LA, Maliga P (1997) The two RNA polymerases encoded by the nuclear and the plastid compartments transcribe distinct groups of genes in tobacco plastids. EMBO J 16:4041–4048

Han CD, Patrie W, Polacco M, Coe EH (1993) Abberations in plastid transcripts and deficiency of plastid DNA in striped and albino mutants in maize. Planta 191:552–563

Hanaoka M, Kanamaru K, Fujiwara M, Takahashi H, Tanaka K (2005) Glutamyl-tRNA mediates a switch in RNA polymerase use during chloroplast biogenesis. EMBO Rep 6:545–550

Hedtke B, Börner T, Weihe A (1997) Mitochondrial and chloroplast phage-type RNA polymerases in *Arabidopsis*. Science 277:809–811

Hedtke B, Wagner I, Börner T, Hess WR (1999a) Inter-organellar crosstalk in higher plants: impaired chloroplast development affects mitochondrial gene and transcript levels. Plant J 19:635–643

Hedtke B, Meixner M, Gillandt S, Richter E, Börner T, Weihe A (1999b) Green fluorescent protein as a marker to investigate targeting of organellar RNA polymerases of higher plants *in vivo*. Plant J 17:557–561

Hedtke B, Börner T, Weihe A (2000) One RNA polymerase serving two genomes. EMBO Rep 1:435–440

Hedtke B, Legen J, Weihe A, Herrmann RG, Börner T (2002) Six active phage-type RNA polymerase genes in *Nicotiana tabacum*. Plant J 30:625–637

Hengeveld R, Fedonkin MA (2004) Causes and consequences of eukaryotization through mutualistic endosymbiosis and compartmentalization. Acta Biotheor 52:105–154

Hess WR, Börner T (1999) Organellar RNA polymerases of higher plants. Int Rev Cytol 190:1–59

Hess WR, Prombona A, Fieder B, Subramanian AR, Börner T (1993) Chloroplast *rps15* and the *rpoB/C1/C2* gene cluster are strongly transcribed in ribosome-deficient plastids: evidence for a functioning non-chloroplast-encoded RNA polymerase. EMBO J 12:563–571

Holec S, Lange H, Canaday J, Gagliardi D (2008) Coping with cryptic and defective transcripts in plant mitochondria. Biochim Biophys Acta 1779:566–573

Hricová A, Quesada V, Micol JL (2006) The SCABRA3 nuclear gene encodes the plastid RpoTp RNA polymerase, which is required for chloroplast biogenesis and mesophyll cell proliferation in *Arabidopsis*. Plant Physiol 141:942–956

Ikeda T, Gray M (1999a) Characterization of a DNA-binding protein implicated in transcription in wheat mitochondria. Mol Cell Biol 19:8113–8122

Ikeda TM, Gray MW (1999b) Identification and characterization of T7/T3 bacteriophage-like RNA polymerase sequences in wheat. Plant Mol Biol 40:567–578

Imamura S, Asayama M (2009) Sigma factors for cyanobacterial transcription. Gene Regul Syst Biol 3:65–87

Imamura S, Yoshihara S, Nakano S, Shiozaki N, Yamada A, Tanaka K, Takahashi H, Asayama M, Shirai M (2003) Purification, characterization, and gene expression of all sigma factors of RNA polymerase in a cyanobacterium. J Mol Biol 325:857–872

Iratni R, Baeza L, Andreeva A, Mache R, Lerbs-Mache S (1994) Regulation of rDNA transcription in chloroplasts: promoter exclusion by constitutive repression. Genes Dev 8:2928–2938

Iratni R, Diederich L, Harrak H, Bligny M, Lerbs-Mache S (1997) Organ-specific transcription of the *rrn* operon in spinach plastids. J Biol Chem 272:13676–13682

Ishihama A (2000) Functional modulation of *Escherichia coli* RNA polymerase. Annu Rev Microbiol 54:499–518

Jishage M, Kvint K, Shingler V, Nyström T (2002) Regulation of sigma factor competition by the alarmone ppGpp. Genes Dev 16:1260–1270

Kabeya Y, Sato N (2005) Unique translation initiation at the second AUG codon determines mitochondrial localization of the phage-type RNA polymerases in the moss *Physcomitrella patens*. Plant Physiol 138:369–382

Kabeya Y, Kobayashi Y, Suzuki H, Itoh J, Sugita M (2007) Transcription of plastid genes is modulated by two nuclear-encoded alpha subunits of plastid RNA polymerase in the moss *Physcomitrella patens*. Plant J 52:730–741

Kaneko T, Sato S, Kotani H, Tanaka A, Asamizu E, Nakamura Y, Miyajima N, Hirosawa M, Sugiura M, Sasamoto S, Kimura T, Hosouchi T, Matsuno A, Muraki A, Nakazaki N, Naruo K, Okumura S, Shimpo S, Takeuchi C, Wada T, Watanabe A, Yamada M, Yasuda M, Tabata S (1996) Sequence analysis of the genome of the unicellular cyanobacterium *Synechocystis* sp. strain PCC6803. II. Sequence determination of the entire genome and assignment of potential protein-coding regions. DNA Res 3:109–136

Kapoor S, Sugiura M (1999) Identification of two essential sequence elements in the nonconsensus Type II P*atpB*-290 plastid promoter by using plastid transcription extracts from cultured tobacco BY-2 cells. Plant Cell 11:1799–1810

Kapoor S, Suzuki JY, Sugiura M (1997) Identification and functional significance of a new class of non-consensus-type plastid promoters. Plant J 11:327–337

Karcher D, Köster D, Schadach A, Klevesath A, Bock R (2009) The *Chlamydomonas* chloroplast HLP protein is required for nucleoid organization and genome maintenance. Mol Plant 2:1223–1232

Kasai K, Usami S, Yamada T, Endo Y, Ochi K, Tozawa Y (2002) A RelA-SpoT homolog (Cr-RSH) identified in *Chlamydomonas reinhardtii* generates stringent factor *in vivo* and localizes to chloroplasts *in vitro*. Nucleic Acids Res 30:4985–4992

Kleine T, Voigt C, Leister D (2009) Plastid signalling to the nucleus: messengers still lost in the mists? Trends Genet 25:185–192

Kobayashi Y, Dokiya Y, Sugita M (2001) Dual targeting of phage-type RNA polymerase to both mitochondria and plastids is due to alternative translation initiation in single transcripts. Biochem Biophys Res Commun 289:1106–1113

Kobayashi Y, Dokiya Y, Kumazawa Y, Sugita M (2002) Non-AUG translation initiation of mRNA encoding plastid-targeted phage-type RNA polymerase in *Nicotiana sylvestris*. Biochem Biophys Res Commun 299:57–61

Krause K (2008) From chloroplasts to "cryptic" plastids: evolution of plastid genomes in parasitic plants. Curr Genet 54:111–121

Krause K, Krupinska K (2000) Molecular and functional properties of highly purified transcriptionally active chromosomes from spinach chloroplasts. Physiol Plant 109:188–195

Krause K, Maier RM, Kofer W, Krupinska K, Herrmann RG (2000) Disruption of plastid-encoded RNA polymerase genes in tobacco: expression of only a distinct set of genes is not based on selective transcription of the plastid chromosome. Mol Gen Genet 263:1022–1030

Krause K, Berg S, Krupinska K (2003) Plastid transcription in the holoparasitic plant genus *Cuscuta*: parallel loss of the *rrn16* PEP-promoter and of the *rpoA* and *rpoB* genes coding for the plastid-encoded RNA polymerase. Planta 216:815–823

Kühn K, Weihe A, Börner T (2005) Multiple promoters are a common feature of mitochondrial genes in *Arabidopsis*. Nucleic Acids Res 33:337–346

Kühn K, Bohne A-V, Liere K, Weihe A, Börner T (2007) *Arabidopsis* phage-type RNA polymerases: accurate *in vitro* transcription of organellar genes. Plant Cell 19:959–971

Kühn K, Richter U, Meyer E, Delannoy E, Falcon de Longevialle A, O'Toole N, Börner T, Millar A, Small I, Whelan J (2009) Phage-type RNA polymerase RPOTmp performs gene-specific transcription in mitochondria of *Arabidopsis thaliana*. Plant Cell 21:2762–2779

Kusumi K, Yara A, Mitsui N, Tozawa Y, Iba K (2004) Characterization of a rice nuclear-encoded plastid RNA polymerase gene *OsRpoTp*. Plant Cell Physiol 45:1194–1201

Lang BF, Burger G, O'Kelly CJ, Cedergren R, Golding GB, Lemieux C, Sankoff D, Turmel M, Gray MW (1997) An ancestral mitochondrial DNA resembling a eubacterial genome in miniature. Nature 387:493–497

Legen J, Kemp S, Krause K, Profanter B, Herrmann RG, Maier RM (2002) Comparative analysis of plastid transcription profiles of entire plastid chromosomes from tobacco attributed to wild-type and PEP-deficient transcription machineries. Plant J 31:171–188

Leino M, Landgren M, Glimelius K (2005) Alloplasmic effects on mitochondrial transcriptional activity and RNA turnover result in accumulated transcripts of *Arabidopsis* ORFs in cytoplasmic male-sterile *Brassica napus*. Plant J 42:469–480

Lerbs-Mache S (2011) Function of plastid sigma factors in higher plants: regulation of gene expression or just preservation of constitutive transcription? Plant Mol Biol 76(3–5):235–249

Lezhneva L, Meurer J (2004) The nuclear factor HCF145 affects chloroplast *psaA-psaB-rps14* transcript abundance in *Arabidopsis thaliana*. Plant J 38:740–753

Li XQ, Zhang M, Brown GG (1996) Cell-specific expression of mitochondrial transcripts in maize seedlings. Plant Cell 8:1961–1975

Liere K, Börner T (2007a) Transcription of plastid genes. In: Grasser KD (ed) Regulation of transcription in plants. Blackwell, Oxford, pp 184–224

Liere K, Börner T (2007b) Transcription and transcriptional regulation in plastids. In: Bock R (ed) Topics in current genetics: cell and molecular biology of plastids, vol 19. Springer, Berlin/Heidelberg, pp 121–174

Liere K, Maliga P (1999a) *In vitro* characterization of the tobacco *rpoB* promoter reveals a core sequence motif conserved between phage-type plastid and plant mitochondrial promoters. EMBO J 18:249–257

Liere K, Maliga P (1999b) Novel *in vitro* transcription assay indicates that the *accD* NEP promoter is contained in a 19 bp fragment. In: Argyroudi-Akoyunoglou JH, Senger H (eds) The chloroplast: from molecular biology to biotechnology. Kluwer, Amsterdam, pp 79–84

Liere K, Maliga P (2001) Plastid RNA polymerases in higher plants. In: Andersson B, Aro E-M (eds) Regulation of photosynthesis. Kluwer, Netherlands, Dordrecht, pp 29–49

Liere K, Kaden D, Maliga P, Börner T (2004) Overexpression of phage-type RNA polymerase RpoTp in tobacco demonstrates its role in chloroplast transcription by recognizing a distinct promoter type. Nucleic Acids Res 32:1159–1165

Liere K, Weihe A, Börner T (2011) The transcription machineries of plant mitochondria and chloroplasts: composition, function, and regulation. J Plant Physiol 168:1345–1360

Link G (1996) Green life: control of chloroplast gene transcription. Bioessays 18:465–471

Link G (2003) Redox regulation of chloroplast transcription. Antioxid Redox Signal 5:79–87

Loschelder H, Homann A, Ogrzewalla K, Link G (2004) Proteomics-based sequence analysis of plant gene expression – the chloroplast transcription apparatus. Phytochemistry 65:1785–1793

Lukens JH, Mathews DE, Durbin RD (1987) Effect of tagetitoxin on the levels of ribulose 1,5-bisphosphate carboxylase, ribosomes, and RNA in plastids of wheat leaves. Plant Physiol 84:808–813

Lupold DS, Caoile AGFS, Stern DB (1999a) Genomic context influences the activity of maize mitochondrial *cox2* promoters. Proc Natl Acad Sci USA 96:11670–11675

Lupold DS, Caoile AG, Stern DB (1999b) The maize mitochondrial *cox2* gene has five promoters in two genomic regions, including a complex promoter consisting of seven overlapping units. J Biol Chem 274:3897–3903

Lysenko EA (2006) Analysis of the evolution of the family of the *Sig* genes encoding plant sigma factors. Russ J Plant Physiol V53:605–614

Lysenko EA (2007) Plant sigma factors and their role in plastid transcription. Plant Cell Rep 26:845–859

Lysenko EA, Kuznetsov VV (2005) Plastid RNA polymerases. Mol Biol V39:661–674

Maier UG, Bozarth A, Funk HT, Zauner S, Rensing SA, Schmitz-Linneweber C, Börner T, Tillich M (2008) Complex chloroplast RNA metabolism: just debugging the genetic programme? BMC Biology 6:36

Maliga P (1998) Two plastid polymerases of higher plants: an evolving story. Trends Plant Sci 3:4–6

Masters BS, Stohl LL, Clayton DA (1987) Yeast mitochondrial RNA polymerase is homologous to those encoded by bacteriophages T3 and T7. Cell 51:89–99

Matsunaga M, Jaehning JA (2004) Intrinsic promoter recognition by a "core" RNA polymerase. J Biol Chem 279:44239–44242

Mayfield SP, Yohn CB, Cohen A, Danon A (1995) Regulation of chloroplast gene expression. Annu Rev Plant Physiol Plant Mol Biol 46:147–166

Minoda A, Nagasawa K, Hanaoka M, Horiuchi M, Takahashi H, Tanaka K (2005) Microarray profiling of plastid gene expression in a unicellular red alga, *Cyanidioschyzon merolae*. Plant Mol Biol 59:375–385

Mochizuki T, Onda Y, Fujiwara E, Wada M, Toyoshima Y (2004) Two independent light signals cooperate in the activation of the plastid *psbD* blue light-responsive promoter in *Arabidopsis*. FEBS Lett 571:26–30

Mochizuki N, Tanaka R, Tanaka A, Masuda T, Nagatani A (2008) The steady-state level of Mg-protoporphyrin IX is not a determinant of plastid-to-nucleus signaling in *Arabidopsis*. Proc Natl Acad Sci U S A 105:15184–15189

Moller SG, Kim YS, Kunkel T, Chua NH (2003) PP7 is a positive regulator of blue light signaling in *Arabidopsis*. Plant Cell 15:1111–1119

Monéger F, Smart CJ, Leaver CJ (1994) Nuclear restoration of cytoplasmic male sterility in sunflower is associated with the tissue-specific regulation of a novel mitochondrial gene. EMBO J 13:8–17

Morikawa K, Shiina T, Murakami S, Toyoshima Y (2002) Novel nuclear-encoded proteins interacting with a plastid sigma factor, Sig1, in *Arabidopsis thaliana*. FEBS Lett 514:300–304

Moulin M, McCormac AC, Terry MJ, Smith AG (2008) Tetrapyrrole profiling in *Arabidopsis* seedlings reveals that retrograde plastid nuclear signaling is not due to Mg-protoporphyrin IX accumulation. Proc Natl Acad Sci USA 105:15178–15183

Muise RC, Hauswirth WW (1995) Selective DNA amplification regulates transcript levels in plant mitochondria. Curr Genet 28:113–121

Mullet JE (1993) Dynamic regulation of chloroplast transcription. Plant Physiol 103:309–313

Mulligan RM, Lau GT, Walbot V (1988) Numerous transcription initiation sites exist for the maize mitochondrial genes for subunit 9 of the ATP synthase and subunit 3 of cytochrome oxidase. Proc Natl Acad Sci U S A 85:7998–8002

Mulligan RM, Leon P, Walbot V (1991) Transcription and posttranscriptional regulation of maize mitochondrial gene expression. Mol Cell Biol 11:533–543

Nagashima A, Hanaoka M, Shikanai T, Fujiwara M, Kanamaru K, Takahashi H, Tanaka K (2004) The multiple-stress responsive plastid sigma factor, SIG5, directs activation of the *psbD* blue light-responsive promoter (BLRP) in *Arabidopsis thaliana*. Plant Cell Physiol 45:357–368

Nayak D, Guo Q, Sousa R (2009) A promoter recognition mechanism common to yeast mitochondrial and phage T7 RNA polymerases. J Biol Chem 284:13641–13647

Newton KJ, Winberg B, Yamato K, Lupold S, Stern DB (1995) Evidence for a novel mitochondrial promoter preceding the *cox2* gene of perennial teosintes. EMBO J 14:585–593

Oelmüller R, Levitan I, Bergfeld R, Rajasekhar VK, Mohr H (1986) Expression of nuclear genes as affected by treatments acting on the plastids. Planta 168:482–492

Okada S, Brennicke A (2006) Transcript levels in plant mitochondria show a tight homeostasis during day and night. Mol Genet Genomics 276:71–78

Onda Y, Yagi Y, Saito Y, Takenaka N, Toyoshima Y (2008) Light induction of *Arabidopsis SIG1* and *SIG5* transcripts in mature leaves: differential roles of cryptochrome 1 and cryptochrome 2 and dual function of SIG5 in the recognition of plastid promoters. Plant J 55:968–978

Paget MS, Helmann JD (2003) The sigma[70] family of sigma factors. Genome Biol 4:203.201–203.206

Parisi MA, Xu B, Clayton DA (1993) A human mitochondrial transcriptional activator can functionally replace a yeast mitochondrial HMG-box protein both *in vivo* and *in vitro*. Mol Cell Biol 13:1951–1961

Pesaresi P, Hertle A, Pribil M, Kleine T, Wagner R, Strissel H, Ihnatowicz A, Bonardi V, Scharfenberg M, Schneider A, Pfannschmidt T, Leister D (2009) *Arabidopsis* STN7 kinase provides a link between short- and long-term photosynthetic acclimation. Plant Cell 21:2402–2423

Pfalz J, Liere K, Kandlbinder A, Dietz K-J, Oelmüller R (2006) pTAC2, -6 and -12 are components of the transcriptionally active plastid chromosome that are required for plastid gene expression. Plant Cell 18:176–197

Pfannschmidt T (2010) Plastidial retrograde signalling – a true "plastid factor" or just metabolite signatures? Trends Plant Sci 15:427–435

Pfannschmidt T, Liere K (2005) Redox regulation and modification of proteins controlling chloroplast gene expression. Antioxid Redox Signal 7:607–618

Pfannschmidt T, Nilsson A, Allen JF (1999a) Photosynthetic control of chloroplast gene expression. Nature 397:625–628

Pfannschmidt T, Nilsson A, Tullberg A, Link G, Allen JF (1999b) Direct transcriptional control of the chloroplast genes *psbA* and *psaAB* adjusts photosynthesis to light energy distribution in plants. IUBMB Life 48:271–276

Pfannschmidt T, Ogrzewalla K, Baginsky S, Sickmann A, Meyer HE, Link G (2000) The multisubunit chloroplast RNA polymerase A from mustard (*Sinapis alba* L.): integration of a prokaryotic core into a larger complex with organelle-specific functions. Eur J Biochem 267:253–261

Pogson BJ, Woo NS, Förster B, Small ID (2008) Plastid signalling to the nucleus and beyond. Trends Plant Sci 13:602–609

Preuten T, Cincu E, Fuchs J, Zoschke R, Liere K, Börner T (2010) Fewer genes than organelles: extremely low and variable gene copy numbers in mitochondria of somatic plant cells. Plant J 64:948–959

Puthiyaveetil S, Allen JF (2008) A bacterial-type sensor kinase couples electron transport to gene expression in chloroplasts. In: Allen JF, Gantt E, Golbeck JH, Osmond B (eds) Photosynthesis. Energy from the sun, Proceedings of the 14th International Congress on Photosynthesis. Springer, Heidelberg, pp 1181–1186

Rapp WD, Lupold DS, Mack S, Stern DB (1993) Architecture of the maize mitochondrial *atp1* promoter as determined by linker-scanning and point mutagenesis. Mol Cell Biol 13:7232–7238

Richter U, Kiessling J, Hedtke B, Decker E, Reski R, Börner T, Weihe A (2002) Two *RpoT* genes of *Physcomitrella patens* encode phage-type RNA polymerases with dual targeting to mitochondria and plastids. Gene 290:95–105

Richter U, Kühn K, Okada S, Brennicke A, Weihe A, Börner T (2010) A mitochondrial rRNA dimethyladenosine methyltransferase in *Arabidopsis*. Plant J 61:558–569

Rodermel S (2001) Pathways of plastid-to-nucleus signaling. Trends Plant Sci 6:471–478

Sato N (2001) Was the evolution of plastid genetic machinery discontinuous? Trends Plant Sci 6:151–155

Sato N, Terasawa K, Miyajima K, Kabeya Y (2003) Organization, developmental dynamics, and evolution of plastid nucleoids. Int Rev Cytol 232:217–262

Sato M, Takahashi K, Ochiai Y, Hosaka T, Ochi K, Nabeta K (2009) Bacterial alarmone, guanosine 5'-diphosphate 3'-diphosphate (ppGpp), predominantly binds the β' subunit of plastid-encoded plastid RNA polymerase in chloroplasts. Chembiochem 10:1227–1233

Scarpulla RC (2008) Transcriptional paradigms in mammalian mitochondrial biogenesis and function. Physiol Rev 88:611–638

Schinkel AH, Groot Koerkamp MJ, Tabak HF (1988) Mitochondrial RNA polymerase of *Saccharomyces cerevisiae*: composition and mechanism of promoter recognition. EMBO J 7:3255–3262

Schmitz-Linneweber C, Small I (2008) Pentatricopeptide repeat proteins: a socket set for organelle gene expression. Trends Plant Sci 13:663–670

Schön A, Krupp G, Gough S, Berry-Lowe S, Kannangara CG, Söll D (1986) The RNA required in the first step of chlorophyll biosynthesis is a chloroplast glutamate tRNA. Nature 322:281–284

Schröter Y, Steiner S, Matthäi K, Pfannschmidt T (2010) Analysis of oligomeric protein complexes in the chloroplast sub-proteome of nucleic acid-binding proteins from mustard reveals potential redox regulators of plastid gene expression. Proteomics 10:2191–2204

Schwacke R, Fischer K, Ketelsen B, Krupinska K, Krause K (2007) Comparative survey of plastid and mitochondrial targeting properties of transcription factors in *Arabidopsis* and rice. Mol Genet Genomics 277:631–646

Schweer J (2010) Plant sigma factors come of age: flexible transcription factor network for regulated plastid gene expression. Endocytobiosis Cell Res 20:1–20

Schweer J, Geimer S, Meurer J, Link G (2009) *Arabidopsis* mutants carrying chimeric sigma factor genes reveal regulatory determinants for plastid gene expression. Plant Cell Physiol 50:1382–1386

Schweer J, Türkeri H, Link B, Link G (2010a) AtSIG6, a plastid sigma factor from *Arabidopsis*, reveals functional impact of cpCK2 phosphorylation. Plant J 62:192–202

Schweer J, Türkeri H, Kolpack A, Link G (2010b) Role and regulation of plastid sigma factors and their functional interactors during chloroplast transcription – recent lessons from *Arabidopsis thaliana*. Eur J Cell Biol 89(12):940–946

Sellaro R, Crepy M, Trupkin SA, Karayekov E, Buchovsky AS, Rossi C, Casal JJ (2010) Cryptochrome as a sensor of the blue/green ratio of natural radiation in *Arabidopsis*. Plant Physiol 154:401–409

Shadel GS, Clayton DA (1993) Mitochondrial transcription. J Biol Chem 268:16083–16086

Shiina T, Tsunoyama Y, Nakahira Y, Khan MS (2005) Plastid RNA polymerases, promoters, and transcription regulators in higher plants. Int Rev Cytol 244:1–68

Shiina T, Ishizaki Y, Yagi Y, Nakahira Y (2009) Function and evolution of plastid sigma factors. Plant Biotechnol 26:57–66

Siemenroth A, Wollgiehn R, Neumann D, Börner T (1981) Synthesis of ribosomal RNA in ribosome-deficient plastids of the mutant "albostrians" of *Hordeum vulgare* L. Planta 153:547–555

Silhavy D, Maliga P (1998) Mapping of the promoters for the nucleus-encoded plastid RNA polymerase (NEP) in the *iojap* maize mutant. Curr Genet 33:340–344

Smart CJ, Moneger F, Leaver CJ (1994) Cell-specific regulation of gene expression in mitochondria during anther development in sunflower. Plant Cell 6:811–825

Smith AC, Purton S (2002) The transcriptional apparatus of algal plastids. Eur J Pharmacol 37:301–311

Sriraman P, Silhavy D, Maliga P (1998) The phage-type PclpP-53 plastid promoter comprises sequences downstream of the transcription initiation site. Nucleic Acids Res 26:4874–4879

Steiner S, Dietzel L, Schröter Y, Fey V, Wagner R, Pfannschmidt T (2009) The role of phosphorylation in redox regulation of photosynthesis genes *psaA* and *psbA* during photosynthetic acclimation of mustard. Mol Plant 2:416–429

Sugiura C, Kobayashi Y, Aoki S, Sugita C, Sugita M (2003) Complete chloroplast DNA sequence of the moss *Physcomitrella patens*: evidence for the loss and relocation of *rpoA* from the chloroplast to the nucleus. Nucleic Acids Res 31:5324–5331

Suzuki JY, Jimmy Ytterberg A, Beardslee TA, Allison LA, Wijk KJ, Maliga P (2004) Affinity purification of the tobacco plastid RNA polymerase and *in vitro* reconstitution of the holoenzyme. Plant J 40:164–172

Swiatecka-Hagenbruch M, Liere K, Börner T (2007) High diversity of plastidial promoters in *Arabidopsis thaliana*. Mol Genet Genomics 277:725–734

Swiatecka-Hagenbruch M, Emanuel C, Hedtke B, Liere K, Börner T (2008) Impaired function of the phage-type RNA polymerase RpoTp in transcription of chloroplast genes is compensated by a second phage-type RNA polymerase. Nucleic Acids Res 36:785–792

Takahashi K, Kasai K, Ochi K (2004) Identification of the bacterial alarmone guanosine 5'-diphosphate 3'-diphosphate (ppGpp) in plants. Proc Natl Acad Sci USA 101:4320–4324

Thum KE, Kim M, Morishige DT, Eibl C, Koop H-U, Mullet JE (2001) Analysis of barley chloroplast *psbD* light-responsive promoter elements in transplastomic tobacco. Plant Mol Biol 47:353–366

Tokuhisa JG, Vijayan P, Feldmann KA, Browse JA (1998) Chloroplast development at low temperatures requires a homolog of DIM1, a yeast gene encoding the 18S rRNA dimethylase. Plant Cell 10:699–712

Topping J, Leaver C (1990) Mitochondrial gene expression during wheat leaf development. Planta 182:399–407

Toulokhonov II, Shulgina I, Hernandez VJ (2001) Binding of the transcription effector ppGpp to *Escherichia coli* RNA polymerase is allosteric, modular, and occurs near the N terminus of the beta'-subunit. J Biol Chem 276:1220–1225

Tracy RL, Stern DB (1995) Mitochondrial transcription initiation: promoter structures and RNA polymerases. Curr Genet 28:205–216

Trifa Y, Privat I, Gagnon J, Baeza L, Lerbs-Mache S (1998) The nuclear *RPL4* gene encodes a chloroplast protein that co-purifies with the T7-like transcription complex as well as plastid ribosomes. J Biol Chem 273:3980–3985

Tsunoyama Y, Morikawa K, Shiina T, Toyoshima Y (2002) Blue light specific and differential expression of a plastid sigma factor, Sig5 in *Arabidopsis thaliana*. FEBS Lett 516:225–228

Tsunoyama Y, Ishizaki Y, Morikawa K, Kobori M, Nakahira Y, Takeba G, Toyoshima Y, Shiina T (2004) Blue light-induced transcription of plastid-encoded *psbD* gene is mediated by a nuclear-encoded transcription initiation factor, AtSig5. Proc Natl Acad Sci USA 101:3304–3309

van der Biezen EA, Sun J, Coleman MJ, Bibb MJ, Jones JD (2000) Arabidopsis RelA/SpoT homologs implicate (p)ppGpp in plant signaling. Proc Natl Acad Sci USA 97:3747–3752

Wagner R, Pfannschmidt T (2006) Eukaryotic transcription factors in plastids – bioinformatic assessment and implications for the evolution of gene expression machineries in plants. Gene 381:62–70

Warmke HE, Lee SL (1978) Pollen abortion in T cytoplasmic male-sterile corn (*Zea mays*): a suggested mechanism. Science 200:561–563

Weihe A (2004) The transcription of plant organelle genomes. In: Daniell H, Chase CD (eds) Molecular biology and biotechnology of plant organelles. Kluwer, Dordrecht, pp 213–237

Weihe A, Börner T (1999) Transcription and the architecture of promoters in chloroplasts. Trends Plant Sci 4:169–170

Wolfe KH, Morden CW, Palmer JD (1992a) Function and evolution of a minimal plastid genome from a nonphotosynthetic parasitic plant. Proc Natl Acad Sci USA 89:10648–10652

Wolfe KH, Morden CW, Ems SC, Palmer JD (1992b) Rapid evolution of the plastid translational apparatus in a nonphotosynthetic plant: loss or accelerated sequence evolution of tRNA and ribosomal protein genes. J Mol Evol 35:304–317

Woloszynska M, Trojanowski D (2009) Counting mtDNA molecules in *Phaseolus vulgaris*: sublimons are constantly produced by recombination via short repeats and undergo rigorous selection during substoichiometric shifting. Plant Mol Biol 70:511–521

Woodson JD, Chory J (2008) Coordination of gene expression between organellar and nuclear genomes. Nat Rev Genet 9:383–395

Xie G, Allison LA (2002) Sequences upstream of the YRTA core region are essential for transcription of the tobacco *atpB* NEP promoter in chloroplasts *in vivo*. Curr Genet 41:176–182

Yan B, Pring DR (1997) Transcriptional initiation sites in *sorghum* mitochondrial DNA indicate conserved and variable features. Curr Genet 32:287–295

Yin C, Richter U, Börner T, Weihe A (2009) Evolution of phage-type RNA polymerases in higher plants: characterization of the single phage-type RNA polymerase gene from *Selaginella moellendorffii*. J Mol Evol 68:528–538

Yin C, Richter U, Börner T, Weihe A (2010) Evolution of plant phage-type RNA polymerases: the genome of the basal angiosperm *Nuphar advena* encodes two mitochondrial and one plastid phage-type RNA polymerases. BMC Evol Biol 10:379

Zer H, Ohad I (2003) Light, redox state, thylakoid-protein phosphorylation and signaling gene expression. Trends Biochem Sci 28:467–470

Zoschke R, Liere K, Börner T (2007) From seedling to mature plant: *Arabidopsis* plastidial genome copy number, RNA accumulation and transcription are differentially regulated during leaf development. Plant J 50:710–722

Zubo YO, Yamburenko MV, Selivankina SY, Shakirova FM, Avalbaev AM, Kudryakova NV, Zubkova NK, Liere K, Kulaeva ON, Kusnetsov VV, Börner T (2008) Cytokinin stimulates chloroplast transcription in detached barley leaves. Plant Physiol 148:1082–1093

Zubo YO, Yamburenko MV, Kusnetsov VV, Börner T (2011) Methyl jasmonate, gibberellic acid and auxin affect transcription and transcript accumulation of chloroplast genes. J Plant Physiol 168(12):1335–1344

Part VI
Evolution of Organelle RNA Processing

Chapter 13
Introns, Mobile Elements, and Plasmids

Georg Hausner

13.1 Introduction

Alpha-proteobacterial and cyanobacterial endosymbionts gave rise to mitochondria and chloroplasts respectively, and their genomes were reconfigured and reduced in size. Some eukaryotes gained chloroplasts by secondary or tertiary endosymbiosis, a process whereby a plastid containing eukaryote was assimilated (Archibald 2009). Most genes controlling the organelles now reside within the nuclear genome. Organellar genomes are usually depicted as circular molecules but linear forms also exist (Burger et al. 2003). Chloroplast and mitochondrial genome sizes are influenced by gene content, noncoding intergenic regions, and in part by the presence or absence of introns. Mitochondria and chloroplasts can also contain autonomously replicating DNA molecules that are, in some cases, derived from the organellar genome or represent true plasmids; the latter show no homology with the organellar chromosome. The endosymbionts that gave rise to the organelles probably carried with them mobile elements such as introns and plasmids. It has been speculated that self-splicing introns and virus-like ancestors to elements such as fungal mtDNA plasmids could have arisen before cellular life forms arose (Koonin et al. 2006).

13.1.1 The Mobilome: Introns, Homing Endonucleases, and Plasmids

There are two types of organellar introns, group I and group II introns, and they are inserted within protein-coding genes and ribosomal RNA genes and, in chloroplast

G. Hausner (✉)
Department of Microbiology, University of Manitoba, Winnipeg, Canada R3T 2N2
e-mail: hausnerg@cc.umanitoba.ca

C.E. Bullerwell (ed.), *Organelle Genetics*,
DOI 10.1007/978-3-642-22380-8_13, © Springer-Verlag Berlin Heidelberg 2012

genomes, introns are frequently found within the tRNA genes (Bonen and Vogel 2001). Group I introns are composite elements that are comprised of two units: the autocatalytic self-splicing intron RNA and often an intron-encoded homing endonuclease gene (HEGs). HEGs are viewed as mobile units that have invaded many genomic niches, including group I and group II introns (Toor and Zimmerly 2002). It is usually thought that HEGs and their intron partners can move independently from each other (Sellem and Belcour 1997), but there are also instances where the composite mobile units persist as a unit for a long time in some evolutionary lineages (Haugen and Bhattacharya 2004).

Mobile group II introns and their reverse transcriptase-like (RT) ORFs in contrast appear to be coevolving, in part driven by the dependence of the group II introns for a maturase that can bind to the various structural variants of group II RNA transcripts plus the inability of the reverse transcriptase ORF to move on its own (Toor et al. 2001). Although group I and II intron RNAs are viewed to be ribozymes, they are in many instances dependent on intron- and nuclear-encoded factors for efficient splicing; this adds another layer of control to organellar gene regulation as splicing deficiency would prevent proper gene function.

Homing endonuclease genes are mobile elements that in order to survive have to insert into neutral sites. In some instances, HEGs appear to mobilize flanking regions to repair the damage they may have caused by inserting into a gene (Paquin et al. 1994; Sethuraman et al. 2009a). This aspect of HE activities can generate new alleles; therefore, HEGs on their own or along with their intron partners can potentially increase the evolvability of organellar genomes (Basse 2010).

Mobile introns and HEGs are neutral elements that go through a cycle of intron invasion, followed by ORF degeneration resulting in both ORF and possibly intron loss (Goddard and Burt 1999). This applies in situations where a HEG-containing intron unit has become fixed within a population and, therefore, the HEG has no biological function that is under selection pressure, thus leading to the accumulation of mutations and eventual deletion. However, recent observations on the long-term persistence of a variety of elements that encoded LAGLIDADG-type HEGs suggest that the relative frequency of such putative neutral elements within a species may depend on the population genetics and biology of the host organisms combined with the ability of the HEGs' and/or introns' ability to coevolve with the host gene/genome or for the HEG to gain a new function (Gogarten and Hilario 2006).

Plasmids, genetic elements that can replicate autonomously from the main genome, have been encountered in various organellar genomes. In general, their biological relevance and mode of transmission or spread within populations are still poorly understood. Plasmids appear to persist by being cryptic (i.e., no phenotype/neutral) and in many instances by encoding at least one component required for their replication.

13.1.2 Distribution of Mobile Introns

With the exception of nuclear rDNA group I introns in some protozoans and fungi, group I and II introns tend to be restricted to organellar genomes. Homing, i.e., moving from an intron-containing allele to a cognate allele that lacks the intron, is efficient when intron/HEG minus and intron/HEG plus alleles exist within the same space; thus, the presence of multiple targets offered by repetitive DNAs (rDNAs) or multi copy genomes (cp and mtDNAs) facilitates homing.

Group I and II introns have been observed within the Fungi, Ameobozoa, Plantae, Chromalveolata, Rhizaria, Excavata, and in a few metazoans. Group I and II introns may have evolved within the preDNA world and group II introns have been proposed as the ancestors for nuclear spliceosomal introns (Koonin et al. 2006; Martin and Koonin 2006). Although some organisms have rather extensive organellar mobilomes, the metazoans are virtually devoid of mobile elements. Self-splicing introns can be quite variable within their primary sequences but they must be able to fold into particular configurations in order to maintain their ribozyme activities.

The persistence and spread of introns within a population is probably favored in: (1) single-celled or coenocytic organisms that readily mate; (2) organisms where cytoplasm can be exchanged; (3) organisms where organelles from both mating partners can fuse and their genomes can recombine; and (4) organisms were germline cells are easily accessible. Fungi can form heterokaryons and, in some instances, hyphal anastomosis can occur between different species which may permit short, transient fusion and the exchange of cytoplasm. Cytoplasm and "germ cells" are easily accessible as many fungi are filamentous, with no true cell walls separating the cells/compartments; instead, various types of pore structures regulate the flow of cytoplasm and nuclei within the hyphal system. Also, some fungi are single celled or truly coenocytic; so are many slime molds and some algae. These factors may partially explain why introns and HEGs have spread so successfully among the fungi and some protozoans. Fungi also appear to have low mtDNA mutation rates and some (albeit limited) mtDNA repair capacities (Palmer et al. 2000; Lynch et al. 2006). Low mutation rates allow introns and their encoded ORFs to have a better chance of avoiding the accumulation of nonadaptive mutations long enough to permit homing into new sites, thus keeping ahead of the "degeneration" part of the HEG and mobile intron life cycle proposed by Goddard and Burt (1999). For mobile elements such as HEGs and their intron partners to efficiently spread and persist within populations and move to other species by homing into conserved sequences, lateral transfers need to occur frequently enough to outpace the degeneration that occurs once an element is fixed.

Metazoans tend to have gene-dense small circular genomes usually lacking introns and noncoding sequences (Gissi et al. 2008). However, unicellular protozoans related to the animals, such as *Monosiga brevicollis* and *Amoebidum parasiticum*, have large mitochondrial genomes suggesting that, as metazoans arose and evolved multicellular body plans, their mitochondrial genomes became streamlined (Burger et al. 2003; Odintsova and Iurina 2005).

Group I introns have been identified in mtDNA genes (*cox* 1 and *nad* 5) within early diverging metazoan lineages such as corals, sponges, and sea anemones (Goddard et al. 2006; Rot et al. 2006; Fukami et al. 2007). In early branching metazoans, germ-line cells segregate continuously during the life span of the organism, whereas in the more complex metazoans germ-line cells segregate early on in development (Juliano and Wessel 2010) and thus these cells are less accessible for invasion by mobile elements. The lack of mtDNA introns among most metazoans may also be due to the higher mutation rates noted in metazoan organellar genomes (Lynch et al. 2006); therefore, maintaining key components for ribozyme activities within group I and II introns would be difficult. Also, HEGs may degenerate too fast where mutations accumulate quickly. The persistence of a HEG-encoding group I intron within the *cox*1 gene of the sea anemone *Metridium senile* was in part explained by low mutation rates and the potential existence of an mtDNA repair system (Goddard et al. 2006).

So far, only one instance of a group II intron within the metazoans has been reported from an annelid worm (*Nephtys* sp.); here, a 1,819-bp group II intron encoding a reverse transcriptase-like ORF is inserted within the *cox*1 gene (Vallès et al. 2008). The authors inferred that this intron is probably the result of a recent horizontal gene transfer (HGT) event from a viral or a bacterial source into the mitochondrial genome of *Nephtys* sp.

Plant mtDNAs can be quite large and harbor many group I introns, and similarly chloroplast genomes tend to contain many group II introns (Palmer et al. 2000; Knoop 2004; Kim and Archibald 2009). The presence of similar introns at identical positions in homologous mtDNA genes among unrelated species suggests that later transfer of introns has occurred frequently in the past (Bergthorsson et al. 2003; Richardson and Palmer 2007). For example, an mtDNA *cox*1 group I intron has been estimated to have been transferred horizontally over 1,000 times during angiosperm evolution (Palmer et al. 2000). Other examples of HGT have been documented between the mtDNAs of flowering plants and *Gnetum* (Gymnosperm) (Won and Renner 2003) and between various land plants to the basal angiosperm *Amborella* (Bergthorsson et al. 2004), suggesting that the potential for extensive lateral transfers exists between plant mitochondrial genomes, but the actual mechanisms of such transfers are elusive (Timmis et al. 2004; Nedelcu et al. 2008). Parasitic plants may promote HGT due to a life style that is almost analogous to fungal parasites where a "haustorial system" invades the host tissues for extracting nutrients (Davis and Wurdack 2004; Davis et al. 2005; Barkman et al. 2007). There is also evidence that sequences have been transferred between chloroplast and mitochondrial genomes among photosynthetic algae and protozoans (Turmel et al. 1995); in particular, mobile introns inserted within organellar rDNAs appear to have evolved according to insertion sites within the rDNA genes not according to the phylogenies of their host species, suggesting later transfers (Haugen and Bhattacharya 2004; Haugen et al. 2005). Incidents of chloroplast sequences being transferred into the mitochondrial genomes have been documented among the land plants (Knoop 2004); thus, mobile elements have the opportunity to move between the different cell compartments. Lateral transfer of plastid group II introns has been noted from euglenid-like species to cryptophyte algae, organisms that gained photosynthesis secondarily by incorporating eukaryotes with plastids (Khan and Archibald 2008). There is evidence that bacteria can be sources for mobile introns; for example, a group II intron found within the chloroplast genome of *Euglena myxocylindracea* appears to be derived from an intron found in the cyanobacterium *Calothrix* (Sheveleva and Hallick 2004).

Fungal sources are frequently cited for the origin of many plant mitochondrial introns (Wolff et al. 1993; Knoop 2004) and for some of the rare metazoan mtDNA introns (Szitenberg et al. 2010). Fungi are good candidates because they tend to have many mobile introns within their mitochondrial genomes and many fungi have close symbiotic associations with plants in parasitic, endophytic, or mycorrhizal associations (Rosewich and Kistler 2000; Kobayashi and Crouch 2009). Many fungi are components of lichens bringing together cyanobacteria or green algae with various fungal species. Fungi are cosmopolitans associated with virtually all forms of life and also have been shown to harbor bacterial endosymbionts (Partida-Martinez and Hertweck 2005; Schmitt and Lumbsch 2009), providing additional potential sources for HGT.

Agents or mechanisms that could facilitate HGT in fungi or protozoans are viruses, the soil bacterium *Agrobacterium tumefaciens*, and transformation. In general, it has been noted that unicellular eukaryotes that engulf other organisms appear to have more HGT. Thus, predatory cells digesting their prey or cells with endosymbionts appear to acquire foreign sequences more readily (Keeling and Palmer 2001; Andersson 2005, 2009).

13.2   Introns

13.2.1   Group I Introns: Structure, Splicing, and Mobility

Historically, short conserved sequence motifs had been defined within group I intron sequences, referred to as the P-, Q-, R-, and S-sequence motifs, each about 10–12 bp in length. These motifs are involved in the formation of paired and unpaired domains that comprise stems and helices within the RNA secondary structure (Michel and Westhof 1990; Cech et al. 1994). However, these sequence motifs cannot be identified in many introns, but the online tool RNAweasel can quickly identify intron type and potential key folds within an intron RNA sequence (Lang et al. 2007). Based on secondary structure characteristics, nucleotide sequences within the conserved core regions, and peculiarities within the secondary structure, group I introns have been classified into classes IA–IE, which can be further subdivided, e.g., IA1, IC3, etc. (Michel and Westhof 1990; Suh et al. 1999; Woodson 2005). Over 1,000 group I introns have been identified in a variety of organisms, and information about group I introns and their secondary structures have been compiled at the comparative RNA Website (R. Gutell; http://www.rna. icmb.utexas.edu/). Other resources that allow for the identification and folding of introns can be found at the RNAweasel Website (http://megasun.bch.umontreal.ca/ RNAweasel/; Lang et al. 2007), a component of AnaBench (Badidi et al. 2003). When ORFs are present within a group I intron, they are usually inserted into any of the several loops that protrude from the core secondary structure (Schäfer 2003), although there are examples where ORF sequences are part of the intron core sequences (Gibb and Edgell 2010).

The introns are removed from the precursor RNA by an autocatalytic RNA splicing event that is mediated by the intron's RNA tertiary structure and proteins; the latter are intron- (Caprara and Waring 2005) and/or nuclear encoded. In group I introns, base-pairing interactions between the 5' end of the intron and flanking exon regions define the location of the 5' and 3' splice sites. Splicing of the ribozymic group I intron RNAs is by transesterification with an external guanosine as an initiating nucleophile; this results in a linear, excised intron (see Fig. 13.1; Bonen and Vogel 2001).

Group I intron mobility is catalyzed by a homing endonuclease (HE), usually encoded by an ORF that is embedded within the mobile intron. Intron-encoded HEs are usually cis-acting and have specific target sites, with some allowance for sequence variation in their homing sites (target-cleavage site). This ensures propagation against the forces of evolutionary drift, which might modify the homing site within its host genome. The actual mechanism for intron mobility facilitated by HEs is described in Sect. 13.3. There is also experimental evidence that group I introns might be able to transpose into new sites in rRNA genes involving RNA intermediates via reverse splicing of RNA (Roman and Woodson 1995; Birgisdottir and Johansen 2005). The resulting recombinant rRNA molecule would then have to be reverse transcribed into DNA and inserted by recombination into the organellar genome. This model of mobility would explain how group I introns that lack ORFs can avoid being lost and can either disperse into new positions or be transferred horizontally between different species. One should however distinguish two terms: intron homing, which refers to events where the intron invades a specific site in a cognate intronless allele and intron transposition, where an intron inserts into a new/different site (alleles); the latter is also referred to as ectopic integration. Loss of an intron can be envisioned by a mechanism that involves reverse transcription of a mature RNA (intron removed) followed by a recombination event that replaces the intron-containing DNA sequence with the cDNA which lacks the intron.

Trans-splicing group I introns have also been identified. For example, the mtDNA *cox*1 mRNA in several unrelated organisms – (1) lycophyte *Isoetes engelmannii* (quillworts) (Grewe et al. 2009), (2) the entomoparasitic green alga *Helicosporidium* (Pombert and Keeling 2010), and (3) within some placozoan animals (the simplest known free animals such as *Trichoplax adhaerens*) – is generated from a discontinuous mtDNA *cox*1 gene and the mRNA is assembled due to the presence of a split group I intron. The intron components that are associated with the exon "fragments" can aggregate in trans and hence mediate trans-splicing, thus joining together exons

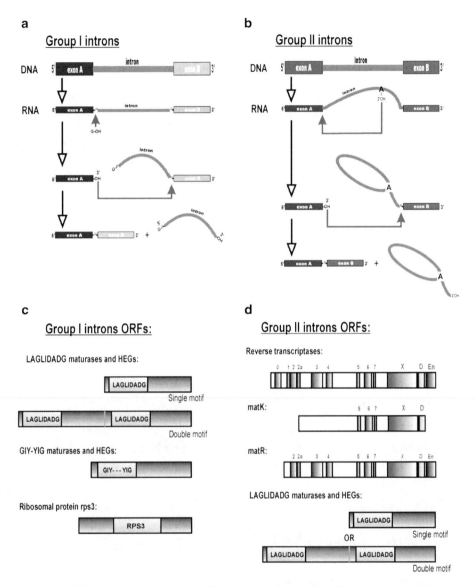

Fig. 13.1 Schematic representation of splicing of group I and group II introns (**a, b**) and intron-encoded ORFs (**c, d**). Conserved amino acid sequences are indicated for group I intron-encoded proteins (**c**). For group II intron-encoded proteins (**d**) the RT conserved sequence blocks are indicated by numbers (0–7), X indicates the maturase domain, and D is the DNA binding domain, and endonuclease activity is associated with the En domain. Some group II introns encode LAGLIDADG-type ORFs (see text for details)

encoded at different loci (Burger et al. 2009). Trans-splicing introns allow for a higher degree of genomic plasticity as disjointed/fragmented genes can be compensated for by split introns (Glanz and Kück 2009).

13.2.2 Group I Intron-Encoded Proteins: RPS3, Maturases, and Homing Endonucleases

Group I intron have been shown to encode site-specific endonucleases, or maturases (Belfort et al. 2002) and, in rare instances in some fungi, essential cellular proteins such as the RPS3 (= S5) ribosomal protein (Burke and RajBhandary 1982; Bullerwell et al. 2000; Sethuraman et al. 2009b) (Fig. 13.1). Most of the maturase-like proteins contain one or two highly conserved symmetrically arranged dodecapeptide motifs, which include conservative variants of the amino-acid sequence LAGLIDADG (Belfort and Roberts 1997). Maturases are thought to facilitate splicing by promoting proper folding of the intron RNA; actual splicing is catalyzed by the ribozyme itself.

Although group I intron maturases are usually cis-acting, there is at least one example of a trans-acting maturase, i.e., the *Saccharomyces cerevisiae cytb* bI4 intron-encoded maturase, which is required for splicing both the *cytb* bI4 intron and the *cox1* aI4α intron (Goguel et al. 1992). The current view is that pre-existing group I introns recruited ORFs that encoded HEs, providing the intron with a mechanism for mobility (Lambowitz et al. 1999). However, once an intron and its ORF have been established within a specific host gene, selection might favor the development of maturase activity over endonuclease activity because correct and efficient splicing would lessen the impact of the intron on the host gene (Bolduc et al. 2003). The reliance of group I introns on host factors for splicing further demonstrates that there is selection pressure on both the group I intron and the host genome for introns to splice accurately and quickly. Group I and II intron-encoded ORFs can be either free-standing within the intron, or be fused in frame to an upstream exon. In the latter case, it has been shown in yeast that such chimeric translation products are proteolytically cleaved to liberate the fused peptides, perhaps by a nucleus-encoded ATP-dependent protease such as PIM1 (van Dyck et al. 1998).

13.2.3 Host Factors Facilitating Group I Intron Splicing

It is now known that many nuclear factors are involved in organellar intron splicing. For example, the *S. cerevisiae cytb* i3 intron RNA recruits for splicing the intron-encoded LAGLIDADG-type maturase (with no DNA cleavage activity) and co-opted nuclear host factors such as Mrs1 (Bassi et al. 2002). Mrs1 is a member of the Rnase H fold superfamily of dimeric DNA junction-resolving enzymes that lacks nuclease activity, and it appears to have lost its original DNA functions and now binds to RNA. This example demonstrates how intron and host factors have been co-opted to new functions to facilitate efficient intron splicing.

Another well-studied system is the mtDNA *rnl* group I intron in the *Neurospora crassa*. Three nuclear mutations (*cyt-4*, *cyt-18*, and *cyt-19*) were recovered by

Bertrand et al. (1982) that showed defective splicing of the *rnl* intron. Cyt-4 was shown to be an RNAse II-like protein that might be involved in the turnover of the excised group I intron (Turcq et al. 1992) and Cyt-18 was revealed to be a tyrosyl-tRNA synthetase that interacts with several group I introns to promote splicing by helping the intron RNA fold into a catalytically active structure (Akins and Lambowitz 1987; Mohr et al. 2001). Cyt-19 appears to be an ATP-dependent RNA chaperone that can recognize and destabilize non-native RNA folds that might arise during the Cyt-18 mediated folding of group I intron RNAs (Mohr et al. 2002; Vicens et al. 2008). What emerges from these studies is that intron RNAs interacted with cellular RNA-binding proteins and fortuitous interactions occurred that promoted RNA folds required for splicing. Such interactions reduced the evolutionary constraint on intron sequences and allowed for the accumulation of nonadaptive mutations.

13.3   Homing Endonuclease Genes

Homing endonuclease genes encode DNA endonucleases which recognize rather long target sites (14–44 bp; Chevalier and Stoddard 2001). HEGs are encoded within group I or II introns, or as in-frame fusions with inteins, and as free-standing ORFs (Belfort et al. 2002; Toor and Zimmerly 2002). HEGs function as mobile elements by introducing a double-strand break (DSB), or nick, in genomes that lack the endonuclease coding sequence. The homing process is completed by host DSB-repair pathways that use the HEG-containing allele as a template to repair the DSB (Dujon 1989). The repair results in the nonreciprocal transfer of the HEG into the HEG-minus allele and is usually associated with co-conversion of markers flanking the HEG insertion site (Belfort et al. 2002). In the case of HEGs encoded within introns or inteins, co-conversion of flanking markers ensures that the self-splicing element is inherited during the mobility process. Once a HEG and its associated intron have invaded an allele, that site cannot be occupied by another HEG as the target site has now been disrupted. Some HEGs can move from an ORF-containing intron to an "ORF-less" intron (Mota and Collins 1988). This supports the notion that the structural group I intron components and the embedded ORFs have evolved independently (Belfort 2003; Bonocora and Shub 2009).

There are four major families of HEs, with naming based on conserved amino acid motifs: the H-N-H, HIS-CYS, LAGLIDADG, and GIY-YIG families (Kowalski and Derbyshire 2002; Stoddard 2006). The LAGLIDADG family of HEs is the most frequently encountered among group I introns. However GIY-YIG endonucleases have been identified within numerous group I introns, and LAGLIDADG HEs and the H-N-H domain are present within the ORFs of group II introns (Chevalier et al. 2005). Recently, additional HE-like proteins have been discovered: the PD-(D/E)XK HEs found in bacterial tRNA group I introns (Stoddard 2006), the Vsr (very short patch repair) endonucleases (a predicted family of HEs found in phages based on environmental metagenomic data)

(Dassa et al. 2009), and the Holliday junction resolvase-like HEs found in phages (Zeng et al. 2009).

LAGLIDADG-type HEGs can encode proteins with either one or two LAG-LIDADG dodecapeptide domains. A single-motif ORF is presumed to be the ancestral version, which after a gene duplication event followed by fusion of the duplication products yielded the double-motif versions of the LAGLIDADG HEGs (Haugen and Bhattacharya 2004). Single-motif LAGLIDADG HEs are active as homodimers, whereas the double-motif HEs are active as monomers (Stoddard 2006). The latter appears to have greater allowance for sequence degeneracy at the homing site as the two parts of the protein can evolve slightly new functions or specificities; thus, double-motif LAGLIDADG HEGs have been more successful in invading new sites (Haugen and Bhattacharya 2004).

13.4 Group II Introns

13.4.1 Structure, ORFs, Mobility

Group II introns have conserved secondary structures at the RNA level, which can be visualized as six stem-loop domains (domains I to VI) emerging from a central wheel (Michel and Ferat 1995). When reverse transcriptase-like ORFs are present, they tend to be embedded within domain IV. However, in some bacterial group II introns, RT-type ORFs have been observed in domain II (Simon et al. 2008) and some LAGLIDADG-type ORFs are inserted in domains 3 or 4 (Toor and Zimmerly 2002). Some group II intron-encoded proteins extend upstream and are fused to the upstream exon; this results in the generation of a fusion protein upon translation, which probably is resolved by proteolysis (Michel and Ferat 1995). Domain 1 is the largest domain and it is involved in assembling the molecular scaffold needed for the intron to assume its active structure, while domain 5 is the phylogenetically most conserved part that comprises the active site of the group II intron ribozyme (reviewed in Kelchner 2002; Lehmann and Schmidt 2003; Fedorova and Zingler 2007; Michel et al. 2009; Toor et al. 2009; Pyle 2010).

Primary sequence conservation among group II introns is minimal except at the intron boundaries, with GUGYG and AY (Y = pyrimidines) defining the 5' and 3' ends, respectively. However, the most reliable diagnostic approach for confirming the presence of a group II intron is to search for the domain V consensus structure (Toor and Zimmerly 2002) and by analyzing a putative group II sequence by RNA folding via MFOLD (http://www.bioinfo.rpi.edu/applications/mfold/old/rna/) or by submitting sequences to the RNAweasel Website (Lang et al. 2007).

Group II intron RNAs found in organellar genomes can be classified into two major subgroups based on specific structural features: subgroups IIA and IIB, which can be further segregated into IIA1, IIA2, IIB1, and IIB2 (Michel and Ferat 1995). In general, many fungal–mitochondrial group II introns can be assigned to subgroup IIA1 and many chloroplast group II introns can be assigned to subgroup IIB. A recent analysis of ORF-containing group II intron RNA structures from prokaryotic and eukaryotic sources found a total of six groups of intron structures: three were conventional forms of group IIA1, B1, and B2 secondary structures (Toor et al. 2001). There are additional subgroups in the bacteria, possibly associated with the most primitive ORFs (Simon et al. 2008, 2009).

Mobile group II introns typically encode a multifunctional protein with three activities. First is a segment homologous to reverse transcriptases (RTs), which is followed by a region referred to as domain X and has been implicated in maturase activity (Lambowitz and Zimmerly 2004). The third activity is provided by the En (previously referred to as Zn) domain, which contains a potential zinc finger and has endonuclease activity; but the En domain is absent in some fungal group II introns (Zimmerly et al. 2001; Dai et al. 2003).

Chloroplast genomes tend to be rich in group II introns that lack ORFs (reviewed in Kelchner 2002). In general, plant mtDNA and cpDNA group II introns tend to deviate from the standard group II intron structures suggesting that these introns rely more on host-encoded factors for splicing (Lambowitz and Zimmerly 2004).

13.4.2 Splicing of Group II Introns

Both splicing and mobility activity of group II introns require the catalytic activity of the intron RNA, the intron-encoded protein, and possibly host factors (Zimmerly et al. 1995a, b). During expression of the host gene, the intronic ORF is translated and a ribonucleoprotein (RNP) particle is formed between the intron lariat and the intron-encoded protein. The RNP recognizes a target homing site, typically an intron-less cognate alleles, and the first cut is made by the 3' end of the intron RNA. This initiates a reverse splicing reaction whereby the intron RNA is inserted into the sense DNA strand. The En domain cleaves the antisense DNA strand, generating a free 3'-OH that serves as a primer for the RT. Eventually, the host DNA repair machinery will remove the RNA and fill in any gaps. This process of "retrohoming" is mediated by a process termed target DNA-primed reverse transcription and has been reviewed in detail by Belfort et al. (2002). Group II introns have also been shown to retrotranspose by reverse splicing into RNA molecules; see Bonen and Vogel (2001) for the various models of RT-mediated group II intron mobility.

Several group II introns in yeast, algae, and bacteria have been shown to catalyze their own removal from primary transcripts (Peebles et al. 1986; Schmelzer and Schweyen 1986; van der Veen et al. 1986; Costa et al. 1997). Intron-catalyzed splicing proceeds by two transesterification reactions and leads to the excision of the intron as a branched, or lariat (Fig. 13.1) molecule with a characteristic 2'–5' phosphodiester bond, as in the branched pathway, or in a linear form in the hydrolytic pathway (Daniels et al. 1996; Vogel and Börner 2002).

A requirement for the group II intron splicing reaction is that the 5′ and 3′ exon sequences flanking the splice site are bound at the active site by base-pair interactions involving sequence elements embedded within domain I (Michel et al. 2009). These exon-binding sites (EBS 1 and 2) bind to the corresponding intron-binding sites (IBS1 and 2) located within the 5′ exon directly upstream of the 5′ intron/exon junction. The 3′ exon sequence is sequestered into the splicing complex by sequence elements referred to as δ (adjacent to EBS1) or EBS3 for group IIA and IIB introns, respectively. These base-pairing interactions are required for splicing, reverse splicing, and for insertion into DNA target sites during intron homing. Mutations within group II intron EBS motifs allow for changing the intron's target specificities (Lambowitz et al. 2005).

13.4.3 Variants of Group II Introns

13.4.3.1 Trans-splicing Introns and Group III Introns

Trans-splicing group II introns are found in some plant chloroplast and mitochondrial genomes, where trans-split genes have been noted. The disruption of the genes also resulted in rearrangements within their group II introns, in turn resulting in exons plus the adjoining partial intron being dispersed within the genome (Bonen 2008; Glanz and Kück 2009; Elina and Brown 2010). These segments are transcribed independently and the mRNAs are generated in trans by the intron components assembling into the proper configuration, facilitating intron splicing and joining of the exons.

Group III introns can be considered group II introns with extremely degenerated structures, where only the D1 and D6 domains have been retained (Robart and Zimmerly 2005). Other variants are so-called twintrons (intron-within-intron) first described in the *Euglena* chloroplast genome, where a group II intron inserted into another group II intron (Copertino and Hallick 1991); here, splicing of the components via lariat intermediates proceeds in a manner where the internal intron splices before the external intron. Twintrons composed of group III introns and/or group II with group III introns have also been described (Drager and Hallick 1993) and twintrons have also been reported from crytomonad alga such as *Rhodomonas salina* (=*Pyrenomonas salina*) (Maier et al. 1995). In the latter organism, twintrons were identified where the internal intron lost its splicing capacity, essentially merging with the outer intron forming "one" splicing unit (Khan and Archibald 2008). This illustrates the potential of generating new intron variants by introns invading other introns.

13.4.3.2 Group II Introns that Encode LAGLIDADG ORFs

A set of IIB1 group II introns has been detected within fungal mitochondrial ribosomal genes that encode LAGLIDADG ORFs, which typically are associated with group I introns (Toor and Zimmerly 2002). LADLIDADG-type ORFs are quite invasive and it was recently shown that these novel composite elements can encode active LAGLIDADG HEs, which may not participate in the splicing of the host introns but could potentially mobilize the group II intron (Mullineux et al. 2010). Phylogenetic analysis showed that different lineages of HEG-like elements have invaded rDNA group II introns on numerous (at least three) occasions (Monteiro-Vitorello et al. 2009). These "intron host jumps" of the LAGLIDADG HEGs are another example of the pervasive nature of HEGs and the pressure on HEGs to continuously invade new niches or acquire new functions in order to avoid extinction.

13.4.4 Maturases and Nuclear Cofactors that Facilitate Splicing of Group II Introns

As for group I introns, there has been great interest in intron and host factors that are either moonlighting or have been co-opted to serve as cofactors for splicing group II introns (Schmidt et al. 2002; Huang et al. 2005). The nuclear-encoded DEAD-box protein Mss116p (a homolog of the *Neurospora* CYT-19 protein) has been identified in yeast to serve as a potential RNA chaperon that promotes splicing and reverse splicing of the *cox* al5γ and *cob* bI1 group II introns (Halls et al. 2007). Mss116p may actually be important in stabilizing group II ribozyme structure by binding to the flanking exon sequences that may compete for folding interactions with intron sequences (Fedorova et al. 2010). The same protein has been shown to also promote the splicing of some group I introns (Huang et al. 2005).

Nuclear-encoded splicing factors have been identified that facilitate the splicing of chloroplast group II introns (Ostheimer et al. 2003; Ostersetzer et al. 2005). As most cp group II introns have lost their ORFs, these introns most likely are highly dependent on host factors for forming splicing competent RNPs. The nuclear gene nMat-2, related to group II intron-encoded reverse transcriptase/maturases (Mohr and Lambowitz 2003), has been shown in *Arabidopsis* to be required for splicing of several mitochondrial group II introns (Keren et al. 2009).

It has also been shown that the cpDNA-encoded matK protein can associate with its own intron RNA and that of six other group II introns, all belonging the group II A structural category (Zoschke et al. 2010). It has been speculated for a long time that matK might be a trans-acting splicing factor and represents an essential gene in the cpDNA that has been co-opted from being an RT protein to a maturase (Mohr et al. 1993; Hausner et al. 2006a). A similar scenario might be present within the plant mtDNAs where the intron-encoded matR protein (*nad 1* i4) has been proposed to be a splicing factor.

13.4.5   Mobile Introns and HEGs: Applications and Biotechnology

Guo et al. (2000) showed that genetically manipulated group II introns can be programmed to retrohome into desired sites. Programming of group II introns can be achieved by mutating the EBS sequences and thus altering the exon-binding specificities of the ribonucleoprotein complex that comprises the active mobile unit of the group II intron (Cui and Davis 2007; Lambowitz and Zimmerly 2011). The current model system for developing group II intron-based gene-targeting vectors is the *Lactococcus lactis* Ll.LtrB group II intron, but other group II introns might eventually be genetically manipulated for biotechnological applications (Zhuang et al. 2009a). So called "targetron" systems (Karberg et al. 2001; Zhong et al. 2003) have been developed and are commercially available; these allow for gene-targeted mutagenesis in a variety of bacteria (reviewed in Yao et al. 2005). Here, the mobile intron is introduced into a bacterial cell by means of a compatible plasmid vector and the group II intron has been programmed to insert into a specific target site/gene. Besides insertional mutagenesis, targetron-like systems are being developed as gene delivery systems, whereby genes are incorporated into domain IV, and the intron-encoded protein (RT) has been relocated either onto a second vector or into a different position within the same vector. If successful, these types of systems would allow for site-specific DNA insertions and provide new tools in genetically manipulating economically important bacteria. Gene replacement strategies are also being developed whereby a modified RT-deficient group II intron is used to introduce targeted site-specific double-stranded breaks. These breaks will induce the host system's DSB DNA repair system involving homologous recombination. A co-transformed DNA fragment can be engineered to be the template for homologous recombination and thus replacing the "damaged" segment (Karberg et al. 2001; Jones et al. 2005).

The above concepts are currently being utilized for developing group II intron-based gene-targeting systems in eukaryotic cells, with some success in the *Xenopus laevis* oocyte and *Drosophila melanogaster* embryo systems. However, some obstacles have to be resolved, such as the requirement of high Mg^{2+} concentrations, host cofactors needed for retrohoming, the presence of lariat debranching enzymes, and the somewhat inhibitory effects of eukaryotic chromatin composition (Nam et al. 1994; Mastroianni et al. 2008; Zhuang et al. 2009b).

Homing endonucleases are studied as they require long DNA recognition sites and therefore cut infrequently within a genome; this makes them useful for DNA engineering and genomics (Stoddard 2006). Some commercially available HEs are used for linearizing large insert-type cloning vectors that have been engineered to include a specific HE target site, or for genomic studies or pulse field electrophoretic studies where large DNA fragments are desired in contrast to the relative small DNA fragments that would be generated by applying type II restriction enzymes. HEs can be engineered to cleave at desired locations and therefore HEs can become site-specific tools that can be used to target specific genes (Gimble 2005; Marcaida et al. 2010; Siegl et al. 2010). Analogous to the strategy employed for group II introns, HEs can be employed to generate a double-stranded break in a gene to be modified or replaced. At the same time, co-transforming the cells with a segment of DNA that shares homology with the target sequence would allow for genes cut by the HE to be replaced via homologous recombination. This strategy would allow for therapeutic applications of HEs that target human diseases caused by one gene (Marcaida et al. 2010; Stoddard 2011).

HEGs have been proposed as tools in managing pest populations (Deredec et al. 2008). For example, it has been shown that, in *Anopheles gambiae* (vector for malaria), HEs can be introduced into *A. gambiae* cells and embryos and the HE I-PpoI can cut genomic rDNA located on the X chromosome. This strategy could eliminate X-carrying spermatozoa and favor a severe male-biased sex ratio (Windbichler et al. 2007).

13.5 Plasmids

Plasmids can be defined as optional, autonomously replicating circular or linear double-stranded extrachromosomal DNA molecules (Griffiths 1995). Organellar plasmids were at one time actively studied as promising candidates for engineering cloning and transformation vectors in eukaryotic systems (Samac and Leong 1989); however, the lack of organellar transformation systems has essentially stopped this line of research. Plasmids however represent small replicons that can be easily studied with standard molecular biology methods and thus are good model systems to resolve DNA replication machineries that operate within the organelles (Fangman et al. 1989; Backert et al. 1998; Hausner et al. 2006b).

13.5.1 Plasmid-Like Elements

Plasmid-like elements (plMEs) that are derived from regions of the organellar genome have been found in a variety of organisms, with the best-studied examples found among the fungi (Hausner 2003). Circular mtDNA-derived plMEs that exist in multimeric forms have been associated with mitochondrial instabilities in *S. cereviseae* (Dujon and Belcour 1989), *Aspergillus amstelodami* (Lazarus et al. 1980), *Podospora anserina* (Begel et al. 1999; Albert and Sellem 2002), and in plant-pathogenic fungi such as *Ophiostoma novo-ulmi* (Abu-Amero et al. 1995) and *Cryphonectria parasitica* (Monteiro-Vitorello et al. 2009). The mechanisms involved in the initial formation of mitochondrial plMEs, their amplification, mode of inheritance, and physiological effects are still poorly understood, although it was recently shown that some pLMEs in *N. crassa* replicate via a rolling circle-type mechanism (Hausner et al. 2006b). There are instances where plMEs can have important functions in mtDNA maintenance; for example, in some yeasts (e.g., *Candida parapsilosis*) that have linear mitochondrial chromosomes, specialized plMEs, referred to as telomeric circles (t-circles), are involved in telomere maintenance (Tomaska et al. 2004). These mitochondrial t-circles can be generated by recombination between randomly repeated sequences present at the telomeric region, and they are propagated by a rolling circle-dependent amplification mechanism, which in turn provides substrates for recombinational telomere elongation (Tomaska et al. 2009).

13.5.2   True Plasmids

Among the true plasmids, at least three broad categories can be recognized (Fig. 13.2): (1) circular plasmids usually encoding a DNA polymerase (Griffiths 1995); (2) linear plasmids with terminal inverted repeats encoding either a DNA or an RNA polymerase or both (Klassen and Meinhardt 2007); and (3) retroplasmids, which are linear or circular plasmids that usually encode an RT (Kennel and Cohen 2004).

Although true plasmids are mostly cryptic in nature, some plasmids have been associated with mitochondrial instabilities and senescence, the latter usually due to insertion of the plasmid into the mtDNA (Griffiths, 1992; Bertrand 2000; Maheshwari and Navaraj 2008; van Diepeningen et al. 2008; Nargang and Kennell 2010).

13.5.2.1   Retroplasmids and Possible Variants

Retroplasmids

The best known retroplasmids (RPs) are the small, circular *mauriceville* (3.6 kb) and *varkud* (3.7 kb) mitochondrial plasmids of *Neurospora* species; these encode functional RTs and replicate via an RNA intermediate (Kennell et al. 1994; Galligan and Kennell 2007). Serial transfer of *Neurospora* strains harboring these plasmids frequently results in erratic colony growth, respiratory dysfunction, mtDNA rearrangement due to integration of these plasmids and, eventually, senescence (Griffiths 1992; Chiang et al. 1994). Variant forms of both the *mauriceville* and *varkud* RPs have been detected; these appear to have arisen due to recombination events that result in deletions of plasmid sequences and insertion of mtDNA segments or segments of plMEs (Mohr et al. 2000; Stevenson et al. 2000; Fox and Kennell 2001). These forms can induce senescence in *N. crassa* by over- replicating and/or by inserting into the mtDNA, but actual phenotypes are host-specific and dependent on the nuclear background of a particular *Neurospora* strain (Fox and Kennell 2001).

Linear mitochondrial plasmids encoding RTs have been found in *Fusarium oxysporum* (Kistler et al. 1997) and in *Rhizoctonia solani* (Katsura et al. 2001). The *R. solani* linear RP (pRS224) consists of 4 986 nucleotides, encodes an ORF for a putative RT, and both termini are covalently closed "hairpin-like" structures of

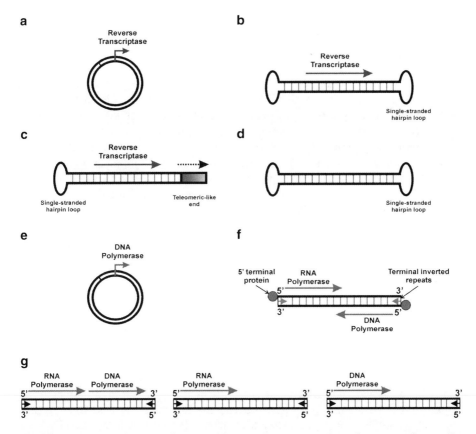

Fig. 13.2 Schematic diagrams of structural features that represent the various types of organellar plasmids. Representatives of the various types of retroplasmids and possible derived forms are shown: (**a**) circular RT-encoding forms; (**b**) linear forms; (**c**) linear "clothespin" forms; and (**d**) "hairpin"-type forms that lack defined ORFs. Circular DNA polymerase-encoding plasmids are represented in (**e**), linear invertron-type plasmids encoding either RNA or DNA polymerases represented in (**f**), and invertron plasmids shown in various ORF configurations are depicted in (**g**)

236 and 264 nucleotides (Katsura et al. 2001). This plasmid forms a single-stranded circle when denatured. The linear RPs of *F. oxysporum* (pFOXC2, and pFOXC3) replicate via their encoded RTs and one of the linear ends of these RPs has structural features that resemble eukaryotic telomeres (Walther and Kennell 1999). The *Fusarium* RPs are 1.9 kb in length and their structures can be described as "clothespin-like." One terminus has a hairpin configuration (covalently closed) and the other terminus has a telomere-like iteration of a 5-bp sequence.

"Hairpin" plasmids

A type of mitochondrial linear plasmids, referred to as hairpin plasmids, has been noted in various vegetative incompatibility groups of *R. solani*. These plasmids (pRS 64-1, pRS64-2, pRS64-3, pRS104, and pRS188) are resistant to both 3′ and 5′ exonuclease treatments as they have covalently closed termini (i.e., hairpin plasmids), and the hairpin loops differ in size (113 and 105 bp), shape, and sequence (Katsura et al. 1997; Hashiba and Nagasaka 2007). In general, these plasmids share regions of sequence homology and have hairpin-like structures that can be folded into cruciform base-paired regions. While no biologically significant ORFs have been detected, these plasmids are transcribed, and some contain a small putative ORF encoding a potential 68- or 91-amino acid peptide that has been implicated in vegetative incompatibility (Hashiba and Nagasaka 2007). So far, these plasmids have not been shown to be associated with pathogenicity. From a structural view-point, these hairpin plasmids share many features with the 4.98-kbp pRS224 hairpin RP that has been found in *R. solani*. It is therefore possible that, in *R. solani*, the hairpin plasmids are derived from linear RPs. A plasmid that could also be described as a "hairpin" plasmid has been described from the chloroplast of a green alga (La Claire and Wang 2004), but little is known about the biology of this element.

13.5.2.2   Circular Plasmids Encoding a DNA Polymerase

The best-characterized circular DNA polymerase encoding plasmids are the *fiji* and *laBelle* plasmids of *Neurospora intermedia*. These elements encode a B-family DNA polymerase (Li and Nargang 1993); and so far, neither the *laBelle* plasmid nor the related *fiji* element has been found to integrate into mtDNA or induce senescence in *N. intermedia*. A circular plasmid, pCRY1, has been described from *C. parasitica* (chestnut blight); it also encodes a B-family DNA polymerase, and studies have shown that this mitochondrial plasmid is an infectious agent (can move horizontally) and it is capable of reducing pathogenicity in some strains of this fungus (Monteiro-Vitorello et al. 2000).

13.5.2.3   Circular Plasmids in Plant Mitochondria

Circular plasmids have been noted in some plant mitochondria, but their function and origins are unknown. They do not appear to have significant ORFs, but there is some evidence that parts of these plasmids might be transcribed (Backert et al. 1996; Homs et al. 2008). These elements appear to be replicated by rolling circle-type mechanisms that may be similar to mechanisms observed for bacterial conjugative plasmids (Backert et al. 1998). In general, these plasmids do not appear to be directly related to viruses or show evidence of having been derived from the mtDNA genome, but some plant mtDNA plasmids contain segments that share similarities with sequences found in plant nuclear and chloroplast genomes. The latter led to suggestions that some of these plasmids might be involved in mechanisms that permit inter-organellar transfer of genetic information (Backert et al. 1998).

13.5.2.4   Linear "Invertron" Plasmids

Linear plasmids are present in both the cytoplasm and organelles of many lower and higher eukaryotes (Klassen and Meinhardt 2007). The most common linear forms of organellar plasmids are invertron-like elements that encode a DNA and an RNA polymerase, have terminal inverted repeats (TIR), and have proteins covalently attached to both 5' ends Kim et al. 2000. Phylogenetic analysis based on the ORFs of the mitochondrial linear plasmids suggests that these plasmids share a common ancestor with some phages (Pöggeler and Kempken 2004). TIRs are important for the formation of replication intermediates and these TIRs also contain sequence motifs required for both transcription and replication (Klassen and Meinhardt 2007). Although the current view is that linear plasmids do not produce obvious physiological or phenotypic effects in their hosts, some linear plasmids such as the *kalilo* and *maranhar* plasmids of *Neurospora* species (Court and Bertrand 1992) induce senescence by integrating into the mitochondrial chromosome. Plasmid integration appears to be a rare event; senescence presumably is the result of accumulation of suppressive defective mtDNAs, possibly generated by a single integration event (Chan et al. 1991). In contrast, the life span of the fungus *Podospora anserina* is prolonged substantially by the integration of the pAL2-1 linear plasmid into the mitochondrial chromosome (Hermanns et al. 1994). Linear plasmids have been found in several plant pathogenic fungi such as *Glomerella musae*, *Tilletia* spp., *Fusarium* spp., *Cochliobolus heterostrophus*, *Gaeumannomyces graminis* var. *tritici*, and *Claviceps purpurea* (Freeman et al. 1997; Meinhardt et al. 1997; Láday et al. 2008), but reports of the effect of these plasmids on virulence and pathogenicity are conflicting.

The presence of TIRs and terminal proteins bound to the 5' ends of these linear plasmids indicates that they likely replicate via a protein-primed mechanism similar to that observed in adenovirus (Klassen and Meinhardt 2007). Linear invertron-type plasmids have been found in plant mitochondria and they resemble the well-characterized fungal counterparts with many encoding DNA and/or RNA polymerase ORFs (Handa 2008). As within the fungi, these linear plasmids have been noted to insert into the mitochondrial genome, potentially introducing mutations or promoting genome rearrangements, and there are many examples of plasmid remnants contributing toward organellar genomes. In plants, mitochondrial linear plasmids were at one time noted to be associated with cytoplasmic male sterility (CMS), but recent research from maize and sugar beet suggests that the presence of these elements in CMS plants might be fortuitous and the plasmids themselves are not the causative agents of the CMS phenotype (Chase 2007; McDermott et al. 2008).

An invertron-type plasmid in the slime mold *Physarum polycephalum*, mF, encodes several putative ORFs including one that appears to encode a function that encourages mitochondrial fusion (Takano et al. 2010). This allows for mitochondrial gnomes to recombine during a sexual cross if one parent carries the plasmid and thus ensures that the plasmid is preferentially passed on to the progeny.

Lateral transfer from various plant-pathogenic fungi has been implicated in the origin of plant linear invertron-type plasmids, although the mechanism of transfer is unknown (Handa 2008). Phylogenetic analysis, however, suggests that plant invertron-type plasmids, although related to fungal plasmids, form their own clade (Klassen and Meinhardt 2007), arguing against a recent lateral transfer; thus, the origin of plant linear plasmids is still an enigma.

In *Candida subhashii,* which has a linear mtDNA, an invertron-type plasmid may have been domesticated to serve as the telomeres. The mtDNA has a protein covalently attached to the 5′ terminus, two ORFs are present that potentially encode DNA polymerases that resemble those found in linear invertron-type plasmids, and the mtDNA termini consist of long inverted repeats (Fricova et al. 2010). Overall, this genome could have arisen by recombination between a presumable circular ancestral mtDNA and an invertron-like plasmid. There are numerous reports of integrated plasmid segments within mitochondrial genomes but they appear to be neutral or cryptic, although they may promote genome rearrangements (Cahan and Kennell 2005; Ferandon et al. 2008).

13.6   Concluding Statements

Organellar genomes are a rich source of catalytic RNA molecules, mobile introns, plasmids, and DNA endonucleases. Group I and II intron RNAs provide a range of catalytic RNAs that could be of value to biotechnology as ribozymes that can be designed to cleave RNA molecules. Mobile introns can be manipulated to target genes or facilitate gene replacements. Intron-encoded HEs offer an almost untapped reserve of novel and rare cutting endonucleases.

The ability of HEGs to move independently of their ribozyme counterparts (group I or II introns) to form new composite mobile units along with the some allowance for degeneracy at their DNA target sites provides the flexibility needed for HEGs and mobile introns to invade new sites. The "ecology" of mobile introns is quite complex, as most introns require cis- and trans-acting factors for their splicing and mobility reactions. These elements have to be nontoxic to the host genome but, for their long-term survival, they have to invade new niches or they face extinction. Therefore, these elements are continuously challenged with the potential accumulation of nonadaptive mutations and, during lateral transfers into a new site/host they have to adapt to their new genomic environment (genetic code, codon biases, etc.), including the need to recruit host factors to facilitate efficient splicing or mobility. Yet despite all these challenges, they have been successful by being invasive and persistent within many organisms. There are also instances where these elements have been domesticated such as some intron-encoded proteins serving as trans-acting splicing factors and some plasmids being involved in telomere maintenance.

Acknowledgments G.H.'s research on mtDNA mobile elements is supported by a Discovery Grant from the Natural Sciences and Engineering Research Council of Canada. I would like to thank Dr. E.A. Gibb for providing critical comments on this manuscript and Mohamed Hafez for help with the figures. This work is dedicated to my late father Georg Hausner Senior.

References

Abu-Amero SN, Charter NW, Buck KW, Brasier CM (1995) Nucleotide-sequence analysis indicates that a DNA plasmid in a diseased isolate of *Ophiostoma novo-ulmi* is derived by recombination between two long repeat sequences in the mitochondrial large subunit ribosomal RNA gene. Curr Genet 28:54–59

Akins RA, Lambowitz AM (1987) A protein required for splicing group I introns in *Neurospora* mitochondria is mitochondrial tyrosyl-tRNA synthetase or derivative thereof. Cell 50:331–345

Albert B, Sellem CH (2002) Dynamics of the mitochondrial genome during *Podospora anserina* aging. Curr Genet 40:365–373

Andersson JO (2005) Lateral gene transfer in eukaryotes. Cell Mol Life Sci 62:1182–1197

Andersson JO (2009) Horizontal gene transfer between microbial eukaryotes. Methods Mol Biol 532:473–487

Archibald JM (2009) The puzzle of plastid evolution. Curr Biol 19:R81–R88

Backert S, Dörfel P, Lurz R, Börner T (1996) Rolling-circle replication of mitochondrial DNA in the higher plant *Chenopodium album* (L.). Mol Cell Biol 16:6285–6294

Backert S, Kunnimalaiyaan M, Börner T, Nielsen BL (1998) In vitro replication of mitochondrial plasmid mp1 from the higher plant *Chenopodium album* (L.): a remnant of bacterial rolling circle and conjugative plasmids? J Mol Biol 284:1005–1015

Badidi E, De Sousa C, Lang BF, Burger G (2003) AnaBench: a Web/CORBA-based workbench for biomolecular sequence analysis. BMC Bioinformatics 4:63

Barkman TJ, McNeal JR, Lim SH, Coat G, Croom HB, Young ND, Depamphilis CW (2007) Mitochondrial DNA suggests at least 11 origins of parasitism in angiosperms and reveals genomic chimerism in parasitic plants. BMC Evol Biol 7:248

Basse CW (2010) Mitochondrial inheritance in fungi. Curr Opin Microbiol 13:712–719

Bassi GS, de Oliveira DM, White MF, Weeks KM (2002) Recruitment of intron-encoded and co-opted proteins in splicing of the bI3 group I intron RNA. Proc Natl Acad Sci USA 99:128–133

Begel O, Boulay J, Albert B, Dufour E, Sainsard-Chanet A (1999) Mitochondrial group II introns, cytochrome c oxidase, and senescence in *Podospora anserina*. Mol Cell Biol 19:4093–4100

Belfort M (2003) Two for the price of one: a bifunctional intron-encoded DNA endonuclease-RNA maturase. Genes Dev 17:2860–2863

Belfort M, Roberts RJ (1997) Homing endonucleases: keeping the house in order. Nucleic Acids Res 25:3379–3388

Belfort M, Derbyshire V, Parker MM, Cousineau B, Lambowitz AM (2002) Mobile introns: pathways and proteins. In: Craig NL, Craigie R, Gellert M, Lambowitz AM (eds) Mobile DNA II. ASM Press, Washington DC, pp 761–783

Bergthorsson U, Adams KL, Thomason B, Palmer JD (2003) Widespread horizontal transfer of mitochondrial genes in flowering plants. Nature 424:197–201

Bergthorsson U, Richardson AO, Young GJ, Goertzen LR, Palmer JD (2004) Massive horizontal transfer of mitochondrial genes from diverse land plant donors to the basal angiosperm *Amborella*. Proc Natl Acad Sci USA 101:17747–17752

Bertrand H (2000) Role of mitochondrial DNA in the senescence and hypovirulence of fungi and potential for plant disease control. Annu Rev Phytopathol 38:397–422

Bertrand H, Bridge P, Collins RA, Garriga G, Lambowitz AM (1982) RNA splicing in *Neurospora* mitochondria. Characterization of new nuclear mutants with defects in splicing the mitochondrial large rRNA. Cell 29:517–526

Birgisdottir AB, Johansen S (2005) Site-specific reverse splicing of a HEG-containing group I intron in ribosomal RNA. Nucleic Acids Res 33:2042–2051

Bolduc JM, Spiegel PC, Chatterjee P, Brady KL, Downing ME, Caprara MG, Waring RB, Stoddard BL (2003) Structural and biochemical analyses of DNA and RNA binding by a bifunctional homing endonuclease and group I intron splicing factor. Genes Dev 17:2875–2888

Bonen L (2008) Cis- and trans-splicing of group II introns in plant mitochondria. Mitochondrion 8:26–34

Bonen L, Vogel J (2001) The ins and outs of group II introns. Trends Genet 17:322–331

Bonocora RP, Shub DA (2009) A likely pathway for formation of mobile group I introns. Curr Biol 19:23–28

Bullerwell CE, Burger G, Lang BF (2000) A novel motif for identifying rps3 homologs in fungal mitochondrial genomes. Trends Biochem Sci 25:363–365

Burger G, Gray MW, Lang BF (2003) Mitochondrial genomes: anything goes. Trends Genet 19:709–716

Burger G, Yan Y, Javadi P, Lang BF (2009) Group I-intron trans-splicing and mRNA editing in the mitochondria of placozoan animals. Trends Genet 25:381–386

Burke JM, RajBhandary UL (1982) Intron within the large rRNA gene of *N. crassa* mitochondria: a long open reading frame and a consensus sequence possibly important in splicing. Cell 31:509–520

Cahan P, Kennell JC (2005) Identification and distribution of sequences having similarity to mitochondrial plasmids in mitochondrial genomes of filamentous fungi. Mol Genet Genomics 273:462–473

Caprara MG, Waring RB (2005) Group I introns and their maturases: uninvited, but welcome guests. In: Belfort M, Derbyshire V, Stoddard BL, Wood DL (eds) Homing endonucleases and inteins. Springer, New York, NY, pp 103–119

Cech TR, Damberger SH, Gutell ER (1994) Representation of the secondary and tertiary structure of group I introns. Nat Struct Biol 1:273–280

Chan BS, Court DA, Vierula PJ, Bertrand H (1991) The kalilo linear senescence-inducing plasmid of *Neurospora* is an invertron and encodes DNA and RNA polymerases. Curr Genet 20:225–237

Chase CD (2007) Cytoplasmic male sterility: a window to the world of plant mitochondrial-nuclear interactions. Trends Genet 23:81–90

Chevalier BS, Stoddard BL (2001) Homing endonucleases: structural and functional insight into the catalysts of intron/intein mobility. Nucleic Acids Res 29:3757–3774

Chevalier B, Monnat RJ Jr, Stoddard BL (2005) The LAGLIDADG homing endonuclease family. In: Belfort M, Derbyshire V, Stoddard BL, Wood DL (eds) Homing endonucleases and inteins. Springer, New York, NY, pp 33–47

Chiang CC, Kennell JC, Wanner LA, Lambowitz AM (1994) A mitochondrial retroplasmid integrates into mitochondrial DNA by a novel mechanism involving the synthesis of a hybrid cDNA and homologous recombination. Mol Cell Biol 14:6419–6432

Copertino DW, Hallick RB (1991) Group II twintron: an intron within an intron in a chloroplast cytochrome b-559 gene. EMBO J 10:433–442

Costa M, Fontaine JM, Loiseaux-de Goër S, Michel F (1997) A group II self-splicing intron from the brown alga *Pylaiella littoralis* is active at unusually low magnesium concentrations and forms populations of molecules with a uniform conformation. J Mol Biol 274:353–364

Court DA, Bertrand H (1992) Genetic organization and structural features of *maranhar*, a senescence-inducing linear mitochondrial plasmid of *Neurospora crassa*. Curr Genet 22:385–397

Cui X, Davis G (2007) Mobile group II intron targeting: applications in prokaryotes and perspectives in eukaryotes. Front Biosci 12:4972–4985

Dai L, Toor N, Olson R, Keeping A, Zimmerly S (2003) Database for mobile group II introns. Nucleic Acids Res 31:424–426

Daniels DL, Michels WJ Jr, Pyle AM (1996) Two competing pathways for self-splicing by group II introns: a quantitative analysis of in vitro reaction rates and products. J Mol Biol 256:31–49

Dassa B, London N, Stoddard BL, Schueler-Furman O, Pietrokovski S (2009) Fractured genes: a novel genomic arrangement involving new split inteins and a new homing endonuclease family. Nucleic Acids Res 37:2560–2573

Davis CC, Wurdack KJ (2004) Host-to-parasite gene transfer in flowering plants: phylogenetic evidence from Malpighiales. Science 305:676–678

Davis CC, Anderson WR, Wurdack KJ (2005) Gene transfer from a parasitic flowering plant to a fern. Proc Biol Sci 272:2237–2242

Deredec A, Burt A, Godfray HC (2008) The population genetics of using homing endonuclease genes in vector and pest management. Genetics 179:2013–2026

Drager RG, Hallick RB (1993) A complex twintron is excised as four individual introns. Nucleic Acids Res 21:2389–2394

Dujon B (1989) Group I introns as mobile genetic elements: facts and mechanistic speculation - a review. Gene 82:91–114

Dujon B, Belcour L (1989) Mitochondrial DNA instabilities and rearrangements in yeasts and fungi. In: Berg DE, Howe MM (eds) Mobile DNA. AMS Press, Washington DC, pp 861–878

Elina H, Brown GG (2010) Extensive mis-splicing of a bi-partite plant mitochondrial group II intron. Nucleic Acids Res 38:996–1008

Fangman WL, Henly JW, Churchill G, Brewer B (1989) Stable maintenance of a 35-base-pair yeast mitochondrial genome. Mol Cell Biol 9:1917–1921

Fedorova O, Zingler N (2007) Group II introns: structure, folding and splicing mechanism. Biol Chem 388:665–678

Fedorova O, Solem A, Pyle AM (2010) Protein-facilitated folding of group II intron ribozymes. J Mol Biol 397:799–813

Ferandon C, Chatel Sel K, Castandet B, Castroviejo M, Barroso G (2008) The *Agrocybe aegerita* mitochondrial genome contains two inverted repeats of the *nad4* gene arisen by duplication on both sides of a linear plasmid integration site. Fungal Genet Biol 45:292–301

Fox AN, Kennell JC (2001) Association between variant plasmid formation and senescence in retroplasmid-containing strains of *Neurospora* spp. Curr Genet 39:92–100

Freeman S, Redman RS, Grantham G, Rodriguez RJ (1997) Characterization of a linear DNA plasmid from the filamentous fungal pathogen *Glomerella musae* [Anamorph: *Colletotrichum musae* (Berk. & Curt.) Arx.]. Curr Genet 32:152–156

Fricova D, Valach M, Farkas Z, Pfeiffer I, Kucsera J, Tomaska L, Nosek J (2010) The mitochondrial genome of the pathogenic yeast *Candida subhashii*: GC-rich linear DNA with a protein covalently attached to the 5′ termini. Microbiology 156:2153–2163

Fukami H, Chen CA, Chiou CY, Knowlton N (2007) Novel group I introns encoding a putative homing endonuclease in the mitochondrial cox1 gene of *Scleractinian corals*. J Mol Evol 64:591–600

Galligan JT, Kennell JC (2007) Retroplasmids: linear and circular plasmids that replicate via reverse transcription. In: Meinhardt F, Klassen R (eds) Microbial linear plasmids. Springer, Berlin, Germany, pp 164–185

Gibb EA, Edgell DR (2010) Better late than early: delayed translation of intron-encoded endonuclease I-TevI is required for efficient splicing of its host group I intron. Mol Microbiol 78:35–46

Gimble FS (2005) Engineering homing endonucleases for genomic applications. In: Belfort M, Derbyshire V, Stoddard BL, Wood DL (eds) Homing endonucleases and inteins. Springer, New York, NY, pp 177–192

Gissi C, Iannelli F, Pesole G (2008) Evolution of the mitochondrial genome of Metazoa as exemplified by comparison of congeneric species. Heredity 101:301–320

Glanz S, Kück U (2009) Trans-splicing of organelle introns – a detour to continuous RNAs. Bioessays 31:921–934

Goddard MR, Burt A (1999) Recurrent invasion and extinction of a selfish gene. Proc Natl Acad Sci USA 96:13880–13885

Goddard MR, Leigh J, Roger AJ, Pemberton AJ (2006) Invasion and persistence of a selfish gene in the Cnidaria. PLoS One 1:e3

Gogarten JP, Hilario E (2006) Inteins, introns, and homing endonucleases: recent revelations about the life cycle of parasitic genetic elements. BMC Evol Biol 6:94

Goguel V, Delahodde A, Jacq C (1992) Connections between RNA splicing and DNA intron mobility in yeast mitochondria: RNA maturase and DNA endonuclease switching experiments. Mol Cell Biol 12:696–705

Grewe F, Viehoever P, Weisshaar B, Knoop V (2009) A trans-splicing group I intron and tRNA-hyperediting in the mitochondrial genome of the lycophyte *Isoetes engelmannii*. Nucleic Acids Res 37:5093–5104

Griffiths AJF (1992) Fungal senescence. Annu Rev Genet 26:351–372

Griffiths AJF (1995) Natural plasmids of filamentous fungi. Microbiol Rev 59:673–685

Guo H, Karberg M, Long M, Jones JP, Sullenger B, Lambowitz AM (2000) Group II introns designed to insert into therapeutically relevant DNA target sites in human cells. Science 289:452–457

Halls C, Mohr S, Del Campo M, Yang Q, Jankowsky E, Lambowitz AM (2007) Involvement of DEAD-box proteins in group I and group II intron splicing. Biochemical characterization of Mss116p, ATP hydrolysis-dependent and -independent mechanisms, and general RNA chaperone activity. J Mol Biol 365:835–855

Handa H (2008) Linear plasmids in plant mitochondria: peaceful coexistences or malicious invasions? Mitochondrion 8:15–25

Hashiba T, Nagasaka A (2007) Hairpin plasmids from the plant pathogenic fungi *Rhizoctonia solani* and *Fusarium oxysporum*. In: Meinhardt F, Klassen R (eds) Microbial linear plasmids. Springer, Berlin, Germany, pp 227–245

Haugen P, Bhattacharya D (2004) The spread of LAGLIDADG homing endonuclease genes in rDNA. Nucleic Acids Res 32:2049–2057

Haugen P, Simon DM, Bhattacharya D (2005) The natural history of group I introns. Trends Genet 21:111–119

Hausner G (2003) Fungal mitochondrial genomes, introns and plasmids. In: Arora DK, Khachatourians GG (eds) Applied mycology and biotechnology, vol III, Fungal genomics. Elsevier Science, New York, pp 101–131

Hausner G, Olson R, Simon D, Johnson I, Sanders ER, Karol KG, McCourt RM, Zimmerly S (2006a) Origin and evolution of the chloroplast trnK (matK) intron: a model for evolution of group II intron RNA structures. Mol Biol Evol 23:380–391

Hausner G, Nummy KA, Stoltzner S, Hubert SK, Bertrand H (2006b) Biogenesis and replication of small plasmid-like derivatives of the mitochondrial DNA in *Neurospora crassa*. Fungal Genet Biol 43:75–89

Hermanns J, Asseburg A, Osiewacz HD (1994) Evidence for a life span-prolonging effect of a linear plasmid in a longevity mutant of *Podospora anserina*. Mol Gen Genet 243:297–307

Homs M, Kober S, Kepp G, Jeske H (2008) Mitochondrial plasmids of sugar beet amplified via rolling circle method detected during curtovirus screening. Virus Res 136:124–129

Huang HR, Rowe CE, Mohr S, Jiang Y, Lambowitz AM, Perlman PS (2005) The splicing of yeast mitochondrial group I and group II introns requires a DEAD-box protein with RNA chaperone function. Proc Natl Acad Sci U S A 102:163–168

Jones JP III, Kierlin MN, Coon RG, Perutka J, Lambowitz AM, Sullenger BA (2005) Retargeting mobile group II introns to repair mutant genes. Mol Ther 11:687–694

Juliano C, Wessel G (2010) Developmental biology. Versatile germline genes. Science 329:640–641

Karberg M, Guo H, Zhong J, Coon R, Perutka J, Lambowitz AM (2001) Group II introns as controllable gene targeting vectors for genetic manipulation of bacteria. Nat Biotechnol 19:1162–1167

Katsura K, Suzuki F, Miyashita S-I, Nishi T, Hirochika H, Hashiba T (1997) The complete nucleotide sequence and characterization of the linear DNA plasmid pRS64-2 from the plant pathogenic fungus *Rhizoctonia solani*. Curr Genet 32:431–435

Katsura K, Sasaki A, Nagaska A, Fuji M, Miyake Y, Hashiba T (2001) Complete nucleotide sequence of the linear DNA plasmid pRS224 with hairpin loops from *Rhizoctonia solani* and its unique transcriptional form. Curr Genet 40:195–202

Keeling PJ, Palmer JD (2001) Lateral transfer at the gene and subgenic levels in the evolution of eukaryotic enolase. Proc Natl Acad Sci USA 98:10745–10750

Kelchner SA (2002) Group II introns as phylogenetic tools: structure, function, and evolutionary constraints. Am J Bot 89:1651–1669

Kennel JC, Cohen SM (2004) Fungal mitochondria: genomes, genetic elements and gene expression. In: Arora DK (ed) The handbook of fungal biotechnology, 2nd edn. Marcel Dekker Inc, New York, pp 131–143

Kennell JC, Wang H, Lambowitz AM (1994) The Mauriceville plasmid of *Neurospora* spp. uses novel mechanisms for initiating reverse transcription in vivo. Mol Cell Biol 14:3094–3107

Keren I, Bezawork-Geleta A, Kolton M, Maayan I, Belausov E, Levy M, Mett A, Gidoni D, Shaya F, Ostersetzer-Biran O (2009) AtnMat2, a nuclear-encoded maturase required for splicing of group-II introns in *Arabidopsis* mitochondria. RNA 15:2299–2311

Khan H, Archibald JM (2008) Lateral transfer of introns in the cryptophyte plastid genome. Nucleic Acids Res 36:3043–3053

Kim E, Archibald JM (2009) Diversity and evolution of plastids and their genomes. Plant Cell Monogr. doi:10.1007/7089_2008_17

Kim E-K, Jeong J-H, Youn HS, Koo YB, Roe J-H (2000) The terminal protein of a linear mitochondrial plasmid is encoded in the N-terminus of the DNA polymerase gene in white-rot fungus *Pleurotus ostreatus*. Curr Genet 38:283–290

Kistler HC, Benny U, Powell WA (1997) Linear mitochondrial plasmids of *Fusarium oxysporum* contain genes with sequence similarity to genes encoding a reverse transcriptase from *Neurospora* spp. Appl Environ Microbiol 63:3311–3313

Klassen R, Meinhardt F (2007) Linear protein primed replicating plasmids in eukaryotic microbes. In: Meinhardt F, Klassen R (eds) Microbial linear plasmids. Springer, Berlin, Germany, pp 188–226

Knoop V (2004) The mitochondrial DNA of land plants: peculiarities in phylogenetic perspective. Curr Genet 46:123–139

Kobayashi DY, Crouch JA (2009) Bacterial/Fungal interactions: from pathogens to mutualistic endosymbionts. Annu Rev Phytopathol 47:63–82

Koonin EV, Senkevich TG, Dolja VV (2006) The ancient Virus World and evolution of cells. Biol Direct 1:29

Kowalski JC, Derbyshire V (2002) Characterization of homing endonucleases. Methods 28:365–373

La Claire JW, Wang J (2004) Structural characterization of the terminal protein domains of linear plasmid-like DNA from the green alga *Ernodesmis* (Chlorophyta). J Phycol 40:1089–1097

Láday M, Stubnya V, Hamari Z, Hornok L (2008) Characterization of a new mitochondrial plasmid from *Fusarium proliferatum*. Plasmid 59:127–133

Lambowitz AM, Zimmerly S (2004) Mobile group II introns. Annu Rev Genet 38:1–35

Lambowitz AM, Zimmerly S (2011) Group II Introns: Mobile ribozymes that invade DNA. Cold Spring Harb Perspect Biol 3(8). doi: 10.1101/cshperspect.a003616

Lambowitz AM, Caprara MG, Zimmerly S, Perlman PS (1999) Group I and group II ribozymes as RNPs: clues to the past and guides to the future. In: Gesteland RF, Cech TR, Atkins JF (eds) The RNA world. Cold Spring Harbor Laboratory Press, New York, NY, pp 451–485

Lambowitz AM, Mohr G, Zimmerly S (2005) Group II intron homing endonucleases: ribonucleo-protein complexes with programmable target specificity. In: Belfort M, Derbyshire V, Stoddard BL, Wood DL (eds) Homing endonucleases and inteins. Springer, New York, NY, pp 121–145

Lang BF, Laforest MJ, Burger G (2007) Mitochondrial introns: a critical view. Trends Genet 23:119–125

Lazarus CM, Earl AJ, Turner G, Kuntzel H (1980) Amplification of a mitochondrial DNA sequence in the cytoplasmically inherited "ragged" mutant of *Aspergillus amstelodami*. Eur J Biochem 106:633–641

Lehmann K, Schmidt U (2003) Group II introns: structure and catalytic versatility of large natural ribozymes. Crit Rev Biochem Mol Biol 38:249–303

Li Q, Nargang FE (1993) Two *Neurospora* mitochondrial plasmids encode DNA polymerases containing motifs characteristic of family B DNA polymerases but lack the sequence asp-thr-asp. Proc Natl Acad Sci USA 90:4299–4303

Lynch M, Koskella B, Schaack S (2006) Mutation pressure and the evolution of organelle genomic architecture. Science 311:1727–1730

Maheshwari R, Navaraj A (2008) Senescence in fungi: the view from *Neurospora*. FEMS Microbiol Lett 280:135–143

Maier UG, Rensing SA, Igloi GL, Maerz M (1995) Twintrons are not unique to the Euglena chloroplast genome: structure and evolution of a plastome cpn60 gene from a cryptomonad. Mol Gen Genet 246:128–131

Marcaida MJ, Muñoz IG, Blanco FJ, Prieto J, Montoya G (2010) Homing endonucleases: from basics to therapeutic applications. Cell Mol Life Sci 67:727–748

Martin W, Koonin EV (2006) Introns and the origin of nucleus-cytosol compartmentalization. Nature 440:41–45

Mastroianni M, Watanabe K, White TB, Zhuang F, Vernon J, Matsuura M, Wallingford J, Lambowitz AM (2008) Group II intron-based gene targeting reactions in eukaryotes. PLoS One 3:e3121

McDermott P, Connolly V, Kavanagh TA (2008) The mitochondrial genome of a cytoplasmic male sterile line of perennial ryegrass (*Lolium perenne* L.) contains an integrated linear plasmid-like element. Theor Appl Genet 117:459–470

Meinhardt F, Schaffrath R, Larsen M (1997) Microbial linear plasmids. Appl Microbiol Biotechnol 47:329–336

Michel F, Ferat JL (1995) Structure and activities of group II introns. Annu Rev Biochem 64:435–461

Michel F, Westhof E (1990) Modelling of the three-dimensional architecture of group I catalytic introns based on comparative sequence analysis. J Mol Biol 216:585–610

Michel F, Costa M, Westhof E (2009) The ribozyme core of group II introns: a structure in want of partners. Trends Biochem Sci 34:189–199

Mohr G, Lambowitz AM (2003) Putative proteins related to group II intron reverse transcriptase/maturases are encoded by nuclear genes in higher plants. Nucleic Acids Res 31:647–652

Mohr G, Perlman PS, Lambowitz AM (1993) Evolutionary relationships among group II intron-encoded proteins and identification of a conserved domain that may be related to maturase function. Nucleic Acids Res 21:4991–4997

Mohr S, Wanner LA, Bertrand H, Lambowitz AM (2000) Characterization of an unusual tRNA-like sequence found inserted in a *Neurospora* retroplasmid. Nucleic Acids Res 28:1514–1524

Mohr G, Rennard R, Cherniack AD, Stryker J, Lambowitz AM (2001) Function of the *Neurospora crassa* mitochondrial tyrosyl-tRNA synthetase in RNA splicing. Role of the idiosyncratic N-terminal extension and different modes of interaction with different group I introns. J Mol Biol 307:75–92

Mohr S, Stryker JM, Lambowitz AM (2002) A DEAD-box protein functions as an ATP-dependent RNA chaperone in group I intron splicing. Cell 109:769–779

Monteiro-Vitorello CB, Baidyaroy D, Bell JA, Hausner G, Fulbright DW, Bertrand H (2000) A circular mitochondrial plasmid incites hypovirulence in some strains of *Cryphonectria parasitica*. Curr Genet 37:242–256

Monteiro-Vitorello CB, Hausner G, Searles DB, Gibb EA, Fulbright DW, Bertrand H (2009) The *Cryphonectria parasitica* mitochondrial *rns* gene: plasmid-like elements, introns and homing endonucleases. Fungal Genet Biol 46:837–848

Mota EM, Collins RA (1988) Independent evolution of structural and coding regions in a *Neurospora* mitochondrial intron. Nature 332:654–656

Mullineux ST, Costa M, Bassi GS, Michel F, Hausner G (2010) A group II intron encodes a functional LAGLIDADG homing endonuclease and self-splices under moderate temperature and ionic conditions. RNA 16:1818–1831

Nam K, Hudson RH, Chapman KB, Ganeshan K, Damha MJ, Boeke JD (1994) Yeast lariat debranching enzyme. Substrate and sequence specificity. J Biol Chem 269:20613–20621

Nargang FE, Kennell JC (2010) Mitochondria and respiration. In: Borkovich KA, Ebbole DJ (eds) Cellular and molecular biology of filamentous fungi. ASM Press, Washington DC, pp 155–178

Nedelcu AM, Miles IH, Fagir AM, Karol K (2008) Adaptive eukaryote-to-eukaryote lateral gene transfer: stress-related genes of algal origin in the closest unicellular relatives of animals. J Evol Biol 21:1852–1860

Odintsova MS, Iurina NP (2005) [Genomics and evolution of cellular organelles]. Genetika 41:1170–1182 [Russian]

Ostersetzer O, Cooke AM, Watkins KP, Barkan A (2005) CRS1, a chloroplast group II intron splicing factor, promotes intron folding through specific interactions with two intron domains. Plant Cell 17:241–255

Ostheimer GJ, Williams-Carrier R, Belcher S, Osborne E, Gierke J, Barkan A (2003) Group II intron splicing factors derived by diversification of an ancient RNA-binding domain. EMBO J 22:3919–3929

Palmer JD, Adams KL, Cho Y, Parkinson CL, Qiu YL, Song K (2000) Dynamic evolution of plant mitochondrial genomes: mobile genes and introns and highly variable mutation rates. Proc Natl Acad Sci USA 97:6960–6966

Paquin B, Laforest M-J, Lang F (1994) Interspecific transfer of mitochondrial genes in fungi and creation of a homologous hybrid gene. Proc Natl Acad Sci U S A 91:11807–11810

Partida-Martinez LP, Hertweck C (2005) Pathogenic fungus harbours endosymbiotic bacteria for toxin production. Nature 437:884–888

Peebles CL, Perlman PS, Mecklenburg KL, Petrillo ML, Tabor JH, Jarrell KA, Cheng HL (1986) A self-splicing RNA excises an intron lariat. Cell 44:213–223

Pöggeler S, Kempken F (2004) Mobile genetic elements in mycelial fungi. In: Kück U (ed) The mycota II: genetics and biotechnology, 2nd edn. Springer, Berlin, Heidelberg, pp 165–197

Pombert JF, Keeling PJ (2010) The mitochondrial genome of the entomoparasitic green alga *Helicosporidium*. PLoS One 5:e8954

Pyle AM (2010) The tertiary structure of group II introns: implications for biological function and evolution. Crit Rev Biochem Mol Biol 45:215–232

Richardson AO, Palmer JD (2007) Horizontal gene transfer in plants. J Exp Bot 58:1–9

Robart AR, Zimmerly S (2005) Group II intron retroelements: function and diversity. Cytogenet Genome Res 110:589–597

Roman J, Woodson SA (1995) Reverse splicing of the *Tetrahymena* IVS: evidence for multiple reaction sites in the 23S rRNA. RNA 1:478–490

Rosewich UL, Kistler HC (2000) Role of horizontal gene transfer in the evolution of fungi. Annu Rev Phytopathol 38:325–363

Rot C, Goldfarb I, Ilan M, Huchon D (2006) Putative cross-kingdom horizontal gene transfer in sponge (Porifera) mitochondria. BMC Evol Biol 6:71

Samac DA, Leong SA (1989) Mitochondrial plasmids of filamentous fungi: characteristics and use in transformation vectors. Mol Plant Microbe Interact 2:155–159

Schäfer B (2003) Genetic conservation versus variability in mitochondria: the architecture of the mitochondrial genome in the petite-negative yeast *Schizosaccharomyces pombe*. Curr Genet 43:311–326

Schmelzer C, Schweyen RJ (1986) Self-splicing of group II introns in vitro: mapping of the branch point and mutational inhibition of lariat formation. Cell 46:557–565

Schmidt U, Lehmann K, Stahl U (2002) A novel mitochondrial DEAD box protein (Mrh4) required for maintenance of mtDNA in *Saccharomyces cerevisiae*. FEMS Yeast Res 2:267–276

Schmitt I, Lumbsch HT (2009) Ancient horizontal gene transfer from bacteria enhances biosynthetic capabilities of fungi. PLoS One 4:e4437

Sellem CH, Belcour L (1997) Intron open reading frames as mobile elements and evolution of a group I intron. Mol Biol Evol 14:518–526

Sethuraman J, Majer A, Friedrich NC, Edgell DR, Hausner G (2009a) Genes-within-genes: multiple LAGLIDADG homing endonucleases target the ribosomal protein S3 gene encoded within a *rnl* group I intron of *Ophiostoma* and related taxa. Mol Biol Evol 26:2299–2315

Sethuraman J, Majer A, Iranpour M, Hausner G (2009b) Molecular evolution of the mtDNA encoded *rps3* gene among filamentous ascomycetes fungi with an emphasis on the ophiostomatoid fungi. J Mol Evol 69:372–385

Sheveleva EV, Hallick RB (2004) Recent horizontal intron transfer to a chloroplast genome. Nucleic Acids Res 32:803–810

Siegl T, Petzke L, Welle E, Luzhetskyy A (2010) I-SceI endonuclease: a new tool for DNA repair studies and genetic manipulations in streptomycetes. Appl Microbiol Biotechnol 87:1525–1532

Simon DM, Clarke NA, McNeil BA, Johnson I, Pantuso D, Dai L, Chai D, Zimmerly S (2008) Group II introns in eubacteria and archaea: ORF-less introns and new varieties. RNA 14:1704–1713

Simon DM, Kelchner SA, Zimmerly S (2009) A broadscale phylogenetic analysis of group II intron RNAs and intron-encoded reverse transcriptases. Mol Biol Evol 26:2795–2808

Stevenson CB, Fox AN, Kennell JC (2000) Senescence associated with the over-replication of a mitochondrial retroplasmid in *Neurospora crassa*. Mol Gen Genet 263:433–444

Stoddard BL (2006) Homing endonuclease structure and function. Q Rev Biophys 38:49–95

Stoddard BL (2011) Homing endonucleases: from microbial genetic invaders to reagents for targeted DNA modification. Structure 19:7–15

Suh S-Q, Jones KG, Blackwell M (1999) A group I intron in the nuclear small subunit rRNA gene of *Cryptendoxyla hypophloia*, an ascomycetes fungus: evidence for a new major class of group I introns. J Mol Evol 48:493–500

Szitenberg A, Rot C, Ilan M, Huchon D (2010) Diversity of sponge mitochondrial introns revealed by *cox 1* sequences of *Tetillidae*. BMC Evol Biol 10:288

Takano H, Onoue K, Kawano S (2010) Mitochondrial fusion and inheritance of the mitochondrial genome. J Plant Res 123:131–138

Timmis JN, Ayliffe MA, Huang CY, Martin W (2004) Endosymbiotic gene transfer: organelle genomes forge eukaryotic chromosomes. Nat Rev Genet 5:123–135

Tomaska L, McEachern MJ, Nosek J (2004) Alternatives to telomerase: keeping linear chromosomes via telomeric circles. FEBS Lett 567:142–146

Tomaska L, Nosek J, Kramara J, Griffith JD (2009) Telomeric circles: universal players in telomere maintenance? Nat Struct Mol Biol 16:1010–1015

Toor N, Zimmerly S (2002) Identification of a family of group II introns encoding LAGLIDADG ORFs typical of group I introns. RNA 8:1373–1377

Toor N, Hausner G, Zimmerly S (2001) Coevolution of the group II intron RNA structure with its intron-encoded reverse transcriptase. RNA 7:1142–1152

Toor N, Keating KS, Pyle AM (2009) Structural insights into RNA splicing. Curr Opin Struct Biol 19:260–266

Turcq B, Dobinson KF, Serizawa N, Lambowitz AM (1992) A protein required for RNA processing and splicing in *Neurospora* mitochondria is related to gene products involved in cell cycle protein phosphatase functions. Proc Natl Acad Sci USA 89:1676–1680

Turmel M, Côté V, Otis C, Mercier JP, Gray MW, Lonergan KM, Lemieux C (1995) Evolutionary transfer of ORF-containing group I introns between different subcellular compartments (chloroplast and mitochondrion). Mol Biol Evol 12:533–545

Vallès Y, Halanych KM, Boore JL (2008) Group II introns break new boundaries: presence in a bilaterian's genome. PLoS One 3:e1488

Van der Veen R, Arnberg AC, van der Horst G, Bonen L, Tabak HF, Grivell LA (1986) Excised group II introns in yeast mitochondria are lariats and can be formed by self-splicing in vitro. Cell 44:225–234

Van Diepeningen AD, Debets AJ, Slakhorst SM, Hoekstra RF (2008) Mitochondrial pAL2-1 plasmid homologs are senescence factors in *Podospora anserina* independent of intrinsic senescence. Biotechnol J 3:791–802

Van Dyck L, Neupert W, Langer T (1998) The ATP-dependent PIM1 protease is required for the expression of intron-containing genes in mitochondria. Genes Dev 12:1515–1524

Vicens Q, Paukstelis PJ, Westhof E, Lambowitz AM, Cech TR (2008) Toward predicting self-splicing and protein-facilitated splicing of group I introns. RNA 14:2013–2029

Vogel J, Börner T (2002) Lariat formation and a hydrolytic pathway in plant chloroplast group II intron splicing. EMBO J 21:3794–3803

Walther TC, Kennell JC (1999) Linear mitochondrial plasmids of *F. oxysporum* are novel, telomere-like retroelements. Mol Cell 4:229–238

Windbichler N, Papathanos PA, Catteruccia F, Ranson H, Burt A, Crisanti A (2007) Homing endonuclease mediated gene targeting in *Anopheles gambiae* cells and embryos. Nucleic Acids Res 35:5922–5933

Wolff G, Burger G, Lang BF, Kück U (1993) Mitochondrial genes in the colourless alga *Prototheca wickerhamii* resemble plant genes in their exons but fungal genes in their introns. Nucleic Acids Res 21:719–726

Won H, Renner SS (2003) Horizontal gene transfer from flowering plants to *Gnetum*. Proc Natl Acad Sci USA 100:10824–10829

Woodson SA (2005) Structure and assembly of group I introns. Curr Opin Struct Biol 15:324–330

Yao J, Zhong J, Lambowitz AM (2005) Gene targeting using randomly inserted group II introns (targetrons) recovered from an *Escherichia coli* gene disruption library. Nucleic Acids Res 33:3351–3362

Zeng Q, Bonocora RP, Shub DA (2009) A free-standing homing endonuclease targets an intron insertion site in the psbA gene of cyanophages. Curr Biol 19:218–222

Zhong J, Karberg M, Lambowitz AM (2003) Targeted and random bacterial gene disruption using a group II intron (targetron) vector containing a retrotransposition-activated selectable marker. Nucleic Acids Res 31:1656–1664

Zhuang F, Karberg M, Perutka J, Lambowitz AM (2009a) EcI5, a group IIB intron with high retrohoming frequency: DNA target site recognition and use in gene targeting. RNA 15:432–449

Zhuang F, Mastroianni M, White TB, Lambowitz AM (2009b) Linear group II intron RNAs can retrohome in eukaryotes and may use nonhomologous end-joining for cDNA ligation. Proc Natl Acad Sci USA 106:18189–18194

Zimmerly S, Guo H, Perlman PS, Lambowitz AM (1995a) Group II intron mobility occurs by target DNA-primed reverse transcription. Cell 82:545–554

Zimmerly S, Guo H, Eskes R, Yang J, Perlman PS, Lambowitz AM (1995b) A group II intron RNA is a catalytic component of a DNA endonuclease involved in intron mobility. Cell 83:529–538

Zimmerly S, Hausner G, Wu X-C (2001) Phylogenetic analysis of group II intron ORFs. Nucleic Acids Res 29:1238–1250

Zoschke R, Nakamura M, Liere K, Sugiura M, Börner T, Schmitz-Linneweber C (2010) An organellar maturase associates with multiple group II introns. Proc Natl Acad Sci USA 107:3245–3250

Chapter 14
tRNA Modification, Editing, and Import in Mitochondria

Mary Anne T. Rubio and Juan D. Alfonzo

14.1 Introduction

Once upon a time, it was widely accepted that despite having a genome of reduced size, compared to their bacterial ancestor, mitochondria still encoded a full set of structural RNAs necessary for protein synthesis. These included copies of the small- and large-subunit ribosomal RNA (rRNA) and a minimally required set of tRNAs. In this simple scenario, the number of tRNA genes, harbored in various mitochondrial genomes, was sufficient to decode all the predicted organellar codons if some tRNA isoacceptors (e.g., those containing a U at nucleotide position 34) were allowed to pair with a G in the third position of the codon by wobbling (Agris 2004; Agris et al. 2007; Crick 1966). In turn, the mitochondrial ribosomes were composed of a majority of nucleus-encoded proteins that, following synthesis in the cytoplasm, were targeted to the mitochondria via the protein-import machinery. These proteins together with the mitochondria-encoded rRNAs assembled into fully functional ribosomes dedicated to mitochondrial translation. This view, however, was challenged by early observations from Suyama's group working with *Tetrahymena* (Suyama 1967), which raised the possibility that in some organisms a number of nucleus-encoded tRNAs were in fact imported into the mitochondria from the cytoplasm, suggesting that some organellar genomes encoded a less than complete set of tRNAs needed for organellar translation. Discoveries of other examples of RNA import into mitochondria soon followed, effectively expanding the mitochondrial RNA import world (Chiu et al. 1975; Dorner et al. 2001; Esseiva et al. 2004; Glover et al. 2001; Magalhaes et al. 1998; Marechal-Drouard et al. 1988; Martin et al. 1979; Putz et al. 2007; Simpson et al. 1989).

M.A.T. Rubio • J.D. Alfonzo (✉)
Department of Microbiology and OSU Center for RNA Biology, The Ohio State University,
Columbus, OH 43210 USA
e-mail: Alfonzo.1@osu.edu

C.E. Bullerwell (ed.), *Organelle Genetics*,
DOI 10.1007/978-3-642-22380-8_14, © Springer-Verlag Berlin Heidelberg 2012

A priori the notion that tRNAs are imported from the cytoplasm raises a number of important issues. For example, given that the mitochondrial translation system predictably resembles its bacterial ancestor, one must wonder how domain-specific nuances in translation are accommodated following the import of tRNAs intended for a eukaryotic system. Differences between the two systems are especially noticeable with the numerous post-transcriptional modifications that occur after tRNA synthesis. These modifications range in chemical complexity from simple methyl group additions to very complicated multi-step reactions involving multiple enzymes and/or protein complexes (Czerwoniec et al. 2009; Dunin-Horkawicz et al. 2006). Importantly for our argument, although some modifications are highly conserved at a given position, such as pseudouridine at position 55 in most tRNAs, modification content greatly varies between different tRNAs within a single organism and even within the same tRNAs from different organisms. These differences have become especially apparent with the advent of genomic databases and the blossoming field of bioinformatics, which have led to the discovery of missing enzymes between bacterial and eukaryotic genomes, highlighting marked differences in tRNA function between these two domains of life (Bishop et al. 2002).

Despite early interest in both the process of tRNA localization within cells and its implications for protein synthesis, the study of tRNA transport into organelles remained more of a curiosity. Perhaps because the original observations involved a few selected protists, these findings were thought to be anecdotal and having little impact in the biology of most organisms. However, in recent years, numerous discoveries have expanded the number of organisms that import RNAs into their organelles, including marsupials (Dorner et al. 2001) and placental mammals (Rubio et al. 2008), suggesting that this process is more widespread than previously thought (Alfonzo and Soll 2009). Almost simultaneously, interest in tRNA modifications has skyrocketed, partly driven by the growing reality that some modifications may be associated with diseased states in human cells and in particular human mitochondria (Putz et al. 2007).

In this chapter, we will review various aspects of the mechanism(s) for tRNA import into mitochondria and discuss a few examples of post-transcriptional modifications that, although not related to tRNA import, may have tremendous impact on mitochondrial function. We will also briefly discuss tRNA editing but since this topic will also be covered in Chap. 17, we will limit our discussions of editing to a specific example and how it relates to other post-transcriptional modifications and their effect on editing specificity. Lastly, we will emphasize recent discoveries that suggest a connection between mitochondrial tRNA maturation and the most primordial mitochondrial function, the process of iron–sulfur cluster assembly. We review these topics with the proviso that our goal is not to produce a compendium of all known tRNA import, editing, and modification systems but rather to emphasize a few examples which, in our humble opinion, provide small beacons that may shed light on new insights and approaches to these important topics.

14.2 tRNA Import into Mitochondria: Many Organisms But Not so Many Mechanisms

Over 40 years have passed since Suyama made the original observation that nucleus-encoded tRNAs were found in *Tetrahymena pyriformis* mitochondria (Suyama 1967). Following import, these tRNAs were proposed to complement the additional subset that was still transcribed from mitochondrial tRNA genes. Mitochondrial import of tRNA has now been found in many evolutionarily divergent organisms but the number and the identity of the imported tRNAs vary greatly from organism to organism (Mirande 2007; Schneider and Marechal-Drouard 2000). By far, the most extreme case, in terms of numbers, is found in trypanosomatid parasites where the mitochondrial genome has a complete lack of tRNA genes (Hancock et al. 1992; Mottram et al. 1991; Simpson et al. 1989). In these organisms, tRNA import is essential for mitochondrial protein synthesis. Likewise, despite the mitochondrial genome of higher plants still encoding a handful of tRNAs (Schneider and Marechal-Drouard 2000) (Salinas et al. 2008), this subset is not sufficient for mitochondrial translation and tRNA import is still essential.

In *Tetrahymena*, only one of three nucleus-encoded glutamine tRNA isoacceptors is imported into mitochondria (Rusconi and Cech 1996a, b). Moreover, in *Tetrahymena*, the anticodon sequence tRNA$_{UUG}^{Gln}$ functions as a mitochondrial localization signal that was both necessary and sufficient for tRNA import (Rusconi and Cech 1996b). The imported tRNA, tRNA$_{UUG}^{Gln}$, decodes the normal glutamine codons CAA and CAG used in both mitochondrial and cytosolic translation. Although the cellular localization of tRNA$_{UUG}^{Gln}$ is mainly cytosolic, 10% was found in mitochondria. The two nonimported glutamine tRNAs, tRNA$_{UUA}^{Gln}$ and tRNA$_{CUA}^{Gln}$, decode the codons UAA and UAG as glutamine, respectively. These codons are termination codons in the universal genetic code but specify glutamine in cytosolic translation while they are still used as stop codons in *Tetrahymena* mitochondria (Horowitz and Gorovsky 1985). Thus, these glutaminyl tRNAs are exclusively cytosolic, as expected, since import into the mitochondria could pose a problem for mitochondrial translation termination.

In yeast, tRNA import of a tRNA$_{CUU}^{Lys}$ isoacceptor for which the gene already exists in mitochondria was found (Martin et al. 1979). This raised the inevitable question regarding the function of the seemingly redundant tRNA. However, it was demonstrated that this tRNA did function in mitochondrial translation (Kolesnikova et al. 2004). It has been suggested that the reason for the import of this isoacceptor may rest on the set of post-transcriptional modifications that occur in the mitochondrial-encoded version, which could restrict base pairing and may limit the use of the mitochondria-encoded tRNA under certain growth conditions (Kamenski et al. 2007). Most recently, our group demonstrated that a second set of tRNAs, tRNA$_{CUG}^{Gln}$ and tRNA$_{UUG}^{Gln}$, are also imported into the yeast mitochondria but by a totally different mechanism (discussed below) (Rinehart et al. 2005). In this particular case, our group and collaborators set out to study the pathway for the synthesis of Gln-tRNA^{Gln} in yeast mitochondria. It was expected that mitochondria, due to its bacterial ancestry and the observed lack of a direct route for Gln-tRNA^{Gln} synthesis in bacteria, would mis-charge tRNA^{Gln} with glutamate where a transamidase would convert glutamate into glutamine (the so-called indirect pathway) (Wu et al. 2009). Indeed, this pathway had been previously described in organelles (Schon et al. 1988). We, however, encountered difficulties when trying to aminoacylate mitochondrial tRNAs with glutamate and could in fact detect a fairly robust glutaminyl-tRNA synthetase activity. This led to the proposal that both the nucleus-encoded tRNA^{Gln} and cognate synthetase were imported into the yeast mitochondria. This hypothesis was confirmed following rigorous localization experiments, which included the isolation of mitochondria devoid of extra-mitochondrial RNA contamination. Together, we could demonstrate, by a combination of northern analysis, RT-PCR sequencing, and in vitro import assays, that two (out of a possible three) tRNA^{Gln} isoacceptors were imported into yeast mitochondria. Furthermore, we could also show that an amber suppressor tRNA^{Gln} variant could rescue mitochondrial expression of cox III from a mutant bearing an amber codon in the middle of its reading frame, demonstrating that the imported tRNA was functional in protein translation in mitochondria in vivo (Rinehart et al. 2005). Currently, however, the reason for tRNA^{Gln} import in yeast has become once again not so clear. Recent reports showed that the indirect pathway for Gln-tRNA^{Gln} does exist in mitochondria (Frechin et al. 2009; Nagao et al. 2009), which would presumably obviate the need for direct aminoacylation of tRNA^{Gln} with glutamine for protein translation; however, it is possible that post-transcriptional modifications would have a say as to what and when a particular tRNA is needed for mitochondrial translation.

With increasing examples of tRNA import systems, questions have been raised as to whether similar import mechanisms occur in mammals. Based on the presence of 22 mitochondria-encoded tRNAs sufficient for translation and on tRNA population studies, it has long been believed that human mitochondria do not import tRNAs. However, our studies with yeast made us realize that, in fact, the gene for tRNA$_{CUG}^{Gln}$ is never found in any known mitochondrial genome and led us to the logical search for tRNA import in mammals. We showed that the same set of tRNA^{Gln} isoacceptors was imported into rat liver and HeLa cell mitochondria (Rubio et al. 2008), suggesting that the import of at least tRNA^{Gln} may be common to all mitochondria-containing organisms (Alfonzo and Soll 2009).

Clearly, when a significant number of tRNA genes are missing from a mitochondrial genome, it becomes fairly straightforward to suspect the import of nucleus-encoded tRNAs as a viable alternative for mitochondrial decoding. However, often, inspection of mitochondrial genomes reveals a limited but predictably sufficient number of tRNAs, if one follows the time-honored decoding rules where wobbling potentially obviates the need for missing tRNA isoacceptors. In these cases, tRNA import could go on undetected. A corollary of these recent findings is that one cannot simply infer intracellular distribution of tRNAs based on decoding rules. This is especially true of mitochondria where an ever-growing number of examples of nonuniversal decoding and unique modifications are rapidly accruing. We thus emphasize that new import studies should provide rigorous controls for localization and purity of organellar fractions used in localization. Import should also be recapitulated in vitro with the now well-described in vitro import assays. Additionally, two possible detection pitfalls may be considered: (1) the possible effect of modifications in decoding especially when this involves tRNA editing, which directly changes the tRNA sequence, and (2) the existence of permuted tRNA genes, split tRNA genes, or tRNA genes with multiple introns, which also pose a problem in that tRNA genes may easily escape detection by conventional means like northern hybridization or RT-PCR (Randau and Soll 2008). In summary, although discovering new examples of tRNA import into mitochondrial is not exactly trivial, the number of organisms that import tRNAs into their mitochondria is expanding. The imminent questions that remain include the contribution of the imported tRNAs for mitochondrial biogenesis and what the import machinery and its mechanisms entail.

14.2.1   Two Main Mechanisms for tRNA Import

Two general mechanisms have been described for tRNA import, differing mainly in the number and types of factors associated with the imported tRNA in a given organism. One mechanism described for import of yeast tRNA^{Lys} (Tarassov et al. 1995) requires an electrochemical potential across the mitochondrial inner membrane, and with the exception of the mitochondrial outer membrane protein MOM72, all other protein import components play a role in tRNA^{Lys} import (Martin et al. 1979; Tarassov et al. 1995; Tarassov and Martin 1996). Prior to import, however, a portion of the tRNA is aminoacylated by the cytosolic lysyl-tRNA synthetase and it is this version of the tRNA that is eventually recognized by the precursor mitochondrial lysyl-tRNA synthetase, pre-MSK1p, which only binds aminoacylated tRNA$_{CUU}$^{Lys}. The current hypothesis is that the glycolytic enzyme enolase, Eno2p, acts as an RNA chaperone by inducing conformational changes in tRNA^{Lys} and enables the tRNA to escape the protein synthesis machinery in the cytoplasm (Brandina et al. 2006; Entelis et al. 2006). In this scheme, the tRNA-enolase complex transits to the mitochondrial surface where the aminoacylated tRNA^{Lys} is bound by the precursor mitochondria-targeted lysyl-tRNA synthetase, preMSK1p. pre-MSK1p serves as the carrier for tRNA$_{CUU}$^{Lys} translocation into the mitochondrial matrix using the canonical protein import pathway. Meanwhile, at the mitochondrial outer membrane, enolase is proposed to partition toward the glycolytic multiprotein complex, consistent with the observation of the glycolytic enzymes activities on the surface of mitochondria in yeast (Brandina et al. 2006). Additional proteins that participate in the process have been found by three- and two-hybrid genetic screenings. Three proteins belonged to the ubiquitin/26S-proteasome system (UPS). Two of them are subunits of the 19S regulatory particle of proteasome – Rpn8p and Rpn13p – while the third, Doa1p, is implicated in cellular ubiquitin metabolism (Brandina et al. 2007). Rpn8p was identified as interacting with tRNA^{Lys}, Rpn13p with the full-size pre-Msk1p, and Doa1p with the N-terminal domain of preMsk1p, shown to be essential for tRNA^{Lys} import. Both full-length

Rpn8p and Rpn13p interact with pre-Msk1 in a two-hybrid system and able to bind the imported tRNA more efficiently than the phage MS2 RNA used as a negative control for binding. Altogether, the case of tRNA^{Lys} import into yeast mitochondria involves multiple cytoplasmic protein factors which acting in concert may help coordinate the specific recognition of a single tRNA out of a pool of nonimportable ones and may explain the delivery mechanism to the mitochondrial protein import complexes. Mechanistically, however, the actual transport involves the protein import machinery and thus has the same bioenergetic requirements (i.e., a membrane potential).

A second, and possibly the most common, tRNA import mechanism acts independently of the protein import pathway and has energetic requirements that are distinct from that described for tRNA^{Lys} in yeast. This mechanism was first described in *Leishmania* but has now been found in many other organisms ranging from other kinetoplastids to humans (Salinas et al. 2008; Schneider and Marechal-Drouard 2000). This protein import-independent pathway can be efficiently reproduced in vitro in the absence of cytoplasmic factors (Mahapatra et al. 1994; Rinehart et al. 2005; Rubio et al. 2000, 2008; Yermovsky-Kammerer and Hajduk 1999); however, it may still be influenced by their presence. For example, in *Trypanosoma brucei*, cytosolic translation elongation factor 1a (eEF1a) plays a key role as a specificity determinant for some imported tRNAs (Bouzaidi-Tiali et al. 2007). Significantly, the import of tRNA^{Gln} into yeast mitochondria occurs in vitro in the absence of added cytosolic factors (Rinehart et al. 2005). Therefore, *Saccharomyces cerevisiae* contains two pathways for tRNA import: one utilizes the protein import machinery and the second is independent of protein import, a feature that is thus far unique to this organism (Rinehart et al. 2005). Regardless of the organism, the proteins involved in the actual transport of the tRNAs across the mitochondrial membrane in the protein import-independent pathway remain elusive (Fig. 14.1).

In *Leishmania tropica*, an RNA import complex (RIC) comprised of an 11-protein complex is assembled at the mitochondrial inner membrane with a stoichiometry adding up to a total mass of ~580 kDa. The import complex requires ATP and membrane potential to import tRNAs. Within the complex, there are three mitochondrion and eight nucleus-encoded subunits. Analyses by knockdown and in vitro reconstitution experiments indicated that six of the eight nucleus-encoded subunits, RIC 1, 4A, 6, 8A, 8B, and 9, are essential for import (Mukherjee et al. 2007). The RIC has been obtained by affinity procedure and has been resolved from other mitochondrial complexes by native gel electrophoresis (Goswami et al. 2006). Functional complexes could be reconstituted with recombinant subunits expressed in *Escherichia coli*. Several essential RIC subunits are identical to specific subunits of respiratory complexes. The two nonessential subunits were identified as RIC3, a M16 metalloproteinases, and RIC5, a trypanosomatid-specific protein. It is proposed that RIC1 and RIC8A are the two receptors involved in initial tRNA binding. Then, trimeric RIC6 and RIC9 form the translocation pore, while RIC4A and RIC8B anchor the complex to the membrane. Membrane-embedded mitochondrion-encoded subunits 2 (dimeric), 4B (sub-stoichiometric), and 7 interact with RIC4A. The dispensable subunits RIC3 and RIC5 are assembled peripherally (Mukherjee et al. 2007).

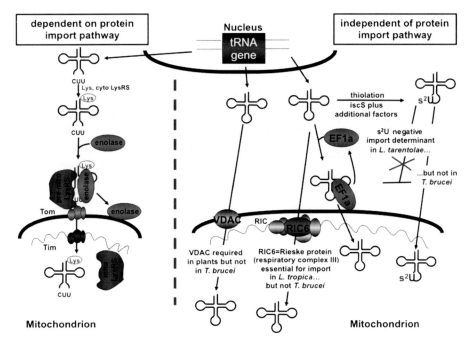

Fig. 14.1 Two general mechanisms of transfer RNA (tRNA) import into mitochondria. One mechanism of tRNA import is strictly dependent on the protein import pathway as found in *S. cerevisiae*, where tRNALys(CUU) (tRK1) is aminoacylated, recognized by enolase and the precursor form of the mitochondrial lysyl-tRNA synthetase followed by delivery to the mitochondrial surface. Another mechanism of tRNA import occurs independently from the protein import pathway, as found for *S. cerevisae* and mammalian tRNAGln (not shown). Additional protein components involved in tRNA import include the VDAC of plant, the RIC of *L. tropica*, EF1a of *T. brucei* mitochondria. Indirectly, the IscS protein involved in tRNA thiolation is a negative determinant in *L. tarentolae*, but not in *T. brucei*

To date despite many reports, the relevance of the *Leishmania* import complex is still controversial. For instances, one of its essential components is the Rieske protein but downregulation of its expression by RNA interference had no effect on tRNA import (Paris et al. 2009). This was despite having the predicted effects on membrane potential and mitochondrial function traditionally ascribed to Rieske function. Thus in the kinetoplastid system, the true nature of the tRNA import machinery is not yet clear.

In the plant system, inhibition of the VDAC (voltage-activated anion channel) by the addition of ruthenium red impaired tRNA import into mitochondria (Salinas et al. 2006), suggesting that the VDAC plays a critical role for tRNA transport across the outer membrane. One major difference from the plant and *T. brucei* systems, however, is the fact that a knockout of the gene encoding the VDAC protein in *T. brucei* had no effect on tRNA transport (Pusnik et al. 2009). Still, just like in the kinetoplastid system, it is not clear what factors transport a tRNA across the inner membrane, the ultimate mitochondrial permeability barrier. In passing, one must recognize that the disparate nature of the various import systems should not be at all surprising given the proposed polyphyletic origin of import (Schneider and Marechal-Drouard 2000). Significantly, what all protein-import-independent mechanisms have in common is the requirement for ATP (although the exact role of ATP is still not clear) and lack of requirement for membrane potential (Alfonzo and Soll 2009), but inasmuch as the actual transporters have not been identified it is difficult to start looking for commonalities in the actual mechanism. Nuances indeed may exist among the ancillary factors used but the ultimate test of common mechanisms still rests in secret in the nature of the import machinery.

14.2.2   The Role of tRNA Modifications on tRNA Function in Organelles

In eukarya, two places for tRNA synthesis exist: the nucleus and the genome-containing organelles (chloroplast and mitochondria). Regardless of the site of synthesis, tRNAs are transcribed with extra sequences at their 5' and 3' ends and undergo processing that trims them to their functional unit length. Nuclear tRNAs are exported to the cytoplasm where they engage in translation of nucleus-encoded mRNAs and likewise organello-encoded tRNAs are used exclusively for organellar protein synthesis. It is now widely accepted that many eukaryotes actively import tRNAs from the cytoplasm into their mitochondria to supplement mitochondrial tRNA pools. At each step of synthesis and processing, tRNAs undergo post-transcriptional modifications. Although currently not much is known about the intracellular localization of most modification enzymes, discoveries over the past 20 years have highlighted a few interesting facts. For example, some modifications occur in the nucleus, indicating that the enzymes involved are actively transported to that compartment following their synthesis in the cytoplasm (Colonna and Kerr 1980; Nishikura and De Robertis 1981). However, what role, if any, nuclear modifications play in tRNA processing and/or transport is not exactly clear. Other modifications occur in the cytoplasm; these are presumed to affect tRNA function by affecting folding, aminoacylation, decoding, or a combination of all of these functions (Hopper and Phizicky 2003; Phizicky 2005). Most recently, new information has provided a glimpse of how particular modifications may even play a role in tRNA stability (Chernyakov et al. 2008).

In the case of mitochondrial modifications, some enzymes are targeted to mitochondria and only operate in that compartment. Alternatively, a single gene may encode two products, by utilizing alternative translation initiation codons, one with a mitochondria-targeting signal and the other without, where the resulting enzyme is identical in function but still it exerts its activity in two different compartments (Dihanich et al. 1987; Hopper et al. 1982; Martin and Hopper 1994). In the end, we can think of at least two factors that may then affect the modification content of a given tRNA: the localization of modification enzymes and the localization of the tRNA.

Despite some advances, like the development of mitochondrial transformation systems in yeast and *Chlamydomonas* (Johnston et al. 1988; Remacle et al. 2006), it is still very difficult to genetically manipulate mitochondria and to directly link a particular mitochondrial modification to a specific effect on tRNA function. Similarly, for reasons that are not obvious, no efficient in vitro translation system exists for mitochondria. Therefore, most studies of the role that particular modifications play in mitochondrial function have been limited to general correlations between mitochondrial tRNA mutations and the existence of a particular mitochondrial defect. In the following sections, we will discuss a few cases where genetic and biochemical data have provided strong insights into the role of modifications in mitochondrial tRNA function far and beyond the realm of mere correlations.

14.2.2.1 Deciphering the Function of 5-Formyl Cytosine (f^5C) and the Importance of a Mitochondrial In Vitro Translation System

The mitochondrial genetic code is far from universal and a number of codons are inferred to have a different meaning in mitochondria from that of the nuclear genome. In fact, the same codon may have different meanings depending on the organism (Jukes and Osawa 1990, 1993; Osawa and Jukes 1988, 1989), for example, the use of stop codons for glutamine in *Tetrahymena* nuclear genes as discussed earlier. Most conserved among these is perhaps the use of UGA codons as tryptophan, which will be discussed in Sect. 14.2.2.4 and also in Chap. 17. Another interesting example is provided by the use of UAU codons as methionine in the mitochondria of a number or organisms including frogs, mammals (rat and humans), fruit flies, squid, etc. (Matsuyama et al. 1998). In these mitochondria, two codons are used as methionine: AUG, the standard codon, and AUA, normally coding for isoleucine in the universal code. Initially it was thought that this codon reassignment could be achieved by lysidine modification of $tRNA_{CAU}^{Met}$. This modification had been described in bacteria and it followed logically that mitochondrial decoding should use the same strategy. However, as mitochondrial genome sequences became available, it was apparent that no $tRNA^{Ile}$ with anticodon UAU existed in animal mitochondria. Thus, it was proposed that the mitochondria of most organisms (with the exception of plants) may use a different strategy. It was suggested that a single $tRNA^{Met}$ with anticodon CAU was responsible for decoding these codons as methionine (Matsuyama et al. 1998). Sequencing of the native tRNA from animal mitochondria then revealed that the first position of the anticodon of mitochondrial $tRNA^{Met}$ was post-transcriptionally modified to 5-formyl cytosine (f^5C) (Matsuyama et al. 1998) (Fig. 14.2). This observation suggested that this modification could allow for an unusual wobble pair involving the modified C_{34} in the first position of the anticodon and an adenosine in the third codon position. However, as explained earlier the inability to genetically manipulate mitochondria and the lack of a robust in vitro translation system left this question unanswered for several years. Recently, however, an in vitro translation system has been described for bovine liver mitochondria (Takemoto et al. 2009).

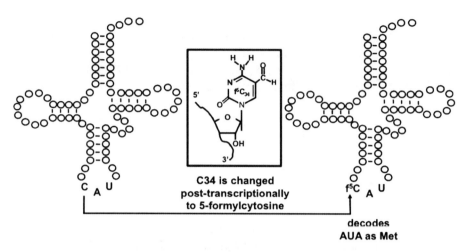

Fig. 14.2 Formylation of the first position of the anticodon reassigns a leucine codon to methionine

Using this system, the unusual decoding of the AUA codons has finally been put through the test. Mitochondrial ribosomes were programmed with poly-ribonucleotides consisting of either AUG or AUA codons and the synthesis of poly-methionine measured with either fully modified (f^5C-containing) or tRNA^{Met} with various degrees of modification. All tRNA species tested could decode the standard AUG codons, however, f^5C_{34} was absolutely required for the decoding of AUA as methionine. Significantly, the presence of this modification did not interfere with AUG decoding but simply expanded the ability of the tRNA to use the additional AUA codons (Takemoto et al. 2009). The availability of this in vitro system should help clarify a number of observations that implicate modifications in mitochondrial defect in a more direct manner. Interestingly, the f^5C system, which does not occur in bacteria, also highlights the fact that despite their origin, years of evolution have led mitochondria to unique solutions to the decoding problem.

14.2.2.2 "tRNA Surgery," Taurine, and Mitochondrial Diseases

Not only modifications play important roles in expanding codon recognition, but mutations that lead to ablation of some conserved modifications can have serious consequences on mitochondrial physiology. An interesting example has been tRNA mutations that show strong correlation with the onset of two mitochondrial defects: (1) mitochondrial encephalophathy, lactic acidosis, and stroke-like episodes (MELAS) and (2) mitochondrial encephalophathy and myoclonous epilepsy with ragged-red fibers (MERRF). MELAS is caused by a single base replacement in the mitochondria-encoded tRNA$_{UAA}^{Leu}$, which decodes UUR codons (where R stands for a purine) (Kobayashi et al. 1991). The mutation can occur at two places either in the dihydrouridine loop (D-loop) or in the anticodon stem (A to G 3243 and U to C

3271 using the mitochondrial genome numbering). Significantly, both mutations cause the same phenotype. In MERRF, a single A to G change in the TΨC loop (A to G 8344) of tRNA$_{UUU}^{Lys}$, responsible for decoding AAA and AAG codons as lysine, leads to the mitochondrial defect (Fukuhara et al. 1980).

In the case of MELAS, it was initially shown that both mutations led to substantially reduced levels of aminoacylation, suggesting that the mitochondrial defect may be caused, in part, by reduced rates of protein synthesis (Yasukawa et al. 2000). However, this initial idea was questioned given that at least in the U3271 to C mutation no reduction in overall protein synthesis was observed. This discrepancy led to a series of experiments involving the use of cybrid cells, where cells from tissues or even patients with a mitochondrial defect can be fused with cells grown in culture but lacking a mitochondrial genome (ρ^0 cells). This technique has been invaluable in clearly pinpointing mutations that directly lead to a mitochondrial defect. Remarkably, it was noticed that even when the protein synthesis levels were normal, a defect in complex I was observed with the MERRF mutations (Yasukawa et al. 2001). Thus, it was hypothesized that the defect was caused by mis-incorporation of the wrong amino acid by the mutant tRNAs. This idea however also fell by the wayside with the fact that only the correct amino acid was found attached to native tRNA^{Leu} despite the mutation. Similar controversies appear in the case of the MERRF mutation, where some reports showed a decrease in aminoacylation of the mutant tRNA^{Lys} (Enriquez et al. 1995) while other investigators showed no differences whatsoever. This led to the proposal that maybe these differences were due to some kind of tissue specificity, granted that these various groups were indeed comparing different tissues (Borner et al. 2000). The first hint of what could reconcile all the disparate observations came from the analysis of the native tRNAs in question and the realization that all of these mutations shared in common the ablation of taurine (τ) (Fig. 14.3) a fairly unique modification found at the first position of the anticodon for all three mutant tRNAs (Yasukawa et al. 2000, 2001). The challenge remained, however, in separating the effects caused by the mutation from that of the lack of taurine, as opposed to other modifications in the tRNA. To settle the argument, the Suzuki group reported a series of technically challenging but elegant experiments. This group purified native tRNAs from human placental mitochondria and performed "surgery" on the tRNA (Kirino et al. 2004). They used a hammerhead ribozyme to specifically split the tRNAs in question at the anticodon, replaced the normally taurine-containing uridine by an unmodified uridine version, and finally ligated back various versions of the tRNA backbone. The resulting constructs were either wild type and fully modified or alternatively had the mutation while fully modified. These manipulations were effectively uncoupling the mutations from the presence or absence of the modifications. They used these various versions of the tRNA to test both ribosome binding and translational efficiency using the mitochondrial in vitro translation systems previously described. They found that the lack of taurine affected the reading of the UUG codons, but even the mutations themselves, despite being at a site distal from the anticodon, affected UUA decoding (Kirino et al. 2004, 2006). These specific effects in the decoding of some, but not all, codons

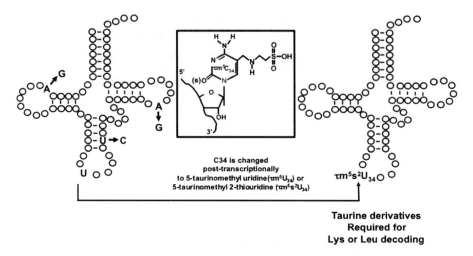

Fig. 14.3 Two modified uridines were identified in mammalian mitochondrial (mt) tRNAs. Mass spectrometric analysis revealed modified uridines possessing a sulfonic acid group derived from taurine; 5-taurinomethyl-uridine from mitochondrial tRNATrp and tRNALeuUUR, and 5-taurinomethyl-2-thiouridine from mitochondrial tRNALys, tRNAGln and tRNAGlu. Taurine modification was absent in mutant mitochondrial tRNALeuUUR and tRNALys from cells derived from patients with encephalomyopathies, MELAS and MERRF, respectively

finally explain why, despite the mitochondrial defect and the presence of the mutations, no effect in overall translation rates was observed by various laboratories as discussed above.

Again, we would like to emphasize that elucidating the specific role that either the lack of a modification or the presence of tRNA mutations plays in mitochondrial physiology inevitably requires the uncoupling of the effects caused by the mutations themselves from indirect effects caused by affecting tRNA modification content. In the case of MELAS and MERRF, as described above, this is currently possible by two significant technological advances: (1) the development of an in vitro translation system that, although it is still far from perfect, is sufficiently robust to measure codon specificity and (2) the tRNA dissection maneuver described by the Watanabe laboratory, coupled with the ability to purify specific tRNAs from native populations, finally permits the assessment of the effect of removing a specific modification in the context of and otherwise fully modified tRNA (Fig. 14.3).

14.2.2.3   tRNA Thiolation: Two Compartments, a Single Modification and Two Separate Enzyme Systems

Information on RNA modifications has steadily increased in the last few years, facilitated by improvements in the application of highly sensitive mass spectrometry approaches (Crain and McCloskey 1998; Limbach et al. 1995; Rozenski et al. 1999) as well as methods for tRNA purification from natural sources (Morl et al.

1995) (Suzuki 2007). Typically, following identification and mapping of a given modification to a particular nucleotide in a tRNA molecule, a search for the modification activity ensues. This eventually leads to the identification of the gene and the recombinant expression and characterization of the enzyme in question. Through a number of studies, however, at least two major points have become clear about modification enzymes: (1) many enzymes do not efficiently modified their substrates in vitro and (2) often their in vitro specificity does not necessarily represent that observed in vivo. The first point has led to the suggestion that although strong genetic evidence may pin a particular modification to a cellular component (i.e., an enzyme), in vitro, one may require additional factors to fully reconstitute the activity (Auxilien et al. 2007). Formation of 2-thiouridine (s^2U) elegantly illustrates this point. This modification is commonly found at U_{34} (the first position of the anticodon) in $tRNA^{Glu}$, $tRNA^{Gln}$, and $tRNA^{Lys}$ in bacteria, eukarya, and possibly in archaea. Several laboratories showed that bacterial s^2U_{34} formation could be reconstituted in vitro by simply incubating the tRNA substrate with two recombinant proteins, iscS and mnmA (Ikeuchi et al. 2006; Lauhon et al. 2004; Nakai et al. 2004). iscS is the universal desulfurase in organisms from all domains of life and partakes in transferring a sulfur group from cysteine to a partner protein. Depending on the partner, the sulfur may be taken into different metabolic routes, namely iron–sulfur cluster assembly or tRNA thiolation. In the specific case of s^2U_{34}, mnmA takes the sulfur from iscS and catalyzes its incorporation into tRNAs. Early, however, it was appreciated that in vitro s^2U_{34} incorporation into tRNA with only these two proteins was extremely inefficient and the levels of thiolated tRNA formed were by no means representative of the in vivo situation. This of course suggested the necessity for some missing factor(s) to efficiently catalyze the reaction and therefore explain the differences between the in vivo and in vitro reactions. The use of a bioinformatics approach then led to the identification of the missing components for bacterial s^2U synthesis. In bacteria, s^2U_{34} formation requires a sulfur relay system that starts with the removal of sulfur from cysteine and its transfer to a series of relatively small proteins (ranging in sizes from ~8 to 14 kDa), starting with TusA which then transfers the sulfur to TusD with the help of two additional factors (TusB and C); this is followed by transfer to TusE and finally to MnmA which ultimately deposits the sulfur group in tRNA (Fig. 14.4) (Ikeuchi et al. 2006). Importantly, the Tus proteins form a multi-protein complex that, in conjunction with iscS and mnmA, mediates efficient s^2U formation (Ikeuchi et al. 2006).

In eukarya, the chemistry is similar in principle, but the components are clearly different. Due to intracellular compartmentalization, two separate locales for s^2U formation exist in eukarya: (1) the cytoplasm, for tRNAs used in the translation of nucleus-encoded mRNAs, and (2) the mitochondria, where s^2U_{34} formation should presumably resemble the bacterial sulfur-relay system and thiolation is exclusively dedicated to tRNAs used in mitochondrial translation. Both cytoplasm and mitochondria share a common need for the desulfurase Nisf1 (the eukaryotic homolog of bacterial iscS) to initiate the sulfur transfer reaction (Shi et al. 2010). However, the cytoplasmic thiolation pathway differs from that in mitochondria (and bacteria) in the factors that mediate sulfur transfer from Nisf1 to the tRNA. A

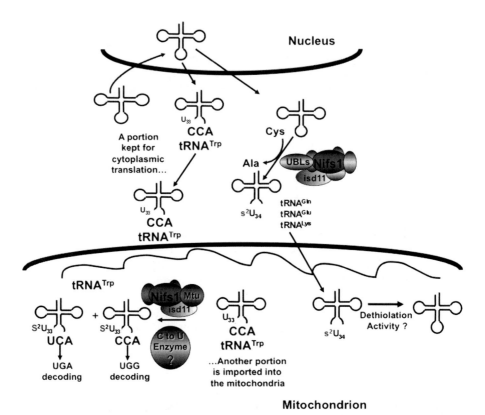

Fig. 14.4 Two separate pathways for tRNA thiolation in *T. brucei*. The cytosolic pathway is important for the stability of U34-containing tRNAGln, tRNAGlu and tRNALys. The cytosolic thiolation complex (UBLs/Nifs1/isd11) is unable to thiolate tRNATrp. This tRNA undergoes C to U editing following mitochondrial import, where thiolation act as a negative determinant for editing. In addition, cytosolically thiolated tRNAs become dethiolated by yet to be discovered activity following mitochondrial import

sulfur-relay-type mechanism was recently discovered in yeast (Leidel et al. 2009; Schlieker et al. 2008; Shigi et al. 2008). Interestingly, this sulfur-relay system has its own set of eukaryotic factors unrelated to the bacterial TusA-E proteins. Thus in the cytoplasm, Nisf1 (the eukaryotic homolog of bacterial iscS) takes the sulfur group from cysteine but then donates it to a series of ubiquitin-like proteins (UBLs) starting with Uba4 which then donates it to Urm1 and further down the cascade to Ncs6/Ncs2. These proteins together, in an analogous manner to the Tus proteins, transfer the sulfur to the tRNA. The route to tRNA thiolation in mitochondria is less clear and at least Nisf1 and Mtu1 (the eukaryotic homolog of mnmA) are required for activity in vivo (Leidel et al. 2009; Noma et al. 2009). Genomic database searches do reveal that no recognizable Tus homologs exist in eukarya and suggest that mitochondria may get by with fewer thiolation components than the rest of the cell. Interestingly, our laboratory in collaboration with the Lukes group have shown that in the early divergent eukaryote *Trypanosoma brucei*, Nisf is undetectable in the cytoplasm and only found (by Western blot analysis) in the mitochondria (Wohlgamuth-Benedum et al. 2009). Currently, it is not clear how an enzyme that is quantitatively imported into mitochondria may mediate thiolation of cytoplasmic tRNAs, but in support of our arguments stated here, it suggests that what may seem like sub-stoichiometric quantities of Nisf in the *T. brucei* cytoplasm may exhibit very high specific activity in vivo with the aid of as yet unidentified factor(s). These may include, but are not limited to, the trypanosome homologs of the yeast Urm1, Uba4, Ncs6, and Ncs2 proteins. Along these lines, we recently discover that in the trypanosome system, isd11, a small protein shown to play a key function in iron–sulfur assembly, is also essential for both cytoplasmic and mitochondrial tRNA thiolation (Paris et al. 2010). This last point once again highlights the fact that through evolution genome dynamics have force organisms to what appear to be evolutionary-domain-specific solutions to identical problems. Such is the case of the Tus proteins in bacteria and Urm–Uba–Ncs system in the cytoplasm of eukarya. Furthermore, in the mitochondria the use of isd11 (a uniquely eukaryotic protein) then emphasizes that at the end of the day the "what ever works" approach has played heavily into the development of tRNA modification systems in organelles where pre-existing activities have been recruited to target new substrate as cellular demands change.

14.2.2.4   Interesting Connection Between Cytoplasmic Thiolation, Mitochondrial Editing, and Iron–Sulfur Cluster Assembly in Trypanosomes

As mentioned above in *T. brucei*, like in most eukaryotes, there are two places where tRNAs can be thiolated: the cytoplasm and the mitochondria (Fig. 14.4). Cytoplasmic thiolation seems to require the same components as in yeast, but the specific contributions of factors such as Urm1, Uba, Ncs, and Ncs have not been formally tested. These are identifiable by genomic searches and are expected to provide similar functions as in the yeast system. Likewise, the use of isd11 has been proposed for other systems, but thus far has only been demonstrated in *T. brucei*. The differentiating feature between trypanosomatids (*T. brucei*, *Leishmania*, etc.) and other eukaryotes in the thiolation systems is the nature of the tRNAs used for mitochondrial thiolation. In the *T. brucei* cytoplasm, $tRNA_{UUG}^{Gln}$, $_{UUC}^{-Glu}$, and $_{UUU}^{-Lys}$ are the only known targets for thiolation, but because of tRNA import, these tRNAs go into the mitochondria already containing the thiol group added by the cytoplasmic thiolation system. So far, the only tRNA known to undergo mitochondrial thiolation in these organisms is $tRNA^{Trp}$, which exists in kinetoplastid mitochondria in two forms: 50% of the tRNAs having CCA anticodon and the remaining 50% UCA (Alfonzo et al. 1999) (Fig. 14.4). A tRNA with a UCA anticodon is not encoded in either the nuclear or mitochondrial genome but is formed by tRNA editing (Alfonzo et al. 1999). Although the editing enzyme is still at large, this reaction presumably occurs by deamination in line with C to U changes observed in other systems. We hypothesized that because of the presence of thiolation at an unusual position (U_{33}) (Crain et al. 2002), the two versions of tRNA^{Trp} are strictly dedicated to UGG and UGA decoding, respectively (Fig. 14.4), suggesting that in fact the edited tRNA could not wobble with the G at the third position of UGG codons. Surprisingly, in our studies of thiolation, we found that downregulation by RNA interference of any of the mitochondrial thiolation factors (including Nisf) led to upregulation of tRNA editing to almost 100% (Bruske et al. 2009; Wohlgamuth-Benedum et al. 2009). This observation implies that s^2U_{33} acts as a negative determinant for tRNA editing and helps maintain the levels of the two isoacceptors as required for UGG and UGA decoding. Notably, tRNA^{Trp} is not thiolated in the cytoplasm in transit. This raised the question of how this tRNA avoids cytoplasmic thiolation. We showed that editing is not required for thiolation at U_{33} in *L. tarentolae*, a close relative of *T. brucei* (Crain et al. 2002). Therefore, the only viable explanation is that the cytoplasmic and mitochondrial tRNA thiolation systems differ in their substrate recognition and that in fact there are features common to tRNA$^{Gln, -Glu}$, and $^{-Lys}$ required for thiolation that are not present in tRNA^{Trp}. Recently, it was shown that following import tRNA$^{Gln, -Glu}$, and $^{-Lys}$ become de-thiolated by a yet-unidentified activity raising the possibility that the mitochondrial thiolation may play a "repair" role for this tRNA set, but again this has not been formally tested (Bruske et al. 2009).

An additionally surprising discovery with the *T. brucei* system involves the fate of the cytoplasmic tRNAs in the absence of thiolation. We showed that tRNA$^{Gln, -Glu}$, and $^{-Lys}$ become unstable and quickly degraded if thiolation is impaired (Wohlgamuth-Benedum et al. 2009). This instability was specific only to the thiolated tRNA set and, in this respect, is different from a more general rapid tRNA degradation pathway described (Alexandrov et al. 2006; Engelke and Hopper 2006). The nature of the enzymes or factors mediating this degradation is however currently unknown. Overall, the thiolation system of *T. brucei* shows not only how intracellular compartmentalization affects tRNA modification but it even exemplifies how location may affect modification enzyme substrate specificity.

A curious corollary of the thiolation story is the remarkable finding that the same desulfurase required for iron–sulfur assembly is also required for tRNA thiolation in both cytoplasm and mitochondria. In the mitochondria, subunits of respiratory complex III require a FeS cluster; therefore, downregulation of Nisf could lead to downregulation of respiratory rates. We suggest a model by which the divergence of the two pathways (FeS assembly and tRNA editing/thiolation) from a common key enzyme may be exploited by these cells to carefully match respiratory rates to mitochondrial translation, perhaps by offsetting the 50/50 ratio for edited and unedited tRNA^{Trp}. It is also worth mentioning that in fact a number of cytoplasmic modification enzymes also require FeS clusters for activity; thus, this hypothesis may even include cytoplasmic modification systems in connection with FeS-cluster assembly for global metabolic regulation (Lill and Muhlenhoff 2006). These proposals are of course largely speculative but their exploration may reveal important aspects of a higher order in the coordination of these aspects of cellular metabolism.

14.3   Concluding Remarks: The Full Impact of Intracellular Compartmentalization on tRNA Function

In the previous sections, we tried to address the issue of specificity by listing some, by no means exhaustive, examples of various factors that may affect modification activity. These include "missing" protein factor(s) as well as modifications themselves. Eukaryotic systems, because of their level of intracellular compartmentalization, pose a real challenge to the use of in vitro data to establish in vivo specificity. Operationally for this argument, we have divided the eukaryotic cell into three compartments in which tRNA modifications can take place: the nucleus, cytoplasm, and the mitochondria. We can think of at least two factors that may then affect the modification content of a given tRNA: the localization of modification enzymes and the localization of the tRNA. Some modification enzymes are imported into the nucleus following their synthesis in the cytoplasm, some are strictly cytoplasmic, yet others are imported into the mitochondria. What this then creates is a situation where a particular enzyme, because of its intracellular localization, may never encounter a particular substrate. For example, the tRNA C to U editing described above takes place in mitochondria and as far as we know only affects tRNA^{Trp}. This raises the possibility that, when studying such reactions, one may observe efficient editing or modification activity in vitro with synthetic tRNA substrates, but in vivo the observed activity is in fact irrelevant given that the substrate simply never comes in contact with the given modification enzyme. These examples thus represent very simple scenarios to highlight the point that in fact intracellular localization in eukarya plays a major role in the decision making of who gets modified.

Perhaps, a more complicated and understudied effect is that of how the rates of transport across membranes may affect modification levels. We know that tRNAs are not only exported from the nucleus after transcription but they are also actively imported into the mitochondria from the cytoplasm in eukarya, including yeast, protozoans, mammals, plants, etc. In cases where the rate of transport exceeds the rate of catalysis by a particular enzyme, populations of differentially modified tRNAs must exist in cells. By "populations," we not only refer here to a particular modification in different tRNAs, but more importantly to multiple modifications within a single tRNA. Given the widely accepted view that modifications are great modulators of RNA structure and thus function, it is then entirely possible that changing modification contents among tRNA populations may in fact have a greater impact on protein synthesis than previously appreciated. In support of this argument, it has been shown that under steady-state conditions of growth, tRNAs are not fully modified and their modification content does vary with growth conditions in bacteria. Toss into the formula the proposed effect on substrate availability created by intracellular partitioning and indeed one can easily envision scenarios where even changing environmental conditions can impact tRNA modification and function. Indeed, recent work has shown that at least in yeast not only can tRNAs travel to the cytoplasm but they can also undergo retrograde transport into the nucleus prompted by certain conditions of starvation (Shaheen and Hopper 2005; Takano et al. 2005). Thus in this particular case even a tRNA that had earlier escape nuclear modification due to a fast rate of export to the cytoplasm can then be further modified following retrograde transport. Likewise, although not proven yet, it is within the realm of possibilities that even imported tRNAs may be exported from the mitochondria under conditions of stress.

References

Agris PF (2004) Decoding the genome: a modified view. Nucleic Acids Res 32:223–238

Agris PF, Vendeix FA, Graham WD (2007) tRNA's wobble decoding of the genome: 40 years of modification. J Mol Biol 366:1–13

Alexandrov A, Chernyakov I, Gu W, Hiley SL, Hughes TR, Grayhack EJ, Phizicky EM (2006) Rapid tRNA decay can result from lack of nonessential modifications. Mol Cell 21:87–96

Alfonzo JD, Soll D (2009) Mitochondrial tRNA import – the challenge to understand has just begun. Biol Chem 390:717–722

Alfonzo JD, Blanc V, Estevez AM, Rubio MA, Simpson L (1999) C to U editing of the anticodon of imported mitochondrial tRNA(Trp) allows decoding of the UGA stop codon in *Leishmania tarentolae*. EMBO J 18:7056–7062

Auxilien S, El Khadali F, Rasmussen A, Douthwaite S, Grosjean H (2007) Archease from *Pyrococcus abyssi* improves substrate specificity and solubility of a tRNA m5C methyltransferase. J Biol Chem 282:18711–18721

Bishop AC, Xu J, Johnson RC, Schimmel P, de Crecy-Lagard V (2002) Identification of the tRNA-dihydrouridine synthase family. J Biol Chem 277:25090–25095

Borner GV, Zeviani M, Tiranti V, Carrara F, Hoffmann S, Gerbitz KD, Lochmuller H, Pongratz D, Klopstock T, Melberg A et al (2000) Decreased aminoacylation of mutant tRNAs in MELAS but not in MERRF patients. Hum Mol Genet 9:467–475

Bouzaidi-Tiali N, Aeby E, Charriere F, Pusnik M, Schneider A (2007) Elongation factor 1a mediates the specificity of mitochondrial tRNA import in *T. brucei*. EMBO J 26:4302–4312

Brandina I, Graham J, Lemaitre-Guillier C, Entelis N, Krasheninnikov I, Sweetlove L, Tarassov I, Martin RP (2006) Enolase takes part in a macromolecular complex associated to mitochondria in yeast. Biochim Biophys Acta 1757:1217–1228

Brandina I, Smirnov A, Kolesnikova O, Entelis N, Krasheninnikov IA, Martin RP, Tarassov I (2007) tRNA import into yeast mitochondria is regulated by the ubiquitin-proteasome system. FEBS Lett 581:4248–4254

Bruske EI, Sendfeld F, Schneider A (2009) Thiolated tRNAs of *Trypanosoma brucei* are imported into mitochondria and dethiolated after import. J Biol Chem 284:36491–36499

Chernyakov I, Whipple JM, Kotelawala L, Grayhack EJ, Phizicky EM (2008) Degradation of several hypomodified mature tRNA species in *Saccharomyces cerevisiae* is mediated by Met22 and the 5′-3′ exonucleases Rat1 and Xrn1. Genes Dev 22:1369–1380

Chiu N, Chiu A, Suyama Y (1975) Native and imported transfer RNA in mitochondria. J Mol Biol 99:37–50

Colonna A, Kerr SJ (1980) The nucleus as the site of tRNA methylation. J Cell Physiol 103:29–33

Crain PF, McCloskey JA (1998) Applications of mass spectrometry to the characterization of oligonucleotides and nucleic acids. Curr Opin Biotechnol 9:25–34

Crain PF, Alfonzo JD, Rozenski J, Kapushoc ST, McCloskey JA, Simpson L (2002) Modification of the universally unmodified uridine-33 in a mitochondria-imported edited tRNA and the role of the anticodon arm structure on editing efficiency. RNA 8:752–761

Crick FH (1966) Codon–anticodon pairing: the wobble hypothesis. J Mol Biol 19:548–555

Czerwoniec A, Dunin-Horkawicz S, Purta E, Kaminska KH, Kasprzak JM, Bujnicki JM, Grosjean H, Rother K (2009) MODOMICS: a database of RNA modification pathways. 2008 update. Nucleic Acids Res 37:D118–D121

Dihanich ME, Najarian D, Clark R, Gillman EC, Martin NC, Hopper AK (1987) Isolation and characterization of MOD5, a gene required for isopentenylation of cytoplasmic and mitochondrial tRNAs of *Saccharomyces cerevisiae*. Mol Cell Biol 7:177–184

Dorner M, Altmann M, Paabo S, Morl M (2001) Evidence for import of a lysyl-tRNA into marsupial mitochondria. Mol Biol Cell 12:2688–2698

Dunin-Horkawicz S, Czerwoniec A, Gajda MJ, Feder M, Grosjean H, Bujnicki JM (2006) MODOMICS: a database of RNA modification pathways. Nucleic Acids Res 34:D145–D149

Engelke DR, Hopper AK (2006) Modified view of tRNA: stability amid sequence diversity. Mol Cell 21:144–145

Enriquez JA, Chomyn A, Attardi G (1995) MtDNA mutation in MERRF syndrome causes defective aminoacylation of tRNA(Lys) and premature translation termination. Nat Genet 10:47–55

Entelis N, Brandina I, Kamenski P, Krasheninnikov IA, Martin RP, Tarassov I (2006) A glycolytic enzyme, enolase, is recruited as a cofactor of tRNA targeting toward mitochondria in *Saccharomyces cerevisiae*. Genes Dev 20:1609–1620

Esseiva AC, Naguleswaran A, Hemphill A, Schneider A (2004) Mitochondrial tRNA import in *Toxoplasma gondii*. J Biol Chem 279:42363–42368

Frechin M, Senger B, Braye M, Kern D, Martin RP, Becker HD (2009) Yeast mitochondrial Gln-tRNA(Gln) is generated by a GatFAB-mediated transamidation pathway involving Arc1p-controlled subcellular sorting of cytosolic GluRS. Genes Dev 23:1119–1130

Fukuhara N, Tokiguchi S, Shirakawa K, Tsubaki T (1980) Myoclonus epilepsy associated with ragged-red fibres (mitochondrial abnormalities): disease entity or a syndrome? Light-and electron-microscopic studies of two cases and review of literature. J Neurol Sci 47:117–133

Glover KE, Spencer DF, Gray MW (2001) Identification and structural characterization of nucleus-encoded transfer RNAs imported into wheat mitochondria. J Biol Chem 276:639–648

Goswami S, Dhar G, Mukherjee S, Mahata B, Chatterjee S, Home P, Adhya S (2006) A bifunctional tRNA import receptor from *Leishmania* mitochondria. Proc Natl Acad Sci USA 103:8354–8359

Hancock K, LeBlanc AJ, Donze D, Hajduk SL (1992) Identification of nuclear encoded precursor tRNAs within the mitochondrion of *Trypanosoma brucei*. J Biol Chem 267:23963–23971

Hopper AK, Phizicky EM (2003) tRNA transfers to the limelight. Genes Dev 17:162–180

Hopper AK, Furukawa AH, Pham HD, Martin NC (1982) Defects in modification of cytoplasmic and mitochondrial transfer RNAs are caused by single nuclear mutations. Cell 28:543–550

Horowitz S, Gorovsky MA (1985) An unusual genetic code in nuclear genes of *Tetrahymena*. Proc Natl Acad Sci USA 82:2452–2455

Ikeuchi Y, Shigi N, Kato J, Nishimura A, Suzuki T (2006) Mechanistic insights into sulfur relay by multiple sulfur mediators involved in thiouridine biosynthesis at tRNA wobble positions. Mol Cell 21:97–108

Johnston SA, Anziano PQ, Shark K, Sanford JC, Butow RA (1988) Mitochondrial transformation in yeast by bombardment with microprojectiles. Science 240:1538–1541

Jukes TH, Osawa S (1990) The genetic code in mitochondria and chloroplasts. Experientia 46:1117–1126

Jukes TH, Osawa S (1993) Evolutionary changes in the genetic code. Comp Biochem Physiol B 106:489–494

Kamenski P, Kolesnikova O, Jubenot V, Entelis N, Krasheninnikov IA, Martin RP, Tarassov I (2007) Evidence for an adaptation mechanism of mitochondrial translation via tRNA import from the cytosol. Mol Cell 26:625–637

Kirino Y, Yasukawa T, Ohta S, Akira S, Ishihara K, Watanabe K, Suzuki T (2004) Codon-specific translational defect caused by a wobble modification deficiency in mutant tRNA from a human mitochondrial disease. Proc Natl Acad Sci USA 101:15070–15075

Kirino Y, Yasukawa T, Marjavaara SK, Jacobs HT, Holt IJ, Watanabe K, Suzuki T (2006) Acquisition of the wobble modification in mitochondrial tRNALeu(CUN) bearing the G12300A mutation suppresses the MELAS molecular defect. Hum Mol Genet 15:897–904

Kobayashi Y, Momoi MY, Tominaga K, Shimoizumi H, Nihei K, Yanagisawa M, Kagawa Y, Ohta S (1991) Respiration-deficient cells are caused by a single point mutation in the mitochondrial tRNA-Leu (UUR) gene in mitochondrial myopathy, encephalopathy, lactic acidosis, and strokelike episodes (MELAS). Am J Hum Genet 49:590–599

Kolesnikova OA, Entelis NS, Jacquin-Becker C, Goltzene F, Chrzanowska-Lightowlers ZM, Lightowlers RN, Martin RP, Tarassov I (2004) Nuclear DNA-encoded tRNAs targeted into mitochondria can rescue a mitochondrial DNA mutation associated with the MERRF syndrome in cultured human cells. Hum Mol Genet 13:2519–2534

Lauhon CT, Skovran E, Urbina HD, Downs DM, Vickery LE (2004) Substitutions in an active site loop of Escherichia coli IscS result in specific defects in Fe-S cluster and thionucleoside biosynthesis in vivo. J Biol Chem 279:19551–19558

Leidel S, Pedrioli PG, Bucher T, Brost R, Costanzo M, Schmidt A, Aebersold R, Boone C, Hofmann K, Peter M (2009) Ubiquitin-related modifier Urm1 acts as a sulphur carrier in thiolation of eukaryotic transfer RNA. Nature 458:228–232

Lill R, Muhlenhoff U (2006) Iron-sulfur protein biogenesis in eukaryotes: components and mechanisms. Annu Rev Cell Dev Biol 22:457–486

Limbach PA, Crain PF, McCloskey JA (1995) Characterization of oligonucleotides and nucleic acids by mass spectrometry. Curr Opin Biotechnol 6:96–102

Magalhaes PJ, Andreu AL, Schon EA (1998) Evidence for the presence of 5S rRNA in mammalian mitochondria. Mol Biol Cell 9:2375–2382

Mahapatra S, Ghosh T, Adhya S (1994) Import of small RNAs into Leishmania mitochondria in vitro. Nucleic Acids Res 22:3381–3386

Marechal-Drouard L, Weil JH, Guillemaut P (1988) Import of several tRNAs from the cytoplasm into the mitochondria in bean Phaseolus vulgaris. Nucleic Acids Res 16:4777–4788

Martin NC, Hopper AK (1994) How single genes provide tRNA processing enzymes to mitochondria, nuclei and the cytosol. Biochimie 76:1161–1167

Martin RP, Schneller JM, Stahl AJ, Dirheimer G (1979) Import of nuclear deoxyribonucleic acid coded lysine-accepting transfer ribonucleic acid (anticodon C-U-U) into yeast mitochondria. Biochemistry 18:4600–4605

Matsuyama S, Ueda T, Crain PF, McCloskey JA, Watanabe K (1998) A novel wobble rule found in starfish mitochondria. Presence of 7-methylguanosine at the anticodon wobble position expands decoding capability of tRNA. J Biol Chem 273:3363–3368

Mirande M (2007) The ins and outs of tRNA transport. EMBO Rep 8:547–549

Morl M, Dorner M, Paabo S (1995) C to U editing and modifications during the maturation of the mitochondrial tRNA(Asp) in marsupials. Nucleic Acids Res 23:3380–3384

Mottram JC, Bell SD, Nelson RG, Barry JD (1991) tRNAs of Trypanosoma brucei. Unusual gene organization and mitochondrial importation. J Biol Chem 266:18313–18317

Mukherjee S, Basu S, Home P, Dhar G, Adhya S (2007) Necessary and sufficient factors for the import of transfer RNA into the kinetoplast mitochondrion. EMBO Rep 8:589–595

Nagao A, Suzuki T, Katoh T, Sakaguchi Y (2009) Biogenesis of glutaminyl-mt tRNAGln in human mitochondria. Proc Natl Acad Sci USA 106:16209–16214

Nakai Y, Umeda N, Suzuki T, Nakai M, Hayashi H, Watanabe K, Kagamiyama H (2004) Yeast Nfs1p is involved in thio-modification of both mitochondrial and cytoplasmic tRNAs. J Biol Chem 279:12363–12368

Nishikura K, De Robertis EM (1981) RNA processing in microinjected Xenopus oocytes. Sequential addition of base modifications in the spliced transfer RNA. J Mol Biol 145:405–420

Noma A, Sakaguchi Y, Suzuki T (2009) Mechanistic characterization of the sulfur-relay system for eukaryotic 2-thiouridine biogenesis at tRNA wobble positions. Nucleic Acids Res 37:1335–1352

Osawa S, Jukes TH (1988) Evolution of the genetic code as affected by anticodon content. Trends Genet 4:191–198

Osawa S, Jukes TH (1989) Codon reassignment (codon capture) in evolution. J Mol Evol 28:271–278

Paris Z, Rubio MA, Lukes J, Alfonzo JD (2009) Mitochondrial tRNA import in *Trypanosoma brucei* is independent of thiolation and the Rieske protein. RNA 15:1398–1406

Paris Z, Changmai P, Rubio MA, Zikova A, Stuart KD, Alfonzo JD, Lukes J (2010) The Fe/S cluster assembly protein Isd11 is essential for tRNA thiolation in *Trypanosoma brucei*. J Biol Chem 285:22394–22402

Phizicky EM (2005) Have tRNA, will travel. Proc Natl Acad Sci USA 102:11127–11128

Pusnik M, Charriere F, Maser P, Waller RF, Dagley MJ, Lithgow T, Schneider A (2009) The single mitochondrial porin of *Trypanosoma brucei* is the main metabolite transporter in the outer mitochondrial membrane. Mol Biol Evol 26:671–680

Putz J, Dupuis B, Sissler M, Florentz C (2007) Mamit-tRNA, a database of mammalian mitochondrial tRNA primary and secondary structures. RNA 13:1184–1190

Randau L, Soll D (2008) Transfer RNA genes in pieces. EMBO Rep 9:623–628

Remacle C, Cardol P, Coosemans N, Gaisne M, Bonnefoy N (2006) High-efficiency biolistic transformation of *Chlamydomonas* mitochondria can be used to insert mutations in complex I genes. Proc Natl Acad Sci USA 103:4771–4776

Rinehart J, Krett B, Rubio MA, Alfonzo JD, Soll D (2005) *Saccharomyces cerevisiae* imports the cytosolic pathway for Gln-tRNA synthesis into the mitochondrion. Genes Dev 19:583–592

Rozenski J, Crain PF, McCloskey JA (1999) The RNA modification database: 1999 update. Nucleic Acids Res 27:196–197

Rubio MA, Liu X, Yuzawa H, Alfonzo JD, Simpson L (2000) Selective importation of RNA into isolated mitochondria from *Leishmania tarentolae*. RNA 6:988–1003

Rubio MA, Rinehart JJ, Krett B, Duvezin-Caubet S, Reichert AS, Soll D, Alfonzo JD (2008) Mammalian mitochondria have the innate ability to import tRNAs by a mechanism distinct from protein import. Proc Natl Acad Sci USA 105:9186–9191

Rusconi CP, Cech TR (1996a) Mitochondrial import of only one of three nuclear-encoded glutamine tRNAs in *Tetrahymena thermophila*. EMBO J 15:3286–3295

Rusconi CP, Cech TR (1996b) The anticodon is the signal sequence for mitochondrial import of glutamine tRNA in *Tetrahymena*. Genes Dev 10:2870–2880

Salinas T, Duchene AM, Delage L, Nilsson S, Glaser E, Zaepfel M, Marechal-Drouard L (2006) The voltage-dependent anion channel, a major component of the tRNA import machinery in plant mitochondria. Proc Natl Acad Sci USA 103:18362–18367

Salinas T, Duchene AM, Marechal-Drouard L (2008) Recent advances in tRNA mitochondrial import. Trends Biochem Sci 33:320–329

Schlieker CD, Van der Veen AG, Damon JR, Spooner E, Ploegh HL (2008) A functional proteomics approach links the ubiquitin-related modifier Urm1 to a tRNA modification pathway. Proc Natl Acad Sci USA 105:18255–18260

Schneider A, Marechal-Drouard L (2000) Mitochondrial tRNA import: are there distinct mechanisms? Trends Cell Biol 10:509–513

Schon A, Kannangara CG, Gough S, Soll D (1988) Protein biosynthesis in organelles requires misaminoacylation of tRNA. Nature 331:187–190

Shaheen HH, Hopper AK (2005) Retrograde movement of tRNAs from the cytoplasm to the nucleus in *Saccharomyces cerevisiae*. Proc Natl Acad Sci USA 102:11290–11295

Shi R, Proteau A, Villarroya M, Moukadiri I, Zhang L, Trempe JF, Matte A, Armengod ME, Cygler M (2010) Structural basis for Fe–S cluster assembly and tRNA thiolation mediated by IscS protein-protein interactions. PLoS Biol 8:e1000354

Shigi N, Sakaguchi Y, Asai S, Suzuki T, Watanabe K (2008) Common thiolation mechanism in the biosynthesis of tRNA thiouridine and sulphur-containing cofactors. EMBO J 27:3267–3278

Simpson AM, Suyama Y, Dewes H, Campbell DA, Simpson L (1989) Kinetoplastid mitochondria contain functional tRNAs which are encoded in nuclear DNA and also contain small minicircle and maxicircle transcripts of unknown function. Nucleic Acids Res 17:5427–5445

Suyama Y (1967) The origins of mitochondrial ribonucleic acids in *Tetrahymena pyriformis*. Biochemistry 6:2829–2839

Suzuki T (2007) Chaplet column chromatography: isolation of a large set of individual RNAs in a single step. Methods Enzymol 425:231–239

Takano A, Endo T, Yoshihisa T (2005) tRNA actively shuttles between the nucleus and cytosol in yeast. Science 309:140–142

Takemoto C, Spremulli LL, Benkowski LA, Ueda T, Yokogawa T, Watanabe K (2009) Unconventional decoding of the AUA codon as methionine by mitochondrial tRNAMet with the anticodon f5CAU as revealed with a mitochondrial in vitro translation system. Nucleic Acids Res 37:1616–1627

Tarassov IA, Martin RP (1996) Mechanisms of tRNA import into yeast mitochondria: an overview. Biochimie 78:502–510

Tarassov I, Entelis N, Martin RP (1995) An intact protein translocating machinery is required for mitochondrial import of a yeast cytoplasmic tRNA. J Mol Biol 245:315–323

Wohlgamuth-Benedum JM, Rubio MA, Paris Z, Long S, Poliak P, Lukes J, Alfonzo JD (2009) Thiolation controls cytoplasmic tRNA stability and acts as a negative determinant for tRNA editing in mitochondria. J Biol Chem 284:23947–23953

Wu J, Bu W, Sheppard K, Kitabatake M, Kwon ST, Soll D, Smith JL (2009) Insights into tRNA-dependent amidotransferase evolution and catalysis from the structure of the *Aquifex aeolicus* enzyme. J Mol Biol 391:703–716

Yasukawa T, Suzuki T, Ueda T, Ohta S, Watanabe K (2000) Modification defect at anticodon wobble nucleotide of mitochondrial tRNAs(Leu)(UUR) with pathogenic mutations of mitochondrial myopathy, encephalopathy, lactic acidosis, and stroke-like episodes. J Biol Chem 275:4251–4257

Yasukawa T, Suzuki T, Ishii N, Ohta S, Watanabe K (2001) Wobble modification defect in tRNA disturbs codon-anticodon interaction in a mitochondrial disease. EMBO J 20:4794–4802

Yermovsky-Kammerer AE, Hajduk SL (1999) In vitro import of a nuclearly encoded tRNA into the mitochondrion of *Trypanosoma brucei*. Mol Cell Biol 19:6253–6259

Chapter 15
Why Do Plants Edit RNA in Plant Organelles?

Toshiharu Shikanai

15.1 Introduction

RNA editing is a process of modifying the genetic information on RNA molecules (Shikanai 2006). U insertion/deletion-type RNA editing has been extensively studied in the kinetoplastid mitochondria of trypanosomes (Stuart et al. 2005). RNA editing, which converts a specific nucleotide, is likely to have an evolutionally distinct origin from the U insertion/deletion-type RNA editing, and its machinery has been well characterized in mammalian cells (Keegan et al. 2001; Wedekind et al. 2003). As in mammals, RNA editing converts C into U residues in land plants, and this was initially discovered in mitochondria (Covello and Gray 1989; Gualberto et al. 1989; Hiesel et al. 1989) and later in plastids (Hoch et al. 1991). In mitochondria (and probably in plastids), the reaction is a deamination or trans-amination of a specific C residue rather than a nucleotide substitution (Rajasekhar and Mulligan 1993; Blanc et al. 1995; Yu and Schuster 1995), but the enzyme involved in the reaction is still unclear. In addition to this C-to-U conversion, U-to-C conversion frequently occurs in *Adiantum capillus-veneris* (fern) and *Anthoceros formosae* (hornwort) (Kugita et al. 2003; Wolf et al. 2004). In *Arabidopsis*, 34 and 441 sites are edited in plastids (Table 15.1, Chateigner-Boutin and Small 2007) and mitochondria, respectively (Giegé and Brennicke 1999).

RNA editing challenges the central dogma and has attracted the attention of many plant molecular biologists, who have aimed to clarify its molecular mechanism. Despite the similarity of the reaction, RNA editing machinery in plants is not very similar to that in mammals, and, at the very least, the site-recognition factor is unique to plants. It was recently shown that a PPR (pentatricopeptide repeat) protein is involved in site recognition (Kotera et al. 2005). This discovery was followed by many reports on PPR proteins involved in RNA editing in both plastids

T. Shikanai (✉)
Faculty of Science, Department of Botany, Kyoto University, Kyoto 606-8502 Japan
e-mail: shikanai@pmg.bot.kyoto-u.ac.jp

C.E. Bullerwell (ed.), *Organelle Genetics*,
DOI 10.1007/978-3-642-22380-8_15, © Springer-Verlag Berlin Heidelberg 2012

and mitochondria (summarized in Hammani et al. 2009). These reports have facilitated discussion on the link between site-recognition factors and their target sites at the genome level. In addition to the mystery of an editing enzyme, we were repeating the same question as to why plants edit so many sites. What is the physiological function of RNA editing? Even with the recent progress in the field, these questions are still unknown. I discuss my answers to these questions in this review.

15.2   Discovery of the *cis*-Element

Pioneering studies on RNA editing in plant organelles have asked how the RNA editing machinery recognizes the target C. There are numerous C residues in transcripts, and recognition of these residues should be precise and efficient so as not to express a mutant protein. The plastid transformation technique facilitated the introduction into plastids of foreign gene-encoding RNA with editing sites. This in vivo analysis revealed 22 nucleotides, 16 nucleotides upstream and five nucleotides downstream of the target C residue in the *psbL* transcript of tobacco plastids as regions required and sufficient for precise RNA editing (Chaudhuri and Maliga 1996). A similar analysis was reported in ndhB-6 and ndhB-7 sites in tobacco plastids (Bock et al. 1996). This idea was also confirmed by a complementary approach using the in vitro RNA editing system, in which an assay with competitor RNA is possible (Hirose and Sugiura 2001). Short sequences surrounding the target C residue are recognized by RNA editing machinery; these sequences are called *cis*-elements. The similar mechanism is also likely involved in RNA editing in mitochondria (Farré et al. 2001; Takenaka et al. 2004).

When the target C is in the protein-coding region, *cis*-elements also carry the information for the encoding peptide sequences. Consequently, *cis*-elements are not highly conserved among RNA editing sites, and, theoretically, each RNA editing site is independently recognized by a distinct site-specific *trans*-factor. This idea was supported by the observation that the *cis*-element of the psbL site competes with endogenous transcripts for a rate-limiting *trans*-factor but does not compete with the endogenous ndhD-1 site that also creates the translational initiation codon (Chaudhuri and Maliga 1996). At present, we have to slightly modify this original idea based on the observation that the over-expression of transgenes carrying the RNA editing sites causes cross-competition with the endogenous RNA editing events (Chateigner-Boutin and Hanson 2002). This study reported the cross-competition among the ndhF-2, ndhB-3, and ndhD-1 sites. The sequences surrounding the editing sites are weakly conserved. However, it became clear later that the ndhF-2 site is recognized by OTP84 (Hammani et al. 2009), and the ndhD-1 site is recognized by CRR4 (Kotera et al. 2005) in *Arabidopsis* (Table 15.1). RNA editing of the ndhB-3 site does not require OTP84 or CRR4. Instead, OTP84 is required for the RNA editing of the ndhB-10 and psbZ sites (Hammani et al. 2009). The story may be more complicated than first imagined, and over-accumulation of endogenous RNA in

Table 15.1 RNA editing sites in Arabidopsis plastids

Locus[a]	AA change	*trans*-Factor	Remarks	References
ndhB-1 (50)	S > L			
ndhB-2 (156)	P > L	CRR28 (DYW)		Okuda et al. (2009)
ndhB-3 (196)	H > Y			
ndhB-7 (249)	S > F	CRR22 (DYW)		Okuda et al. (2009)
ndhB-8 (277)	S > L			
ndhB-9 (279)	S > L	OTP82 (DYW)	Partial[b]/silent[c]	Okuda et al. (2010)
ndhB-11 (291)	S > L			
ndhB-12 (419)	S > L			
ndhB-10 (494)	P > L	OTP84 (DYW)		Hammani et al. (2009)
ndhD-1 (1)	T > first M	CRR4 (E)	Partial	Kotera et al. (2005)
ndhD-2 (128)	S > L	CRR21 (E)		Okuda et al. (2007)
ndhD-4 (225)	S > L	OTP85 (DYW)	Silent	
ndhD-3 (293)	S > L	CRR28 (DYW)		Okuda et al. (2009)
ndhD-5 (296)	P > L	CRR22 (DYW)		Okuda et al. (2009)
ndhF-2 (97)	S > L	OTP84 (DYW)		Hammani et al. (2009)
		VAC1 (DYW)[d]		Tseng et al. (2010)
ndhG-1 (17)	S > F	OTP82 (DYW)	Partial/silent	Okuda et al. (2010)
accD-1 (265)	S > L	RARE1 (DYW)		Robbins et al. (2009)
		ECB2/VAC1 (DYW)[b]		Ching-chih et al. (2010)
accD-2	3′UTR			
atpF-1 (31)	P > L			
clp-2 (187)	H > Y	CLB19 (E)		Chateigner-Boutin et al. (2008)
matK-2 (236)	H > Y			
petL-2 (2)	P > L			
psbE-1 (72)	P > S			
psbZ (17)	S > L	OTP84 (DYW)	Silent	Hammani et al. (2009)
psbF-1 (26)	S > F	LPA66 (DYW)		
rpoA-1 (67)	S > F	CLB19 (E)		Chateigner-Boutin et al. (2008)
rpoB-1 (113)	S > F	YS1 (DYW)		Zhou et al. (2008)
rpoB-3 (184)	S > L	CRR22 (DYW)	Silent	Okuda et al. (2009)
rpoB-7 (811)	S > L			
rpoC-1 (170)	S > L			
rps12	Intron	OTP81 (DYW)	Partial/silent	
rps14-1 (27)	S > L	OTP86 (DYW)	Silent	Hammani et al. (2009)
rps14-2 (50)	P > L			
rpl23 (30)	S > L	OTP80 (E)	Silent	Hammani et al. (2009)

[a]Locus names are based on Tsudzuki et al. (2001) with codon numbers in parentheses
[b]Partially edited in the wild type
[c]The amino acid alteration caused by the RNA editing is not required for the protein function
[d]VAC1 may be secondarily required for the two RNA editing events

chloroplasts may cause these artificial effects. However, it is true that some *trans*-factors are required for multiple RNA editing sites in both plastids (Table 15.1) and mitochondria. This means that the same number of *trans*-factors is not necessarily needed to manage approximately 500 RNA editing sites.

15.3   Discovery of the *trans*-Factor

A *trans*-factor involved in recognizing a *cis*-element was first discovered in a study of photosynthetic electron transport (Kotera et al. 2005). Eleven plastid *ndh* genes encode subunits of chloroplast NADH dehydrogenase-like complex (NDH), which is involved in photosystem I cyclic electron transport and chlororespiration (Munekage et al. 2004; Shikanai 2007). Among 34 RNA editing sites in *Arabidopsis*, 16 sites are concentrated in four *ndh* genes (*ndhB*, *ndhD*, *ndhF*, and *ndhG*) (Tillich et al. 2005). The *Arabidopsis chlororespiratory reduction 4* (*crr4*) mutants were identified based on their phenotypes lacking NDH activity, by chlorophyll fluorescence imaging (Kotera et al. 2005). The *crr4* mutants are specifically defective in RNA editing of the ndhD-1 site, which generates a translational initiation codon of the *ndhD* gene encoding an NDH subunit. Because the chloroplast NDH complex is dispensable under growth chamber conditions, even its knockout mutants do not show any strong mutant phenotypes of growth or photosynthesis, except for a minor alteration in the specific chlorophyll fluorescence (Shikanai et al. 1998; Kotera et al. 2005). Many defects in chloroplast function lead to common phenotypes, such as drastic reductions in growth and photosynthesis that are often accompanied by low levels of pigments, and it would be difficult to find the mutant specifically defective in RNA editing in the large mutant pool exhibiting a similar phenotype. To identify the RNA editing mutant, the key was focusing on NDH activity.

Map-based cloning clarified that *crr4* mutants are defective in a gene encoding a PPR family protein (Kotera et al. 2005). A PPR motif is a highly degenerate unit of 35 amino acids that usually appears as tandem repeats in the family members (Small and Peeters 2000). A PPR protein became a candidate for the *trans*-factor in RNA editing in plastids because (1) it was considered to be a sequence-specific RNA-binding protein involved in various RNA maturation processes in plastids and mitochondria and (2) it forms an extraordinarily large family, especially in higher plants, where RNA editing is prevalent (Lurin et al. 2004). This characteristic of the PPR protein is critical for a *trans*-factor because approximately 500 editing sites are independently recognized by these factors. All the *trans*-factors identified so far belong to the E and DYW subclasses of the PPR family, and the *Arabidopsis* genome encodes approximately 280 members of these subclasses (Lurin et al. 2004). Given that some PPR proteins recognize multiple sites, this number is sufficient to explain all the RNA editing sites present in both plastid and mitochondrial genomes.

Because an E subclass member of the PPR protein family is also involved in intergenic RNA cleavage (Hashimoto et al. 2003), it is also possible that CRR4 is involved in the intergenic RNA cleavage between *ndhD* and the upstream gene *psaC*. Therefore, the mutation may secondarily influence the RNA editing that creates the translational initiation codon of *ndhD* (ndhD-1). To eliminate this possibility, we performed an RNA protection assay to show that the intergenic RNA cleavage is unaffected in *crr4* (Kotera et al. 2005). Subsequently, we showed that the recombinant CRR4 expressed in and purified from *Escherichia coli* binds to 36 nucleotides in vitro (25 upstream and 10 downstream nucleotides of the ndhD-1 site), confirming that a PPR protein is a *trans*-factor required for the site recognition of RNA editing in plastids (Okuda et al. 2006). From the discovery of CRR4, many PPR proteins were shown to be specifically involved in RNA editing in plastids (Okuda et al. 2007, 2009, 2010; Chateigner-Boutin et al. 2008; Zhou et al. 2008; Cai et al. 2009; Robbins et al. 2009; Yu et al. 2009). This is similar in RNA editing in higher plant mitochondria (Kobayashi et al. 2007; Kim et al. 2009; Zehrmann et al. 2009; Sung et al. 2010; Takenaka 2010; Takenaka et al. 2010; Verbitskiy et al. 2010) and also in the mitochondria of *Physocomitrella patens* (Tasaki et al. 2010). For the past 5 years, 15 PPR proteins have been shown to be involved in 21 RNA editing events among the total of 34 RNA editing sites discovered in *Arabidopsis* plastids (Table 15.1). A more comprehensive understanding of the relationship between RNA editing sites and PPR proteins is needed for the genome-level discussion, but it is likely that almost all of the RNA editing events are mediated by PPR proteins in both plastids and mitochondria. Even after the genome-wide reverse genetics focused on the E and DWY subclasses (Hammani et al. 2009), *trans*-factors for the remaining 13 RNA editing sites are still unclear. If a single *cis*-element is alternatively recognized by multiple *trans*-factors, the mutant phenotype is detected only in the multiple mutant backgrounds. If this is the case, the forward genetics would not be enough to identify the mutants. The time-consuming study of double mutants may be necessary to analyze the overlapping function of PPR proteins in recognizing the single target site.

15.4   PPR Protein is Involved in Multiple Steps of RNA Maturation in Plant Organelles

The PPR protein family was first recognized by an *in silico* analysis of the *Arabidopsis* genome (Small and Peeters 2000). Prior to this discovery, the function of few members was analyzed, and all of them are involved in various RNA maturation steps in organelles (Manthey and McEwen 1995; Coffin et al. 1997; Fisk et al. 1999). The majority of members are predicted to target to plastids or mitochondria (Lurin et al. 2004), and consistently all the defined functions of PPR proteins are restricted to two organelles (Schmitz-Linneweber and Small 2008). The PPR protein family is classified into the P and PLS subfamilies (Lurin et al. 2004). The authentic PPR proteins consist of the 35-amino-acid unit that belong to the P subfamily, and the members of the PLS subfamily contain motifs related to the PPR motif and the PPR-like S and PPR-like L motifs. In addition to the N-terminal plastid or mitochondrial targeting signal and a tandem array of PPR and PPR-related motifs, some PLS members also have a C-terminal extension. Due to the presence of these C-terminal motifs, the PLS subfamily is further classified into three subclasses: PLS without C-terminal motifs, E with the E motif and DYW with the E, and DYW motifs (Lurin et al. 2004). All the *trans*-factors discovered in plastids and mitochondria belong to the E or DYW subclasses, implying the involvement of these C-terminal motifs in RNA editing reaction.

15.5   Function of C-Terminal Motifs

What is the function of the C-terminal E and DYW motifs conserved in *trans*-factors? Recently, the DYW motif was proposed to possess C deaminase activity as an RNA editing enzyme (Salone et al. 2007). The hypothesis is based on the phylogenetic distribution of the DYW members being strictly correlated with RNA editing in plants. Furthermore, the DYW motif contains the amino acid residues that are conserved in C deaminase. Several lines of experimental evidence do not support this hypothesis. First, CRR4, CRR21, and CLB19 are involved in RNA editing in plastids, belong to the E subgroup, and do not have the DYW motif (Kotera et al. 2005; Okuda et al. 2007; Chateigner-Boutin et al. 2008). Second, the RNA editing activity of *crr22*, *crr28* and *otp82* mutants was fully complemented by the introduction of mutant versions of CRR22, CRR28, and OTP82, where the DYW motif is truncated, indicating that the DYW motif is not essential for RNA editing (Okuda et al. 2009, 2010). In contrast, the E domain is essential for the function of CRR4, CRR21, CRR22, CRR28, and OTP82, although we do not rule out the possibility that the lack of the E motif destabilizes the PPR protein (Okuda et al. 2007, 2009, 2010). Based on these results, we proposed a two-component model of RNA editing machinery in which the E domain of the PPR protein helps recruit an unknown editing enzyme (Okuda et al. 2007). Lastly, a DYW member, CRR2, is involved in intergenic RNA cleavage between *rps7* and *ndhB* (Hashimoto et al. 2003). Consistent with this fact, the DYW domain of CRR2 has endonuclease activity (Okuda et al. 2009), which was also reported in some other DYW members (Nakamura and Sugita 2008). CRR2 is a site-specific endonuclease in which the PPR motifs determine the substrate specificity, and the DYW motif has catalytic function. The first and second reasons discussed above do not necessarily eliminate the possibility that the DYW motif has C deaminase activity if the domain can be provided by another molecule that is included in the same machinery. Both ECB2 and RARE1 are required for editing the accD-1 site (Robbins et al. 2009; Yu et al. 2009), suggesting that the PPR protein forms a heterodimer. However, in the *vanilla cream 1* (*vac1*) mutant, which is allelic to *ecb2*, RNA editing of the accD-1 site is only partially affected, and pleiotropic defects in plastid function were reported. This suggests that VAC1/ECB2 secondarily affect the RNA editing of accD-1 (Tseng et al. 2010). Outside the DYW domain, there are no other candidates for the editing enzyme; however, why the DYW domain is related to two distinct activities, C deaminase and endonuclease, should be further assessed.

15.6   Partial RNA Editing

One of the mysteries in plant RNA editing is the presence of incomplete RNA editing (Shikanai and Obokata 2008). While the ndhD-1 site is partially edited even in wild-type *Arabidopsis* (42%), the ndhD-2 site present in the same transcript is almost completely edited (99%) (Okuda et al. 2007). CRR21 is a PPR protein that mediates RNA editing of the ndhD-2 site, altering it from Ser128 to Leu, which is not required for stabilizing the NDH complex but essential for its activity (Okuda et al. 2007). To not express the inactive NDH complex with NdhD originated from unedited transcript, the ndhD-2 site should be edited perfectly. However, the RNA editing of the ndhD-1 site generates the translational initiation codon, and unedited transcripts do not interfere with the expression of active NDH. The system looks intriguing; however, the physiological meaning of the difference in RNA editing efficiency should be interrogated.

The efficiency of the RNA editing at the ndhD-1 site is variable among species. Similar to *Arabidopsis*, the site is partially edited in *Nicotiana tabacum* (42%) and *N. sylvestris* (37%), but the efficiency is significantly lower in *N. tomentosiformis* (15%) (Okuda et al. 2008). What is the molecular mechanism for determining this species-specific editing efficiency? The *cis*-region for the CRR4 binding is completely conserved among three *Nicotiana* species, and it is likely that a *trans*-factor, CRR4, is a determinant for this efficiency. To test this possibility, we cloned *CRR4* orthologs from three *Nicotiana* species and introduced them into the null allele of the *Arabidopsis crr4* mutant (Okuda et al. 2008). The transformation fully complemented the NDH activity in *crr4* and provided direct evidence that the *trans*-factor is conserved between distantly related species. CRR4, isolated from *N. sylvestris* (NsylCRR4) and *N. tomentosiformis* (NtomCRR4), showed 60% and 57% amino acid identity, respectively, to *Arabidopsis* CRR4 (Okuda et al. 2008). In *Arabidopsis crr4*, 40% of the *ndhD* transcripts were edited at the ndhD-1 site by the introduction of *NsylCRR4*, whereas only 21% were edited by the introduction of *NtomCRR4*. The efficiency was similar to that in original plants with 37% in *N. sylvestris* and 15% in *N. tomentosiformis* (Okuda et al. 2008). It is likely that the PPR protein CRR4 is a determinant for species-specific efficiency in RNA editing. Although the efficiency of ndhD-1 editing is significantly lower in *N. tomentosiformis* compared to that in other species, *N. tomentosiformis* accumulates a similar level of the NDH complex to *N. tabacum* and *N. sylvestris* (Okuda et al. 2008). These results suggest that the 15% level of RNA editing does not limit the translation of *ndhD*. The ndhD-1 site is partially edited, and the editing creates the translational initiation codon, providing the possibility of the regulatory function of RNA editing as a rare example of among approximately 500 RNA editing events. However, I do not find any physiological meaning for the variation in RNA editing efficiency between species even at this site.

The translation of *ndhD* is regulated by the intergenic RNA cleavage between *psaC* and *ndhD* (del Campo et al. 2002). The protein level of PsaC is approximately 100 times higher than that of NdhD, although the precursor RNA is transcribed from the identical promoter and the level of the primary transcript is the same. These facts suggest the importance of post-transcriptional regulation. The RNA editing efficiency of the ndhD-1 site also depends on developmental and environmental conditions (Hirose and Sugiura 1997). In tobacco, the extent of RNA editing at this site is highest in young leaves (56%) and is very low in nongreen tissue (<5%), suggesting that RNA regulates translation (Hirose and Sugiura 1997). However, this difference may simply reflect the status of plastid conditions rather than the result of regulation. Notably, even with the low level of RNA editing at the ndhD-1 site (15%), *N. tomentosiformis* accumulates a similar NDH complex level to *N. sylvestris*, suggesting that the 15% level of RNA editing does not limit the translation under the greenhouse conditions (Okuda et al. 2008). I do not eliminate the possibility that the alteration in RNA editing efficiency affects the translation under certain conditions. Furthermore, it is possible that less than 5% of RNA editing limits the translation of *ndhD* in nongreen tissues, which is consistent with the fact that chloroplast NDH is absent in nongreen plastids, with the exception of etioplasts (Peng et al. 2008). Most likely, the ndhD-1 site is partially edited because it is not necessary that it be edited completely. The extent may be variable in the range of 15–60% among species and among green tissues (lower in nongreen tissues), but even the 15% level of RNA editing does not limit the translation. This site is substituted by T in the genomes of monocots, suggesting that the translational regulation by RNA editing is dispensable. It is possible that dicots are in the process of losing the ndhD-1 site by substituting the genomic sequence, and partial editing at the site does not have any physiological meaning. As suggested (Tillich et al. 2006), RNA editing is unlikely to have any regulatory roles, at least in plastids.

15.7 Closely Located Editing Sites Are Recognized by Independent *trans*-Factors

In our previous review (Shikanai and Obokata 2008), we discussed the mechanism of editing a pair of sites that are closely located to each other. A question remained as to whether a single *trans*-factor recognizes the common *cis*-element to recruit the editing machinery to both sites. In mitochondria, two sites are often edited in a codon, and it would seem economical that a single *trans*-factor could manage both sites. After the previous review, many *trans*-acting factors were discovered, and it became possible to discuss the mechanism on the basis of experimental results.

In *Arabidopsis*, the *ndhD* transcript has five RNA editing sites, and the ndhD-3 (Ser293Leu) site and ndhD-5 site (Pro296Leu) are separated by only eight nucleotides. Our nomenclature is essentially based on Tsudzuki et al. (2001), where only three editing sites are listed in *Arabidopsis*. The ndhD-4 site (Ser225Leu) was later discovered between the ndhD-2 (Ser128Leu) and ndhD-3 sites (Table 15.1). Although the ndhD-3 site is not edited in the *crr28* mutants, the ndhD-5 site is edited as in the wild type. The ndhD-5 site is not edited in the *crr22* mutants, but the ndhD-3 site is edited in these mutants (Okuda et al. 2009). These results indicate the ndhD-3 and ndhD-5 sites are recognized by distinct *trans*-factors, CRR28 and CRR22, respectively. These results suggest that the PPR protein recruits the RNA editing enzyme exactly to the target site. This is so that any C residue adjacent to the target site is not edited erroneously. Although CRR22 and CRR28 cannot edit the site that is separated by eight nucleotides from the target C, both PPR proteins are involved in the RNA editing of multiple sites, which are located in the different transcripts (Okuda et al. 2009). A similar story is true for the ndhB-8 (Ser277Leu) and ndhB-9 sites, which are separated by only five nucleotides. Although OTP82 is essential for RNA editing of the ndhB-9 site (Ser279Leu), the ndhB-8 site is edited in the *otp82* mutants (Okuda et al. 2010). The factor involved in the RNA editing of the ndhB-8 site is still unclear. More information is needed for the mitochondrial *trans*-factors, but the PPR protein is likely to recognize the distance between the *cis*-element and the target C, and this is probably required to avoid erroneous RNA editing. Plants took the strategy of a single PPR protein that recognizes multiple *cis*-elements, which are often not highly conserved, to save the number of PPR proteins involved in RNA editing.

15.8   Physiological Function of RNA Editing

RNA editing is prevalent in plant organelles, but why do plants edit so many sites? It is evident from the mutant phenotypes that many sites are required to be edited to express functional proteins; thus, RNA editing is believed to be an essential process. Plants have increased the number of RNA editing sites during their evolution to correct mutations that occurred in the genome. However, it raises the question of why plants did not correct the genome information directly.

Identification of a mutant defective in a specific RNA editing event enables the evaluation of the physiological significance of each editing event. The first RNA editing mutant was isolated by forward genetics based on its phenotypes in photo-synthetic electron transport (Kotera et al. 2005). Once the E and DYW subclasses of the PPR protein family became candidates for *trans*-factors, however, reverse genetics were also a powerful tool for identifying the mutants more efficiently (Hammani et al. 2009; Okuda et al. 2010). Based on the extensive survey of knockout lines of the subclasses of PPR protein genes, one surprising discovery is the absence of particular mutant phenotypes in the protein function that is encoded by the target RNAs.

The *Arabidopsis otp82* mutants are defective in RNA editing of ndhB-9 and ndhG-1, and the mutants were identified by the direct analysis of RNA editing in plastids (Okuda et al. 2010). The ndhB-9 editing alters a Ser279 to Leu in *Arabidopsis*. Ser279 is conserved in dicots as well as in maize and wheat, but the site is substituted by Leu at the genome level in rice and sorghum. Because the putative OTP82 ortholog is discovered in the rice genome, the RNA editing site may have been created prior to the division of dicots and monocots and may have been lost in maize and sorghum. In contrast, the ndhG-1 editing that alters Ser17 to Phe is conserved in only dicots, and Phe is encoded in the genomes of monocots. Originally, OTP82 may have been a *trans*-factor for the ndhB-9 site but later may have obtained the novel function of recognizing the ndhG-1 site in dicots. The putative *cis*-elements are not highly conserved between the ndhB-9 and ndhG1 sites (Okuda et al. 2010), and it is unclear how the ndhG1 was selected as the alternative target by OTP82.

The most surprising features of the *otp82* mutant are the absence of any mutant phenotype in NDH activity, the accumulation of the NDH complex, and the supercomplex formation between NDH and PSI (Okuda et al. 2010). Furthermore, both sites are unedited in 30% and 20% of transcripts in wild-type Nössen and wild-type Columbia accessions, respectively. These results suggest the possibility that the NDH complex consists of two versions of NdhB and NdhD subunits that originate from edited and unedited transcripts if both peptides are equally stable and translated with the same efficiency. Unfortunately, our mass analysis failed to detect peptides that specifically originated from unedited transcripts even in the mutants, probably due to a technical problem. Thus, it is still unclear whether two versions of the peptide originating from edited and unedited RNA are functionally identical, but the subunits originating from unedited RNAs behave like the wild-type proteins in the mutants. I do not find any physiological reason for editing the ndhB-9 and ndhG-1 sites in *Arabidopsis*.

In tobacco, the E III site of the *ndhB* transcript is edited to convert Ser204 to Leu in green tissues, but the site is not edited in nongreen mutants. This suggests the possibility that the expression of two versions of NdhD is developmentally regulated (Karcher and Bock 2002). This site corresponds to ndhB-4 (Tsuzuki et al. 2001), and, surprisingly, the site is not edited in *Arabidopsis* (Tillich et al. 2005). This fact indicates that both Ser and Leu are acceptable for the 204 residue of NdhB, and the RNA editing of the site occurring in green tobacco tissues is not essential.

These examples are not exceptional; defects in the RNA editing of ndhD-4, psbZ, rpoB-3, rps12, rps14-1, and rpl23 do not lead to any mutant phenotypes in protein function (Okuda et al. 2009; Hammani et al. 2009). Among 34 RNA editing events present in *Arabidopsis* plastids, at least seven events are unlikely to alter the protein function drastically. We cannot overlook this fact. I agree that RNA editing corrects the T-to-C mutations, but in the absence of any mutant phenotype (i.e., any selection pressure), how did plants recognize the C residues that required correction via RNA editing?

In the *Arabidopsis* Cvi-0 ecotype, the T-to-G transition is located immediately upstream of the ndhG-1site, resulting in the codon change from UCC (Ser) to GCC (Ala) (Tillich et al. 2005). As in other ecotypes, the ndhG-1 site (underlined) is edited in RNA, and, consequently, the GCC codon is converted to GUC (Val). As in Col and Nös (Okuda et al. 2010), the ndhG-1 site is partially edited in Cvi-0. These results imply that, for expressing the functional NdhG, the 17th codon of ndhG can be Ser, Phe, Ala, or Val. It is likely that OTP82 recognizes the target C present in the variant codon in Cvi-0. As mentioned above, I do not find any physiological reason why OTP82 edits the ndhG-1 site in Col-0, and it is surprising that plants do not stop editing the site via OTP82 after the variation in the corresponding codon. It may be true that plants continue to edit sites even when there is no physiological advantage because a *trans*-factor recognizing the site is available.

15.9 Speculation

RNA editing is so mysterious in plant organelles that it is difficult for any one discussion to sufficiently explain its physiological function. The most acceptable hypothesis is that RNA editing is a process for correcting the T-to-C mutations that occurred in the genome at the RNA level (Covello and Gray 1993; Tillich et al. 2006; Shikanai and Obokata 2008), but this simply describes what we observe in plants and does not explain why plants prefer this strategy. Rapid evolution of the PPR protein may have restored the protein function, which was once disturbed by a T-to-C mutation. The story successfully explains the co-evolution of mitochondrial genes causing cytoplasmic male sterility (CMS) and fertility restorer genes (*Rf*) encoded by the nuclear genome (Fujii and Toriyama 2008). Some *Rf* loci contain several copies of PPR protein genes, suggesting that the rapid gene duplication followed by the amino acid alteration occurred to suppress the activity of mito-chondrial CMS genes in the very recent evolution of each species (Akagi et al. 2004; Bentolila et al. 2002; Desloire et al. 2003; Fujii and Toriyama 2009; Komori et al. 2004). The CMS trait is deleterious for plant reproduction; thus, there was the strong pressure for the evolution of *Rf* genes. As discussed in this review, however, the evolution of some PPR proteins occurred in the absence of strong selection pressure. By the simple analogy of *Rf* genes, we may not be able to fully explain the evolution of *trans*-factors in RNA editing.

An alternative idea is that some PPR proteins began recognizing extra sites by chance. This allowed the T-to-C mutation to take place at the novel target site and to back up the amino acid correction via RNA editing. In this scenario, the editing site is conserved not only when the resulting amino acid alteration is beneficial for plants against the selection pressure but also when it is neutral. This may be the reason why plants edit many sites even when the resulting amino acid alteration is not essential, as in the case of ndhB-10 and ndhG-1 (Okuda et al. 2010). Rip-chip assay of PPR proteins PPR4 and PPR10 selectively enriched the target sites, and this suggests that the PPR protein rather strictly recognizes the target sequences

(Schmitz-Linneweber et al. 2006; Pfalz et al. 2009). The same story is also likely for *trans*-factors of RNA editing to edit the target C residues exactly. For my hypothesis, it is necessary to explain how a PPR protein specifically recognizes the multiple *cis*-elements, which are not necessarily highly conserved. PROTON GRADIENT REGULATION 3 (PGR3) is a member of the P subfamily and possesses dual targets, *petL* and *ndhA* (Yamazaki et al. 2004; Cai et al. 2011). The binding to *petL* mRNA is specifically affected in the *pgr3-2* mutant allele, which has an amino acid alteration in a PPR motif (Yamazaki et al. 2004; Cai et al. 2011). This result suggests that an amino acid alteration in a PPR motif makes a PPR protein interact with an extra target. For further discussion, the exact molecular mechanism of how PPR proteins recognize the target RNA sequence should be clarified.

Although the number of RNA editing sites is roughly constant in angiosperms, the sites are variable among species. Some sites are highly conserved among plants, but some sites are not conserved even between closely related species (Tsudzuki et al. 2001). These results suggest that the RNA editing sites were gained during the recent evolution of each species. The comparison of editing sites among species also suggests that many sites were lost during the recent evolution of land plants, and this is consistent with the fact that RNA editing is not required if the genomic information is corrected. Consequently, we cannot find any physiological meaning of RNA editing in the present plant species. To discuss the function of RNA editing, however, we must consider the process whereby plants repeatedly gained and lost the RNA editing sites. The process may have been beneficial for plants to accelerate the evolution of organelle genes, as suggested (Tillich et al. 2006).

In angiosperm plastids, RNA editing sites are concentrated in some *ndh* genes that encode subunits of the chloroplast NDH complex, and 16 sites among the 34 sites are in four *ndh* genes (*ndhB*, *ndhD*, *ndhF*, and *ndhG*) in *Arabidopsis* plastids. What is the reason for this biased distribution of RNA editing sites in the genome? The chloroplast NDH complex is a machine for stress resistance (Endo et al. 1999; Horváth et al. 2000; Wang et al. 2006), and the knockout mutants can grow like the wild type in greenhouse conditions (Shikanai et al. 1998). A reasonable explanation is that the nonlethal phenotypes of NDH-less plants allowed the accumulation of RNA editing sites in *ndh* genes. Recently, we clarified that the structure of the NDH complex altered drastically during the evolution of land plants (Peng et al. 2009). RNA editing may have assisted this drastic evolution of the chloroplast NDH complex. With the aid of *trans*-factors and PPR proteins, some T-to-C mutations were fixed in the genome. Many sites were lost by correcting the genome information; however, some C residues may have been more suitable for the protein function than the original T, and the C was finally fixed in the genome. This story is probable especially if another mutation occurred in the same gene or a different gene encoding protein, which interacted with the protein affected by the original T-to-C mutation. In this case, the site is not edited, but the RNA editing promoted the evolution of the gene by stabilizing the T-to-C mutation.

15.10   Future Aspects

The discovery of *trans*-factors has improved our knowledge of RNA machinery in plant organelles, but the editing enzyme is still unclear. The most important objective is the clarification of the RNA editing enzyme, and this is related to identifying the function of the C-terminal motifs, E and DYW, in *trans*-factors. It should be clarified as to whether the same enzyme catalyzes both C-to-U and U-to-C RNA editing. It is also important to determine the molecular mechanism of how a tandem array of PPR motifs can specifically recognize a target RNA sequence. To understand the co-evolution of *trans*-factors and RNA editing sites, it is necessary to clarify how a PPR protein recognizes multiple target sequences, which are not highly conserved. The experimental results would make it possible to discuss the function of RNA editing more rigorously and further consider the possibility that RNA editing was and is a driving force for the organelle genome evolution.

References

Akagi H, Nakamura A, Yokozeki-Misono Y, Inagaki A, Takahashi H, Mori K, Fujimura T (2004) Positional cloning of the rice *Rf-1* gene, a restorer of BT-type cytoplasmic male sterility that encodes a mitochondria-targeting PPR protein. Theor Appl Genet 108:1449–1457

Bentolila S, Alfonso AA, Hanson MR (2002) A pentatricopeptide repeat-containing gene restores fertility to cytoplasmic male-sterile plants. Proc Natl Acad Sci USA 99:10887–10892

Blanc V, Litvak S, Araya A (1995) RNA editing in wheat mitochondria proceeds by a deamination mechanism. FEBS Lett 373:56–60

Bock R, Hermann M, Kössel H (1996) *In vivo* dissection of *cis*-acting determinants for plastid RNA editing. EMBO J 15:5052–5059

Cai W, Ji D, Peng L, Guo J, Ma J, Zou M, Lu C, Zhang L (2009) LPA66 is required for editing *psbF* chloroplast transcripts in Arabidopsis. Plant Physiol 150:1260–1271

Cai W, Okuda K, Peng L Shikanai T (2011) PROTON GRADIENT REGULATION 3 recognizes multiple targets with limited similarity and mediates translation and RNA stabilization in plastids. Plant J 67:318–327

Chateigner-Boutin AL, Hanson MR (2002) Cross-competition in transgenic chloroplasts expressing single editing sites reveals shared *cis* elements. Mol Cell Biol 22:8448–8456

Chateigner-Boutin AL, Small I (2007) A rapid high-throughput method for the detection and quantification of RNA editing based on high-resolution melting of amplicons. Nucleic Acids Res 35:e114

Chateigner-Boutin AL, Ramos-Vega M, Guevara-García A, Andrés C, de la Luz Gutiérrez-Nava M, Cantero A, Delannoy E, Jiménez LF, Lurin C, Small I, León P (2008) CLB19, a pentatricopeptide repeat protein required for editing of *rpoA* and *clpP* chloroplast transcripts. Plant J 56:590–602

Chaudhuri S, Maliga P (1996) Sequences directing C to U editing of the plastid *psbL* mRNA are located within a 22 nucleotide segment spanning the editing site. EMBO J 15:5958–5964

Coffin JW, Dhillon R, Ritzel RG, Nargang FE (1997) The *Neurospora crassa cya-5* nuclear gene encodes a protein with a region of homology to the *Saccharomyces cerevisiae* PET309 protein and is required in a post-transcriptional step for the expression of the mitochondrially encoded COXI protein. Curr Genet 32:273–280

Covello PS, Gray MW (1989) RNA editing in plant mitochondria. Nature 341:662–666

Covello PS, Gray MW (1993) On the evolution of RNA editing. Trends Genet 9:265–268

del Campo EM, Sabater B, Martin M (2002) Post-transcriptional control of chloroplast gene expression. Accumulation of stable psaC mRNA is due to downstream RNA cleavages in the ndhD gene. J Biol Chem 277:36457–36464

Desloire S, Gherbi H, Laloui W, Marhadour S, Clouet V, Cattolico L, Falentin C, Giancola S, Renard M, Budar F, Small I, Caboche M, Delourme R, Bendahmane A (2003) Identification of the fertility restoration locus, *Rfo*, in radish, as a member of the pentatricopeptide-repeat protein family. EMBO Rep 4:588–594

Endo T, Shikanai T, Takabayashi A, Asada K, Sato F (1999) The role of chloroplastic NAD(P)H dehydrogenase in photoprotection. FEBS Lett 457:5–8

Farré JC, Leon G, Jordana X, Araya A (2001) *cis* Recognition elements in plant mitochondrion RNA editing. Mol Cell Biol 21:6731–6737

Fisk DG, Walker MB, Barkan A (1999) Molecular cloning of the maize gene *crp1* reveals similarity between regulators of mitochondrial and chloroplast gene expression. EMBO J 18:2621–2630

Fujii S, Toriyama K (2008) Genome barriers between nuclei and mitochondria exemplified by cytoplasmic male sterility. Plant Cell Physiol 49:1484–1494

Fujii S, Toriyama K (2009) Suppressed expression of Retrograde-Regulated Male Sterility restores pollen fertility in cytoplasmic male sterile rice plants. Proc Natl Acad Sci USA 106:9513–9518

Giegé P, Brennicke A (1999) RNA editing in Arabidopsis mitochondria effects 441 C to U changes in ORFs. Proc Natl Acad Sci USA 96:15324–15329

Gualberto JM, Lamattina L, Bonnard G, Weil JH, Grienenberger JM (1989) RNA editing in wheat mitochondria results in the conservation of protein sequences. Nature 341:660–662

Hammani K, Okuda K, Tanz S, Chateigner-Boutin AL, Shikanai T, Small I (2009) General features of chloroplast RNA editing factors and their target sites gained from a study of new *Arabidopsis* editing mutants. Plant Cell 21:3686–3699

Hashimoto M, Endo T, Peltier G, Tasaka M, Shikanai T (2003) A nucleus-encoded factor, CRR2, is essential for the expression of chloroplast *ndhB* in Arabidopsis. Plant J 36:541–549

Hiesel R, Wissinger B, Schuster W, Brennicke A (1989) RNA editing in plant mitochondria. Science 246:1632–1634

Hirose T, Sugiura M (1997) Both RNA editing and RNA cleavage are required for translation of tobacco chloroplast *ndhD* mRNA: a possible regulatory mechanism for the expression of a chloroplast operon consisting of functionally unrelated genes. EMBO J 16:6804–6811

Hirose T, Sugiura M (2001) Involvement of a site-specific *trans*-acting factor and a common RNA-binding protein in the editing of chloroplast mRNAs: development of a chloroplast *in vitro* RNA editing system. EMBO J 20:1144–1152

Hoch B, Maier RM, Appel K, Igloi GL, Kössel H (1991) Editing of a chloroplast mRNA by creation of an initiation codon. Nature 353:178–180

Horváth EM, Peter SO, Joët T, Rumeau D, Cournac L, Horváth GV, Kavanagh TA, Schäfer C, Peltier G, Medgyesy P (2000) Targeted inactivation of the plastid *ndhB* gene in tobacco results in an enhanced sensitivity of photosynthesis to moderate stomatal closure. Plant Physiol 123:1337–1350

Karcher D, Bock R (2002) The amino acid sequence of a plastid protein is developmentally regulated by RNA editing. J Biol Chem 277:5570–5574

Keegan LP, Gallo A, O'Connell MA (2001) The many roles of an RNA editor. Nat Rev Genet 2:869–878

Kim SR, Yang JI, Moon S, Ryu CH, An K, Kim KM, Yim J, An G (2009) Rice *OGR1* encodes a pentatricopeptide repeat-DYW protein and is essential for RNA editing in mitochondria. Plant J 59:738–749

Kobayashi K, Suzuki M, Tang J, Nagata N, Ohyama K, Seki H, Kiuchi R, Kaneko Y, Nakazawa M, Matsui M, Matsumoto S, Yoshida S (2007) Muranaka T (2007) Lovastatin insensitive 1, a novel pentatricopeptide repeat protein, is a potential regulatory factor of isoprenoid biosynthesis in Arabidopsis. Plant Cell Physiol 48:322–331

Komori T, Ohta S, Murai N, Takakura Y, Kuraya Y, Suzuki S, Hiei Y, Imaseki H, Nitta N (2004) Map-based cloning of a fertility restorer gene, *Rf-1*, in rice (*Oryza sativa* L.). Plant J 37:315–325

Kotera E, Tasaka M, Shikanai T (2005) A pentatricopetide repeat protein is essential for RNA editing in chloroplasts. Nature 433:326–330

Kugita M, Yamamoto Y, Fujikawa T, Matsumoto T, Yoshinaga K (2003) RNA editing in hornwort chloroplasts makes more than half the genes functional. Nucleic Acids Res 31:2417–2423

Lurin C, Andrés C, Aubourg S, Bellaoui M, Bitton F, Bruyère C, Caboche M, Debast C, Gualberto J, Hoffmann B, Lecharny A, Le Ret M, Martin-Magniette ML, Mireau H, Peeters N, Renou JP, Szurek B, Taconnat L, Small I (2004) Genome-wide analysis of Arabidopsis pentatricopeptide repeat proteins reveals their essential role in organelle biogenesis. Plant Cell 16:2089–2103

Manthey GM, McEwen JE (1995) The product of the nuclear gene PET309 is required for translation of mature mRNA and stability or production of intron-containing RNAs derived from the mitochondrial COX1 locus of *Saccharomyces cerevisiae*. EMBO J 14:4031–4043

Munekage Y, Hashimoto M, Miyake C, Tomizawa K, Endo T, Tasaka M, Shikanai T (2004) Cyclic electron flow around photosystem I is essential for photosynthesis. Nature 429:579–582

Nakamura T, Sugita M (2008) A conserved DYW domain of the pentatricopeptide repeat protein possesses a novel endoribonuclease activity. FEBS Lett 582:4163–4168

Okuda K, Nakamura T, Sugita M, Shimizu T, Shikanai T (2006) A pentatricopeptide repeat protein is a site-recognition factor in chloroplast RNA editing. J Biol Chem 281:37661–37667

Okuda K, Myouga F, Motohashi R, Shinozaki K, Shikanai T (2007) Conserved domain structure of pentatricopeptide repeat proteins involved in chloroplast RNA editing. Proc Natl Acad Sci USA 104:8178–8183

Okuda K, Habata Y, Kobayashi Y, Shikanai T (2008) Amino acid sequence variations in *Nicotiana* CRR4 orthologs determine the species-specific efficiency of RNA editing in plastids. Nucleic Acids Res 36:6155–6164

Okuda K, Chateigner-Boutin A-L, Nakamura T, Delannoy E, Sugita M, Myouga F, Motohashi R, Shinozaki K, Small I, Shikanai T (2009) Pentatricopeptide repeat proteins with the DYW motif have distinct molecular functions in RNA editing and RNA cleavage in *Arabidopsis* chloroplasts. Plant Cell 21:147–156

Okuda K, Hammani K, Tanz SK, Peng L, Fukao Y, Myouga F, Motohashi R, Shinozaki K, Small I, Shikanai T (2010) The pentatricopeptide repeat protein OTP82 is required for RNA editing of plastid *ndhB* and *ndhG* transcripts. Plant J 61:339–349

Peng L, Shimizu H, Shikanai T (2008) The chloroplast NAD(P)H dehydrogenase complex interacts with photosystem I in Arabidopsis. J Biol Chem 283:34873–34879

Peng L, Fukao Y, Fujiwara M, Takami T, Shikanai T (2009) Efficient operation of NAD(P)H dehydrogenase requires supercomplex formation with photosystem I via minor LHCI in Arabidopsis. Plant Cell 21:3623–3640

Pfalz J, Bayraktar OA, Prikryl J, Barkan A (2009) Site-specific binding of a PPR protein defines and stabilizes 5' and 3' mRNA termini in chloroplasts. EMBO J 28:2042–2052

Rajasekhar VK, Mulligan RM (1993) RNA editing in plant mitochondria: α-phosphate is retained during C-to-U conversion in mRNAs. Plant Cell 5:1843–1852

Robbins JC, Heller WP, Hanson MR (2009) A comparative genomics approach identifies a PPR-DYW protein that is essential for C-to-U editing of the Arabidopsis chloroplast *accD* transcript. RNA 15:1142–1153

Salone V, Rüdinger M, Polsakiewicz M, Hoffmann B, Groth-Malonek M, Szurek B, Small I, Knoop V, Lurin C (2007) A hypothesis on the identification of the editing enzyme in plant organelles. FEBS Lett 581:4132–4138

Schmitz-Linneweber C, Small I (2008) Pentatricopeptide repeat proteins: a socket set for organelle gene expression. Trends Plant Sci 13:663–670

Schmitz-Linneweber C, Williams-Carrier RE, Williams-Voelker PM, Kroeger TS, Vichas A, Barkan A (2006) A pentatricopeptide repeat protein facilitates the trans-splicing of the maize chloroplast *rps12* pre-mRNA. Plant Cell 18:2650–3663

Shikanai T (2006) RNA editing in plant organelles: machinery, physiological function and evolution. Cell Mol Life Sci 63:698–708

Shikanai T (2007) Cyclic electron transport around photosystem I: genetic approaches. Annu Rev Plant Biol 58:199–217

Shikanai T, Obokata J (2008) Machinery of RNA editing in plant organelles. In: Smith HC (ed) RNA and DNA editing: molecular mechanisms and their integration into biological systems. Wiley-Interscience, Hoboken, pp 99–119

Shikanai T, Endo T, Hashimoto T, Yamada Y, Asada K, Yokota A (1998) Directed disruption of the tobacco *ndhB* gene impairs cyclic electron flow around photosystem I. Proc Natl Acad Sci USA 95:9705–9709

Small ID, Peeters N (2000) The PPR motif – a TPR-related motif prevalent in plant organellar proteins. Trends Biochem Sci 25:46–47

Stuart KD, Schnaufer A, Ernst NL, Panigrahi AK (2005) Complex management: RNA editing in trypanosomes. Trends Biochem Sci 30:97–105

Sung TY, Tseng CC, Hsieh MH (2010) The SLO1 PPR protein is required for RNA editing at multiple sites with similar upstream sequences in Arabidopsis mitochondria. Plant J 63:499–511

Takenaka M (2010) MEF9, an E-subclass pentatricopeptide repeat protein, is required for an RNA editing event in the *nad7* transcript in mitochondria of Arabidopsis. Plant Physiol 152:939–947

Takenaka M, Neuwirt J, Brennicke A (2004) Complex *cis*-elements determine an RNA editing site in pea mitochondria. Nucleic Acids Res 32:4137–4144

Takenaka M, Verbitskiy D, Zehrmann A, Brennicke A (2010) Reverse genetic screening identifies five E-class PPR proteins involved in RNA editing in mitochondria of *Arabidopsis thaliana*. J Biol Chem 285:27122–27129

Tasaki E, Hattori M, Sugita M (2010) The moss pentatricopeptide repeat protein with a DYW domain is responsible for RNA editing of mitochondrial *ccmFc* transcript. Plant J 62:560–570

Tillich M, Funk HT, Schmitz-Linneweber C, Poltnigg P, Sabater B, Martin M, Maier RM (2005) Editing of plastid RNA in *Arabidopsis thaliana* ecotypes. Plant J 43:708–715

Tillich M, Lehwark P, Morton BR, Maier UG (2006) The evolution of chloroplast RNA editing. Mol Biol Evol 23:1912–1921

Tseng CC, Sung TY, Li YC, Hsu SJ, Lin CL, Hsieh MH (2010) Editing of *accD* and *ndhF* chloroplast transcripts is partially affected in the Arabidopsis *vanilla cream1* mutant. Plant Mol Biol 73:309–323

Tsudzuki T, Wakasugi T, Sugiura M (2001) Comparative analysis of RNA editing sites in higher plant chloroplasts. J Mol Evol 53:327–332

Verbitskiy D, Zehrmann A, van der Merwe JA, Brennicke A, Takenaka M (2010) The PPR protein encoded by the *LOVASTATIN INSENSITIVE 1* gene is involved in RNA editing at three sites in mitochondria of *Arabidopsis thaliana*. Plant J 61:446–455

Wang P, Duan W, Takabayashi A, Endo T, Shikanai T, Ye JY, Mi H (2006) Chloroplastic NAD(P)H dehydrogenase in tobacco leaves functions in alleviation of oxidative damage caused by temperature stress. Plant Physiol 141:465–474

Wedekind JE, Dance GS, Sowden MP, Smith HC (2003) Messenger RNA editing in mammals: new members of the APOBEC family seeking roles in the family business. Trends Genet 19:207–216

Wolf PG, Rowe CA, Hasebe M (2004) High levels of RNA editing in a vascular plant chloroplast genome: analysis of transcripts from the fern *Adiantum capillus-veneris*. Gene 339:89–97

Yamazaki H, Tasaka M, Shikanai T (2004) PPR motifs of the nucleus-encoded factor, PGR3, function in the selective and distinct steps of chloroplast gene expression in Arabidopsis. Plant J 38:152–163

Yu W, Schuster W (1995) Evidence for a site-specific cytidine deamination reaction involved in C to U RNA editing of plant mitochondria. J Boil Chem 270:18227–18233

Yu QB, Jiang Y, Chong K, Yang ZN (2009) AtECB2, a pentatricopeptide repeat protein, is required for chloroplast transcript *accD* RNA editing and early chloroplast biogenesis in *Arabidopsis thaliana*. Plant J 59:1011–1023

Zehrmann A, Verbitskiy D, van der Merwe JA, Brennicke A, Takenaka M (2009) A DYW domain-containing pentatricopeptide repeat protein is required for RNA editing at multiple sites in mitochondria of *Arabidopsis thaliana*. Plant Cell 21:558–567

Zhou W, Cheng Y, Yap A, Chateigner-Boutin AL, Delannoy E, Hammani K, Small I, Huang J (2008) The Arabidopsis gene *YS1* encoding a DYW protein is required for editing of *rpoB* transcripts and the rapid development of chloroplasts during early growth. Plant J 58:82–96

Part VII
Evolution of Organelle Translation, tRNAs and the Genetic Code

Chapter 16
Conserved and Organelle-Specific Molecular Mechanisms of Translation in Mitochondria

Kirsten Kehrein and Martin Ott

16.1 Introduction

Mitochondria are the power plants of eukaryotic cells that use oxidative phosphorylation as a highly efficient way to synthesize ATP. This oxidative phosphorylation is catalyzed by a number of large protein complexes in the inner membrane, the respiratory chain and the ATP-synthase. As detailed in the endosymbiotic theory, the ancestor of mitochondria, presumably of α-proteobacterial origin, joined a primitive eukaryotic cell, and this union laid the foundation to the evolutionary success of eukaryotes (Sagan 1967) (see Chap. 1). In the course of mitochondrial evolution, most of the ancestral genes were transferred to the nucleus (see Chap. 7). This was accompanied by the invention of dedicated machineries allowing post-translational import of most mitochondrial proteins (Neupert and Herrmann 2007; Chacinska et al. 2009) (see Chap. 8). Although small in number, the mitochondrially encoded proteins represent the reactions centers of the respiratory chain and are essential for oxidative phosphorylation. The effort to express these few genes is immense: In simple eukaryotes such as *Saccharomyces cerevisiae*, almost 250 different proteins are required to maintain the genetic system and to allow synthesis of the mitochondrially encoded proteins (Sickmann et al. 2003). The genes which are (almost) consistently present in mitochondrial genomes encode the central membrane-embedded reaction centers of the respiratory chain subunits (Fig. 16.1): ND1, ND2, ND3, ND4, ND4L, ND5, and ND6 of NADH-dehydrogenase (complex I); cytochrome *b* of cytochrome *c* reductase (complex III); Cox1, Cox2, and Cox3 of cytochrome *c* oxidase (complex IV); and Atp6 and Atp8 of the ATP-synthase

K. Kehrein • M. Ott (✉)
Center for Biomembrane Research, Department for Biochemistry and Biophysics, University of Stockholm, Stockholm, Sweden
e-mail: martin.ott@dbb.su.se

C.E. Bullerwell (ed.), *Organelle Genetics*,
DOI 10.1007/978-3-642-22380-8_16, © Springer-Verlag Berlin Heidelberg 2012

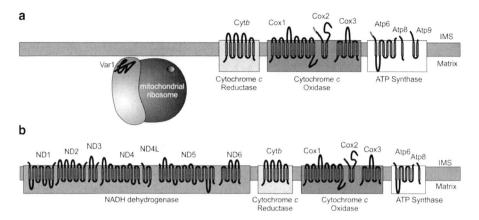

Fig. 16.1 Mitochondrially encoded proteins. (**a**) Mitochondrially encoded proteins of *S. cerevisiae*. The mitochondrial genome of baker's yeast encodes seven subunits of the oxidative phosphorylation system as well as one protein of the small ribosomal subunit. All eight proteins are synthesized on mitochondrial ribosomes close to the inner membrane. Cox2 and Atp6 are processed after insertion. (**b**) Mitochondrially encoded proteins of mammalian mitochondria. All 13 mitochondrially encoded proteins are exclusively membrane proteins. In contrast to *S. cerevisiae*, subunit 9 of the ATP-synthase and the ribosomal protein Var1 are not encoded in mammalian mitochondrial genomes

(complex V). Only in a few examples some of these genes are absent. For instance, the mitochondrial DNA of baker's yeast, the model system most commonly used to investigate molecular mechanisms of translational control and respiratory chain assembly in mitochondria, lacks complex I and hence the mitochondrial genes encoding its subunits. Instead, this mitochondrial DNA contains additionally the genes coding for Atp9 of ATP-synthase and Var1, a constituent of the small ribosomal subunit (Borst and Grivell 1978).

16.1.1 Why Do Mitochondria Still Synthesize Proteins?

It remains a puzzling question why – after the transfer of hundreds of genes – the last few mitochondrial genes obviously had to be retained within the organelle despite the enormous effort for the cell. Three mutually nonexclusive explanations for the presence of DNA in mitochondria were put forward which are supported by experimental evidence: (1) the organelle-specific codon usage that deviated substantially from the standard genetic code inhibits gene transfer to the nucleus. In most mitochondria, the codon TAG is translated to a tryptophan while it is normally used as a stop codon to terminate translation. Tryptophan is a typical amino acid of the proteins encoded in mitochondrial genomes, because most of them are highly hydrophobic membrane proteins. A transfer of these genes to the nucleus will not result in a functional nuclear gene because they are punctuated with stop codons.

Instead, many rounds of evolutionary modifications are necessary to recode the TAG codons to tryptophan codons used in the cytoplasm. A successful gene transfer to the nucleus followed by proper random recoding is an unlikely event, leading to a trapping of the genes in the mitochondrial genetic system. (2) The extreme hydrophobicity of some proteins might prevent their efficient import from the cytosol to the inner membrane. This clearly could explain why mitochondrial genomes encode primarily hydrophobic proteins. Experiments in which a recoded version of cytochrome *b*, a protein normally encoded by mitochondrial DNA, was fused to a mitochondrial presequence and synthesized in the cytosol indeed revealed an unproductive aggregation of this protein that impaired import (Claros et al. 1995). However, experiments with Atp6 expressed allotopically from a nuclear gene in human cells suggest that import of this hydrophobic protein into mitochondria is possible, though extremely inefficient (Manfredi et al. 2002). Likewise, screening of mutations in a recoded version of *COX2* (a mitochondrially gene encoding a subunit of cytochrome *c* oxidase) resulted in the identification of a version of Cox2 that can be post-translationally imported into mitochondria and assembled into a functional cytochrome *c* oxidase (Supekova et al. 2010). In these versions, a tryptophan in the first transmembrane segment of the polypeptide was exchanged to arginine, thus decreasing significantly the hydrophobicity of the protein. (3) An alternative explanation for the retention of mitochondrial DNA is the possibility to couple synthesis and assembly in order to regulate gene expression. Such an elegant regulatory circuit exists for expression of *COX1* and ATP-synthase genes in yeast mitochondria which is explained in Sect. 16.1.4.2.

16.1.2 The General Mechanism of Protein Synthesis

To date, no robust system could be established by which molecular mechanisms of protein synthesis by mitochondrial ribosomes can be studied in vitro. Hence, much of our current understanding stems from the characterization of protein synthesis in prokaryotes, the system most closely related to that of mitochondria. Because of the evolutionary relationship between bacteria and mitochondria, a series of complementation experiments could be performed, allowing to shed light on conserved and diverged aspects of mitochondrial translation.

The ribosomes of the model bacterium *Escherichia coli* consist of small and large subunits that sediment as 30 S and 50 S particles, respectively. Both subunits are composed of protein and RNA elements. Three essential binding sites for tRNAs are located at the surface of the small subunit that contacts the large subunit: The amino acyl-tRNA-site (A-site), the peptidyl-tRNA-site (P-site) and the exit-site (E-site). These serve as docking sites for the tRNAs and help to decode repeatedly the mRNA in a highly accurate manner. During elongation of the polypeptide, alternate rounds of decoding, peptide bond formation, and translocation of the mRNA proceed until the stop codon signals termination of translation. These reactions are controlled and mediated by a number of conserved translation factors.

Table 16.1 Translation factors of bacteria and their mitochondrial homologs

Eubacteria	Saccharomyces cerevisiae mitochondria	Mammalian mitochondria
Initiation factors		
IF1	Not identified; IF2 might perform IF1 function	Not identified; IF2 might perform IF1 function
IF2	MIF2	IF2(mt)
IF3	Not identified	IF3(mt)
Elongation factors		
EF-G	Mef1/Mef2	TUF1
EF-TU	Tuf1	mtEF-TU
EF-Ts	Not identified; Tuf1 has GDP exchange function	mtEF-Ts
EF4 (LepA)	Guf1	Guf1
Release factors		
RF1	MRF1	mtRF1a
RF2	Not identified	Not identified
RF3	Not identified	Not identified
Functional homologs not identified	Yol114cp (putative homolog of Ict1)	ICT1

In the following, the molecular mechanisms of those factors in bacterial protein synthesis will be shortly explained and these insights will be used to compare them to so-far determined mechanistic details of mitochondrial translation and the so-far identified factors (Table 16.1). A detailed review on the pioneering work of Linda Spremulli on mammalian mitochondrial translation factors has been published (Spremulli et al. 2004).

16.1.2.1 Translation Initiation

In bacteria, interactions of most mRNAs with ribosomes are established by the help of complementary RNA sequences known as Shine–Dalgarno sequences. These are present in the 5′-untranslated region (UTR) of the mRNA and in 3′-end of the rRNA of the small subunit. By the help of these sequences, the mRNA interacts with the small subunit and the distance between the Shine–Dalgarno sequence and the Start-AUG is important for the correct loading of the mRNA onto the ribosome. The interaction of the mRNA with the small subunit is controlled by three initiation factors (IF1-3) and utilizes the initiator formylmethionine-tRNA, (fMet-tRNA) (Fig. 16.2). IF-1 blocks the A-site to inhibit binding of fMet-tRNA to this site. IF-3 binds to the small subunit and prevents it from interacting prematurely with the large subunit. Binding of the mRNA to this complex is facilitated by the help of the Shine–Dalgarno interaction. Next, the ribosome-bound GTPase IF-2 positions fMet-tRNA on the AUG codon of the bound mRNA. The large subunit is recruited to this complex and the initiation factors are released. The ribosome is now assembled and can start protein synthesis.

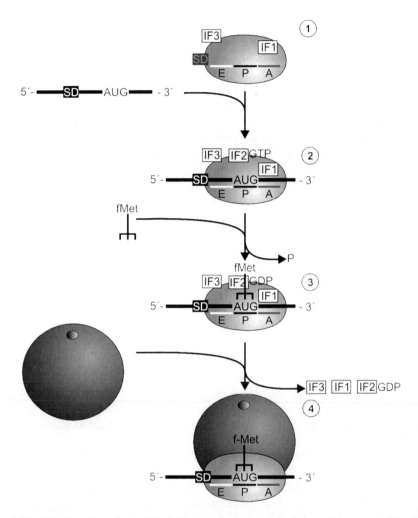

Fig. 16.2 Initiation of translation. The initiation of protein synthesis in prokaryotes requires three initiation factors (IF1, IF2, IF3; *white boxes*). IF1 and IF3 bind to a free small subunit (1). The mRNA is bound and the start codon is properly placed at the P-site of the ribosome by interaction of the Shine–Dalgarno sequence (SD) at the 5′ end of the mRNA and the 16S rRNA (2). mRNA binding to this complex recruits IF2GTP and this factor positions fMet-tRNA onto the AUG at the P-site (3). After GTP hydrolysis by IF2, the initiation factors are released and the large ribosomal subunit joins the initiation complex (4)

In mitochondria, the scenario is slightly different. Mitochondrial mRNAs do not contain sequences similar to the Shine–Dalgarno sequences and it is unclear how initiation of translation on mitochondrial ribosomes is accomplished. Mitochondria possess homologs of bacterial IF2 and IF3 (Koc and Spremulli 2002; Liao and Spremulli 1990) and are termed IF2(mt) and IF3(mt), respectively, but so far no homolog to IF1 has been found. Like in the bacterial system, the molecular function of IF2(mt) is to promote the binding of fMet-tRNA to the AUG codon at the P-site (Spencer and Spremulli 2004). When expressed in yeast, the mammalian mitochondrial factor can also function without a formylated methionine (Tibbetts et al. 2003). Consequently, methionine formylation is not required for yeast mitochondrial translation as deletion of the gene encoding mitochondrial formyltransferase does not abolish respiratory growth (Li et al. 2000). An accessory protein for the binding of not-formylated Met-tRNA to IF-2 might be the factor Aep3 (Lee et al. 2009), but this has not been directly demonstrated.

The mammalian IF3(mt) contains additional N- and C-terminal extensions. Analyses of variants of IF3(mt) where these domains are deleted suggest that they have evolved to ensure the proper dissociation of IF3(mt) from the small subunit upon joining of the large subunit during initiation (Haque et al. 2008). Another function of IF3(mt) appears to be to promote the dissociation of fMet-tRNA that has bound to a small subunit in the absence of a correctly loaded mRNA (Bhargava and Spremulli 2005; Christian and Spremulli 2009).

While many steps of mitochondrial initiation are still unknown, it is clear that a random initiation on mRNAs does not occur. Mammalian mitochondria contain mRNAs with only very few, if any, additional nucleotides at the 5′-end. A recent report indicated that an initiation complex can be formed with isolated bovine ribosomes, initiation factors, and fMet-tRNA only when an mRNA was included that contained AUG at or very close to the 5′-end (Christian and Spremulli 2010). Similarly, experiments in yeast showed that a re-initiation on mRNAs containing two functional coding sequences on one transcript is not possible (Bonnefoy and Fox 2000). To date, no experiments have been reported that allow establishing an initiation complex in vitro that can proceed to elongation. Many attempts have been undertaken to establish a manipulatable system of mitochondrial translation. In one of these, McGregor et al. used electroporation to deliver mRNAs for translation into isolated mitochondria. This method did indeed allow to transfer mRNAs into the mitochondrial matrix; the mRNAs, however, were not translated (McGregor et al. 2001). These results could suggest that in organello additional factors and mechanisms might be involved in transfer of the mRNA from transcription to initiation of translation (Rodeheffer and Shadel 2003).

16.1.2.2   Elongation of the Peptide Chain

The elongation of the polypeptide chain in bacteria occurs by a repetitive set of reactions, termed the elongation cycle (Fig. 16.3). These reactions involve three elongation factors, termed EF-Tu, EF-G, and EF-Ts. The elongation cycle starts with the binding of the correct amino acyl-tRNA to the A-site (decoding). Next, the growing polypeptide is transferred to the A-site-bound tRNA (peptidyl transfer), and the mRNA and the interacting tRNAs are moved by a base-triplet in a reaction termed translocation. The GTPase EF-Tu delivers the amino acyl-tRNAs to the A-site. To do so, EF-Tu binds GTP and this GTP is hydrolyzed when the amino acyl-tRNA is accommodated correctly in the A-site. This step is crucial for translation,

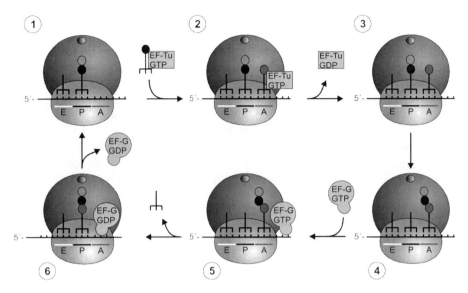

Fig. 16.3 Elongation of the polypeptide during translation. Elongation of the polypeptide occurs in a repetitive reaction. The GTPase EF-TU delivers the amino acyl-tRNA to the A-site of the ribosome (1, 2). When the tRNA is properly attached to the codon at the A-site, EF-Tu detaches after GTP hydrolysis, leaving an aminoacyl-tRNA at the A-site (3). The polypeptide chain of the P-site bound tRNA is transferred onto the A-site tRNA (peptidyl transfer, 4). The mRNA and the bound tRNAs are moved through the ribosomes by the help of the EF-G that hydrolyzes GTP (5), and the empty tRNA at the E-site exits the ribosome (6). Next, EF-G (with bound GDP) is released and the EF-Tu can deliver another aminoacyl-tRNA to the empty A-site (1)

as it directly determines the accuracy of translation. The interaction of the anticodon with the amino acid-specifying codon on the mRNA is proofread thermodynamically (Rodnina and Wintermeyer 2001) aided by complex conformational changes that finally result in the hydrolysis of GTP and the dissociation of EF-Tu from the ribosome. Next, the polypeptide is transferred by the peptidyltransferase center from the P-site-bound tRNA to the amino acyl-tRNA of the A-site, leaving an uncharged tRNA in the P-site. EF-Tu is recharged with GTP by the help of the nucleotide exchange factor EF-Ts and forms another trimeric complex by binding an amino acyl-tRNA.

Next, the mRNA and the bound tRNAs are moved through the ribosome by a reaction catalyzed by the GTPase EF-G. This factor induces, by the help of GTP hydrolysis, a large conformational change of the ribosome resulting in a rotation of the small subunit relative to the large subunit. After GTP hydrolysis, EF-G dissociates from the ribosome and the ribosome rotates back. The movement of the ribosomal subunits resembles a ratchet-like mechanism (Frank and Agrawal 2000) that has to be accomplished very precisely to ensure movement of the mRNA by only three nucleotides. To rescue noncorrectly translocated ribosomes, bacterial ribosomes have a fourth elongation factor, termed LepA (or EF4) (Qin et al. 2006). This factor recognizes ribosomes that underwent a defective translocation reaction and induces back-translocation, thus giving EF-G a second chance to translocate the ribosome correctly. Recent analyses of this factor in *E. coli* suggest that it might play an important role in controlling ribosomal dynamics and timing without having a direct effect on fidelity of translation (Shoji et al. 2010).

Mitochondria contain homologs to EF-Tu, EF-Ts (in some fungal mitochondria, EF-Ts is not present because fungal mitochondrial EF-Tus have a high affinity for GTP) and EF-G and it is assumed that the general mechanism of the elongation cycle is well conserved, albeit small changes in ribosome interaction might occur (Piechulla and Kuntzel 1983; Nagata et al. 1983; Eberly et al. 1985; Schwartzbach and Spremulli 1989; Schwartzbach et al. 1996; Chung and Spremulli 1990). Mitochondria also contain the Guf1-protein that is homologous to LepA and it was shown that it is important for mitochondrial protein synthesis under suboptimal conditions (Bauerschmitt et al. 2008).

16.1.2.3 Termination and Recycling

Translation is terminated when a stop codon is present in the A-site (Fig. 16.4). In bacteria, the three different stop codons are recognized by the release factors RF-1 and RF-2. The binding of both factors by the help of their anticodon-like domains leads to the dissociation of the polypeptide from the P-site-bound tRNA. RF-1 and RF-2 are released from the ribosome by the help of the GTPase RF-3. Post-translational ribosomes are split into small and large subunits by the concerted activity of EF-G and the GTPase RRF (ribosome recycling factor); the individual subunits can be used for another round of translation when translation is initiated on a free small subunit.

Baker's yeast mitochondria have one release factor, termed mRF1, that is required for respiratory growth (Towpik et al. 2004). In contrast, mammalian mitochondria contain two release factors that are closely related to RF1. Analyses

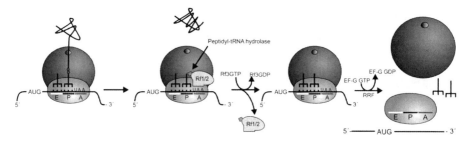

Fig. 16.4 Termination of translation and recycling of ribosomes. Termination of translation is initiated when a stop codon is present at the A-site. This codon is recognized by class I release (RF1 and RF2) factors by protein–RNA interactions. The bound release factors hydrolyze the peptide-tRNA bond to release the newly synthesized protein from the ribosome. The class II release factor RF3 induces liberation of Rf1/2. The resulting post-translational complex is split by the action of ribosome recycling factor and EF-G

using a heterologous system of fission yeast and mammalian factors showed that mtRF1a acts as the general termination release factor, decoding the two major stop codons UAA and UAG of human mitochondria (Soleimanpour-Lichaei et al. 2007). Until recently, however, it was unclear how the other two stop codons used in mammalian mitochondria, AGA and AGG, respectively, are recognized to induce termination. Recent work by Chrzanowska-Lightowlers and co-workers demonstrated how this works: Directly in front of these codons are uracil bases. They are important because by 3'-located secondary structures that both block the ribosome and induce a tension on the mRNA, the reading frame slides by −1 from AGA (or AGG) to UAG, thus allowing mtRF1 to terminate translation by interacting with the standard stop codon (Temperley et al. 2010). In addition, the codons AGA and AGG are hardly used, thus giving the system additional pausing time to allow frameshifting.

Mitochondria have at least one additional release factor, termed ICT1. This protein appears to play a dedicated role in rescuing translation complexes that contain mRNAs lacking stop codons (Fig. 16.5) (Haque and Spremulli 2010; Richter et al. 2010). In the bacterial case, such a scenario is resolved by tmRNA.

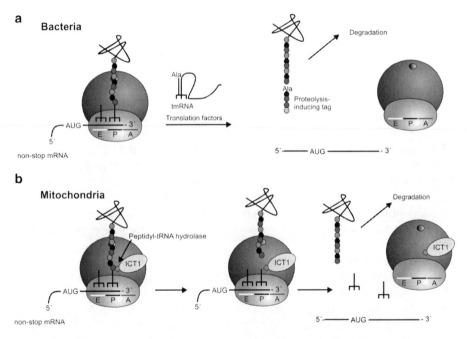

Fig. 16.5 Different strategies to deal with ribosomes that are stalled on mRNAs lacking a stop codon. (**a**) Stalled ribosomes are rescued by tmRNAs in bacteria. Bacteria use trans-translation to release ribosomes trapped on a nonstop mRNA. During this process a tmRNA allows to continue translation by supplying a short template with a stop codon that adds a proteolysis-inducing tag to the broken polypeptide. Thus, the ribosome can be recycled and the broken polypeptide is degraded rapidly. (**b**) Rescue of stalled ribosomes by ICT1 in mammalian mitochondria. The ribosomal protein ICT1 acts as a peptidyl-tRNA hydrolase that cleaves the peptidyl-tRNA when no intact codon is present at the A-site. This cleavage releases the nascent chain and allows recycling the ribosome for further translation

This tmRNA recognizes ribosomes that are stalled on the 3'-end of an mRNA. The tmRNA then binds to the A-site, allowing to transfer the polypeptide onto the bound alanine. Next, the ribosome uses sequence information of the tmRNA to finish protein synthesis employing a stop codon in this sequence. The released polypeptide is rapidly removed because of the presence of a proteolysis promoting tag at the C terminus, which has been introduced by translation of the tmRNA. Mitochondria do not have tmRNAs but instead employ ICT1 that appears to be permanently bound to the large subunit of mitochondrial ribosomes. When mitochondrial ribosomes are stalled on an mRNA lacking a stop codon, ICT1 can hydrolyze polypeptides from peptidyl-tRNAs, thus liberating nascent chains and ribosomal subunits that can subsequently be recycled (Richter et al. 2010). The recycling of mitochondrial ribosomes is mediated by the mitochondrial homolog of RRF, termed mtRRF (Rorbach et al. 2008; Teyssier et al. 2003; Zhang and Spremulli 1998). This factor can associate with mitochondrial ribosomes and is essential for respiratory growth.

16.1.3   Accuracy of Translation

The quality of newly synthesized proteins is determined by speed and accuracy of translation. Under in vivo conditions, translation is a highly precise process with amino acid mis-incorporation occurring at a frequency of 1 in 10^3–10^4 (Rodnina and Wintermeyer 2001). The two steps that are responsible for this fidelity are, on the one hand, the aminoacylation of tRNAs with the cognate amino acid and, on the other hand, the accurate selection of the correct aminoacyl-tRNAs by the ribosome. Aminoacylation by aa-tRNA synthetases is a remarkably precise process with only 1 error in 10^4–10^5 (Soll 1990; Swanson et al. 1988), indicating that accuracy of translation in vivo is mainly determined by the frequency of mistakes during decoding of the mRNA by the ribosome. The precision of decoding by ribosomes is mainly achieved by a kinetic proofreading of the codon–anticodon interaction (Rodnina and Wintermeyer 2001; Rodnina et al. 2005). In vitro, the preferential selection of cognate over near-cognate tRNAs is highly dependent on Mg^{2+} and polyamines. Polyamines reduce the error frequency, while high concentrations of Mg^{2+} decrease fidelity by reducing the efficiency of the initial selection (Pape et al. 1999). This tRNA selection process can also be impaired by several drugs. One class of drugs that is especially important in this context is the group of the aminoglycoside antibiotics such as streptomycin, neomycin, kanamycin, paromomycin, or gentamycin. These antibiotics reduce the fidelity of translation by binding to the 16S rRNA in vicinity to the decoding site of bacterial ribosomes (Davies and Davis 1968). In addition, accurate translation in bacteria is influenced by three ribosomal proteins, namely S12, S4, and S5 (Piepersberg et al. 1975; Zaher and Green 2010; Yaguchi et al. 1975). Certain mutations in S4 and S5 lead to the so-called *ram* (for *r*ibosome *am*biguity) mutations that reduce the level of accuracy in a way similar to streptomycin, whereas some S12 mutations have an opposite effect and increase fidelity (Rosset and Gorini 1969; Ozaki et al. 1969; Biswas and Gorini 1972). Furthermore, Noller and colleagues could show that mutations in S12, S5, and S4 result in structural alterations of the RNA components of the decoding center, which might directly change accuracy of translation (Allen and Noller 1989; Powers and Noller 1991). Although the accuracy of translation is mainly a feature of the 30 S subunit, certain mutations of the large ribosomal subunit L6 could be identified that cause resistance to aminoglycosides, notably gentamycin (Buckel et al. 1977).

In contrast to the situation for bacterial ribosomes, only little is known about translation fidelity in mitochondria. Because of the importance of mitochondrial translation for the assembly of the respiratory chain, mistakes in this process can have fatal consequences for the cell. Consequently, a series of mitochondrial diseases are linked to translation accuracy in mitochondria such as MELAS and MERRF (Sasarman et al. 2008). Both disorders are caused by the absence of a functional tRNA due to a mutation in the mitochondrial genome (Kobayashi et al. 1990; Goto et al. 1994; Shoffner et al. 1990). In the case of MERRF, a decrease in translational accuracy has been reported (Sasarman et al. 2008).

Mitochondrial ribosomes are sensitive to aminoglycoside antibiotics. Similar to the bacterial system, they decrease fidelity of translation (Zagorski et al. 1987). In addition to general effects of these drugs on mitochondrial translation, certain mutations in the mitochondrial genome are found to cause hypersensitivity against aminoglycosides, which results in the loss of cochlear neurons and, consequently, in deafness (Kokotas et al. 2007). These mutations in the mitochondrial DNA typically affect the 12S rRNA (Prezant et al. 1993; Matthijs et al. 1996; Rydzanicz et al. 2010). Hence, it was suggested that the susceptibility to aminoglycosides in patients with nonsydromic hearing loss due to mutations in the 12S rRNA are caused by alterations of the secondary structure of this rRNA molecule (Ballana et al. 2006). However, and in contrast to the bacterial system, the mechanistical insights into the accuracy of translation in mitochondria are scarce and not much is known how this parameter of mitochondrial biogenesis is influenced by the cell and its surroundings.

16.1.4   Translational Control in Mitochondria

Biogenesis of mitochondrially encoded proteins starts with their transcription. As detailed in Sect. V, the general features of mitochondrial transcription differ significantly between species. In mammalian mitochondria, the RNA polymerase is directed to the start of transcription by a specificity factor that recognizes only three distinct sites on the mitochondrial DNA. Next, the RNA polymerase synthesizes large polycistronic RNA precursors that are matured by a series of subsequent reactions to release tRNAs, rRNAs, as well as mRNAs (Ojala et al. 1981; Montoya et al. 1981). The mRNAs of mammalian mitochondria contain only very short 5'- and 3'-untranslated regions. In contrast, fungal mRNAs contain large

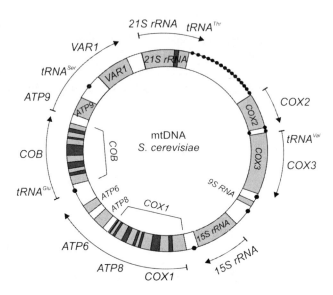

Fig. 16.6 Transcription of mitochondrially encoded genes in *S. cerevisiae*. The mitochondrial genome of *S. cerevisiae* encodes 8 proteins, 2 rRNAs, 1 subunit of RNAseP (*light gray boxes*) and 24 tRNAs (*black dots*). *COB, COX1*, and 21S rRNA are interrupted by introns (*dark gray boxes*). Almost all genes are co-transcribed with a tRNA or a proximate gene. Transcription units and the directions of RNA synthesis are indicated by *arrows*

5′- and 3′-UTRs that are important to regulate protein synthesis. In contrast to mammalian mitochondria, the yeast specificity factor recognizes 19 transcription sites in the mitochondrial DNA (Fig. 16.6). In some cases, a primary transcript is produced that can be used without further maturation. Other transcripts have to undergo complex processing to either liberate the functional entities such as tRNAs and mRNAs or remove a number of introns by group I as well as group II intron splicing (Fig. 16.6).

Expression of specific sets of mitochondrially encoded proteins is not controlled at the level of transcription. Instead, post-transcriptional mechanisms directly control levels of translation in mitochondria. This central role is governed by a class of messenger-specific mitochondrial proteins, the so-called translational activators (Table 16.2). In yeast, and presumably also in other species (Weraarpachai et al. 2009; Sasarman et al. 2010), each mitochondrial mRNA requires other factors for translation. While it is clear that the long 5′-UTRs of mitochondrial transcripts are the target of these factors in yeast, no information exists on how translational regulation might be exerted on the small, almost leaderless mRNAs of mammalian mitochondria. In the following, we will summarize the current knowledge on translational control in yeast, because this system is best understood owing to the powerful genetics that can be applied to the analysis of mitochondrial biogenesis.

Table 16.2 Translational activators of mitochondrially encoded proteins in yeast

Protein	Function	Acts on 5'UTR of mRNA	Localisation	mRNA stability impaired	Mutant phenotype
Cbs1	Translational activator of *COB*	+	Associated with the inner membrane	−	Accumulation of *COB* splicing intermediates; accumulation of mature *COX1* mRNA abnormal
Cbs2	Translational activator of *COB*	+	Associated with the inner membrane	−	Accumulation of *COB* splicing intermediates; accumulation of mature *COX1* mRNA abnormal
Cbp6	Enhances *COB* translation	?	?	−	Reduced Cob levels
Cbp1	Stabilization of *COB* mRNA	+	Peripheral associated with the inner membrane	+	Reduced levels of pre-*COB* mRNA; mature *COB* mRNA missing; tRNA^{Glu} level normal
Pet309	Translational activator of *COX1*; involved in stabilization of *COX1* mRNA	+	Inner membrane spanning Protein with domains facing the IMS	+	Cox1 missing ; mature *COX1* and pre-*COX1* mRNA missing in case of intron containing *COX1*; mature and pre-*COX1* mRNA unaffected in case of intronless COX1
Mss51	Translational activator of *COX1*; couples Cox1 synthesis to COX assembly	+	associated with the inner membrane	−	Cox1 missing
Pet111	Translational activator of *COX2*	+	Inner membrane bound	+	Defective in Cox2 synthesis, mature *COX2* mRNA levels slightly reduced

(continued)

Table 16.2 (continued)

Protein	Function	Acts on 5'UTR of mRNA	Localisation	mRNA stability impaired	Mutant phenotype
Pet54	Translational activator of *COX3*	+	Peripheral inner membrane protein	−	Cox3 missing ; Cox1 reduced; splicing of *COX1* I5β blocked
Pet122	Translational activator of *COX3*	+	Integral inner membrane protein	−	Cox3 missing
Pet494	Translational activator of *COX3*	+	Integral inner membrane protein	−	Cox3 missing
Atp22	Translational activator of *ATP6*	+	Component of the inner membrane	−	Defective in Atp6 synthesis, defective in F_o assembly; high frequency of ρ^- and ρ^0 cells
Aep3	Stabilization of *ATP6/ATP8* mRNA	?	Inner membrane protein facing the matrix	+	*ATP8/6* mRNA processing impaired; high frequency of ρ^- and ρ^0 cells
Aep1/ Nca1	Required for expression of *ATP9*	+	No predicted membrane spanning domains; probably soluble protein	−	Defective in Atp9 synthesis, mature *ATP9* mRNA present
Aep2/ Atp13	Likely involved in *ATP9* translation and stabilization of mRNA	+	Predicted soluble mitochondrial protein	+	Defective in Atp9 synthesis, mature *ATP9* mRNA missing
Atp25	Involved in stabilization of *ATP9* mRNA and oligomerization of the Atp9 ring	?	Mitochondrial inner membrane protein	+	Deficit of *ATP9* mRNA and Atp9 protein; defective in F_o assembly

16.1.4.1 Synthesis of Apocytochrome *b*

Complex III of *S. cerevisiae* is composed of ten subunits that are all encoded in the nucleus with the exception of cytochrome *b* (Cob). Synthesis of Cob requires three nuclear encoded translational activators called Cbs1, Cbs2, and Cbp6 (Rödel 1986; Dieckmann and Tzagoloff 1985). The *COB* gene is one of three intron-containing genes in the yeast mitochondrial genome and is co-transcribed with the upstream tRNA^{Glu} (Christianson et al. 1983). Hence, the primary transcription unit is a bicistronic RNA molecule composed of tRNA^{Glu} and *COB*. For maturation, it undergoes multiple processing steps, including the removal of five introns (bI1-bI5). The introns bI2, bI3, and bI4 encode so-called maturases, proteins required for the excision of the introns that encode them. The ORFs of these maturases are in frame with the upstream exon and therefore dependent on functional and accurate translation of *COB* (Lazowska et al. 1980; Nobrega and Tzagoloff 1980).

The first protein identified that appeared to be essential for Cob synthesis was Cbs1. It is a mitochondrial protein with a size of about 27 kDa including a 3.5 kDa presequence that is removed after mitochondrial import (Korte et al. 1989). It tightly associates with the inner membrane but lacks transmembrane segments (Krause-Buchholz et al. 2000). Cbs2 was also identified as a membrane-associated protein but this interaction seems not to be as firm as that of Cbs1 (Michaelis et al. 1991). Cbs2 has a molecular weight of 45 kDa and has no cleavable mitochondrial targeting signal.

A first indication that Cbs1 and Cbs2 act on the 5'-UTR of *COB* mRNA came from the analyses of suppressor mutants lacking *CBS1* and *CBS2*. In these spontaneous occurring mutants, the original 5'-UTR of *COB* was exchanged by that of *ATP9*. This restored respiratory growth and synthesis of Cob (Rödel 1986; Rödel et al. 1985). A similar experiment indicated that the synthesis of Cox3 is dependent on Cbs1 when the protein is synthesized from an mRNA containing the 5'-UTR of *COB* but not when bearing the authentic *COX3* 5'-UTR (Rödel and Fox 1987). However, it is still unknown whether Cbs1 and Cbs2 interact directly with the 5'-UTR of *COB* or whether they activate other proteins. The 954-bp-long 5'-UTR of *COB* has no obvious features that could suggest possible binding sites. Using mitochondrial genetics, it has been shown that sequence elements between -232 and -4 are important for translation of *COB* and thereby possible targets of Cbs1 and Cbs2 (Mittelmeier and Dieckmann 1995).

Besides Cbs1 and Cbs2, another protein, Cbp6, has been identified to be involved in Cob synthesis (Dieckmann and Tzagoloff 1985). Interestingly, a replacement of the *COB* 5'-UTR cannot rescue the phenotype as seen for Cbs1 and Cbs2 (Tzagoloff et al. 1988). This observation would argue for a possible dual function. An involvement of Cbp6 in Cob synthesis could be mediated by a participation in the formation of the initiation complex or by assisting the ribosome to find the right start codon (Dieckmann and Tzagoloff 1985).

As described above, the expression of *COB* depends on a set of nuclear genes. Among these genes some are required for translation (Cbs1, Cbs2, Cbp6), others for excision of introns and intervening sequences. *CBP1* (for cytochrom *b* processing) is a further nuclear gene involved in Cob biogenesis. Cbp1 is synthesized in the cytosol as a 76-kDa pre-protein that is matured after import into mitochondria (Weber and Dieckmann 1990), where it is peripherally associated with the membrane by hydrophobic interactions (Krause et al. 2004) (Islas-Osuna et al. 2003). Cbp1 is positively charged (pI = 10.36) and especially the N- and C-terminal ends contain many arginine and lysine residues pointing to a possible function in binding to nucleic acids (Dieckmann et al. 1984). Likewise, Cbp1 is required for accumulation of *COB* mRNA by preventing the degradation of unprocessed *COB* transcripts produced by endonucleolytic cleavage at the 3'-end of tRNA^{Glu} (Dieckmann et al. 1984; Weber and Dieckmann 1990; Dieckmann et al. 1982). Δ*cbp1* mutants have tRNA^{Glu} levels close to that of the wild type, whereas pre-*COB* mRNA is reduced to about 25% of wild-type levels. Mature *COB* mRNA is not present in these cells, which accounts for the respiration-deficient phenotype. A CCG triplet located near the 5'-end of *COB* mRNA (−944 to 942) is essential for Cbp1-dependent stability (Chen and Dieckmann 1997) because mutations of any of these nucleotides result in degradation of the mRNA. Hence, it appears that Cbp1 acts through a sequence near the 5'- end of the *COB* mRNA (−961 to −898), whereas Cbs1 and Cbs2 act on a sequence between −232 and −4 (Chen and Dieckmann 1997; Mittelmeier and Dieckmann 1995). One possible model for the function of Cbp1 is that it associates with the *COB* mRNA already during transcription and then delivers the mRNA to the membrane-bound Cbs1 and Cbs2, allowing both proteins to promote the association of the mRNA with mitochondrial ribosomes (Islas-Osuna et al. 2002).

16.1.4.2   Synthesis of Cox1, Cox2, and Cox3

The mechanisms regulating translation of the three mRNAs specifying mitochondrially encoded subunits of cytochrome *c* oxidase are quite well understood. The first factors acting as translational activators were identified already during the early work on the characterization of yeast mutants unable to respire. Two of these mutants were especially interesting because they completely lacked synthesis of Cox2 and Cox3 and this absence of translation could be explained by the absence of functional versions of two different nuclear genes (*PET111* and *PET494*, respectively) (Cabral and Schatz 1978). Subsequent analyses of both gene products by Tom Fox and co-workers revealed a number of different aspects of translational control regulating the synthesis of mitochondrially encoded cytochrome *c* oxidase subunits.

Pet111 is a membrane-bound protein with an apparent molecular mass of 94 kDa that is present in very low quantities in the mitochondrial matrix (Green-Willms et al. 2001). The absence of this protein destabilizes the *COX2* mRNA. A direct interaction of Pet111 with the mRNA was suggested by the analysis of spontaneous revertants of Δ*pet111* cells. These revertants were heteroplasmic for mitochondrial DNA and carried, in addition to a wild type mitochondrial genome, genomes where 5'-portions of *COX1* or *ATP9* were fused in frame to the coding sequence of *COX2* (Poutre and Fox 1987). Subsequent analyses indicated that Pet111 acts specifically on the 5'-UTR of *COX2* (Mulero and Fox 1993b) because a point mutation in this 5'-UTR (that results in a respiratory growth defect) can be suppressed by a point mutation in *PET111* (Mulero and Fox 1993a). This 5'-UTR is, in comparison to other 5'-UTR of mRNAs of yeast mitochondria, relatively short and comprises 54 nucleotides. It contains a stem loop formed by the sequence UAGACAAAAGAGUCUA. Genetic modifications of this region impair Cox2 synthesis in mitochondria, suggesting that this structure might in fact be the element that is recognized by Pet111 (Dunstan et al. 1997). The exact molecular functions that Pet111 exerts on the mRNA to control its translation on mitochondrial ribosomes have not yet been identified.

The situation for *COX3* mRNA is a little bit more complex. Cox3 synthesis requires three gene products, namely Pet54, Pet122, and Pet494. Pet122 and Pet494 interact tightly with the inner membrane (McMullin and Fox 1993) while Pet54 is a peripheral membrane protein with an additional role in the maturation of *COX1* (Valencik and McEwen 1991). Pair-wise yeast two hybrid analyses suggest that the three proteins form a complex in which Pet54 interacts with both Pet122 and Pet494 while both membrane-embedded factors do not show a direct interaction (Brown et al. 1994). They all act on the 5'-UTR of *COX3* mRNA that is 613 nucleotides long. Within this domain, a stretch of 159 nucleotides is apparently the target of translational activation (Costanzo and Fox 1993). *PET494* is expressed at very low levels and the expression is modulated significantly according to the metabolic state of the cell (Marykwas and Fox 1989). Hence, it was suggested that tuning the amounts of Pet494 (and probably also of other translational activators that are expressed at similarly limiting levels) might allow adjusting the expression of the mitochondrially encoded *COX3* by the help of general regulation of nuclear genes. By this mechanism, nuclear gene control would directly modulate gene expression in mitochondria.

Cox1 is the largest of the three mitochondrially encoded COX subunits. It spans the membrane 12 times, with the N- and C-terminus facing the matrix. To function in the context of cytochrome *c* oxidase, it has to be equipped with redox cofactors. Synthesis of Cox1 is regulated by two specific proteins, Mss51 and Pet309. Mss51 was initially identified as a factor involved in the splicing necessary for the maturation of *COX1* mRNA (Simon and Faye 1984b). Mss51 is a peripheral membrane protein residing in the mitochondrial matrix. Subsequent experiments revealed that this protein does not play a direct role in splicing of *COX1* but is instead required for efficient synthesis of Cox1 (Decoster et al. 1990). Yeast-three hybrid analyses suggest that Mss51 indeed interacts with the 5'-UTR of *COX1* (Zambrano et al. 2007). Interestingly, exchange of the 5'-UTR of *COX1* does not allow circumventing the requirement of Mss51 for Cox1 biogenesis (Perez-Martinez et al. 2003). The other Cox1 translational activator, Pet309, was identified as a factor that stabilizes *COX1* mRNA and at the same time is required for translational activation of the messenger (Manthey and McEwen 1995). Pet309 contains at least seven pentatricopeptide repeats (PPRs), the signature of the class of PPR proteins that are RNA-binding proteins of the genetic systems of mitochondria and chloroplasts. Like most translational activators, Pet309 strongly interacts with the inner membrane (Manthey et al. 1998).

Mss51 has an additional function that links the protein to the assembly of cytochrome *c* oxidase. The first hints to such a role were obtained from analyzing yeast mutants lacking *SHY1*, the yeast homolog of a human gene implicated in Leigh's syndrome (Barrientos et al. 2002), which is a fatal mitochondrial disorder. In these mutants, a number of repressor mutations were identified that allowed respiratory growth in the absence of Shy1. Genetic analyses of these revertants showed that the mutation responsible for the reversion mapped to the *MSS51* gene. While newly synthesized Cox1 is hardly detectable and unstable in the absence of Shy1, the mutated forms of Mss51 allowed a robust expression of *COX1* as well as an increased stability to the protein (Barrientos et al. 2002). Subsequent analyses by the group of Tom Fox showed that Mss51 does not only activate translation of mRNAs containing the 5'-UTR of *COX1*, but is also required for the biogenesis of Cox1 when Cox1 is synthesized from an mRNA containing the 5'-UTR of *COX2* (Perez-Martinez et al. 2003). This direct function of Mss51 on Cox1 apparently involves a direct interaction of both proteins, because newly synthesized Cox1 can be efficiently crosslinked to Mss51 (Perez-Martinez et al. 2003). This interaction of Mss51 with newly synthesized Cox1 occurs in a complex that contains Cox14 (Barrientos et al. 2004). Deletion of *COX14* allows robust synthesis of Cox1 even in the absence of Cox assembly, while this is normally not possible as deletion of structural genes of cytochrome *c* oxidase or its assembly factors normally inhibit *COX1* translation. Because Cox14 does also interact with newly synthesized Cox1, Barrientos and co-workers concluded that synthesis of this mitochondrially encoded protein is regulated by a feedback loop employing Mss51 and Cox14 (Fig. 16.7) (Barrientos et al. 2004). Binding of Mss51 to newly synthesized Cox1 sequesters Mss51 until the newly synthesized Cox1 is assembled into cytochrome *c* oxidase. Only then Mss51 is liberated and can activate a new round of Cox1 synthesis. By this feedback mechanism, the amounts of newly synthesized Cox1 are adjusted to levels that can be successfully incorporated into a functional cytochrome *c* oxidase (Mick et al. 2011). Such a feedback loop possibly inhibits unwanted accumulation of nonassembled Cox1 in the membrane that is potentially harmful because it might give rise to reactive oxygen species due to its redox-active cofactors

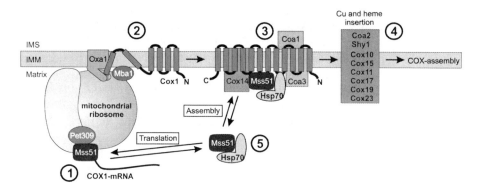

Fig. 16.7 Translational control on *COX1*. Mss51 activates Cox1 synthesis by interaction with the *COX1* 5'-UTR (1) and Cox1 is co-translationally inserted into the inner membrane (2). Here, Mss51 (in a complex with Hsp70), Cox14, Coa1, and Coa3 bind newly synthesized Cox1 (3). Cox1 is further assembled with the help of a variety of assembly factors (orange box). This assembly liberates Mss51 from Cox1 (5), thus allowing a new round of Cox1 translation (1). In the case of a defect in assembly of the cytochrome *c* reductase, Mss51 is sequestered in a complex with Cox1 and thereby prevented to activate translation of Cox1

(Khalimonchuk et al. 2007). Recent work indicated that in addition to Cox14, three other proteins are part of the Mss51 regulatory complex, namely Coa1, Shy1, and Cox25/Coa3 (Pierrel et al. 2007; Mick et al. 2007, 2010; Fontanesi et al. 2011). The mitochondrial Hsp70, Ssc1, interacts with Mss51 to support the biogenesis of Cox1 (Fontanesi et al. 2010). The sequence of Cox1 that is recognized by the regulatory complex is the C-terminal region of Cox1, because genetic ablation of this region intersects the translational control by Mss51 (Shingu-Vazquez et al. 2010).

16.1.4.3   Controlling the Synthesis of the Membrane-Embedded F_o-Part of ATP-Synthase

Only little is known about the regulatory mechanisms involved in the biogenesis of the mitochondrial encoded subunits of ATP-synthase. One reason for this could be that ATP-synthase-deficient mutants have unstable mitochondrial genomes resulting in the accumulation of ρ^- and ρ^0 cells. In yeast the ATP-synthase is composed of 17 different subunit proteins of which three (Atp6, Atp8, and Atp9) are encoded in the mitochondrial genome. These three components are part of the membrane-embedded F_o-portion of the enzyme. *ATP6* and *ATP8* are transcribed as a polycistronic mRNA together with *COX1* (Figs. 16.6 and 16.8). The individual mRNAs are released through endonucleotic cleavage (Simon and Faye 1984a). A set of nuclear proteins has been shown to be involved in the stabilization of the resulting bicitronic and/or the single mRNAs, respectively, namely Nca3, Nca2, Nam1, and Aep3 (Ellis et al. 2004; Camougrand et al. 1995; Pelissier et al. 1995; Groudinsky et al. 1993).

Atp22 was first identified as a protein with an essential function in the assembly of the F_o sector of the ATP-synthase (Helfenbein et al. 2003). Some years later, an additional role as an *ATP6*-specific translation factor could be attributed to Atp22 (Zeng et al. 2007). This 71-kDa protein has overall a hydrophilic character but it was found to be a component of the inner membrane (Helfenbein et al. 2003). $\Delta atp22$ mutants are defective in F_o and impaired in the synthesis of Atp6 (Helfenbein et al. 2003; Zeng et al. 2007). A function of Atp22 as a translational activator of *ATP6* was indicated by analyzing a hybrid gene consisting of the 5'-UTR, first exon, and first intron of *COX1* fused to the sequence of *ATP6* beginning with the fourth codon, which was able to rescue the $\Delta atp22$ mutant by restoring Atp6 synthesis. This fusion between the *COX1* parts and *ATP6* causes a substitution of the *COX1* 5'-UTR for the normal 5'-UTR of *ATP6*, thus allowing the *COX1*-specific translational activators to regulate translation in the absence of Atp22 (Zeng et al. 2007). This assumption was confirmed by the observation that Atp22 is able to regulate translation of the mitochondrial reporter *ARG8m* when the coding sequence of this recoded gene is inserted in mitochondrial DNA in place of that of *ATP6* (Rak and Tzagoloff 2009).

Recently, Rak et al. showed that the translation of *ATP6* and *ATP8* is directly regulated by F_1. In the absence of α and β subunits of F_1 or of their chaperones

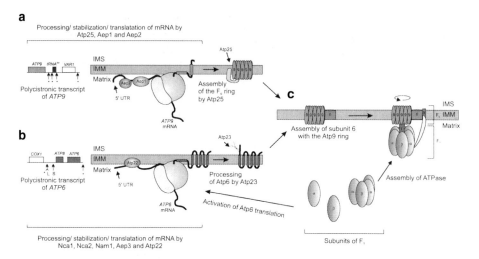

Fig. 16.8 Expression of the mitochondrially encoded subunits *ATP6* and *ATP9* and their assembly in the F$_o$ part of the ATPase in *Saccharomyces cerevisiae*. (**a**) Atp9 is co-transcribed as polycistronic precursor transcripts with *VAR1* and tRNA^{Ser}. Cleavage sites of the polycictronic transcripts are shown by *asterisks*. Three genes *(ATP25, AEP1* and *AEP2)* have been identified that influence the stability and translation of the *ATP9* mRNA. After synthesis of Atp9, an oligomeric ring structure is assembled by Atp25. (**b**) *ATP6* is co-transcribed together with *COX1* and *ATP8*. This primary transcript is cleaved, resulting in *COX1* mRNA and the bicistronic *ATP8/ATP6* mRNA. Cleavage sites are indicated by *asterisks*. L and S mark the cleavage sites of the *ATP6* precursor molecule that produce the long (L) and short (S) form of the bicistronic *ATP8/ATP6* messenger. Nca1, Nca2, and Aep3 stabilize the *ATP6* mRNA during maturation. Synthesis of Atp6 depends on the interaction of Atp22 with the *ATP6* 5'-UTR. Atp6 is inserted into the membrane and subsequently processed by the metalloprotease Atp23. (**c**) Atp6 is assembled with the Atp9 ring and subsequently inserted into ATP-synthase. The presence of subunits of the F$_1$ part is required for synthesis of Atp6 and Atp8. This feedback mechanism adjusts expression of F$_o$ subunits to levels that can be successfully incorporated into ATP-synthase [see also review (Rak et al. 2009)]

Atp11 and Atp12, the synthesis of Atp6 and Atp8 is strongly reduced. This mechanism allows controlling translation of the mitochondrially encoded ATP-synthase genes during assembly. This control prevents the premature assembly of Atp6/Atp9 intermediates in the membrane that would dissipate the membrane potential by unregulated proton leak through Atp6 and thus create a dangerous scenario for the cell (Rak and Tzagoloff 2009). Similar to *ATP6*, also *ATP9* is transcribed as a polycistronic precursor RNA molecule encompassing tRNA^{Ser} and *VAR1* (Figs. 16.6 and 16.8) (Zassenhaus et al. 1984). Atp9 (or subunit c) forms the proton translocating sector of the ATP-synthase rotor. Its expression requires two nuclear genes called *AEP1/NCA1* and *AEP2/ATP13* (Payne et al. 1991, 1993) Mutations in *AEP1* result in the failure to produce Atp9 but the mature mRNA is detectable (Payne et al. 1991). Defects in Aep2 likewise cause failure in Atp9 synthesis but in contrast to *AEP1* mutations there are no detectable levels of mature *ATP9* mRNA (Ellis et al. 1999). Ellis et al. proposed that the inability to detect mature *ATP9* mRNA in the Δ*aep2* mutant is not caused by an influence of Aep2 on the stability of the *ATP9* mRNA but rather a consequence of impaired translation (Ellis et al. 1999).

Another nuclear gene implicated in expression of *ATP9* is *ATP25*. Atp25 is a component of the inner membrane and seems to be involved in both stabilization of the *ATP9* mRNA and the post-translational assembly of Atp9 into the oligomeric ring structure (Zeng et al. 2008). Interestingly, Atp25 is first synthesized as a 70-kDa protein that is then cleaved in two parts. The resulting 35 kDa C-terminal domain is sufficient to stabilize the *ATP9* mRNA and to restore Atp9 synthesis, but the regeneration of respiratory growth depends on both C-and N-terminal domain, indicating that the N-terminal domain has another function, presumably in the assembly of the Atp9 ring (Zeng et al. 2008).

Atp8 is, besides Var1, the only mitochondrially encoded protein for which no translation factors have been identified so far. The evidence that translation of *ATP8* is regulated by F_1 (see above) and that Atp22 is able to partially restore synthesis of Atp8 (Rak and Tzagoloff 2009) indicates that a still missing factor might be involved in regulating also the expression of this mitochondrially encoded gene.

16.1.4.4   What is the Molecular Function of Translational Activators in Mitochondrial Translation?

It is currently not known in which step of protein synthesis translational activators fulfill their function. Genetic evidence suggests that at least certain translational activators might play a role in the initiation of translation (Green-Willms et al. 1998; Nouet et al. 2007; Williams et al. 2007). Because Shine–Dalgarno-like sequences are not present in mitochondria, it is conceivable that mRNAs are loaded by other mechanisms onto mitochondrial ribosomes. A likely scenario is that this is mediated by the translational activators. Specifically, interactions of the translational activator with both the mRNA and the ribosome could allow aligning the start codon onto the P-site of the ribosome in a spatially correct way. An alternative, yet not exclusive, function might be to localize mitochondrial mRNAs to the inner face of the inner membrane, thus facilitating interaction with the membrane-bound ribosomes (Fox 1996).

Pioneering work of Tom Fox and co-workers demonstrated that the translational activators appear to also play a role in the general organization of mitochondrial translation. Such a function has been suggested by the observation that the translational activators controlling the synthesis of cytochrome *c* oxidase subunits can directly interact with each other and therefore allow the synthesis of subunits destined for the same complex in close proximity (Naithani et al. 2003). This spatial organization at the level of the inner membrane might increase the efficacy of respiratory chain biogenesis by channeling the assembly process.

16.2   Conclusion

Due to the lack of a robust system to study mitochondrial translation in vitro, only some molecular mechanisms of mitochondrial protein synthesis have been clarified. It will therefore be important to address this fundamental process with biochemically and genetically well-defined strategies. Clearly, many challenging question are open that await methodological as well as conceptual advances.

Acknowledgments   We would like to thank all members of our group and Jan Riemer (University of Kaiserslautern) for stimulating discussions. We thank Flavia Fontanesi and Toni Barrientos (University of Miami, USA) for comments and suggestions on the manuscript. Work in our laboratory is supported by grants from the Swedish research council (VR), the Center for Biomembrane Research (CBR) at Stockholm University, Stiftung Rheinland-Pfalz für Innovation, Deutscher akademischer Austauschdienst (DAAD), Deutsche Forschungsgemeinschaft and the Rheinland-Pfalz Research Initiative Membrane Transport. Kirsten Kehrein is supported by a PhD-grant from the Carl Zeiss Stiftung.

References

Allen PN, Noller HF (1989) Mutations in ribosomal proteins S4 and S12 influence the higher order structure of 16 S ribosomal RNA. J Mol Biol 208(3):457–468

Ballana E, Morales E, Rabionet R, Montserrat B, Ventayol M, Bravo O, Gasparini P, Estivill X (2006) Mitochondrial 12S rRNA gene mutations affect RNA secondary structure and lead to variable penetrance in hearing impairment. Biochem Biophys Res Commun 341 (4):950–957

Barrientos A, Korr D, Tzagoloff A (2002) Shy1p is necessary for full expression of mitochondrial COX1 in the yeast model of Leigh's syndrome. EMBO J 21(1–2):43–52

Barrientos A, Zambrano A, Tzagoloff A (2004) Mss51p and Cox14p jointly regulate mitochondrial Cox1p expression in *Saccharomyces cerevisiae*. EMBO J 23(17):3472–3482

Bauerschmitt H, Funes S, Herrmann JM (2008) The membrane-bound GTPase Guf1 promotes mitochondrial protein synthesis under suboptimal conditions. J Biol Chem 283(25):17139–17146

Bhargava K, Spremulli LL (2005) Role of the N- and C-terminal extensions on the activity of mammalian mitochondrial translational initiation factor 3. Nucleic Acids Res 33 (22):7011–7018

Biswas DK, Gorini L (1972) Restriction, de-restriction and mistranslation in missense suppression. Ribosomal discrimination of transfer RNA's. J Mol Biol 64(1):119–134

Bonnefoy N, Fox TD (2000) In vivo analysis of mutated initiation codons in the mitochondrial *COX2* gene of *Saccharomyces cerevisiae* fused to the reporter gene *ARG8m* reveals lack of downstream reinitiation. Mol Gen Genet 262(6):1036–1046

Borst P, Grivell LA (1978) The mitochondrial genome of yeast. Cell 15:705–723

Brown NG, Costanzo MC, Fox TD (1994) Interactions among three proteins that specifically activate translation of the mitochondrial COX3 mRNA in *Saccharomyces cerevisiae*. Mol Cell Biol 14(2):1045–1053

Buckel P, Buchberger A, Bock A, Wittmann HG (1977) Alteration of ribosomal protein L6 in mutants of *Escherichia coli* resistant to gentamicin. Mol Gen Genet 158(1):47–54

Cabral F, Schatz G (1978) Identification of cytochrome c oxidase subunits in nuclear yeast mutants lacking the functional enzyme. J Biol Chem 253(12):4396–4401

Camougrand N, Pelissier P, Velours G, Guerin M (1995) NCA2, a second nuclear gene required for the control of mitochondrial synthesis of subunits 6 and 8 of ATP synthase in *Saccharomyces cerevisiae*. J Mol Biol 247(4):588–596

Chacinska A, Koehler CM, Milenkovic D, Lithgow T, Pfanner N (2009) Importing mitochondrial proteins: machineries and mechanisms. Cell 138(4):628–644

Chen W, Dieckmann CL (1997) Genetic evidence for interaction between Cbp1 and specific nucleotides in the 5′ untranslated region of mitochondrial cytochrome b mRNA in *Saccharomyces cerevisiae*. Mol Cell Biol 17(11):6203–6211

Christian BE, Spremulli LL (2009) Evidence for an active role of IF3mt in the initiation of translation in mammalian mitochondria. Biochemistry 48(15):3269–3278

Christian BE, Spremulli LL (2010) Preferential selection of the 5′-terminal start codon on leaderless mRNAs by mammalian mitochondrial ribosomes. J Biol Chem 285(36):28379–28386

Christianson T, Edwards JC, Mueller DM, Rabinowitz M (1983) Identification of a single transcriptional initiation site for the glutamic tRNA and COB genes in yeast mitochondria. Proc Natl Acad Sci USA 80(18):5564–5568

Chung HK, Spremulli LL (1990) Purification and characterization of elongation factor G from bovine liver mitochondria. J Biol Chem 265(34):21000–21004

Claros MG, Perea J, Shu YM, Samatey FA, Popot JL, Jacq C (1995) Limitations to in vivo import of hydrophobic proteins into yeast mitochondria – the case of a cytoplasmically synthesized apocytochrome b. Eur J Biochem 228(3):762–771

Costanzo MC, Fox TD (1993) Suppression of a defect in the 5′ untranslated leader of mitochondrial COX3 mRNA by a mutation affecting an mRNA-specific translational activator protein. Mol Cell Biol 13(8):4806–4813

Davies J, Davis BD (1968) Misreading of ribonucleic acid code words induced by aminoglycoside antibiotics. The effect of drug concentration. J Biol Chem 243(12):3312–3316

Decoster E, Simon M, Hatat D, Faye G (1990) The *MSS51* gene product is required for the translation of the *COX1* mRNA in yeast mitochondria. Mol Gen Genet 224(1):111–118

Dieckmann CL, Tzagoloff A (1985) Assembly of the mitochondrial membrane system. *CBP6*, a yeast nuclear gene necessary for synthesis of cytochrome b. J Biol Chem 260(3):1513–1520

Dieckmann CL, Pape LK, Tzagoloff A (1982) Identification and cloning of a yeast nuclear gene (*CBP1*) involved in expression of mitochondrial cytochrome b. Proc Natl Acad Sci USA 79(6):1805–1809

Dieckmann CL, Koerner TJ, Tzagoloff A (1984) Assembly of the mitochondrial membrane system. *CBP1*, a yeast nuclear gene involved in 5′ end processing of cytochrome b pre-mRNA. J Biol Chem 259(8):4722–4731

Dunstan HM, Green-Willms NS, Fox TD (1997) In vivo analysis of *Saccharomyces cerevisiae* COX2 mRNA 5′-untranslated leader functions in mitochondrial translation initiation and translational activation. Genetics 147(1):87–100

Eberly SL, Locklear V, Spremulli LL (1985) Bovine mitochondrial ribosomes. Elongation factor specificity. J Biol Chem 260(15):8721–8725

Ellis TP, Lukins HB, Nagley P, Corner BE (1999) Suppression of a nuclear aep2 mutation in *Saccharomyces cerevisiae* by a base substitution in the 5′-untranslated region of the mitochondrial oli1 gene encoding subunit 9 of ATP synthase. Genetics 151(4):1353–1363

Ellis TP, Helfenbein KG, Tzagoloff A, Dieckmann CL (2004) Aep3p stabilizes the mitochondrial bicistronic mRNA encoding subunits 6 and 8 of the H+−translocating ATP synthase of *Saccharomyces cerevisiae*. J Biol Chem 279(16):15728–15733

Fontanesi F, Soto IC, Horn D, Barrientos A (2010) Mss51 and Ssc1 facilitate translational regulation of cytochrome c oxidase biogenesis. Mol Cell Biol 30(1):245–259

Fontanesi F, Clemente P, Barrientos A (2011) Cox25 teams up with Mss51, Ssc1 and Cox14 to regulate mitochondrial cytochrome C oxidase subunit 1 expression and assembly in *Saccharomyces cerevisiae*. J Biol Chem 286(1):555–566

Fox TD (1996) Translational control of endogenous and recoded nuclear genes in yeast mitochondria: regulation and membrane targeting. Experientia 52(12):1130–1135

Frank J, Agrawal RK (2000) A ratchet-like inter-subunit reorganization of the ribosome during translocation. Nature 406(6793):318–322

Goto Y, Tsugane K, Tanabe Y, Nonaka I, Horai S (1994) A new point mutation at nucleotide pair 3291 of the mitochondrial tRNA(Leu(UUR)) gene in a patient with mitochondrial myopathy, encephalopathy, lactic acidosis, and stroke-like episodes (MELAS). Biochem Biophys Res Commun 202(3):1624–1630

Green-Willms NS, Fox TD, Costanzo MC (1998) Functional interactions between yeast mitochondrial ribosomes and mRNA 5' untranslated leaders. Mol Cell Biol 18(4):1826–1834

Green-Willms NS, Butler CA, Dunstan HM, Fox TD (2001) Pet111p, an inner membrane-bound translational activator that limits expression of the *Saccharomyces cerevisiae* mitochondrial gene COX2. J Biol Chem 276(9):6392–6397

Groudinsky O, Bousquet I, Wallis MG, Slonimski PP, Dujardin G (1993) The NAM1/MTF2 nuclear gene product is selectively required for the stability and/or processing of mitochondrial transcripts of the atp6 and of the mosaic, cox1 and cytb genes in *Saccharomyces cerevisiae*. Mol Gen Genet 240(3):419–427

Haque ME, Spremulli LL (2010) ICT1 comes to the rescue of mitochondrial ribosomes. EMBO J 29(6):1019–1020

Haque ME, Grasso D, Spremulli LL (2008) The interaction of mammalian mitochondrial translational initiation factor 3 with ribosomes: evolution of terminal extensions in IF3mt. Nucleic Acids Res 36(2):589–597

Helfenbein KG, Ellis TP, Dieckmann CL, Tzagoloff A (2003) *ATP22*, a nuclear gene required for expression of the F0 sector of mitochondrial ATPase in *Saccharomyces cerevisiae*. J Biol Chem 278(22):19751–19756

Islas-Osuna MA, Ellis TP, Marnell LL, Mittelmeier TM, Dieckmann CL (2002) Cbp1 is required for translation of the mitochondrial cytochrome *b* mRNA of *Saccharomyces cerevisiae*. J Biol Chem 277(41):37987–37990

Islas-Osuna MA, Ellis TP, Mittelmeier TM, Dieckmann CL (2003) Suppressor mutations define two regions in the Cbp1 protein important for mitochondrial cytochrome b mRNA stability in *Saccharomyces cerevisiae*. Curr Genet 43(5):327–336

Khalimonchuk O, Bird A, Winge DR (2007) Evidence for a pro-oxidant intermediate in the assembly of cytochrome oxidase. J Biol Chem 282(24):17442–17449

Kobayashi Y, Momoi MY, Tominaga K, Momoi T, Nihei K, Yanagisawa M, Kagawa Y, Ohta S (1990) A point mutation in the mitochondrial tRNA(Leu)(UUR) gene in MELAS (mitochondrial myopathy, encephalopathy, lactic acidosis and stroke-like episodes). Biochem Biophys Res Commun 173(3):816–822

Koc EC, Spremulli LL (2002) Identification of mammalian mitochondrial translational initiation factor 3 and examination of its role in initiation complex formation with natural mRNAs. J Biol Chem 277(38):35541–35549

Kokotas H, Petersen MB, Willems PJ (2007) Mitochondrial deafness. Clin Genet 71(5):379–391

Korte A, Forsbach V, Gottenof T, Rodel G (1989) In vitro and in vivo studies on the mitochondrial import of CBS1, a translational activator of cytochrome b in yeast. Mol Gen Genet 217(1):162–167

Krause K, Lopes de Souza R, Roberts DG, Dieckmann CL (2004) The mitochondrial message-specific mRNA protectors Cbp1 and Pet309 are associated in a high-molecular weight complex. Mol Biol Cell 15(6):2674–2683

Krause-Buchholz U, Tzschoppe K, Paret C, Ostermann K, Rodel G (2000) Identification of functionally important regions of the *Saccharomyces cerevisiae* mitochondrial translational activator Cbs1p. Yeast 16(4):353–363

Lazowska J, Jacq C, Slonimski PP (1980) Sequence of introns and flanking exons in wild-type and box3 mutants of cytochrome b reveals an interlaced splicing protein coded by an intron. Cell 22(2 Pt 2):333–348

Lee C, Tibbetts AS, Kramer G, Appling DR (2009) Yeast AEP3p is an accessory factor in initiation of mitochondrial translation. J Biol Chem 284(49):34116–34125

Li Y, Holmes WB, Appling DR, RajBhandary UL (2000) Initiation of protein synthesis in *Saccharomyces cerevisiae* mitochondria without formylation of the initiator tRNA. J Bacteriol 182(10):2886–2892

Liao HX, Spremulli LL (1990) Identification and initial characterization of translational initiation factor 2 from bovine mitochondria. J Biol Chem 265(23):13618–13622

Manfredi G, Fu J, Ojaimi J, Sadlock JE, Kwong JQ, Guy J, Schon EA (2002) Rescue of a deficiency in ATP synthesis by transfer of MTATP6, a mitochondrial DNA-encoded gene, to the nucleus. Nat Genet 30(4):394–399

Manthey GM, McEwen JE (1995) The product of the nuclear gene *PET309* is required for translation of mature mRNA and stability or production of intron-containing RNAs derived from the mitochondrial *COX1* locus of *Saccharomyces cerevisiae*. EMBO J 14(16):4031–4043

Manthey GM, Przybyla-Zawislak BD, McEwen JE (1998) The *Saccharomyces cerevisiae* Pet309 protein is embedded in the mitochondrial inner membrane. Eur J Biochem 255(1):156–161

Marykwas DL, Fox TD (1989) Control of the *Saccharomyces cerevisiae* regulatory gene PET494: transcriptional repression by glucose and translational induction by oxygen. Mol Cell Biol 9(2):484–491

Matthijs G, Claes S, Longo-Mbenza B, Cassiman JJ (1996) Non-syndromic deafness associated with a mutation and a polymorphism in the mitochondrial 12S ribosomal RNA gene in a large Zairean pedigree. Eur J Hum Genet 4(1):46–51

McGregor A, Temperley R, Chrzanowska-Lightowlers ZM, Lightowlers RN (2001) Absence of expression from RNA internalised into electroporated mammalian mitochondria. Mol Genet Genomics 265(4):721–729

McMullin TW, Fox TD (1993) COX3 mRNA-specific translational activator proteins are associated with the inner mitochondrial membrane in *Saccharomyces cerevisiae*. J Biol Chem 268(16):11737–11741

Michaelis U, Körte A, Rödel G (1991) Association of cytochrome *b* translational activator proteins with the mitochondrial membrane: implications for cytochrome *b* expression in yeast. Mol Gen Genet 230:177–185

Mick DU, Wagner K, van der Laan M, Frazier AE, Perschil I, Pawlas M, Meyer HE, Warscheid B, Rehling P (2007) Shy1 couples Cox1 translational regulation to cytochrome *c* oxidase assembly. EMBO J 26(20):4347–4358

Mick DU, Vukotic M, Piechura H, Meyer HE, Warscheid B, Deckers M, Rehling P (2010) Coa3 and Cox14 are essential for negative feedback regulation of *COX1* translation in mitochondria. J Cell Biol 191(1):141–154

Mick DU, Fox TD, Rehling P (2011) Inventory control: cytochrome c oxidase assembly regulates mitochondrial translation. Nat Rev Mol Cell Biol 12(1):14–20

Mittelmeier TM, Dieckmann CL (1995) In vivo analysis of sequences required for translation of cytochrome *b* transcripts in yeast mitochondria. Mol Cell Biol 15(2):780–789

Montoya J, Ojala D, Attardi G (1981) Distinctive features of the 5'-terminal sequences of the human mitochondrial mRNAs. Nature 290(5806):465–470

Mulero JJ, Fox TD (1993a) Alteration of the *Saccharomyces cerevisiae* COX2 mRNA 5'-untranslated leader by mitochondrial gene replacement and functional interaction with the translational activator protein PET111. Mol Biol Cell 4(12):1327–1335

Mulero JJ, Fox TD (1993b) PET111 acts in the 5'-leader of the *Saccharomyces cerevisiae* mitochondrial COX2 mRNA to promote its translation. Genetics 133(3):509–516

Nagata S, Tsunetsugu-Yokota Y, Naito A, Kaziro Y (1983) Molecular cloning and sequence determination of the nuclear gene coding for mitochondrial elongation factor Tu of *Saccharomyces cerevisiae*. Proc Natl Acad Sci USA 80(20):6192–6196

Naithani S, Saracco SA, Butler CA, Fox TD (2003) Interactions among *COX1*, *COX2*, and *COX3* mRNA-specific translational activator proteins on the inner surface of the mitochondrial inner membrane of *Saccharomyces cerevisiae*. Mol Biol Cell 14:324–333

Neupert W, Herrmann JM (2007) Translocation of proteins into mitochondria. Annu Rev Biochem 76:723–749

Nobrega FG, Tzagoloff A (1980) Assembly of the mitochondrial membrane system. DNA sequence and organization of the cytochrome b gene in *Saccharomyces cerevisiae* D273-10B. J Biol Chem 255(20):9828–9837

Nouet C, Bourens M, Hlavacek O, Marsy S, Lemaire C, Dujardin G (2007) Rmd9p controls the processing/stability of mitochondrial mRNAs and its overexpression compensates for a partial deficiency of oxa1p in *Saccharomyces cerevisiae*. Genetics 175(3):1105–1115

Ojala D, Montoya J, Attardi G (1981) tRNA punctuation model of RNA processing in human mitochondria. Nature 290(5806):470–474

Ozaki M, Mizushima S, Nomura M (1969) Identification and functional characterization of the protein controlled by the streptomycin-resistant locus in *E. coli*. Nature 222(5191):333–339

Pape T, Wintermeyer W, Rodnina M (1999) Induced fit in initial selection and proofreading of aminoacyl-tRNA on the ribosome. EMBO J 18(13):3800–3807

Payne MJ, Schweizer E, Lukins HB (1991) Properties of two nuclear pet mutants affecting expression of the mitochondrial oli1 gene of *Saccharomyces cerevisiae*. Curr Genet 19(5):343–351

Payne MJ, Finnegan PM, Smooker PM, Lukins HB (1993) Characterization of a second nuclear gene, AEP1, required for expression of the mitochondrial OLI1 gene in *Saccharomyces cerevisiae*. Curr Genet 24(1–2):126–135

Pelissier P, Camougrand N, Velours G, Guerin M (1995) NCA3, a nuclear gene involved in the mitochondrial expression of subunits 6 and 8 of the Fo-F1 ATP synthase of *S. cerevisiae*. Curr Genet 27(5):409–416

Perez-Martinez X, Broadley SA, Fox TD (2003) Mss51p promotes mitochondrial Cox1p synthesis and interacts with newly synthesized Cox1p. EMBO J 22(21):5951–5961

Piechulla B, Kuntzel H (1983) Mitochondrial polypeptide elongation factor EF-Tu of *Saccharomyces cerevisiae*. Functional and structural homologies to *Escherichia coli* EF-Tu. Eur J Biochem 132(2):235–240

Piepersberg W, Bock A, Wittmann HG (1975) Effect of different mutations in ribosomal protein S5 of *Escherichia coli* on translational fidelity. Mol Gen Genet 140(2):91–100

Pierrel F, Bestwick ML, Cobine PA, Khalimonchuk O, Cricco JA, Winge DR (2007) Coa1 links the Mss51 post-translational function to Cox1 cofactor insertion in cytochrome c oxidase assembly. EMBO J 26(20):4335–4346

Poutre CG, Fox TD (1987) PET111, a *Saccharomyces cerevisiae* nuclear gene required for translation of the mitochondrial mRNA encoding cytochrome c oxidase subunit II. Genetics 115(4):637–647

Powers T, Noller HF (1991) A functional pseudoknot in 16S ribosomal RNA. EMBO J 10(8): 2203–2214

Prezant TR, Agapian JV, Bohlman MC, Bu X, Oztas S, Qiu WQ, Arnos KS, Cortopassi GA, Jaber L, Rotter JI et al (1993) Mitochondrial ribosomal RNA mutation associated with both antibiotic-induced and non-syndromic deafness. Nat Genet 4(3):289–294

Qin Y, Polacek N, Vesper O, Staub E, Einfeldt E, Wilson DN, Nierhaus KH (2006) The highly conserved LepA is a ribosomal elongation factor that back-translocates the ribosome. Cell 127(4):721–733

Rak M, Tzagoloff A (2009) F1-dependent translation of mitochondrially encoded Atp6p and Atp8p subunits of yeast ATP synthase. Proc Natl Acad Sci USA 106(44):18509–18514

Rak M, Zeng X, Briere JJ, Tzagoloff A (2009) Assembly of F0 in *Saccharomyces cerevisiae*. Biochim Biophys Acta 1793(1):108–116

Richter R, Rorbach J, Pajak A, Smith PM, Wessels HJ, Huynen MA, Smeitink JA, Lightowlers RN, Chrzanowska-Lightowlers ZM (2010) A functional peptidyl-tRNA hydrolase, ICT1, has been recruited into the human mitochondrial ribosome. EMBO J 29(6):1116–1125

Rodeheffer MS, Shadel GS (2003) Multiple interactions involving the amino-terminal domain of yeast mtRNA polymerase determine the efficiency of mitochondrial protein synthesis. J Biol Chem 278(20):18695–18701

Rödel G (1986) Two yeast nuclear genes, *CBS1* and *CBS2*, are required for translation of mitochondrial transcripts bearing the 5′-untranslated *COB* leader. Curr Genet 11(1):41–45

Rödel G, Fox TD (1987) The yeast nuclear gene *CBS1* is required for translation of mitochondrial mRNAs bearing the cob 5' untranslated leader. Mol Gen Genet 206(1):45–50

Rödel G, Korte A, Kaudewitz F (1985) Mitochondrial suppression of a yeast nuclear mutation which affects the translation of the mitochondrial apocytochrome *b* transcript. Curr Genet 9(8): 641–648

Rodnina MV, Wintermeyer W (2001) Fidelity of aminoacyl-tRNA selection on the ribosome: kinetic and structural mechanisms. Annu Rev Biochem 70:415–435

Rodnina MV, Gromadski KB, Kothe U, Wieden HJ (2005) Recognition and selection of tRNA in translation. FEBS Lett 579(4):938–942

Rorbach J, Richter R, Wessels HJ, Wydro M, Pekalski M, Farhoud M, Kuhl I, Gaisne M, Bonnefoy N, Smeitink JA, Lightowlers RN, Chrzanowska-Lightowlers ZM (2008) The human mitochondrial ribosome recycling factor is essential for cell viability. Nucleic Acids Res 36(18):5787–5799

Rosset R, Gorini L (1969) A ribosomal ambiguity mutation. J Mol Biol 39(1):95–112

Rydzanicz M, Wrobel M, Pollak A, Gawecki W, Brauze D, Kostrzewska-Poczekaj M, Wojsyk-Banaszak I, Lechowicz U, Mueller-Malesinska M, Oldak M, Ploski R, Skarzynski H, Szyfter K (2010) Mutation analysis of mitochondrial 12S rRNA gene in Polish patients with non-syndromic and aminoglycoside-induced hearing loss. Biochem Biophys Res Commun 395(1):116–121

Sagan L (1967) On the origin of mitosing cells. J Theor Biol 14(3):255–274

Sasarman F, Antonicka H, Shoubridge EA (2008) The A3243G tRNALeu(UUR) MELAS mutation causes amino acid misincorporation and a combined respiratory chain assembly defect partially suppressed by overexpression of EFTu and EFG2. Hum Mol Genet 17(23): 3697–3707

Sasarman F, Brunel-Guitton C, Antonicka H, Wai T, Shoubridge EA (2010) LRPPRC and SLIRP interact in a ribonucleoprotein complex that regulates posttranscriptional gene expression in mitochondria. Mol Biol Cell 21(8):1315–1323

Schwartzbach CJ, Spremulli LL (1989) Bovine mitochondrial protein synthesis elongation factors. Identification and initial characterization of an elongation factor Tu-elongation factor Ts complex. J Biol Chem 264(32):19125–19131

Schwartzbach CJ, Farwell M, Liao HX, Spremulli LL (1996) Bovine mitochondrial initiation and elongation factors. Methods Enzymol 264:248–261

Shingu-Vazquez M, Camacho-Villasana Y, Sandoval-Romero L, Butler CA, Fox TD, Perez-Martinez X (2010) The carboxyl-terminal end of Cox1 is required for feedback-assembly regulation of Cox1 synthesis in *Saccharomyces cerevisiae* mitochondria. J Biol Chem 285(45): 34382–34389

Shoffner JM, Lott MT, Lezza AM, Seibel P, Ballinger SW, Wallace DC (1990) Myoclonic epilepsy and ragged-red fiber disease (MERRF) is associated with a mitochondrial DNA tRNA(Lys) mutation. Cell 61(6):931–937

Shoji S, Janssen BD, Hayes CS, Fredrick K (2010) Translation factor LepA contributes to tellurite resistance in *Escherichia coli* but plays no apparent role in the fidelity of protein synthesis. Biochimie 92(2):157–163

Sickmann A, Reinders J, Wagner Y, Joppich C, Zahedi R, Meyer HE, Schonfisch B, Perschil I, Chacinska A, Guiard B, Rehling P, Pfanner N, Meisinger C (2003) The proteome of *Saccharomyces cerevisiae* mitochondria. Proc Natl Acad Sci USA 100(23):13207–13212

Simon M, Faye G (1984a) Organization and processing of the mitochondrial oxi3/oli2 multigenic transcript in yeast. Mol Gen Genet 196(2):266–274

Simon M, Faye G (1984b) Steps in processing of the mitochondrial cytochrome oxidase subunit I pre-mRNA affected by a nuclear mutation in yeast. Proc Natl Acad Sci USA 81(1):8–12

Soleimanpour-Lichaei HR, Kuhl I, Gaisne M, Passos JF, Wydro M, Rorbach J, Temperley R, Bonnefoy N, Tate W, Lightowlers R, Chrzanowska-Lightowlers Z (2007) mtRF1a is a human mitochondrial translation release factor decoding the major termination codons UAA and UAG. Mol Cell 27(5):745–757

Soll D (1990) The accuracy of aminoacylation-ensuring the fidelity of the genetic code. Experientia 46(11–12):1089–1096

Spencer AC, Spremulli LL (2004) Interaction of mitochondrial initiation factor 2 with mitochondrial fMet-tRNA. Nucleic Acids Res 32(18):5464–5470

Spremulli LL, Coursey A, Navratil T, Hunter SE (2004) Initiation and elongation factors in mammalian mitochondrial protein biosynthesis. Prog Nucleic Acid Res Mol Biol 77:211–261

Supekova L, Supek F, Greer JE, Schultz PG (2010) A single mutation in the first transmembrane domain of yeast COX2 enables its allotopic expression. Proc Natl Acad Sci USA 107(11): 5047–5052

Swanson R, Hoben P, Sumner-Smith M, Uemura H, Watson L, Soll D (1988) Accuracy of in vivo aminoacylation requires proper balance of tRNA and aminoacyl-tRNA synthetase. Science 242(4885):1548–1551

Temperley R, Richter R, Dennerlein S, Lightowlers RN, Chrzanowska-Lightowlers ZM (2010) Hungry codons promote frameshifting in human mitochondrial ribosomes. Science 327(5963):301

Teyssier E, Hirokawa G, Tretiakova A, Jameson B, Kaji A, Kaji H (2003) Temperature-sensitive mutation in yeast mitochondrial ribosome recycling factor (RRF). Nucleic Acids Res 31(14): 4218–4226

Tibbetts AS, Oesterlin L, Chan SY, Kramer G, Hardesty B, Appling DR (2003) Mammalian mitochondrial initiation factor 2 supports yeast mitochondrial translation without formylated initiator tRNA. J Biol Chem 278(34):31774–31780

Towpik J, Chacinska A, Ciesla M, Ginalski K, Boguta M (2004) Mutations in the yeast mrf1 gene encoding mitochondrial release factor inhibit translation on mitochondrial ribosomes. J Biol Chem 279(14):14096–14103

Tzagoloff A, Crivellone MD, Gampel A, Muroff I, Nishikimi M, Wu M (1988) Mutational analysis of the yeast coenzyme QH2-cytochrome c reductase complex. Philos Trans R Soc Lond B Biol Sci 319(1193):107–120

Valencik ML, McEwen JE (1991) Genetic evidence that different functional domains of the PET54 gene product facilitate expression of the mitochondrial genes COX1 and COX3 in *Saccharomyces cerevisiae*. Mol Cell Biol 11(5):2399–2405

Weber ER, Dieckmann CL (1990) Identification of the *CBP1* polypeptide in mitochondrial extracts from *Saccharomyces cerevisiae*. J Biol Chem 265(3):1594–1600

Weraarpachai W, Antonicka H, Sasarman F, Seeger J, Schrank B, Kolesar JE, Lochmuller H, Chevrette M, Kaufman BA, Horvath R, Shoubridge EA (2009) Mutation in TACO1, encoding a translational activator of COX I, results in cytochrome c oxidase deficiency and late-onset Leigh syndrome. Nat Genet 41(7):833–837

Williams EH, Butler CA, Bonnefoy N, Fox TD (2007) Translation initiation in *Saccharomyces cerevisiae* mitochondria: functional interactions among mitochondrial ribosomal protein Rsm28p, initiation factor 2, methionyl-tRNA-formyltransferase and novel protein Rmd9p. Genetics 175(3):1117–1126

Yaguchi M, Wittmann HG, Cabezon T, De Wilde M, Villarroel R, Herzog A, Bollen A (1975) Cooperative control of translational fidelity by ribosomal proteins in *Escherichia coli*. II. Localization of amino acid replacements in proteins S5 and S12 altered in double mutants resistant to neamine. Mol Gen Genet 142(1):35–43

Zagorski W, Boguta M, Mieszczak M, Claisse M, Guiard B, Spyridakis A, Slonimski PP (1987) Phenotypic suppression and nuclear accommodation of the mit- oxi1-V25 mutation in isolated yeast mitochondria. Curr Genet 12(5):305–310

Zaher HS, Green R (2010) Hyperaccurate and error-prone ribosomes exploit distinct mechanisms during tRNA selection. Mol Cell 39(1):110–120

Zambrano A, Fontanesi F, Solans A, de Oliveira RL, Fox TD, Tzagoloff A, Barrientos A (2007) Aberrant translation of cytochrome c oxidase subunit 1 mRNA species in the absence of Mss51p in the yeast *Saccharomyces cerevisiae*. Mol Biol Cell 18(2):523–535

Zassenhaus HP, Martin NC, Butow RA (1984) Origins of transcripts of the yeast mitochondrial var 1 gene. J Biol Chem 259(9):6019–6027

Zeng X, Hourset A, Tzagoloff A (2007) The *Saccharomyces cerevisiae* ATP22 gene codes for the mitochondrial ATPase subunit 6-specific translation factor. Genetics 175(1):55–63

Zeng X, Barros MH, Shulman T, Tzagoloff A (2008) ATP25, a new nuclear gene of *Saccharomyces cerevisiae* required for expression and assembly of the Atp9p subunit of mitochondrial ATPase. Mol Biol Cell 19(4):1366–1377

Zhang Y, Spremulli LL (1998) Identification and cloning of human mitochondrial translational release factor 1 and the ribosome recycling factor. Biochim Biophys Acta 1443(1–2):245–250

Chapter 17
Mitochondrial tRNA Structure, Identity, and Evolution of the Genetic Code

B. Franz Lang, Dennis Lavrov, Natacha Beck, and Sergey V. Steinberg

17.1 Introduction

This chapter provides an overview of organelle (in particular mitochondrial DNA-encoded) tRNAs. We begin with discussing their unorthodox (reduced) structural features, the basis for understanding reassignment of tRNA identity, and evolution of the genetic (translation) code. We will then go on to analyze challenges in predicting unorthodox tRNAs and discuss ideas for the development of better-performing tRNA annotation tools.

17.1.1 Finding Organelle tRNA Genes, Assigning Their Identity Together with the Genetic Code: An Intricate Task

Mitochondria and plastids, descendents of α-Proteobacteria and Cyanobacteria, respectively, possess their own DNA, transcription, and protein translation machineries, and therefore require sets of tRNAs that are sufficient to recognize all codons in protein-coding genes encoded by organelle DNA. These tRNAs are most often encoded by the organelle DNAs themselves, but may also be nucleus-encoded and transported into the organelle, in which case they are usually shared with the cytoplasmic translational machinery. As a first approximation, tRNA import may be predicted from genome sequence (by far the most common and

B.F. Lang (✉) • N. Beck • S.V. Steinberg
Centre Robert Cedergren, Département de Biochimie, Université de Montréal, Boulevard Edouard Montpetit, Montréal, QC, Canada
e-mail: Franz.Lang@Umontreal.ca

D. Lavrov
Department of Ecology, Evolution and Organismal Biology, Iowa State University, Ames, IA, USA

C.E. Bullerwell (ed.), *Organelle Genetics*,
DOI 10.1007/978-3-642-22380-8_17, © Springer-Verlag Berlin Heidelberg 2012

often only available data), when the set of inferred tRNA genes is insufficient for recognizing all codons in coding sequences (e.g., Burger et al. 1995; Gray et al. 1998; Bullerwell et al. 2003a, see also below). Loss of aminoacyl-tRNA synthetases (aaRSs) encoded in the nuclear genome are another indication (Haen et al. 2010). Yet, highly unorthodox structure, extensive tRNA editing, or the presence of genes with introns make detection of tRNA genes difficult. When it comes to predicting tRNA identities and the corresponding translation code, knowledge of the genome sequence alone is often insufficient. This is because tRNA editing and RNA modification may alter tRNA identity, and changes in the genetic code are easily detected only when stop codons are reassigned to specify amino acids (e.g., UGA from stop to tryptophan, or UAG from stop to leucine) but otherwise require sequence comparison and biochemical confirmation [such as re-assignment of AUA from isoleucine to methionine, and CUN from leucine to threonine in *Saccharomyces cerevisiae*, as reviewed in Miranda et al. 2006)]. Finally, there is uncertainty as to the translation code of intron ORFs. Organelle introns are mobile elements that transfer via intron-encoded homing endonuclease (group I) or reverse transcriptase/endonuclease (group II) activities (Michel and Lang 1985; Dujon et al. 1986; Zimmerly et al. 1995; Lucas et al. 2001; Galburt and Stoddard 2002). Yet, intron ORFs that are transferred from evolutionarily distant donors may have a codon usage pattern that differs from that of native organelle genes (e.g., including nonstandard UGA-tryptophan codons). For the descendents of the host cell, it is relevant only if the intron protein has a helper (maturase) function in RNA splicing, but such functions are likely acquired only with time, secondarily. Small amounts of maturase protein may even be produced by (probably inefficient) decoding of nonstandard codons, like UGA codons recognized by tRNAs with a CCA anticodon, as previously proposed for mitochondria of *Schizosaccharomyces* species (Bullerwell et al. 2003b); for more information on this topic see Sect. 17.3.2.

17.1.2   Other Ways to Complicate Organelle Life: Import of tRNAs and tRNA Genes

Only a few mitochondrial proteins are encoded by mtDNA; the majority are nucleus-encoded and transported into mitochondria. That tRNA import exists came with the realization that certain mt genomes lack a substantial number of the required tRNA genes [e.g., land plants, sea anemones, chytrid fungi, *Acanthamoeba castellanii* (Marechal-Drouard et al. 1990; Burger et al. 1995; Laforest et al. 1997; Beagley et al. 1998; Glover et al. 2001; Bullerwell et al. 2003a)] if not all of them [e.g., Trypanosoma, (Yermovsky-Kammerer and Hajduk 1999)]. For more details on this topic, see Sect. 17.4.2.

Another surprising aspect of tRNA import exists in flowering plant mitochondria. They not only actively import cytoplasmic tRNAs but also integrate foreign genomic DNAs into mtDNA – from a variety of sources including nuclear and plastid DNA (Koulintchenko et al. 2003). As a consequence, tRNA genes of obviously plastid DNA (ptDNA) origin have been identified in mtDNA [e.g., (Wintz et al. 1988; Joyce and Gray 1989; Binder et al. 1991)] that are transcribed and apparently functional in mitochondria, a total of five in potato and six in wheat (Marechal-Drouard et al. 1990; Glover et al. 2001). With plant mt tRNAs originating from up to three sources (native, imported as RNA, transferred via genomic DNA), plant mt ribosomes have to be quite flexible with respect to accommodating tRNA identity. Yet, although mtDNAs of bilaterian animals usually encode complete sets of tRNAs, their structural flexibility is even more pronounced (see Sect. 17.2). Evidently, organelle tRNAs have a general tendency to function under relaxed structural constraints.

17.2 Features of Organelle tRNA Structure

Bacterial and cytosolic eukaryotic tRNAs are characterized by a common secondary structure often visualized as a cloverleaf (Fig. 17.1a). Yet, in three-dimensional (3D) space tRNAs take on an L-shape structure (Fig. 17.1b), which is stabilized by tertiary interactions and is required for proper interaction with the ribosome's A- and P-sites. In contrast, mt-tRNAs of bilaterian animals often do not conform to this model, with helical regions varying in length and sometimes even missing [e.g., (Anderson et al. 1981; Wolstenholme et al. 1987; Okimoto and Wolstenholme 1990; Watanabe et al. 1994b)], and loss of otherwise highly conserved tertiary interactions. Because mitochondria originate from a bacterium, the standard bacterial tRNA structure described in the following section will serve as a reference point for analyzing aberrant mitochondrial tRNAs.

17.2.1 The Standard tRNA Structure

The standard tRNA secondary structure, the cloverleaf (Fig. 17.1a), is valid throughout all three domains life, and is characterized by four double helical regions, the acceptor, D-, anticodon, and T-stems. These stems are composed, respectively, of seven, four, six, and five base pairs. On the tertiary structure level, they are combined into two helical domains, acceptor/T and D/anticodon. The perpendicular arrangement of these domains (L-shape; Fig. 17.1b) is responsible for the particular juxtaposition of the two functional centers, the anticodon and the acceptor terminus, and is thus essential for tRNA function. The two domains are linked together by connector regions, between the acceptor and D-stems (connector 1), and between the anticodon and T-stem (connector 2). In addition to the four universal stems, the Leu-, Ser-, and Tyr-tRNAs (Tyr in bacteria, plastids, some fungal, and protist mitochondria) contain a fifth stem-loop named extra arm, located in the RNA sequence between the anticodon and T-stems. The extra arm is

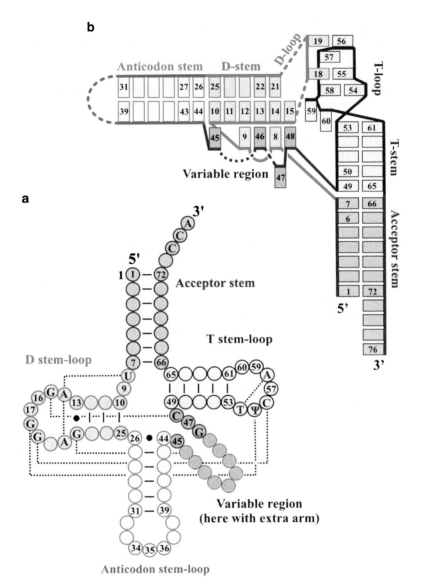

Fig. 17.1 Standard tRNA secondary structure. (**a**) Cloverleaf: Each stem-loop is shown with its own color. *Short lines* connect nucleotides forming Watson–Crick base pairs within stems. *Black filled circles* indicate that base pairs 13-22 and 26-44 in many tRNAs are non-Watson–Crick. *Dotted lines* connect nucleotides involved in conserved tertiary interactions. Identities of these nucleotides are indicated. Nucleotides are numbered in accordance with the standard tRNA nomenclature, which is based on the yeast Phe-tRNA (Rich and RajBhandary 1976). Note that numerous cytosolic tRNAs do not fit this structure. For instance, nucleotide 47 does not exist in some tRNAs, and position 17 may be either empty or additional nucleotides are inserted between positions 17 and 18. Further, some tRNAs have one to two additional nucleotides between positions 20 and 21. All mentioned nucleotides are unstacked and do not have a distinct position in the tertiary structure. Finally, the variable region located between the anticodon and T stem-loop,

not required for basic tRNA functions but serves as a recognition element for aminoacylation [e.g., (Watanabe et al. 1994a)].

The tRNA L-shape is stabilized by various tertiary interactions of the variable region with the D-stem-loop, and interactions between the D- and T-loops (Fig. 17.1b). First, nucleotides of connectors 1 and 2 form several contacts with the D-stem and with the proximate part of the D-loop. The two tertiary contacts that are present in all cytosolic tRNAs are a reverse-Hoogsteen base pair U8-A14 (Fig. 17.2a), and a reverse-Watson–Crick (WC) base pair 15-48. In most sequences, the latter base pair is G15-C48 (Fig. 17.3a). It is also sometimes A15-U48 (Fig. 17.3b) and rarely G15-G48 (Fig. 17.3c). The formation of tertiary base pairs U8-A14 and 15-48 extends the number of stacked nucleotide layers in the D/anticodon domain to 12, when counting the total number of base pairs in anticodon, D-stem, plus tertiary base pairs. As we discuss later, the extension of the D/anticodon domain with tertiary base pairs is essential for the proper interaction of the two helical domains, and formation of the tRNA L-shape. Additional tertiary interactions in this region are formed by nucleotides of the two connectors with base pairs of the D-stem. The standard pattern, in most but not all tRNAs, includes base pairs 9-23, 10-45, and 22-46 (Fig. 17.1b). These interactions make the overall tRNA structure more rigid, which is important for accurate codon–anticodon recognition (Curran and Yarus 1987).

Nucleotide 47 is part of connector 2, is present in most cytosolic tRNAs, and plays a special role in the formation of tertiary interactions in the D-stem and -loop. If present, it is bulged out, so that its base does not interact with other nucleotides (Fig. 17.1b). It is important for the integrity of tertiary interactions by spatially separating nucleotides 46 and 48, thus allowing them to *simultaneously* form the two above-mentioned tertiary base pairs 22-46 (Fig. 17.4a) and 15-48 (Fig. 17.3a) (Steinberg and Ioudovitch 1996). In the few tRNAs where nucleotide 47 is absent, a U13-G22 (Fig. 17.4b) pair allows nucleotide G22 to acquire a position in which it can connect to nucleotide 48, even in the absence of an intervening nucleotide 47 (Steinberg and Ioudovitch 1996). In a few exceptional cytosolic tRNAs, base pair 13-22 is not a U–G, although nucleotide 47 is missing; instead, 13-22 is a

---

Fig. 17.1 (continued) which minimally contains only three nucleotides, can be extended by eight or more nucleotides and form an extra arm. In such tRNAs, nucleotide 45 is usually not involved in base pairing with the opposite strand of this arm, and nucleotides 46 and/or 47 may not exist. (**b**) L-form: *Rectangles* represent individual nucleotides. Nucleotides of the stem-loops are shown in the same color and with the same numbering as in (**a**). Nucleotides of the anticodon loop and nonstacked nucleotides of the D-loop (16, 17 and 20) are not shown (indicated by *dashed lines*). The *dashed gray line* connecting nucleotides 46 and 47 of the variable region indicates that additional nonnumbered nucleotides may be inserted and form the extra arm. Tertiary interactions U8-A14 and 15-48 are found in the D-stem-loop region of all cytosolic tRNAs. The presence of a Watson–Crick or U–G base pair 13-22 and of the tertiary contacts 9-23 and 22-46 constitutes the standard pattern of tertiary interactions. This pattern may also be accompanied by the tertiary contact 10-45. In the DT-region, there are two inter-loop base pairs G18-Ψ55 and G19-C56 separated by a purine-57. The dinucleotide 59-60 bulges between base pairs 53-61 and 54-58. Nucleotide 59 stacks to the tertiary base pair 15-48, which constitutes the last layer of the D/anticodon helical domain. This interaction stabilizes the perpendicular arrangement of the two helical domains of the tRNA L-shape

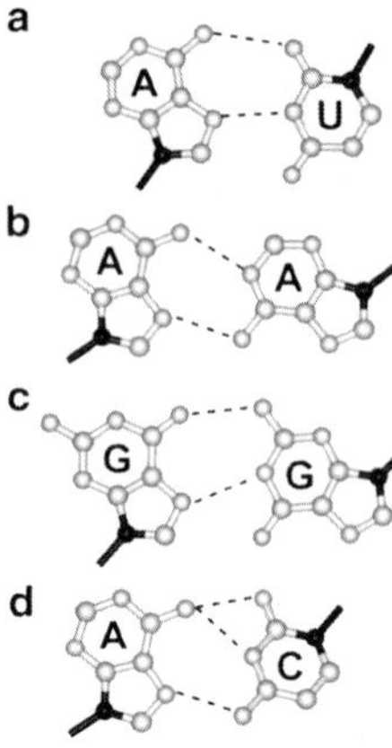

Fig. 17.2 Reverse-
Hoogsteen base pairs found in
positions 8-14 and 54-58 of
different tRNAs. In the
standard tRNA structure, base
pairs 8U-14A and 54T
(U)-58S are formed as in
panel (**a**). In the eukaryotic
initiator tRNA, the 54A-58A
base pair is formed as in panel
(**b**) (Basavappa and Sigler
1991). In various mitochondrial
tRNAs (Fig. 17.5) base pair
8-14 can also be A–A (**b**),
G–G (**c**), or C–A (**d**), while
base pair 54-58 can be A–A
(**b**) or G–G (**c**)

Fig. 17.3 The structure of
base pair 15-48. In the
standard tRNA structure, this
base pair has identity G15-
C48 and is formed as in panel
(**a**). In some cytosolic tRNAs,
this base pair is U–A (**b**) or
G–G (**c**). In many
mitochondrial tRNAs, base
pair 15-48 has alternative
identities, the most frequent
of which are U–A, C–G,
G–U, and A–C. The
juxtapositions of the bases in
such base pairs can be close to
one of those shown in panels
(**a**), (**b**), or (**d**)

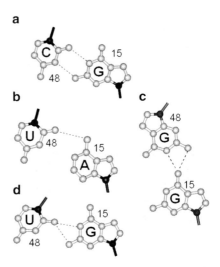

Watson–Crick pair and in most cases, nucleotide 46 is either U or A. A smaller-size
uridine-46 (i.e., compared to purines) would allow a reasonably strong interaction
with base pair 13-22, without affecting nucleotide 48 (Fig. 17.4c). Adenosine-46 is
an alternative as it can form an A-minor interaction with base pair 13-22 (Doherty
et al. 2001; Nissen et al. 2001), as shown in Fig. 17.4d. Although sterically possible,
all alternative interactions without nucleotide 47 will be less stable than standard
tertiary interactions (Fig. 17.4a, b).

The presence of an extra arm in some tRNAs is usually accompanied by
rearrangements in the D-stem, and in tertiary interactions between the D-stem and
connectors 1 and 2. Changes include the replacement of a Watson–Crick or U–G base
pair 13-22 by a sheared base pair 13G-22A or 13A-22A (Fig. 17.4e, f). In addition, the
tertiary base pair 9-23 is replaced by a 9-13 base pair, while tertiary base pairs 10-45
and 22-46 may not exist at all. The specific position of nucleotide 9 allows it to stack to

Fig. 17.4 Tertiary interactions involving base pair 13-22. (**a, b**) Standard patterns of tertiary interaction between guanosine-46 and base pair C13-G22 (**a**) or U9-G22 (**b**). (**c, d**) suggested modifications of the standard pattern when nucleotide 46 is U (**c**) or A (**d**). Tertiary interactions in tRNAs with an extra arm (**e, f**), or an extra-arm-like pattern (**e–g**)

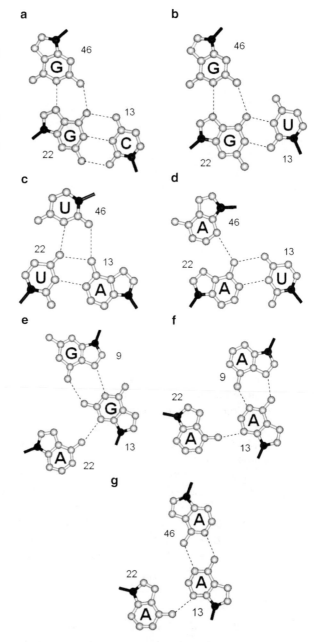

the first base pair of the extra arm, thus contributing to the stabilization of the particular position of the extra arm with respect to the rest of the tRNA structure (Biou et al. 1994; Ioudovitch and Steinberg 1998). In tRNAs containing the extra arm, the number of nucleotides attributed to region 46-47 varies between zero and two. For instance, the region 46–47 of *E. coli* Ser-, Leu-, and Tyr-tRNAs contains, zero, one, and two nucleotides, respectively, and this difference serves as identity element for aminoacyl-tRNA synthetase recognition (Watanabe et al. 1994a). Finally, some Cys-, His-, Glu-, and Gln-tRNAs contain a sheared base pair 13A-22A or 13G-22A, similar to tRNAs with an extra arm, and base pair 13-22 is usually involved in a triple with adenosine in either position 9 or 45 (Fig. 17.4e–g). This pattern of tertiary interactions is referred to as extra-arm-like (Nissen et al. 1999).

The second region of tertiary interactions, the so-called DT-region, is located at the corner of the L-shape, where the D- and T-loops meet (Fig. 17.1b). The structure of the DT region represents a four-layer stack of three tertiary base pairs with nucleotide 57, which is always a purine. One of the base pairs, T54-A58 (T is 5-methyl-uridine), is formed within the T-loop, while the other two, G18-Ψ55 (Ψ is pseudouridine) and G19-C56, are formed between the two loops. Base pairs G18-Ψ55 and G19-C56 are present in all cytosolic tRNAs. Base pair T54-A58 exists in all bacterial elongator and initiator tRNAs, whereas in eukaryotic initiator tRNA it is replaced by A54-A58. This structural variant allows positioning of key nucleotides in the DT-region in a way most similar to the standard one [Fig. 17.2a, b; (Basavappa and Sigler 1991)].

All three tertiary base pairs of the DT-region participate in the fixation of the particular juxtaposition of the acceptor/T and D/anticodon domains within the tRNA L-shape. The fixation proceeds at two levels. First, the formation of the two inter-loop base pairs G18-Ψ55 and G19-C56 attaches the two loops to each other. Then, the formation of base pair T54-A58 results in bulging of nucleotides 59 and 60, which can now stack to the last layer of the D/anticodon helical domain (base pair 15-48, Fig. 17.1b) (Zagryadskaya et al. 2003, 2004; Doyon et al. 2004). The stacking between nucleotide 59 and base pair 15-48 provides for the proper arrangement of the two helical domains, which in turn guarantees the proper juxtaposition of the two functional centers of the tRNA, the acceptor terminus and the anticodon. Note that the exact number of stacked layers in this region is critical for tRNA function: elimination of base pair 15-48 renders the tRNA nonfunctional, but a compensatory extension of bulge 59-60 in the T-loop from two to three nucleotides restores tRNA function (Zagryadskaya et al. 2004). In other words, the total number of stacked layers in the D/anticodon domain and in the bulge of the T-loop should be 14, regardless of how many of these layers come from each of the two moieties.

Note that plant, fungal, and protist mitochondria have canonical tRNA structures, with few exceptions. The majority of structural "aberrations" occur in bilaterian animals that will be discussed in the following (Sects. 17.2.2–17.2.5)

17.2.2   Variations in the Tertiary Interactions of the D-stem and Loop of Bilaterian Animal mt-tRNAs

Analysis of publicly available mitochondrial tRNA sequences shows that although nucleotides 8, 14, 15, and 48 allow the formation of the normal base pairs 8-14 and 15-48 in most cases, the number of exceptions is substantial. For base pair 8-14, the most frequent alternative is AA, which occurs in many Leu- (Fig. 17.5b), Lys-, and Asn-tRNAs. Two adenosines may form a reverse-Hoogsteen base pair that is similar to the standard base pair U8-14A (Fig. 17.2a, b) and would fit into the given structural context. The same applies to G–G (in Val- and Leu-tRNAs) and C–A (Ile-tRNAs) (Fig. 17.2c, d). Yet in other instances (e.g., Fig. 17.5c, d), relatively stable reverse-Hoogsteen base pair substitutes do not exist, pointing to a further deterioration of tRNA structure.

For base pair 15-48, the most frequent alternatives are U–A, C–G, as well as G–U and A–C all of which can be arranged in a reverse-Watson–Crick conformation, as required by the given structural context (Fig. 17.3). In general, base pair 15-48 seems to be more conserved in mitochondrial tRNAs than 8-14, which is reasonable given the fundamental role played by base pair 15-48 in maintenance of

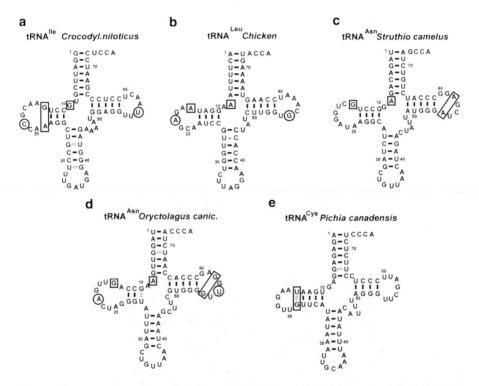

Fig. 17.5 Examples of animal mitochondrial tRNAs with various structural aberrations. (**a**) Ile-tRNA from *Crocodylus niloticus* with the extra-arm-like pattern of tertiary interactions (key nucleotides are shown in *rectangles*, and a mispair C19-U55 (*circled*). (**b**) Chicken Leu-tRNA with mismatch U27-C43, tertiary base pairs A8-A14 (*squared*) and A18-G54 (*circled*). (**c**) Asn-tRNA from *Struthio camelus* with base pair A54-A58 in the T-loop (enclosed in the inclined *rectangle*) and an unusual combination A8-G14 (*squared*). (**d**) Asn-tRNA from *Oryctolagus cuniculus* with base pair G54-G58 in the T-loop (enclosed in the inclined *rectangle*), a Watson–Crick base pair A19-U54 (*circled*), and an unusual combination A8-G14 (*squared*). (**e**) Cys-tRNA from *Pichia canadensis* with an additional base pair U14-G21 (enclosed in the *rectangle*) that compensates for the absence of the standard tertiary base pair 8-14

the tRNA L-shape (Zagryadskaya et al. 2004). A most notable change in bilaterian animal mt-tRNAs is the deletion of the extra arm in Leu-, Ser-, and Tyr-tRNAs, accompanied by various rearrangements of tertiary interactions in the D-stem. In particular in most Tyr-tRNAs, tertiary interactions in the D-stem correspond to the extra-arm-like pattern (Fig. 17.4e–g). Leu-tRNAs follow either the pattern observed in Tyr-tRNAs or they have four base pairs in the D-stem, as in the standard pattern (Figs. 17.4a–d and 17.5b). Finally, most Ser-tRNAs have an extended anticodon stem, which leads to severe deformations of the tertiary interactions in the D-stem (discussed in the next section).

Some animal mt-tRNAs that in bacteria have no extra arm (tRNAs other than Leu, Ser, and Tyr) have undergone rearrangements in the D-stem. For instance, whereas most bacterial Gln-tRNAs have an extra-arm-like pattern, in animal mitochondria they have adopted the standard pattern. The opposite has taken place in Ile-tRNAs, where conversion has occurred from a standard tertiary to an extra-arm-like pattern (Fig. 17.5a). Finally, Cys-tRNAs have the extra-arm-like pattern in bacteria, while in some mitochondria they have converted to the standard pattern (Fig. 17.5c). Alternatively, the anticodon stem of a mitochondrial Cys-tRNA can be extended, leading to major rearrangements of the tertiary interactions in the D-stem (these tRNAs will be discussed in more detail in the following section).

Mitochondrial tRNAs of bilaterian animals with four base pairs in the D-stem and tertiary interactions similar to the standard pattern have a strong tendency to lose nucleotide 47, which remains almost exclusively in Lys- and Asn-tRNAs. In addition, unlike in cytosolic tRNAs where the absence of nucleotide 47 is almost always accompanied by the presence of a 13U-22G base pair, it is either U–A or A–U in mitochondria, and nucleotide 46 is either an A or a U. We suggest that this conformation is stabilized by a weak interaction of nucleotide 46 with base pair 13-22 (Fig. 17.4c, d).

17.2.3 Structural Variations in the Anticodon and D-Stems

Various mitochondrial tRNAs have rearrangements in the anticodon and D-stems, including frequent mismatches in helical regions (Fig. 17.5b), and changes in the length of either the anticodon or D-stem (Figs. 17.6 and 17.5e). Yet, proper spacing and 3D coordination of the anticodon/D domains is essential for positioning of both the anticodon and the acceptor terminus. Therefore, length variation in one domain needs to be compensated by corresponding changes in other domains. As we will see below, this compensation is usually imperfect as it comes at the expense of tertiary interactions in the D-stem and -loop. A simple scheme of such compensation would be extension of the D-stem at the expense of the tertiary base pair 8-14, which is frequently observed in Cys-tRNAs. As shown in Fig. 17.5e, nucleotides A8 and U14 in the mitochondrial Cys-tRNA of *Pichia canadensis* do not allow the formation of the standard tertiary base pair U8-A14. On the other hand, G21 can pair to U14, which would be the fifth base pair in the D-stem. Thus, in the most

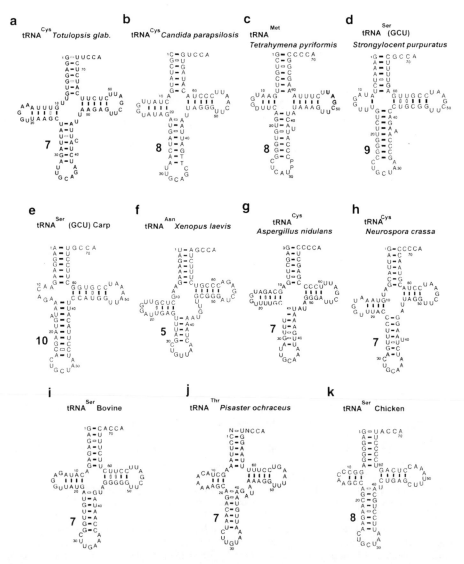

Fig. 17.6 Examples of mitochondrial tRNAs with nonstandard length anticodon and D-stems. For each tRNA, the number of base pairs in the anticodon stem is indicated. For tRNAs (**a**)–(**h**), the structure of the D/anticodon domain is shown in Fig. 17.8. For tRNA (**i**), the structure of this domain is equivalent to that of (**b**). For tRNAs (**j**, **k**), the structure of this domain is unclear due to the absence of the standard tertiary interactions at the DT-region (see text)

probable structure of this molecule, the additional base pair in the D-stem would replace the standard tertiary base pair 8-14. Due to such rearrangement, the total number of stacked layers in the anticodon/D domain remains constant, and the overall tRNA shape and its functionality will not be affected.

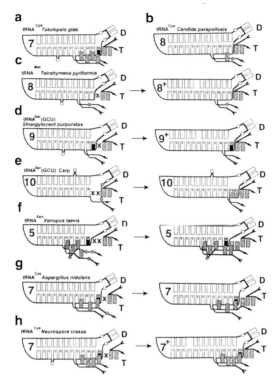

Fig. 17.7 Base-pairing, stacking, and tertiary interaction scheme for the D/anticodon domain in abnormal mitochondrial tRNAs. The corresponding nucleotide sequences are shown in Fig. 17.6 with the same tag. The structures of the D/anticodon domain are represented as in Fig. 17.1b. *Rectangles* in the D-loop represent the universal guanines 18 and 19; *circles* represent bulged and other unstacked nucleotides. For each tRNA, the number of base pairs in the anticodon stem is indicated and each stacked layer of the D-anticodon domain is numbered. In (**a**)–(**f**), the D/anticodon domain contains 12 stacked layers. In (**g**, **h**), due to the short T-stem, the required length of this domain is 13 layers. In (**a**, **b**), the proper length of the D/anticodon domain is achieved through summation of the number of base pairs in the anticodon and D-stems and the formation of one tertiary base pair in the D-loop (e.g., Fig. 17.7, panel (**a**)). In (**c**), (**d**), and (**h**), the proper length of the D/anticodon domain is achieved through intercalation of bulged nucleotides into the anticodon stem (compare figures before (*left*) and after (*right*) intercalation). A plus sign after the number of base pairs indicates that the length of the anticodon stem becomes extended due to intercalation of a bulged nucleotide. In (**e**)–(**g**), the proper length of the D/anticodon domain is achieved through participation of unpaired nucleotides of the D-loop (**e**), of connector 1 (**f**) and of connector 2 (**g**). The *arrows* between the *left* and *right* structures indicate the extension of the D/anticodon domain to the required length

A more complex situation arises with variations of the anticodon stem that contains six base pairs in standard, but five, seven, eight, nine, or even ten base pairs in mitochondrial tRNAs (Steinberg and Cedergren 1994; Steinberg et al. 1994). Examples of such abnormal tRNAs are shown in Fig. 17.6. Regardless of the length of the anticodon stem, one might expect that the normal juxtaposition of the anticodon and the acceptor terminus requires that the D/anticodon domain contains 12 layers of stacked nucleotides. This would allow stacking of the last layer of the D/anticodon domain to nucleotide 59 of the T-loop, i.e., a proper tRNA L-shape (panels a, b, and i of Fig. 17.6 and panels a and b of Fig. 17.7). Surprisingly, this principle does not seem to apply to certain mt-tRNAs (panels c–f in Fig. 17.6 and in left column of Fig. 17.7). To attain the number of required layers in these cases, bulged nucleotides in the anticodon stem (panels c, d in Figs. 17.6 and 17.7) and unpaired nucleotides of both connector regions and of the D-loop (panels e, f) have to be included in this count. Following this principle, a normally L-shaped tRNA can be formed (Steinberg et al. 1997), although the formation of stacked layers in the D/anticodon domain with unpaired nucleotides will compromise overall structural stability.

Further, the requirement of 12 layers in the D/anticodon domain is valuable only in the context of proper positioning of nucleotide 59, which in turn depends on the number of base pairs in the T-stem. Standard tRNAs contain five-base pair T-stems, but mitochondrial tRNAs may have four (panels g, h in Figs. 17.6 and 17.7) or six base pairs (panels d, e). These variations will cause rotation of the T-loop around the axis of the T-stem, in one or the other direction. Correspondingly, nucleotide 59 becomes displaced, potentially affecting its interaction with the 12th stacked layer of the D/anticodon domain. Molecular modeling of these tRNAs shows that the extension of the T-stem from five to six base pairs does not change the distance between nucleotide 59 and the 12th layer of the D/anticodon domain. However, decreasing the T-stem from five to four base pairs displaces nucleotide 59 away from the 12th layer, by about 3 Å. Accordingly, a T-stem containing six base pairs requires a D/anticodon domain with 12 layers, while with four base pairs an additional 13th layer has to be added to the D/anticodon domain. According to these rules, tRNAs^{Cys} of *N. crassa* and *A. nidulans* also fold into the proper L-shape [panels (g, h) in Figs. 17.6 and 17.7; (Steinberg et al. 1997)].

For any given length of the anticodon stem, not only the length of the D/anticodon domain but also the lengths of connectors 1 and 2 have to correspond. If the anticodon stem becomes longer, the anticodon/D-stem junction will be closer to the acceptor/T-stem junction; thus, fewer nucleotides will be required for connectors 1 and 2. Indeed, connectors 1 and 2 in standard tRNAs have at least two and three nucleotides, respectively, but with seven-base-pair anticodon stems these numbers drop to one and two nucleotides. For tRNAs with eight, nine, and ten base pairs, the corresponding numbers are one and one nucleotides, zero and one nucleotides, and finally, zero and zero nucleotides. Molecular modeling confirms that only under these conditions do the anticodon and the acceptor stems fit the standard L-shaped tRNA structure. Finally, when the anticodon stem becomes shorter as in tRNA^{Asn} in *Xenopus laevis* (panel f in Fig. 17.7), the distance between the two junction points increases and longer connector regions are required, three and five nucleotides, respectively.

In summary, there are two types of requirements that together guarantee the normal juxtaposition of the two helical domains. First, adapting the lengths of connectors 1 and 2 allows the anticodon and acceptor stems to occupy standard positions, a requirement that should be satisfied under all conditions. Second, the lengths of the D/anticodon domain and the T-stem have to be adapted in concert, in a way that allows fixation of a proper L-shape tRNA conformation. The key structural aspect here is stacking of nucleotide 59 in the T-loop to the last layer of the D/anticodon domain. Therefore, the second requirement is conditional: it is only valid if the conformation of the T-loop and its interaction with the D-loop is standard. Deviations from the standard pattern are discussed in the following section.

tRNAs conforming to the above rules are likely to function not only in mitochondria. In fact, tRNAs with seven, eight, nine, and even ten base pairs in the anticodon stem are functional in *E. coli* as suppressors of a nonsense mutation (Bourdeau et al. 1998). The efficiency of these suppressors is somewhat lower than that of normally structured tRNA suppressors.

17.2.4 Structural Abnormalities in the D- and T-Loops

As shown above, interaction of D- and T-loop in the standard tRNA structure plays a critical role in fixing the juxtaposition of the two helical domains. However, at some stage of mitochondrial evolution, this DT-conformation is no longer enforced. Correspondingly, conformational rearrangements of the DT-region can be divided into two groups, those that contain elements that preserve the standard tRNA L-shape, and those that do not. Our view is based on results of in vivo expression and molecular modeling of tRNA mutants with abnormalities in the DT region (Zagryadskaya et al. 2003, 2004; Doyon et al. 2004; Kotlova et al. 2007).

The evolutionarily most flexible element of the DT-region is base pair G19-C56 (Fig. 17.1b). Because it is located at the top of the DT-arrangement, its modification would not interfere with other parts of the molecule. Correspondingly, in many mitochondrial tRNAs the G19-C56 base pair is replaced by either another Watson–Crick base pair (e.g., Fig. 17.5d) or other combinations of two nucleotides (e.g., Fig. 17.5a). When expressed in *E. coli*, tRNAs with a 19-56 mispair retain a somewhat lower level of activity than normal tRNAs (Doyon et al. 2004).

The other inter-loop base pair G18-Ψ55 seems to be more critical for the integrity of the standard DT-interaction. In mitochondrial tRNAs, a modification of base pair G18-Ψ55 is practically always associated with a major rearrangement of the DT-region. The only exception is replacement of this base pair by A18-G55, which is present in many mitochondrial tRNAs with otherwise standard DT regions (e.g., Fig. 17.5b). In addition, the A18-G55 combination occurs in the T-loop of some viral tRNA-like particles (Fechter et al. 2001), and in the T-loop-like elements of RNase P (Krasilnikov and Mondragon 2003) and of ribosomal RNA (Nagaswamy and Fox 2002; Lee et al. 2003). Finally, the A18-G55 combination emerges in tRNA nonsense suppressors that are selected in vivo (Doyon et al. 2004). In the known tertiary structures, base pair A18-G55 is arranged as shown in

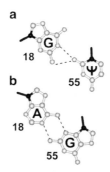

Fig. 17.8 GU–AG base pairs in position 18-55. Nucleotide arrangements for the two inter-loop base pairs G18-U(Ψ)55 (**a**) and A18-G55 (**b**). The structure of base-pair G18-U(Ψ)55 is taken from Shi and Moore (2000), and that of base-pair A18-G55 from Krasilnikov et al. (2003) (PDB entry 1NBS). Although the two base pairs are not completely isosteric, the juxtapositions of the glycosidic bonds between the two nucleotides (shown in *black*) are sufficiently close to guarantee interchangeability of these base pairs in the DT region

Fig. 17.8, which makes it almost isosteric to base pair G18-Ψ55 in the standard tRNA structure. Therefore, base pair A18-G55 should be considered an equivalent to G18-Ψ55.

The reverse-Hoogsteen base pair T54-A58 is equally important for the structural integrity of the DT-region. As mentioned earlier, base pair T54–A58 is conserved in all cytosolic elongator tRNAs, while in the eukaryotic initiator tRNA it is replaced by an A54-A58 base pair (Rich and RajBhandary 1976; Basavappa and Sigler 1991). Despite its somewhat larger size (Fig. 17.2a, b), an A54-A58 base pair does not seem to affect other parts of the T-loop and occurs in many mitochondrial Asn-tRNAs (Fig. 17.5c). The isosteric G54-G58 (Fig. 17.2c) base pair is less frequent; for examples of tRNAs with A54-A58 or G54-G58 base pairs, see Fig. 17.5c, d.

Finally, in the standard tRNA structure, the two inter-loop base pairs G19-C56 and G18-Ψ55 are separated by nucleotide 57 (Fig. 17.1b). As discussed above, the stability of this arrangement requires nucleotide 57 to be a purine. This also applies to mitochondrial tRNAs with a close-to standard DT-conformation, and to functional suppressor tRNAs that were selected from combinatorial gene libraries (Doyon et al. 2004). In other words, a purine in position 57 is an essential aspect of the standard T-loop pattern and cannot be changed without serious damage to the T-loop conformation.

The variations in the DT-region discussed above exhaust almost all possibilities for varying the nucleotide sequences in the D- and T-loops without major structural rearrangement. Only recently, an alternative way to stabilize the juxtaposition of the two helical domains was reported, based on the formation of a two- to three-base-pair Watson–Crick double helix between the D- and T-loops (Kotlova et al. 2007). In rare cases, animal mitochondrial tRNAs follow this pattern as shown in Fig. 17.9a. Most mitochondrial tRNAs with deformed D- and T-loops are highly

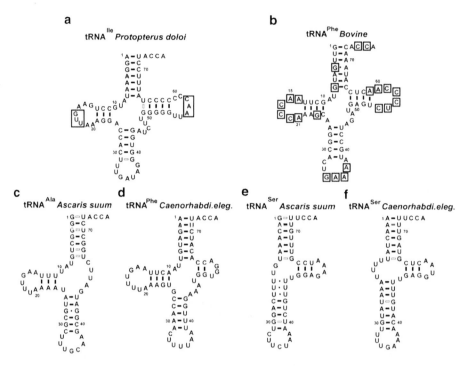

Fig. 17.9 Mitochondrial tRNAs with a major deformation in DT-interactions. (**a**) Ile-tRNA from *Protopterus doloi* with a Watson–Crick double helix between the D- and T-loops (the nucleotides involved in the base pairing are shown in *rectangles*). (**b**) Structure of the bovine mitochondrial Phe-tRNA. In this molecule, the D- and T-loops do not form stable interactions. *Squared* nucleotides are reactive to RNase T2 and to chemical reagents under native conditions (Wakita et al. 1994). (**c**) Example of a nematode mitochondrial tRNA without T-stem. (**d**) Example of a nematode mitochondrial tRNA with a two-base-pair T-stem. (**e, f**) Mitochondrial Ser-tRNA from *A. suum* and *C. elegans*, in which the D-stem has been consumed by the anticodon stem. Unlike other mitochondrial tRNAs, these tRNAs are recognized by a special Tu-factor (Ohtsuki et al. 2002)

variable in sequence, without evident formation of a distinct alternative structure. In a well-studied example of the bovine Phe-tRNA belonging to this class, the D- and T-loops had highly flexible conformations (Fig. 17.9b; Wakita et al. 1994). However, deterioration of the DT-interactions is usually not accompanied by loss of tertiary interactions in the D-stem-loop (Wakita et al. 1994).

The utmost level of structural deterioration of the DT-region consists in the replacement of the T-stem by a short unpaired sequence, providing a direct connection between the anticodon stem and the 3′-strand of the acceptor stem. Such deformations occur in nematode mitochondria, where practically all tRNAs have lost the T-stem-loop, except for occasional one or two base pairs in the T-stem (Fig. 17.9c, d; Wolstenholme et al. 1987). Again, the elimination of the T-stem-loop does not affect the tertiary interactions in the D-stem-loop (Wolstenholme et al. 1987, 1994).

17.2.5   How tRNAs Compensate for Loss of Structural Features

Modeling of mitochondrial tRNA structures shows that many of them perfectly fit standard structures, whereas others deviate substantially. The degree of deviation is minor for mitochondrial tRNAs of most protists, fungi, and plants, but may be so severe in bilaterian animals that even their identification in mtDNA sequences becomes problematic. The evolution of animal mitochondrial tRNAs can be generally described as a deteriorative one, in which elements of the standard tRNA structure are modified or deleted such that the whole molecule becomes less stable and less effective in its basic function.

As we have seen, structural deviations in animal mitochondrial tRNAs come in two distinct flavors. In the first group, helical regions have frequent mismatches, tertiary base pairs in the D-stem-and-loop and the extra arm are lost, anticodon and D-stems vary in length, and D- and T-loops may undergo minor deformations – yet without eliminating DT-interactions. Although these deviations seem to be detrimental to tRNA stability and efficiency, tRNAs are likely to adopt the standard L-shape. They are predicted to function even when expressed in the cytosol, albeit with low efficiency, and they would profit from stabilization by interaction with other molecules.

The second group of structural variations is more extreme, including gradual deterioration of DT-interactions to a point that a juxtaposition of the two helical domains (i.e., the L-shape) is lost. Such tRNAs are unlikely to function in any system unless the L-shape is somehow stabilized. The deterioration of DT interaction itself has distinct variants, with the T-stem remaining to some variable extent (as in most animals) or lost completely [as in nematodes (Okimoto and Wolstenholme 1990), and in certain arachnids (Masta and Boore 2008)]. In a most revealing case of nematode Ser-tRNA, the T-stem still has base pairing but the D-stem no longer exists, having been incorporated into the anticodon stem (Fig. 17.9f). Unlike other nematode mitochondrial tRNAs, Ser-tRNA is delivered to the ribosome by a special EF-Tu protein that contains a new structural domain (Sakurai et al. 2006). This example in which the mitochondrial protein synthesis machinery is adapted to dealing with deteriorated tRNA structure foreshadows a larger trend – gradual takeover of RNA by protein function [other examples include the reductive evolution of mitochondrial rRNA, RNase P and tmRNA; e.g., Schneider 2001; Seif et al. 2003; Jacob et al. 2004; Holzmann et al. 2008]. We suggest that tRNA chaperones other than EF-Tu remain to be detected in bilaterian animal (or even all) mitochondria, and that components directly or indirectly involved with tRNAs or in protein translation are likely candidates.

17.3   Reducing the Minimum Number of mt-tRNAs and Assigning New tRNA Identity

Organelles of most eukaryotes undergo rapid evolutionary change, both at the level of genome sequence (increased mutational rate) and organization (gene content, gene order, and genome architecture). Several bilaterian animal lineages are special in preserving mitochondrial genome organization over long periods of time, which may be related to a lack of effective DNA recombination and repair mechanisms. In most other eukaryotes, recombination plays a major role in shaping organelle genome and genes, such as changes of gene order, segmental genome duplication, intron mobility, insertion of mitochondrial plasmids into mtDNAs, and overall changes of genome architecture (circular, linear concatemers mapping as circles, monomeric linear, multiple chromosomes; for more details, see Chap. 3). Mitochondrial genomes are particularly inventive in dealing with the resulting mutational burden, often leading to radical departures from conventional solutions in order to restore gene function.

In this respect, the mutational drifting of mitochondrial tRNA gene structure and identity is no exception. For instance, 5′ and 3′ tRNA editing in mitochondria has been invented independently several times, in a wide variety of eukaryotes (Lonergan and Gray 1993; Laforest et al. 1997; Lavrov et al. 2000; Gray 2001, 2003; Leigh and Lang 2004; Segovia et al. 2011). Other more radical changes include the replacement of defunct genes with copies derived from other tRNA genes, some of which are created by recombination among tRNAs, and alterations of tRNA identity in conjunction with changes of the genetic code.

17.3.1   Wobble and Super-Wobble Interactions and the Minimum Set of tRNAs

In the Wobble Hypothesis, Crick proposed that nucleotides 35 and 36 of a tRNA's anticodon would form canonical Watson–Crick (WC) pairings at respective codon positions, and that tRNA position 34 could be a WC, G–U, or U–G pair (Crick 1966). This implies a substantial reduction in the number of tRNAs that are required to recognize all codons. In this same publication, Crick interprets the presence of the nucleoside inosine in some tRNAs (Holley et al. 1965a, b), proposing that the number of distinct tRNAs may be even smaller, because this type of post-transcriptional modification allows pairing with A, U, and C (Crick 1966). Based on G–U wobble rules alone [but including a special tRNA(Ile) with a CAU anticodon in which C carries a lysidine modification (Muramatsu et al. 1988; Weber et al. 1990; Moriya et al. 1994)], a minimum of 32 distinct tRNAs is required to recognize all codons, in species with a standard genetic code. However, several genetic systems, including organelles (plastids and mitochondria) and bacteria [e.g., *Mycoplasma* and relatives; for a detailed overview, see de Crecy-Lagard et al. 2007], do not encode all 32 predicted tRNAs. Possible explanations are: tRNA modifications (as in *Mycoplasma*), tRNA import (restricted to organelles), changes in the genetic code, relaxed wobble rules ("super-wobble"), and finally, partial editing of an organelle-encoded tRNA to create two distinct tRNAs. Precedence for partial RNA editing comes from opossum mitochondria, where a portion of the encoded glycine tRNA (GCC) is post-transcriptionally transformed into an aspartate (GUC) tRNA (Borner et al. 1996). All five mechanisms are known to impact the evolution of organelle tRNA sets, but as super-wobble is most common it will be considered in more detail.

When Crick proposed the wobble hypothesis; he postulated that U–C and U–U pairing would be hampered at the wobble position by steric problems (Crick 1966). However, uridine modification in the wobble position of the anticodon does promote reading of all codons by a single tRNA (Nasvall et al. 2004; Weixlbaumer et al. 2007), creating a super-wobble. Interestingly, in both chloroplasts and mitochondria, an unmodified U in the wobble position of the anticodon position is also able to accommodate any of the four standard nucleotides in the codon's third position (Heckman et al. 1980; Rogalski et al. 2008), without creating steric incompatibility as predicted. In fact, it seems that a U in this position neither contributes to the codon-anticodon interaction, nor does it weaken the interaction of the neighboring two base pairs (Lagerkvist 1978). While most plastids and flowering plant mitochondria still follow the wobble rule with few exceptions, apparently for translational efficiency (Rogalski et al. 2008), most mitochondria of animals and fungi have adopted super-wobble. With an unmodified U at a tRNA's wobble position (U34), four-codon families are read using two out of three rules. In two-codon families, an unmodified G34 recognizes only pyrimidines (U or C), and a modified U34 restricts reading to only purines (A or G) (Heckman et al. 1980). The latter type of U modification, together with the lysidine modification of isoleucine tRNA(CAU) and combined with super-wobble rules, reduces the theoretical minimum of distinct tRNAs to 25 [reviewed in (Marck and Grosjean 2002)]. In angiosperm chloroplasts, which like plant mitochondria use the standard genetic code, a minimum of 32 tRNAs would be required if conventional wobble base pairing occurs, but only 30 tRNAs have been identified (Pfitzinger et al. 1990). It has been demonstrated by the same authors that three chloroplast tRNAs are able to read all four codons of the respective amino acid families, apparently employing a super-wobble base pair recognition mechanism that is in an initial phase of being established.

Further changes in the rules of anticodon–codon interaction have been reported in animal mitochondria and are usually associated with changes in the genetic code. These include recognition of serine AGR codons by tRNA(GCU), with modified G34 (Matsuyama et al. 1998; Tomita et al. 1998), and recognition of lysine AAA codons as asparagines by tRNA(GUU), due to a pseudouridine modification at U35 but no modification at the wobble position (Tomita et al. 1999).

17.3.2   Mitochondrial tRNAs Recognizing Stop as a Sense Codon, and Sense as Stop

The reassignment of UGA stop codons to tryptophan is among the most frequent translation code changes in organelles, first identified in yeast mitochondria as early as 1979 (Macino et al. 1979; Fox 1979), and then detected in mtDNAs of many other eukaryotes [for a recent compilation, see Massey and Garey 2007] and certain bacteria [e.g., Inagaki et al. 1996; de Crecy-Lagard et al. 2007]. Reassignment of UGA codons is relatively straightforward in organelle genomes, due to the codon bias that comes with their usually high A + T content. In mtDNAs, stop codons are often exclusively UAA, although in some instances UAG is also used. The corresponding tRNA that recognizes both UGA and UGG as tryptophan has a UCA anticodon that will interact effectively with both codons due to standard U–G wobble [e.g., Martin et al. 1980, 1981; Sibler et al. 1980].

Curiously, a limited number of UGA(Trp) codons may occur in certain fungal mtDNAs (including fission yeasts), although the only available mtDNA-encoded tRNA has a CCA anticodon that would not (effectively) recognize UGA codons, except if the C in the first anticodon position of tRNA(Trp) is post-transcriptionally modified or partially edited to permit interaction with the G or A residues of the respective UGG or UGA codons (Bullerwell et al. 2003b, c; Seif et al. 2005). Indeed in *Trypanosoma*, an imported tRNA(Trp) with a CCA anticodon undergoes C to U editing in the first position of the anticodon, which allows effective decoding of mitochondrial UGA codons as tryptophan (Alfonzo et al. 1999). Alternatively, a less effective UGA suppressor-like recognition of UGA(Trp) codons may as well be envisioned, i.e., a tRNA structure that allows for latent, ambiguous decoding, and that has evolved specifically in the given fungal species (Bullerwell et al. 2003b, c; Seif et al. 2005; Massey and Garey 2007). In fact, according to our unpublished results, the mitochondrial tRNA(Trp)(CCA) of *Schizosaccharomyces pombe* does exhibit particularly effective UGA suppressor activity in a wheat germ in vitro translation system. Yet, to resolve this issue beyond reasonable doubt, sequencing of native tRNA(Trp) and identification of potential nucleotide modification of its anticodon sequence would be required.

Following the logic developed above, reassignment of UAG stop codons may also be expected. In fact, several reports have identified assignment of mitochondrial UAG codons to leucine, in several chytrids and in chlorophycean algae (Hayashi-Ishimaru et al. 1996; Laforest et al. 1997; Kück et al. 2000; Nedelcu et al. 2000). In other chlorophycean algae, UAG appears to be translated as alanine (Hayashi-Ishimaru et al. 1996), suggesting that the availability of "free" UAG codons may lead to different codon reassignments following different evolutionary constraints. The same situation seems to apply to the reassignment of yeast CUN (leucine) codons to either threonine or (according to our analysis) alanine in mitochondria (for more details, see Sect. 17.3.3 and Fig. 17.13).

Stop codons may not only become assigned to amino acids, but sense codons may become stops as well. In the curious, seemingly unique cases of the chlorophytes *Pycnococcus provasolii* and *Scenedesmus obliquus*, both are realized. In *Pycnoccus*, UGA stop codons stand for tryptophan (as in many other mtDNAs), but, in turn, the leucine codons UUA and UUG are used as stop (Turmel et al. 2010). In *Scenedesmus*, UCA serine codons have become translation stops, whereas UAG stop codons are assigned to leucine as also described for other representatives of this algal lineage (Kück et al. 2000; Nedelcu et al. 2000). The conclusion that UCA equals stop was reached with confidence, because these codons (a) do not occur at all in protein-coding genes; (b) occur exactly at or close to stop codons relative to positions of the latter in other species; (c) are not edited at the RNA level; and (d) because a tRNA serine recognizing UCA codons is missing (the existing tRNA(Ser) has a GGU anticodon, recognizing only UCU an UCC codons). It is interesting that both UAG and UCA codons are also absent in mitochondrial genes of close relatives of *S. obliquus* (i.e., *Chlamydomonas* and *Pedinomonas*) (Nedelcu et al. 2000). This suggests that favorable conditions for the introduction of codon reassignments existed, and apparently continue to exist, throughout this green algal lineage. Finally, one of the longest known cases of codon reassignment has been challenged by a recent study. In vertebrate mitochondria, the rarely used AGR triplets were thought to be stop codons (Anderson et al. 1981; Osawa et al. 1989). However, the AGR motif of human mitochondrial *coxl* and *nad6* has been demonstrated to be the site of a −1 translational frameshift, creating standard UAA and UAG stop codons (Temperley et al. 2010).

17.3.3 Assignment of New tRNA Identity and Changes of the Genetic Code

Assignment of new tRNA identity linked with a change of the genetic code has occurred many times during mitochondrial evolution. In the following discussion, we provide a short overview of the underlying questions, and then focus on a well-known example in yeast mitochondria. For a recent comprehensive overview on this topic (including bilaterian animal mitochondria), the reader is referred to Massey and Garey (2007).

Some amino acids, including leucine, serine, and arginine (in the standard translation code), are encoded not by a single codon family, but by two. It was initially assumed that tRNAs accepting the *same* amino acid (isoacceptors) evolved by gene duplication and divergence of the resulting copies. Yet, according to a more recent hypothesis, some tRNAs might have evolved without respecting the genetic code, changing tRNA anticodon and acceptor identities of duplicated alloacceptor (recognizing *different* amino acid) tRNA genes (Cedergren and Lang 1985). This view is supported by unexpected close evolutionary relationships within mitochondrial tRNA sets of *S. pombe* and *Acanthamoeba castellanii* (Cedergren and Lang 1985; Burger et al. 1995) and other mitochondrial systems (Cantatore et al. 1987; Higgs et al. 2003; Rawlings et al. 2003). The feasibility of tRNA identity change (tRNA gene recruitment or "identity theft") was for the first time experimentally demonstrated with *E. coli* mutants (Saks et al. 1998). By using phylogenetic inferences with tRNA sequences, it was further possible to identify several specific cases of mitochondrial alloacceptor and isoacceptor tRNA recruitment that occurred during demosponge evolution (Lavrov and Lang 2005; Wang and Lavrov 2011). In addition, a systematic database search for additional clear-cut instances (at least two standard deviations higher than average) identified tRNA sequence similarities between 73 and 90% in mitochondria of *Scenedesmus obliquus* (*trnI* (uau) and *trnM*(cau) genes) and *Amoebidium parasiticum* [among tRNA genes for methionine, isoleucine, valine, and lysine, and between threonine and alanine (Lavrov and Lang 2005)] as well as nuclear genomes of several primates (Wang and Lavrov 2011). These investigations suggest that the evolution of tRNA multigene families is by far more complex than previously appreciated and that phylogenetic inferences with tRNAs may only be interpreted based on a given species tree, cautioning against reliance on tRNA features alone for inferring evolutionary relationships.

Finally, in a most recent investigation, recruitment of yeast mt-tRNA(Thr) from tRNA(His) was demonstrated by phylogenetic inference, and more directly by biochemical approaches (Su et al. 2011). The mitochondrial tRNA(Thr) in question has a UAG anticodon that would be read as leucine in the standard gene code. This tRNA has a most unorthodox eight-nucleotide anticodon loop, occurs in *S. cerevisiae* and some of its close relatives, and evolves at a time point of yeast evolution when a large number of mitochondrial (*nad*) genes are lost. Reassignment of tRNA identity is not a simple matter of changing a tRNA's sequence and structure, but has to be interpreted in a larger biochemical and evolutionary context. To continue with the yeast example, the creation of a tRNA that recognizes CUN codons ("N" stands for any of the four nucleotides) as threonine will require (a) the availability of an unused tRNA gene copy (in this case from a tRNA(His) alloacceptor) that is transcribed and the resulting tRNA precursor properly matured; (b) an aaRS that charges this tRNA with threonine. These requirements imply either adaptation of an existing threonyl-tRNA synthetase to also recognize this new tRNA, or gene duplication followed by mutational change to create a separate enzyme. In the yeast example, it is known that both the standard and the new unorthodox tRNA(Thr) are charged by the same aaRS (Su et al. 2011), i.e., not implying a second aaRS as previously postulated (Pape et al. 1985); the creation of a tRNA will require (c) mutation of all functionally important CUN(Leu) codon positions in mitochondrial genes to UUA or UUG, i.e., before switching CUN(Leu) to CUN(Thr) and (d) inactivation or loss of the resident tRNA(CUN) leucine. Once all changes have been fully established (co-evolved), threonine positions in proteins may now be assigned to the new CUN codons even in highly conserved amino acid positions of a protein.

Codon reassignment may evolve according to a codon capture mechanism (Brown and Doolittle 1995), assuming complete disappearance of a codon and its corresponding tRNA gene from a genome, followed either by evolution of a tRNA gene with new specificity and codon reintroduction or by an intermediate mechanism that allows codon assignment to more than one amino acid (Schultz and Yarus 1994). According to the codon capture model, the complete disappearance of codons from a genome is much more easily achieved in the small organelle genomes, whereas translation code changes in nuclear genomes almost inevitably proceed via an ambiguous intermediate mechanism. In fact, in the cytoplasm of several *Candida* species, the CUG codon is decoded by a tRNA(CAG) as either leucine or serine (Suzuki et al. 1997; Miranda et al. 2006), whereas the codon reassignment of *S. cerevisiae* mitochondrial CUN codons from leucine to threonine appears to follow a codon capture mechanism (Su et al. 2011).

To resolve details about the evolutionary transitions that occurred in yeast mitochondria as described above, estimated at some time after the divergence of *Pichia canadensis* and close to the emergence of *Kluyveromyces* species, we have extended the analysis. In this case we analyzed additional yeast species, in particular the very rapidly evolving *Hanseniaspora uvarum* and *Ashbya gossypii*. Briefly, a species tree was constructed with PhyloBayes and the CAT model (Lartillot and Philippe 2004), based on the concatenated sequences of mtDNA-encoded proteins (Fig. 17.10). In addition, codon tables for these genes were calculated from the same

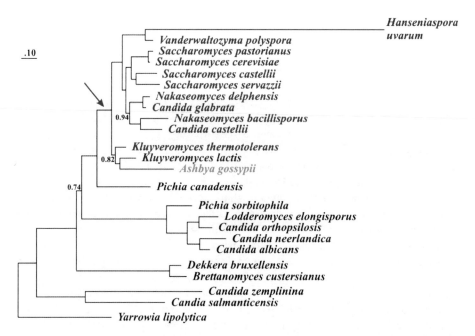

Fig. 17.10 Phylogeny of yeast species based on concatenated mtDNA-encoded protein sequences. The phylogenetic analysis with PhyloBayes and the CAT model is based on 13 mtDNA-encoded proteins. All divergence points are supported by posterior probability values of 1.0, except where indicated. The *red arrow* points to the concomitant loss of all 7 *nad* genes (i.e., respiratory complex I or NADH dehydrogenase) and the start of major mitochondrial codon reassignments, including AUA methionine, CUN threonine (species marked red, *K. lactis* marked magenta as it has no CUN codons and no corresponding tRNA with a UAG anticodon), and CUN alanine (*A. gossypii*, marked blue; see also Figs. 17.12 and 17.13)

species, and multiple protein alignments were analyzed for potential additional codon reassignments. Results are unexpected in several ways. *Hanseniaspora uvarum*, a very rapidly evolving and most unorthodox yeast species with shortened gene sequences (Pramateftaki et al. 2006), belongs within the group of *Saccharomyces* species that recognize CUN codons as threonine, placed with PhyloBayes and the CAT model close to *Vanderwaltozyma* (Fig. 17.10). This tree topology differs from the published one that places *Hanseniaspora* within a lineage of *Candida* species containing *Candida albicans*, which we interpret as a "long-branch attraction artifact" [LBA; (Felsenstein 1978)]. LBA may be identified and overcome by using more realistic phylogenetic models [such as CAT; (Lartillot and Philippe 2004; Lartillot et al. 2007)].

As expected from its phylogenetic position, *Hanseniaspora* encodes a mitochondrial tRNA with a UAG anticodon, and although it clearly belongs to the tRNA$_1$^{Thr} cluster (Fig. 17.11), it has a regular seven nucleotide anticodon loop (Fig. 17.12). An inspection of highly conserved positions in mitochondrial protein alignments (i.e., the data set used for phylogenetic analysis) confirms translation of several UAG codons as threonine also in *Hanseniaspora* (Fig. 17.13). Yet, due to its elevated evolutionary rate, unexpected sequence deviations (even including small deletions) are abundant in this species, sometimes leading to ambiguous interpretation. To resolve with confidence the question of whether UAG is decoded as threonine or leucine, protein sequence data from *Hanseniaspora* will be required. Note that in the original publication of this mtDNA (Pramateftaki et al. 2006), this tRNA was assigned as tRNA^{Thr}, although the phylogenetic *Candida* neighbors (according to the this paper) read CUN codons as leucine, not threonine. Our proposed change of phylogenetic position of *Hanseniaspora* into the *Saccharomyces* yeast complex eliminates the need for a correction.

Finally, according to our interpretation of multiple protein alignments, *A. gossypii* CUN codons are most likely translated as alanine (Fig. 17.13). According to our phylogenetic analysis (Fig. 17.10), *A. gossypii* belongs (with marginal support) to a lineage that includes *Kluyveromyces lactis* and *Kluyveromyces thermotolerans*, diverging directly after the point at which CUN codon reassignments start and where seven mitochondrial *nad* genes were lost (see the arrow in that figure). That CUN codons were free to encode either threonine as in *K. thermotolerans* or alanine as in *A. gossypii* seems to be perfectly in line with the interpretation that CUN codons first disappeared [as it is the case in *K. lactis*; (Sengupta et al. 2007)]. In addition, the putative alanine tRNA with an UAG anticodon in *A. gossypii* has a G3:U70 base pair (Fig. 17.12), which is a well-known key recognition element for alanine tRNA synthetases [e.g., (Beebe et al. 2008)].

Our proposal of a dual tRNA identity shift would benefit from further investigations, including improvement of tree resolution with an extended set of yeast species and proteomics analysis in *A. gossypii* demonstrating translation of CUN codons as alanine. Testing the specificity of alanine tRNA synthase for this tRNA would follow procedures described in a recent publication on the corresponding threonine tRNA in *S. cerevisiae* (Su et al. 2011).

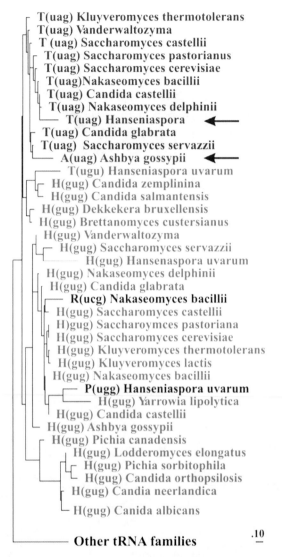

Fig. 17.11 Phylogeny of yeast mitochondrial tRNAs. The phylogenetic analysis with PhyloBayes contained all tRNA sequences from the species shown in Fig. 17.10. Only the sections of the tRNA phylogeny covering the tRNA$_1^{Thr}$ and tRNA^{His} clusters are shown (marked *red* and *blue*, respectively), confirming monophyly of tRNA$_1^{Thr}$ and a sister group relationship to tRNA^{His}. The posterior probability support for the two tRNA groups is 1.0 (note that phylogenetic analysis with tRNA sequences depends on only few informative nucleotide positions, which does not allow one to resolve the branching order within these groups). Interestingly, *H. uvarum* and *A. gossypii* tRNAs with a UAG anticodon and a standard seven-nucleotide anticodon loop cluster with tRNA$_1^{Thr}$ homologs. According to our interpretation of multiple protein alignments, *H. uvarum* CUN codons are translated as threonine (in line with its phylogenetic position, Fig. 17.10), whereas *A. gossypii* CUN codons are most likely translated as alanine (*arrow*, tRNA marked *magenta*). Three tRNA sequences (marked *black*) apparently do not fit within these two clades, have relatively long branches, and are potentially misplaced due to accelerated evolutionary rates

```
                                     [[[[[[[ [[[[        ]]]] [[[[[      ↓      ]]]]]    [[[[[         ]]]]]]]]]]]]
Sacc. cerevisiae  trnT(uag)   GTAAATATAATTTAAT-GGT-AAAATGTATGTTTTTACGTGCATATTATCTAAATTCAAATCTTAGTATTTACA
Sacc. servazzii   trnT(uag)   GTAAATATAATTTAATAGGT-AAAATGTATGTTTCTAGGGATATATTATCTAAGTTCAAGTCTTAGTATTTACA
Sacc. pastor.     trnT(uag)   GTAAATATAATTTAAT-GGT-AAAATGTATGTTTTTAGGTGCATATTATCTAAGTTCAAATCTTAGTATTTACA
Sacc. castellii   trnT(uag)   GTAAATATAATTTAAT-GGTTAAAATATATGTTTTTAGGTGCATATTATCTGAGTTCAAATCTTAGTATTTACA
Nakaseo. bacill.  trnT(uag)   GTAAATATAATTTAAT-GGT-AAAATGTATGTTTTTACGTGCATATTATCTCAAGTTCAAATCTTAGTATTTACA
Nakaseo. delph.   trnT(uag)   GTAAATATAATTTAATCGGTTAAAATGTATGTTTTTAGGTGCATATAATCTAAGTTCAAATCTTAGTATTTACA
Cand. glabrata    trnT(uag)   GTAGATATAATTTAATCGGT-AAAATGTATGTTTTTACGTACATATTATCTAAGTTCAAATCTTAGTATTTACA
Vand. polyspora   trnT(uag)   GTAAATATAATTTAAT-GGT-AAAATATATGTTTTTACGTGCATATTATCAGAGTTCAAATCTCTGTGTTTACA
Kluy. thermotol.  trnT(uag)   GTAAATATAGTTTAAT-GGT-AGAATATATGTTTTTAGGTGCATATGATCTGAGTTCAATTCTCAGTGTTTACA
Ashbya gossypii   trnA(uag)   GTGGATATAGTTTAATTGGT-AAAACATATGTTT-TAGGGACATATATCTTCAGTTCAAAACTGAATATCTACA
Hansen. uvarum    trnT(uag)   GTAATAATAATATAAT-GGTTATTATATAGTGTT-TAGGTCACTATTATCTAGGTTCAATTCCTAGTTATTACA

Sacc. cerevisiae  trnH(gug)   GGTGAATATATTTCAAT-GGT-AGAAATACGCTT-GTGGTGCGTTAAATCTGAGTTCGATTCTCAGTATTCACC
Sacc. castellii   trnH(gug)   GGTGAATATATTTTAAT-GGT-AAAAAGTACGCTT-GTGGTGCGTTAATCTAAGTTCAATTCTTAGTATTCACC
Ashbya gossypii   trnH(gug)   GGTAAATATATTTCAAT-GGT-AGAAAAGATGCTT-GTGTGTGCATTCAATATGAGTTCAATTCTCATTATTTACC
Hansen. uvarum    trnH(gug)   GATTATATTGATTTAAT-GGT-AGAATAAGTACTT-GTGTGCGTACTAAGTCCAGGTTCAATCCCTGGGTATAATC
                                     [[[[[[[ [[[[        ]]]] [[[[[             ]]]]]    [[[[[         ]]]]]]]]]]]]
```

Fig. 17.12 Alignment of yeast mitochondrial tRNAs with UAG and GUG anticodons. Two types of yeast mt-tRNAs are aligned (*upper* and *lower* block of sequences, respectively), with either UAG or GUG anticodons (anticodon marked *orange*). *Square brackets* indicate the four standard helical regions in tRNAs. The *arrow* points to the insertion of a nucleotide leading to an 8-nt anticodon in trnT(uag) of some species. *Hanseniaspora uvarum* has a regular 7 nt anticodon loop and according to our analysis (and consistent with its phylogenetic position, Fig. 17.11) translates CTN codons as threonine (Fig. 17.13). In *Ashbya gossypii*, the anticodon loop has also 7 nt, but although it groups with *Kluyveromyces thermotolerans,* it apparently reads CTN codons as alanine not threonine (Fig. 17.13; note that *K. lactis* has no tRNA with a UAG anticodon and no CTN codon). As trnT(uag) is most likely derived from trnH(gug) by duplication, we have included four histidine tRNAs in the alignment for a comparison [see also (Su et al. 2011)]. Note the characteristic G residue at position −1 (constituting the 5′ terminus of tRNA histidine) that pairs with the C at the 3′ discriminator position

```
Ashbya gossypii   GHAIL-SGAGTG-TTISSLLGA-ITA-DRNFNTS-MIYAMGS-DTRAYF-IIAIPTSIKVFSWLATI-LYALSFLFLFTV-ANASLD
Hanseniaspora     AHAIM-SGAGTG-tTISSLLGA-ItA-DRNFNtS-MIYAMGS-DtRAYF-IIAVPtGIKIFSWLAtI-LFALAFLVLFtA-SNASVD
Sac. cerevisiae   GHAVL-SGAGTG-TSISSLLGA-ITA-DRNFNtS-MVYAMAS-DtRAYF-IIAIPTGIKIFSWLAtI-LYAIAFLFLFTM-ANASLD
Sac. castellii    GHAIL-SGAGTG-TSISSLLGA-ITA-DRNFNTS-MVYAMAS-DTRAYF-IIAIPTGIKIFSWLATI-LFAIAFLFLFTM-ANASLD
Sac. pastorian.   GHAVL-SGAGTG-TSISSLLGA-ITA-DRNFNTS-MVYAMAS-DTRAYF-IIAIPTGIKIFSWLATV-LYAIAFLFLFTM-ANASLD
Sac. servazzii    GHAVL-SGAGTG-TSISSLLGA-ITA-DRNFNTS-MVYAMAS-DTRAYF-IIAIPTGIKIFSWLATI-MFAIAFLFLFTV-ANASLD
Cand. castellii   GHAVA-SGAGTG-TSISSLLGA-ITA-DRNFNTS-MVYAMAS-DTRAYF-IIAIPTGIKIFSWLATI-LYAISFLFLFTI-ANASLD
Cand. glabrata    GHAVL-AGAGTG-TSISSLLGS-ITA-DRNFNTS-MVYAMAS-DTRAYF-IIAIPTGIKIFSWLATI-LYAIAFLFLFTI-ANASLD
Cand. albicans    GHAIA-TGAGTG-TSISSLLGA-FTA-DRNFNTG-MIYAIGS-DSRAYF-VIAIPTGIKIFSWLATI-LFALGFLFLFTI-SNASID
Kluyv. lactis     GHAVL-SGAGTG-TSISSLLGA-ITA-DRNFNTS-MVYAMAS-DTRAYF-IIAIPTGIKIFSWLATI-LYAIAFLFLFTI-ANASLD
Kluyv. thermo.    GHAVL-SGAGTG-TSISSLLGA-ITA-DRNFNTS-MVYAMAS-DTRAYF-IIAIPTGIKIFSWLATI-LYAIAFLFLFTV-ANASLD
Nakas. bacill.    GHAVL-SGAGTG-SSISSLLGA-ITA-DRNFNTS-MVYAMAS-DTRAYF-IIAIPTGIKIFSWLATI-LYAIAFLFLFTI-ANASLD
Nakas. delph.     GHAVL-AGAGTG-TSISSLLGA-ITA-DRNFNTS-MVYAMAS-DTRAYF-IIAIPTGIKIFSWLATI-LYAIAFLFLFTI-ANASLD
Vanderwaltozyma   GHAIL-SGAGTG-TSISSLLGA-ITA-DRNFNTS-MVYAMAS-DTRAYF-IIAIPTGIKIFSWLATI-MYAIAFLFLFTL-ANASLD
Cand. orthops.    GHAIA-SGAGTG-TSISSLLGA-FTA-DRNFNTG-MIYAIAS-DSRAYF-VIAVPTGIKIFSWLATI-LFALGFLFLFTV-SNASID
Dekkera brux.     AHSVL-EGPGTG-TSISSILSS-FTA-DRNFNTA-MVYAMGS-DTRAYF-VIAIPTGIKIFSWLATL-LYAIAFLFIFTI-SNANLD
Pic. canadensis   GHALL-SWAGTG-TSISSLLGA-ITA-DRNFNTS-MVYAMAS-DTRAYF-VIAVPTGIKIFSWLATL-LYAIAFLFLFTI-ANASLD
```

Fig. 17.13 Alignment of derived Cox1 protein sequences demonstrating reading of CUN codons as threonine in *Hanseniaspora* and as alanine in *Ashbya*. Only parts of the Cox1 alignment are shown (blocks of amino acids separated by a space, corresponding to positions 61-65, 106-111, 154-162, 191-193, 213-219, 273-279, 300-305, 311-327, 339-350, and 359-364 in *S. cerevisiae*). CUN codons that are translated into threonine in *Hanseniaspora uvarum* or *S. cerevisiae* are in lower case; so are CUN codons translated into alanine in *Ashbya gossypii*. Corresponding columns are marked *orange* and *yellow*, respectively. A similar pattern of conservation is valid throughout all mtDNA-encoded proteins of these species

17.3.4 Specificity of Translation Initiation

In bacteria, initiation of translation usually depends on a specific methionine initiator tRNA that recognizes AUG, GUG, and UUG codons, in descending order of efficiency. This is also the case in some organelles, whereas in others initiation no longer depends on an extra tRNA. In addition, other initiation codons may extend the given list, such as AUA, UUA, etc. Confident recognition of initiator tRNA genes requires complicated biochemical testing procedures, including formylation of the purified mitochondrial tRNA. To our knowledge, such data are only available for yeast (Canaday et al. 1980). Alternatively, conserved nucleotide signatures known from bacteria (Söll and RajBhandary 1995; Marck and Grosjean 2002) together with phylogenetic analysis of tRNAs may help with identification; however, nucleotide signatures are not necessarily conserved in mitochondria, and inferences may be misled by potential tRNA identity shift. The presence of three distinct tRNAs with CAU anticodon in an organelle genome is another valuable indicator of the presence of separate initiator and elongator tRNAs, plus tRNA(Ile) recognizing specifically AUA codons due to a lysidine modification of the C residue in the anticodon (Muramatsu et al. 1988; Weber et al. 1990). In cases when only one tRNA(Met) is present in an organelle, it most likely serves in both initiation and elongation.

A special case is tRNA(Met) in *S. cerevisiae* that has an unmodified C in the wobble position. A structural irregularity in the T-stem contains an unpaired nucleotide within the base-paired T psi C stem that may account for its ability to decode both AUG and AUA as methionine in a heterologous translation system (Sibler et al. 1985). Finally, a most interesting situation exists in monoblepharidalian fungal mtDNAs, in which almost every protein-coding gene has a guanosine residue upstream of the predicted AUG or GUG start codons (Bullerwell et al. 2003a). The appearance of this conserved G residue correlates with the presence of a nonorthodox cytosine residue at position 37 in the anticodon loop of the assumed mitochondrial initiator tRNAs, suggesting a 4-bp interaction between a CAUC anticodon and quartet GAUG/GGUG codons. A similar interaction may also be involved in mitochondrial translation initiation in the sea anemone *Metridium senile* (Beagley et al. 1998; Bullerwell et al. 2003a).

17.4   tRNA Sets Occurring in Organelle Genomes

17.4.1   tRNAs in Plastid Genomes

The plastids of most photosynthetic algae and plants possess a small genome coding for 100–250 genes, including an extended set of tRNAs with often more than one isoacceptor for decoding the fourfold degenerate codon families for serine (UCN codons), leucine (CUN), threonine (ACN), arginine (CGN), and glycine (GGN), and the twofold degenerate codon family for lysine (AAR) (de Koning and Keeling 2006; Lung et al. 2006). In addition, some genes are located in two copies of a large inverted repeat typical for chloroplast genomes. For example, the chloroplast genome of tobacco (*Nicotiana tabacum*) encodes 30 distinct tRNAs that are encoded by 37 tRNA genes, seven of which are duplicated

(Shinozaki et al. 1986). There is a tendency for tRNA gene loss that is associated with overall genome reduction in parasitic species, such as the parasitic green alga *Helicosporidium* sp. (37.5 kpb, 25 tRNAs) (de Koning and Keeling 2006), the parasitic flowering plant *Epifagus virginiana* (70 kbp, 17 tRNAs) (Wolfe et al. 1992), and the apicoplast genome of apicomplexans such a *Plasmodium* (Preiser et al. 1995). In addition, two species of lycophytes, *Selaginella uncinata* (Tsuji et al. 2007) and *S. moellendorffii* (Smith 2009), contain highly derived chloroplast genomes that encode only 12 and 13 tRNAs, respectively. Finally, some of the tRNA isoacceptors appear to be especially prone to loss. For example, trnR(ccg) has been lost repeatedly in several clades of lycophytes, ferns, and seed plants (GAO et al. 2010). In several of the cases listed above, chloroplast genomes have to import tRNAs; however, the mechanism remains unknown.

17.4.2 tRNA Genes in Mitochondrial Genomes

Compared to plastids, mitochondrial genomes exhibit far more variation in both size and gene content, and the number of tRNAs encoded in mitochondria is usually less than the "minimal" set due to changes in codon recognition rules, base modifications, and tRNA import. Most of the changes in the genetic code have occurred in animals, fungi, and green algae (Knight et al. 2001a, b). However, one change, the reassignment of the UGA codon from termination to tryptophan, has occurred repeatedly throughout eukaryotes. Changes in the genetic code often reduce the number of tRNA genes required for mitochondrial translation. For example in bilaterian animals, the reassignment of the AUA codon from isoleucine to methionine led to the loss of one isoleucine tRNA, and the reassignment in invertebrates of AGA/AGG codons from arginine to serine led to the loss of the corresponding tRNA(Arg) [e.g., (Himeno et al. 1987; Andersson and Kurland 1991; Tomita et al. 1998)]. In addition, the gene for a separate elongator methionine tRNA has been repeatedly lost from mitochondrial genomes, without accompanying changes in the genetic code. In most bilaterian animals, these losses, combined with super-wobble and base modifications in tRNAs, allowed a reduction to 22 in the number of distinct tRNAs required for mitochondrial translation. A further reduction in this number by leaving codon families unassigned by a tRNA is possible (e.g., CUN in the yeast *K. lactis*), but likely occurs only during transitions in codon reassignment as discussed above. This is because mutations that create unassigned codons are likely to result in ribosome stalling and will be deleterious, i.e., codon reassignment should be under positive selection.

Further reduction of the number of mtDNA-encoded tRNA occurs usually only via tRNA import of nucleus-encoded tRNAs into organelles. The systems mediating this import have likely evolved independently in several groups of eukaryotes, with mitochondrial tRNA import having been verified experimentally in plants, protozoa, the yeast *Saccharomyces cerevisiae*, and marsupials (Salinas et al. 2008). There is a clear difference in the extent of tRNA import among different eukaryotic groups. Based on mitochondrial genome analysis, most bilaterian animals and fungi need not import tRNAs into their mitochondria (Hopper and Phizicky 2003; Lung et al. 2006); for import of tRNA-Lys1 into yeast mitochondria, required for compensation of a functional defect in the mtDNA-coded Lys-tRNA under certain physiological conditions, see Kamenski et al. (2007) and Tarassov et al. (2007). In contrast, tRNA import is common in nonbilaterian animals, protists, plants, and several chytrid fungi [e.g., (Beagley et al. 1998; Gray et al. 1998; Bullerwell et al. 2003a; Voigt et al. 2008; Haen et al. 2010)]. Plant mitochondrial genomes typically encode fewer than the 25 tRNA genes minimally required for mitochondrial translation and import the rest from the cytosol (Marechal-Drouard et al. 1990; Glover et al. 2001). In extreme cases, no tRNAs are found in mitochondrial DNA and all of them are inferred to be imported from the cytosol. Complete absence of mtDNA-encoded tRNA genes has been reported in trypanosomatids (such as *Trypanosoma brucei* and *Leishmania* spp.) (Schneider 2001). For a discussion on the various mechanisms by which tRNAs may be imported, the reader is referred to a selected list of recent publications (Schneider and Marechal-Drouard 2000; Tarassov et al. 2007; Rubio et al. 2008; Salinas et al. 2008; Alfonzo and Soll 2009; Berglund et al. 2009; Kamenski et al. 2010).

Remarkably, the remaining tRNA genes in cnidarians (Beagley et al. 1998), chaetognathes (Helfenbein et al. 2004), and some sponges (Wang and Lavrov 2008) include *trnM(cau)* and *trnW(uca)*, supporting the inference of a special role of these tRNAs in animal mitochondria, translation initiation, and recognition of the stop codon UGA as tryptophan. Similarly in chytrid fungi, most mt tRNAs have to be imported, whereas those decoding UGA (Trp) or UAG (Leu) are always present [(Bullerwell et al. 2003a; Laforest et al. 2004), unpublished analyses of *Batrochochytrium* mtDNAs]. An exception to this rule is in *Trypanosoma*, which imports a nucleus-encoded tRNA(Trp), but because this tRNA apparently does not efficiently recognize the UGA codons in mtDNA-encoded genes, it is modified by C-to-U editing (Charriére et al. 2006).

Several studies have investigated the fate of missing tRNAs and their corresponding aaRSs (Glover et al. 2001; Haen et al. 2010). It appears that imported tRNAs are always of eukaryotic evolutionary origin, i.e., there is no known case where tRNA genes of mitochondrial origin are integrated into the nuclear genome, with the corresponding tRNA being re-imported into mitochondria. Furthermore, mtDNA-encoded mitochondrial tRNAs and their corresponding nucleus-encoded AARS are usually lost together. The latter are replaced by cytosolic AARS, which gain a mitochondrial targeting sequence and become dual-targeted to cytosol and mitochondria. Dual targeting of aaRSs has been demonstrated experimentally in several organisms (Rinehart et al. 2004, 2005) [yet recently contested for yeast GlnRS that is not imported into mitochondria as previously believed. Instead, Gln-tRNA(Gln) is generated by transamidation (Frechin et al. 2009)]. Because of the coordinated loss of mt-tRNAs and AARS, in organisms that have lost mtDNA-encoded tRNAs, the nuclear genome encodes fewer distinct aaRSs (Berriman et al. 2005; Haen et al. 2010).

17.5   Bioinformatics Tools for Identification of tRNA Genes

Since the introduction of massive parallel sequencing techniques, genome sequencing has accelerated to an extent that, predictably, only automated analysis tools will be able to keep up with future genome annotation and GenBank submission. This constraint applies also to the relatively small organelle genomes (currently >2,500 at NCBI/GenBank), whose tRNAs up to now have been annotated either manually, or with the help of tRNAscan SE (Lowe and Eddy 1997) followed by expert corrections. These revisions include the elimination of false positives, identification of pseudogenes, and the addition of absent tRNAs (in particular in bilaterian animals). The only dedicated organelle genome annotator [Dogma; Wyman et al. 2004] that is specialized for use on bilaterian animal mtDNAs uses tRNAscan SE with an adapted global tRNA profile (different from the one supplied with tRNAscan SE), which improves predictions to some degree but continues to be incomplete. In our opinion, the reason for the observed shortcomings in automated tRNA prediction is the use of a single, global tRNA model that covers most diverse tRNA sequences and structures, some with missing D and T loops. The resulting *"one-fits-all model"* provides results with relatively low scores (or E-values), together with an increase in false positives. The most serious shortcomings are with A + T-rich genomes (e.g., yeast mtDNAs), in combination with unorthodox and structurally reduced tRNAs and tRNA editing. Finally, tRNAscan SE does not recognize tRNA genes with group I or II introns.

A more recently developed tool for identification of animal mt tRNAs is Arwen (Laslett and Canback 2008). It uses relaxed structural constraints for searching tRNA-like structures, at the cost of high numbers of false positives that are (arguably) easily identified by manual curation. Like tRNAscan SE, Arwen does not consider shifting the $5'$ terminus of tRNA histidine to position -1 of the standard model.

Unfortunately, even if the approach provides a better sensitivity than tRNAscan SE, Arwen perpetuates the burden of manual annotation. In face of the overwhelming number of tRNA annotation errors that have been introduced into GenBank records exactly due to the lack of objective structural criteria, the development of (a) a more precise structure/sequence-based search algorithm in conjunction with (b) training sets that represent the complete range of structural variation as described in this review does remain a high priority. To what extant upgrading of structural search models for use with tRNAscan SE [or the similar but more advanced general-purpose Infernal; (Eddy 2008b)] will provide such a solution remains to be seen. In fact, after initial submission of this review, a publication dealing with animal mtDNA annotation appeared online [Mitos; (Donath et al. 2011)], which claims high-precision identification of tRNA (and rRNA) genes using Infernal, based on a set of updated tRNA structure models. Yet, when comparing tRNA predictions for the centipede *Lithobius forficatus* (NC_002629), which has 22 putative mtDNA-encoded tRNAs some of which undergo $3'$ tRNA editing (Lavrov et al. 2000), tRNAscan-SE finds only one, Arwen seven (plus four false positives), and Mitos 20 out of 22. (a) Yet, tRNA-Ser (positions 9768-9828) of the GenBank record is slightly shifted in its 5′ position compared with Mitos (which arguably found the correct solution); (b) it is the only tRNA found with tRNAscan-SE but identified as tRNA-Phe and (c) becomes tRNA-Trp (and further shifted positions) with Arwen. Evidently, biochemical identification is urgently required in this and probably numerous other cases, in addition to verification of structural models based on the rules developed above.

17.5.1 Pseudo-tRNAs and tRNA-Like Structures

Before entering into a more detailed discussion of automated tRNA prediction, it is imperative to understand rules and "gold standards" that are the basis for distinguishing *bona fide* tRNAs from pseudogenes and "tRNA-like structures." Most importantly, RNA sequencing of complete tRNA sets is the very basis for confirming bioinformatics predictions, and such sequence sets are the optimal training sets for automated predictions. In addition, direct sequencing of tRNAs permits identification of imported tRNAs (i.e., that are not encoded in an organelle's genome), validation of transcription and maturation, pinpointing of nucleotide modification, and detection of nucleotide positions that are edited post-transcriptionally. Ideally (and we are unfortunately far from it), sequences of mitochondrial tRNAs should be available for representatives of every major eukaryotic lineage, in particular those with fast-evolving and/or unusual tRNA features. Finally, purified native or synthetic tRNAs should be tested for their in vitro activity to confirm the functionality of unusual tRNA structures.

The shortage of tRNA sequence information is obviously related to the high cost and effort of classic RNA sequencing, which requires painstaking purification procedures and manual sequencing techniques that require radioactive labeling, and chemical or enzymatic degradation of RNA. The application of new RNA sequencing technology such as Illumina/RNAseq is expected to change this situation, although it does not rival traditional techniques that are able to identify RNA termini other than 5′-phosphate and 3′-OH, and a variety of RNA modifications. In conjunction with inexpensive complete genome sequencing, deep RNA sequencing will still allow mapping of tRNAs and processing intermediates to the genome of a given species, and identify potential tRNA editing.

In turn, as genome sequencing from a wide range of species is now affordable, with minute amounts of total DNA and without the need for further DNA purification, comparative genome information alone will compensate to some degree for the lack of tRNA data. The expected flood of genomic data will permit more precise phylogenetic modeling of tRNA structures, in silico predictions of translation code changes and tRNA identity shifts, and the development of better tRNA sequence profiles for more sensitive searches.

17.5.2   Bioinformatics Tools for Identification of tRNAs and Codon Reassignments

As an illustration, we conduct here a study of mitochondrial tRNA evolution and the evolution of the genetic code, based on all publicly available (close to 40) yeast mtDNAs (see above, Sect. 17.3.3). The challenge starts with GenBank records that contain a multitude of inconsistencies and imprecise information, such as incorrect tRNA coordinates, an indiscriminate use of the "yeast mitochondrial" translation code number 3 that is valid only for a defined subset of yeast species, annotation of pseudogenes and tRNA-like structures as true tRNAs, incorrect "codon recognized" features that ignore wobble and super-wobble recognition rules, etc.

The record describing *Saccharomyces cerevisiae* mtDNA is among the few remarkably high-quality exceptions (apart from a two-nucleotide shift of the tRNA histidine 3′ terminus); it obviously has been curated manually based on expert knowledge (Table 17.1). In contrast, tRNAscan SE predicts seven false positive tRNAs for the same mtDNA (Table 17.1). Incorrect solutions are essentially composed of A and U (folding into "tRNA like structures"), some of which contain short intron insertions of a type occurring in nuclear, but never in mt, genomes. An inspection of the yeast genome tRNAscan SE database (Lowe 1997) reveals additional differences. Some tRNAs on the mitochondrial chromosome (chrM) are correctly labeled "pseudo," yet other (biochemically confirmed) yeast tRNAs are missing that are otherwise predicted with tRNAscan SE (see also Table 17.1). The generally most inconsistent annotations are for tRNA histidine. The mitochondrial *S. cerevisiae* GenBank correctly assumes an eight-base-pair acceptor stem for this tRNA, i.e., the 5′ terminus is extended to position −1 (numbering according to the standard tRNA model; Fig. 17.1a). This position either is a G residue in the genome sequence (Burkard and Soll 1988) or may be edited at the tRNA level (Rao et al. 2011; Cooley et al. 1982; Burkard and Soll 1988; Leigh and Lang 2004; Jackman and Phizicky 2006b; Jackman and Phizicky 2006a). tRNAscan SE does not account for the extra nucleotide at position −1 of tRNA histidine, which may explain the numerous inconsistencies in GenBank records.

In short, for the purpose of this review, tRNAs had to be predicted employing a different bioinformatics approach, and we decided to develop a specific yeast mitochondria tRNA profile to accommodate unorthodox structures such as eight-nucleotide anticodon loops. In addition, given the many gene name inconsistencies and errors in GenBank gene annotations (the basis for calculating codon usage tables and deriving protein sequence), all mt genomes had to be been re-annotated (using MFannot, a tool under development in our lab, followed by a few manual corrections). MFannot is not formally published, but may be used via our webservice (Beck and Lang 2010).

For organelle tRNA identification, our laboratory previously employed tRNAscan SE (Lowe and Eddy 1997) but has now converted to more regular use of Erpin for search model development as outlined below. Both approaches use algorithms that take advantage of the most informative, aligned tRNA

Table 17.1 tRNA predictions with tRNAscan SE versus Erpin, compared to annotations in GenBank records

Yeast species[a]	GenBank[b]	tRNAscan[c]	Erpin[d]	T(ugu)[e]	T(uag)[f]	L(uag)[g]
Ashbya gossypii	23	23	23	+	−	+[g]
Brettanomyces custers	25	25	25	+	−	+
Candida alai	25	25	25	+	−	+
Candida albicans	30	30	30	+	−	+
Candida castellii	23	43(20)	23	+	+	−
Candida glabrata	23	25(2)	23	+	+	−
Candida maltosa	26	26	26	+	−	+
Candida metapsilosis	24	24	24	+	−	+
Candida neerlandica	24	24	24	+	−	+
Candida orthopsilosis	24	24	24	+	−	+
Candida parapsilosis	24	25(1)	24	+	−	+
Candida salmanticensis	25	25	25	+	−	+
Candida sojae	29	29	29	+	−	+
Candida jiufengensis	24	24	24	+	−	+
Candida subhashii	24[h]	24	24	+	−	+
Candida viswanathii	25	25	25	+	−	+
Candida zemplinina	25	25	25	+	−	+
Debaryomyces hansenii	25	25	25	+	−	+
Dekkera bruxellensis	25	29(4)	25	+	−	+
Hanseniaspora uvarum	23	23	22[i]	−	−	+[g]
Kluyveromyces lactis	22	22	22	+	−[j]	−[j]
Kluyveromyces thermotolerans	24	24	24	+	+	−
Nakaseomyces bacilli sp.	23[k]	80(56)	24	+	+	−
Nakoseomyces delphensis	23	27(4)	23	+	+	−
Pichia canadensis	25	27(1)[l]	26[l]	+	−	+
Pichia farinosa[m] (*P. sorbitophila*)	25	25	25	+	−	+
Saccharomyces castellii	23	23	23	+	+	−
Saccharomyces cerevisiae	24	31(7)	24	+	+	−
Saccharomyces pastori	24	24	24	+	+	−
Saccharomyces servazzii	23	24(1)	23	+	+	−
Vanderwaltozyma	23	26(3)	23	+	+	−
Yarrowia lipolytica	27[n]	24	24	+	−	+

[a]Hemiascomycetes with complete mtDNA GenBank records (as of 10 December 2010). Species encoding tRNA threonine (UAG) with an 8-nt anticodon loop are in *bold*

[b]Number of tRNAs in GenBank records; note that incorrect sequence positions are common

[c]Number of tRNAs predicted by tRNAscan SE (Lowe 1997; Lowe and Eddy 1997); number of mistaken predictions (according to expert verification) in *brackets*

[d]Number of tRNAs predicted by Erpin/RNAweasel (using a custom training set based on yeast mt tRNAs)

[e]Occurrence of regular threonine tRNA(UGU)

[f]Occurrence of *S. cerevisiae*-related threonine tRNA(UAG) with an 8-nt anticodon loop; respective species in *bold*

[g]Occurrence of canonically structured tRNA(UAG); recognizes leucine according to codon conservation patterns, except in *Hanseniaspora* where it stands for threonine, and in *Ashbya gossypii* where it recognizes alanine (see also Fig. 17.13)

(continued)

[h]Includes an unorthodox glutamic acid tRNA with a shortened T loop
[i]*Hanseniaspora* tRNA methionine has a three-nucleotide insertion in the T-loop and is therefore not recognized with Erpin; according to our interpretation it is a pseudo-tRNA
[j]*K. lactis* has no UAG codons, and no tRNA recognizing them
[k]GenBank record lacks one tRNA
[l]Includes a potential tRNA that has an anticodon stem-loop region with exclusively A and U residues, of ambiguous structure
[m]Potential misidentification of *Pichia sorbitophila* as *Pichia (Millerozyma) farinosa*
[n]Three extra "tRNA-like structures" in the *Yarrowia* GenBank record that are not identified by tRNAscan SE and Erpin

sequence profiles (sequence training sets) that contain information on secondary structure interactions. Profile-based algorithms for RNAs [e.g., Erpin, Infernal, and its precursor version Cove; (Eddy and Durbin 1994; Lowe and Eddy 1997; Gautheret and Lambert 2001; Griffiths-Jones et al. 2003; Eddy 2008b)] are more sensitive by far than binary sequence comparison methods (such as Blast and Fasta). Because RNA profile searches include primary sequence information together with RNA secondary structure information (i.e., they are based on a "*structural alignment*"), these are also more sensitive than primary sequence profile searches alone [e.g., the Hidden Markov Model – based HMMER3; (Eddy 2008a)]. In our opinion, Infernal (Griffiths-Jones et al. 2003; Eddy 2008b) is the currently most sensitive and sophisticated tool for searching for structured RNAs. However, two unfortunate drawbacks of Infernal are (a) its inability to model pseudoknot interactions and (b) slow execution speed, which becomes limiting even with the small tRNAs, when many genomes are analyzed or when repeated analyses are required during the development of specific tRNA models.

An alternative solution is Erpin (Gautheret and Lambert 2001), which does not make use of the elaborate insertion/deletion and nucleotide transition statistics implemented in Infernal. For searching genomes, Erpin uses only positional nucleotide and base pairing probabilities (calculated from sequence profiles), which make it about two orders of magnitude faster than Infernal. Erpin further allows integration of pseudoknot (and other tertiary) interactions into its structural model, which may be a decisive advantage in given instances [e.g., group I intron structures; Lang et al. 2007)]. As the underlying structural RNA alignments are essentially the same for both programs, training sets developed with Erpin are easily adapted for use with Infernal (or tRNAscan SE). This may be important because Infernal is able to find tRNAs with unusual structural features (in particular positional insertions) in new genome sequences, whereas Erpin may require modification of the search model.

In summary, in our opinion Erpin is best suited for rapid development of alignments and models from genome sequences, for searches of standard-structure RNAs, and for modeling RNAs with pseudoknot structures. Infernal excels with highest sensitivity and with a potential for finding new RNAs that are not perfectly colinear with a given training set alignment.

17.5.3 Development of Species-Specific Trainings Sets for tRNA Identification

As mentioned above, "one-fits-all" global tRNA models are imperfect because inferences based on them have low statistical support, with an elevated risk of false positives, and they may even not find all tRNAs. Examples of such global models are the ones that come with tRNAscan SE, and our own organelle tRNA model previously used in MFannot, which performs perfectly well except for budding yeast and bilaterian animal mtDNAs. To overcome this difficulty, specific tRNA models for these two large groups of fast-evolving species need to be developed, as well as for intron-containing organelle tRNA genes that occur (rarely) in a wide range of species [e.g., plants, red and green algal and jakobid mitochondria; (Oda et al. 1992a, b; Leblanc et al. 1995; Lang et al. 1997; Turmel et al. 2002, 2003)].

Building a yeast mt-specific tRNA search model is straightforward, starting with the collection of well-characterized *S. cerevisiae* mt-tRNAs [e.g., Martin et al. 1977; Canaday et al. 1980; Sibler et al. 1985; Bordonne et al. 1987a, b; Chen and Martin 1988; Kolesnikova et al. 2000] and adding similar tRNA sequences from other yeast mtDNAs. The resulting training set allows rapid finding (with Erpin) of the expected tRNAs in all published yeast mtDNAs [Table 17.1; accessible via MFannot and RNAweasel; (Lang et al. 2007; Beck and Lang 2009, 2010)]. On the other hand, our attempts to proceed in a similar way for bilaterian animal mt-tRNAs were less successful, which is essentially due to major structural differences (in particular, lack of D or T loops) among tRNAs of a given species, as well as to substantial differences across species (e.g., a much more pronounced relaxation of tRNA structure in nematodes). As a potential solution, we have started compiling separate alignments for all 22 specific tRNAs of mammalian species [based in part on a well-curated data set; (Putz et al. 2007)]. Together, these 22 profiles allow identification of mammalian, a large portion of vertebrate, but only a limited fraction of invertebrate mt-tRNAs. Evidently, similar sets of tRNA models will have to be developed for various groups of invertebrates. For that, substantially more tRNA sequence information is required, also because of uncertainty about the degree of tRNA editing. The available literature suggests that tRNA editing occurs more frequently in invertebrates [e.g., Morl et al. 1995; Yokobori and Pääbo 1995, 1997; Tomita et al. 1996; Borner et al. 1997; Lavrov et al. 2000; Segovia et al. 2011], complicating both the interpretation of tRNA structure and identity.

Finally, mt-tRNA genes may contain introns [most of group II, some of group I; e.g., Kuhsel et al. 1990; Manhart and Palmer 1990; Oda et al. 1992a, b; Leblanc et al. 1995; Lang et al. 1997; Vogel et al. 1997; Besendahl et al. 2000; Turmel et al. 2002, 2003]. So far, intron-containing tRNA genes have had to be identified and annotated manually, based on several lines of evidence, in particular (a) lack of one or more of the expected tRNA genes in a given genome (compared to related species); (b) partial tRNA sequences, in genome regions without evident alternative coding capacity; and (c) presence of orphan intron structures, i.e., without known flanking exon sequences. Recognition of such orphan introns is now quite effective due to automated search procedures (Lang et al. 2007; Beck and Lang 2009). Computerized finding of tRNA genes that contain such introns is likewise within reach, as numerous examples have been published in both mtDNAs and ptDNAs, sufficient for building specific training sets. Yet, this task remains complex, as the respective intron group has to be identified to assist in prediction of the varying intron/exon boundaries. We provide a preliminary search option for split tRNA genes at our website (Beck and Lang 2009), but strongly recommend checking (and potential adaptation) of exon/intron boundaries. As more gene sequences of this type become available, the precision of predictions will improve.

17.6   Conclusions

The notion that tRNAs have universal features and properties that are well understood for a long time does not apply to mitochondria. This deficiency is reflected in the slow progress toward the development of bioinformatics tools that predict bilaterian mt-tRNAs from genome sequence alone, i.e., completely, without false positives, with accurate RNA termini, and reliable tRNA identity predictions. We were surprised to learn that the same applies, to a less significant and more-readily-corrected degree, to yeast mitochondria. Clearly, better bioinformatics tools plus training sets have to be developed, and these sets rely on systematic tRNA plus genome sequencing. In addition, currently available search models (including our own) include primary and secondary sequence information but not tertiary interactions (i.e., pseudknots); this represents valuable extra information, which is important for identification and structural verification of bilaterian animal tRNA predictions. Systematic determination of post-transcriptional nucleotide modifications and RNA editing is also lacking for organelle tRNAs, and much more biochemistry has to be employed to investigate tRNA structure and function. Particularly needed is an understanding of unorthodox tRNA structures, and of complex functional scenarios in which tRNA identity or even the genetic code might have changed.

Acknowledgments We are grateful to Dr. Michael W. Gray (Dalhousie University, Canada), Dr. Jiqiang Ling (Yale University), and Dr. Robert Martin (CNRS – Université de Strasbourg, France) for insightful discussion and comments on the manuscript, to Konstantin Bokov (University of Montreal) for help with the preparation of figures, and to Dr. Robert Martin for suggesting and hosting the in vitro translation experiments of mitochondrial *S. pombe* transcripts in his laboratory. This work was supported by the Canadian Research Chair Program (BFL), National Sciences and Engineering Research Council of Canada (SS), and the National Science Foundation, U.S. (DL).

References

Alfonzo JD, Soll D (2009) Mitochondrial tRNA import – the challenge to understand has just begun. Biol Chem 390:717–722

Alfonzo JD, Blanc V, Estevez AM, Rubio MA, Simpson L (1999) C to U editing of the anticodon of imported mitochondrial tRNA(Trp) allows decoding of the UGA stop codon in *Leishmania tarentolae*. EMBO J 18:7056–7062

Anderson S et al (1981) Sequence and organization of the human mitochondrial genome. Nature 290:457–465

Andersson GE, Kurland CG (1991) An extreme codon preference strategy: codon reassignment. Mol Biol Evol 8:530–544

Basavappa R, Sigler PB (1991) The 3 A crystal structure of yeast initiator tRNA: functional implications in initiator/elongator discrimination. EMBO J 10:3105–3111

Beagley CT, Okimoto R, Wolstenholme DR (1998) The mitochondrial genome of the sea anemone *Metridium senile* (Cnidaria): introns, a paucity of tRNA genes, and a near-standard genetic code. Genetics 148:1091–1108

Beck N, Lang BF (2009) RNAweasel, a webserver for identification of mitochondrial, structured RNAs. http://megasun.bch.umontreal.ca/RNAweasel

Beck N, Lang BF (2010) MFannot, organelle genome annotation websever. http://megasun.bch.umontreal.ca/papers/MFannot

Beebe K, Mock M, Merriman E, Schimmel P (2008) Distinct domains of tRNA synthetase recognize the same base pair. Nature 451:90–93

Berglund AK et al (2009) Dual targeting to mitochondria and chloroplasts: characterization of Thr-tRNA synthetase targeting peptide. Mol Plant 2:1298–1309

Berriman M et al (2005) The genome of the African trypanosome *Trypanosoma brucei*. Science 309:416–422

Besendahl A, Qiu YL, Lee J, Palmer JD, Bhattacharya D (2000) The cyanobacterial origin and vertical transmission of the plastid tRNA(Leu) group-I intron. Curr Genet 37:12–23

Binder S, Knoop V, Brennicke A (1991) Nucleotide sequences of the mitochondrial genes trnS (TGA) encoding tRNA(TGASer) in *Oenothera berteriana* and *Arabidopsis thaliana*. Gene 102:245–247

Biou V, Yaremchuk A, Tukalo M, Cusack S (1994) The 2.9 A crystal structure of *T. thermophilus* seryl-tRNA synthetase complexed with tRNA(Ser). Science 263:1404–1410

Bordonne R, Bandlow W, Dirheimer G, Martin RP (1987a) A single base change in the extra-arm of yeast mitochondrial tyrosine tRNA affects its conformational stability and impairs aminoacylation. Mol Gen Genet 206:498–504

Bordonne R, Dirheimer G, Martin RP (1987b) Transcription initiation and RNA processing of a yeast mitochondrial tRNA gene cluster. Nucleic Acids Res 15:7381–7394

Borner GV, Morl M, Janke A, Paabo S (1996) RNA editing changes the identity of a mitochondrial tRNA in marsupials. EMBO J 15:5949–5957

Borner GV, Yokobori S, Morl M, Dorner M, Paabo S (1997) RNA editing in metazoan mitochondria: staying fit without sex. FEBS Lett 409:320–324

Bourdeau V, Steinberg SV, Ferbeyre G, Emond R, Cermakian N, Cedergren R (1998) Amber suppression in *Escherichia coli* by unusual mitochondria-like transfer RNAs. Proc Natl Acad Sci USA 95:1375–1380

Brown JR, Doolittle WF (1995) Root of the universal tree of life based on ancient aminoacyl-tRNA synthetase gene duplications. Proc Natl Acad Sci USA 92:2441–2445

Bullerwell CE, Forget L, Lang BF (2003a) Evolution of monoblepharidalean fungi based on complete mitochondrial genome sequences. Nucleic Acids Res 31:1614–1623

Bullerwell CE, Leigh J, Forget L, Lang BF (2003b) A comparison of three fission yeast mitochondrial genomes. Nucleic Acids Res 31:759–768

Bullerwell CE, Leigh J, Seif E, Longcore JE, Lang BF (2003c) Evolution of the fungi and their mitochondrial genomes. In: Arora D, Khachatourians GG (eds) Applied mycology and biotechnology. Elsevier Science, Amsterdam, pp 133–160

Burger G, Plante I, Lonergan KM, Gray MW (1995) The mitochondrial DNA of the amoeboid protozoon, *Acanthamoeba castellanii*: complete sequence, gene content and genome organization. J Mol Biol 245:522–537

Burkard U, Soll D (1988) The 5'-terminal guanylate of chloroplast histidine tRNA is encoded in its gene. J Biol Chem 263:9578–9581

Canaday J, Dirheimer G, Martin RP (1980) Yeast mitochondrial methionine initiator tRNA: characterization and nucleotide sequence. Nucleic Acids Res 8:1445–1457

Cantatore P, Gadaleta MN, Roberti M, Saccone C, Wilson AC (1987) Duplication and remoulding of tRNA genes during the evolutionary rearrangement of mitochondrial genomes. Nature 329:853–855

Cedergren R, Lang BF (1985) Probing fungal mitochondrial evolution with tRNA. Biosystems 18:263–267

Charriére F, Helgadottir S, Horn E, Söll D, Schneider A (2006) Dual targeting of a single tRNA (Trp) requires two different tryptophanyl-tRNA synthetases in *Trypanosoma brucei*. Proc Natl Acad Sci USA 103:6847–6852

Chen JY, Martin NC (1988) Biosynthesis of tRNA in yeast mitochondria. An endonuclease is responsible for the 3'-processing of tRNA precursors. J Biol Chem 263:13677–13682

Cooley L, Appel B, Soll D (1982) Post-transcriptional nucleotide addition is responsible for the formation of the 5' terminus of histidine tRNA. Proc Natl Acad Sci USA 79:6475–6479

Crick F (1966) Codon–anticodon pairing: the wobble hypothesis. J Mol Biol 19:548–555

Curran JF, Yarus M (1987) Reading frame selection and transfer RNA anticodon loop stacking. Science 238:1545–1550

de Crecy-Lagard V, Marck C, Brochier-Armanet C, Grosjean H (2007) Comparative RNomics and modomics in Mollicutes: prediction of gene function and evolutionary implications. IUBMB Life 59:634–658

de Koning A, Keeling P (2006) The complete plastid genome sequence of the parasitic green alga *Helicosporidium* sp. is highly reduced and structured. BMC Biol 4:12

Doherty EA, Batey RT, Masquida B, Doudna JA (2001) A universal mode of helix packing in RNA. Nat Struct Biol 8:339–343

Donath A et al. (2011) http://www.bioinf.uni-leipzig.de/publications/09–028

Doyon FR, Zagryadskaya EI, Chen J, Steinberg SV (2004) Specific and non-specific purine trap in the T-loop of normal and suppressor tRNAs. J Mol Biol 343:55–69

Dujon B, Colleaux L, Jacquier A, Michel F, Monteilhet C (1986) Mitochondrial introns as mobile genetic elements: the role of intron-encoded proteins. Basic Life Sci 40:5–27

Eddy S (2008a) Hmmer website. http://hmmer.janelia.org

Eddy S (2008b) Infernal website. http://infernal.janelia.org

Eddy SR, Durbin R (1994) RNA sequence analysis using covariance models. Nucleic Acids Res 22:2079–2088

Fechter P, Rudinger-Thirion J, Florentz C, Giege R (2001) Novel features in the tRNA-like world of plant viral RNAs. Cell Mol Life Sci 58:1547–1561

Felsenstein J (1978) Cases in which parsimony and compatibility methods will be positively misleading. Syst Zool 27:27–33

Fox TD (1979) Five TGA "stop" codons occur within the translated sequence of the yeast mitochondrial gene for cytochrome c oxidase subunit II. Proc Natl Acad Sci USA 76:6534–6538

Frechin M, Senger B, Braye M, Kern D, Martin RP, Becker HD (2009) Yeast mitochondrial Gln-tRNA(Gln) is generated by a GatFAB-mediated transamidation pathway involving Arc1p-controlled subcellular sorting of cytosolic GluRS. Genes Dev 23:1119–1130

Galburt EA, Stoddard BL (2002) Catalytic mechanisms of restriction and homing endonucleases. Biochemistry 41:13851–13860

Gautheret D, Lambert A (2001) Direct RNA motif definition and identification from multiple sequence alignments using secondary structure profiles. J Mol Biol 313:1003–1011

Gao L, Yi X, Yang YX, Su YJ, Wang T (2009) Complete chloroplast genome sequence of a tree fern *Alsophila spinulosa*: insights into evolutionary changes in fern chloroplast genomes. BMC Evol Biol 9:130

Glover K, Spencer D, Gray M (2001) Identification and structural characterization of nucleus-encoded transfer RNAs imported into wheat mitochondria. J Biol Chem 276:639–648

Gray MW (2001) Speculations on the origin and evolution of editing. In: Bass BL (ed) RNA editing. Oxford University Press, Oxford, pp 160–184

Gray MW (2003) Diversity and evolution of mitochondrial RNA editing systems. IUBMB Life 55:227–233

Gray MW et al (1998) Genome structure and gene content in protist mitochondrial DNAs. Nucleic Acids Res 26:865–878

Griffiths-Jones S, Bateman A, Marshall M, Khanna A, Eddy SR (2003) Rfam: an RNA family database. Nucleic Acids Res 31:439–441

Haen K, Pett W, Lavrov D (2010) Parallel loss of nuclear-encoded mitochondrial aminoacyl-tRNA synthetases and mtDNA-encoded tRNAs in Cnidaria. Mol Biol Evol 27:2216–2219

Hayashi-Ishimaru Y, Ohama T, Kawatsu Y, Nakamura K, Osawa S (1996) UAG is a sense codon in several chlorophycean mitochondria. Curr Genet 30:29–33

Heckman J, Sarnoff J, Alzner-DeWeerd B, Yin S, RajBhandary U (1980) Novel features in the genetic code and codon reading patterns in *Neurospora crassa* mitochondria based on sequences of six mitochondrial tRNAs. Proc Natl Acad Sci USA 77:3159–3163

Helfenbein K, Fourcade H, Vanjani R, Boore J (2004) The mitochondrial genome of *Paraspadella gotoi* is highly reduced and reveals that chaetognaths are a sister group to protostomes. Proc Natl Acad Sci USA 101:10639–10643

Higgs PG, Jameson D, Jow H, Rattray M (2003) The evolution of tRNA-Leu genes in animal mitochondrial genomes. J Mol Evol 57:435–445

Himeno H et al (1987) Unusual genetic codes and a novel gene structure for tRNA(AGYSer) in starfish mitochondrial DNA. Gene 56:219–230

Holley R et al (1965a) Structure of a ribonucleic acid. Science 147:1462–1465

Holley R, Everett G, Madison J, Zamir A (1965b) Nucleotide sequences in the yeast alanine transfer ribonucleic acid. J Biol Chem 240:2122–2128

Holzmann J, Frank P, Loffler E, Bennett KL, Gerner C, Rossmanith W (2008) RNase P without RNA: identification and functional reconstitution of the human mitochondrial tRNA processing enzyme. Cell 135:462–474

Hopper A, Phizicky E (2003) tRNA transfers to the limelight. Genes Dev 17:162–180

Inagaki Y, Bessho Y, Hori H, Osawa S (1996) Cloning of the *Mycoplasma capricolum* gene encoding peptide-chain release factor. Gene 169:101–103

Ioudovitch A, Steinberg SV (1998) Modeling the tertiary interactions in the eukaryotic selenocysteine tRNA. RNA 4:365–373

Jackman JE, Phizicky EM (2006a) tRNAHis guanylyltransferase adds G-1 to the 5′ end of tRNAHis by recognition of the anticodon, one of several features unexpectedly shared with tRNA synthetases. RNA 12:1007–1014

Jackman JE, Phizicky EM (2006b) tRNAHis guanylyltransferase catalyzes a 3′-5′ polymerization reaction that is distinct from G-1 addition. Proc Natl Acad Sci USA 103:8640–8645

Jacob Y, Seif E, Paquet PO, Lang BF (2004) Loss of the mRNA-like region in mitochondrial tmRNAs of jakobids. RNA 10:605–614

Joyce PB, Gray MW (1989) Chloroplast-like transfer RNA genes expressed in wheat mitochondria. Nucleic Acids Res 17:5461–5476

Kamenski P et al (2007) Evidence for an adaptation mechanism of mitochondrial translation via tRNA import from the cytosol. Mol Cell 26:625–637

Kamenski P et al (2010) tRNA mitochondrial import in yeast: mapping of the import determinants in the carrier protein, the precursor of mitochondrial lysyl-tRNA synthetase. Mitochondrion 10:284–293

Knight R, Freeland S, Landweber L (2001a) Rewiring the keyboard: evolvability of the genetic code. Nat Rev Genet 2:49–58

Knight R, Landweber L, Yarus M (2001b) How mitochondria redefine the code. J Mol Evol 53:299–313

Kolesnikova OA, Entelis NS, Mireau H, Fox TD, Martin RP, Tarassov IA (2000) Suppression of mutations in mitochondrial DNA by tRNAs imported from the cytoplasm. Science 289:1931–1933

Kotlova N, Ishii TM, Zagryadskaya EI, Steinberg SV (2007) Active suppressor tRNAs with a double helix between the D- and T-loops. J Mol Biol 373:462–475

Koulintchenko M, Konstantinov Y, Dietrich A (2003) Plant mitochondria actively import DNA via the permeability transition pore complex. EMBO J 22:1245–1254

Krasilnikov AS, Mondragon A (2003) On the occurrence of the T-loop RNA folding motif in large RNA molecules. RNA 9:640–643

Krasilnikov AS, Yang X, Pan T, Mondragon A (2003) Crystal structure of the specificity domain of ribonuclease P. Nature 421:760–764

Kück U, Jekosch K, Holzamer P (2000) DNA sequence analysis of the complete mitochondrial genome of the green alga *Scenedesmus obliquus*: evidence for UAG being a leucine and UCA being a non-sense codon. Gene 253:13–18

Kuhsel MG, Strickland R, Palmer JD (1990) An ancient group I intron shared by eubacteria and chloroplasts. Science 250:1570–1573

Laforest MJ, Roewer I, Lang BF (1997) Mitochondrial tRNAs in the lower fungus *Spizellomyces punctatus*: tRNA editing and UAG 'stop' codons recognized as leucine. Nucleic Acids Res 25:626–632

Laforest MJ, Bullerwell CE, Forget L, Lang BF (2004) Origin, evolution, and mechanism of 5′ tRNA editing in chytridiomycete fungi. RNA 10:1191–1199

Lagerkvist U (1978) "Two out of three": an alternative method for codon reading. Proc Natl Acad Sci USA 75:1759–1762

Lang BF et al (1997) An ancestral mitochondrial DNA resembling a eubacterial genome in miniature. Nature 387:493–497

Lang BF, Laforest MJ, Burger G (2007) Mitochondrial introns: a critical view. Trends Genet 23:119–125

Lartillot N, Philippe H (2004) A Bayesian mixture model for across-site heterogeneities in the amino-acid replacement process. Mol Biol Evol 21:1095–1109

Lartillot N, Brinkmann H, Philippe H (2007) Suppression of long-branch attraction artefacts in the animal phylogeny using a site-heterogeneous model. BMC Evol Biol 7(Suppl 1):S4

Laslett D, Canback B (2008) ARWEN: a program to detect tRNA genes in metazoan mitochondrial nucleotide sequences. Bioinformatics 24:172–175

Lavrov DV, Lang BF (2005) Transfer RNA gene recruitment in mitochondrial DNA. Trends Genet 21:129–133

Lavrov DV, Brown WM, Boore JL (2000) A novel type of RNA editing occurs in the mitochondrial tRNAs of the centipede *Lithobius forficatus*. Proc Natl Acad Sci USA 97:13738–13742

Leblanc C, Boyen C, Richard O, Bonnard G, Grienenberger JM, Kloareg B (1995) Complete sequence of the mitochondrial DNA of the rhodophyte *Chondrus crispus* (Gigartinales). Gene content and genome organization. J Mol Biol 250:484–495

Lee JC, Cannone JJ, Gutell RR (2003) The lonepair triloop: a new motif in RNA structure. J Mol Biol 325:65–83

Leigh J, Lang BF (2004) Mitochondrial 3′ tRNA editing in the jakobid *Seculamonas ecuadoriensis*: a novel mechanism and implications for tRNA processing. RNA 10:615–621

Lonergan KM, Gray MW (1993) Editing of transfer RNAs in *Acanthamoeba castellanii* mitochondria. Science 259:812–816

Lowe TM (1997) tRNAscan SE website. In: http://gtrnadb.ucsc.edu/Sacc_cere/Sacc_cere-by-locus.html

Lowe TM, Eddy SR (1997) tRNAscan-SE: a program for improved detection of transfer RNA genes in genomic sequence. Nucleic Acids Res 25:955–964

Lucas P, Otis C, Mercier JP, Turmel M, Lemieux C (2001) Rapid evolution of the DNA-binding site in LAGLIDADG homing endonucleases. Nucleic Acids Res 29:960–969

Lung B et al (2006) Identification of small non-coding RNAs from mitochondria and chloroplasts. Nucleic Acids Res 34:3842–3852

Macino G, Coruzzi G, Nobrega FG, Li M, Tzagoloff A (1979) Use of the UGA terminator as a tryptophan codon in yeast mitochondria. Proc Natl Acad Sci USA 76:3784–3785

Manhart JR, Palmer JD (1990) The gain of two chloroplast tRNA introns marks the green algal ancestors of land plants. Nature 345:268–270

Marck C, Grosjean H (2002) tRNomics: analysis of tRNA genes from 50 genomes of Eukarya, Archaea, and Bacteria reveals anticodon-sparing strategies and domain-specific features. RNA 8:1189–1232

Marechal-Drouard L et al (1990) Transfer RNAs of potato (*Solanum tuberosum*) mitochondria have different genetic origins. Nucleic Acids Res 18:3689–3696

Martin RP, Schneller JM, Stahl AJ, Dirheimer G (1977) Study of yeast mitochondrial tRNAs by two-dimensional polyacrylamide gel electrophoresis: characterization of isoaccepting species and search for imported cytoplasmic tRNAs. Nucleic Acids Res 4:3497–3510

Martin NC, Pham HD, Underbrink-Lyon K, Miller D, Donelson JE (1980) Yeast mitochondrial tRNATrp can recognize the nonsense codon UGA. Nature 285:579–581

Martin RP, Sibler AP, Dirheimer G, de Henau S, Grosjean H (1981) Yeast mitochondrial tRNATrp injected with *E. coli* activating enzyme into *Xenopus* oocytes suppresses UGA termination. Nature 293:235–237

Massey SE, Garey JR (2007) A comparative genomics analysis of codon reassignments reveals a link with mitochondrial proteome size and a mechanism of genetic code change via suppressor tRNAs. J Mol Evol 64:399–410

Masta SE, Boore JL (2008) Parallel evolution of truncated transfer RNA genes in arachnid mitochondrial genomes. Mol Biol Evol 25:949–959

Matsuyama S, Ueda T, Crain P, McCloskey J, Watanabe K (1998) A novel wobble rule found in starfish mitochondria. Presence of 7-methylguanosine at the anticodon wobble position expands decoding capability of tRNA. J Biol Chem 273:3363–3368

Michel F, Lang BF (1985) Mitochondrial class II introns encode proteins related to the reverse transcriptases of retroviruses. Nature 316:641–643

Miranda I, Silva R, Santos MA (2006) Evolution of the genetic code in yeasts. Yeast 23:203–213

Moriya J et al (1994) A novel modified nucleoside found at the first position of the anticodon of methionine tRNA from bovine liver mitochondria. Biochemistry 33:2234–2239

Morl M, Dorner M, Paabo S (1995) C to U editing and modifications during the maturation of the mitochondrial tRNA(Asp) in marsupials. Nucleic Acids Res 23:3380–3384

Muramatsu T et al (1988) A novel lysine-substituted nucleoside in the first position of the anticodon of minor isoleucine tRNA from *Escherichia coli*. J Biol Chem 263:9261–9267

Nagaswamy U, Fox GE (2002) Frequent occurrence of the T-loop RNA folding motif in ribosomal RNAs. RNA 8:1112–1119

Nasvall S, Chen P, Bjork G (2004) The modified wobble nucleoside uridine-5-oxyacetic acid in tRNAPro(cmo5UGG) promotes reading of all four proline codons in vivo. RNA 10:1662–1673

Nedelcu AM, Lee RW, Lemieux C, Gray MW, Burger G (2000) The complete mitochondrial DNA sequence of *Scenedesmus obliquus* reflects an intermediate stage in the evolution of the green algal mitochondrial genome. Genome Res 10:819–831

Nissen P, Thirup S, Kjeldgaard M, Nyborg J (1999) The crystal structure of Cys-tRNACys-EF-Tu-GDPNP reveals general and specific features in the ternary complex and in tRNA. Structure 7:143–156

Nissen P, Ippolito JA, Ban N, Moore PB, Steitz TA (2001) RNA tertiary interactions in the large ribosomal subunit: the A-minor motif. Proc Natl Acad Sci USA 98:4899–4903

Oda K et al (1992a) Gene organization deduced from the complete sequence of liverwort *Marchantia polymorpha* mitochondrial DNA. A primitive form of plant mitochondrial genome. J Mol Biol 223:1–7

Oda K et al (1992b) Transfer RNA genes in the mitochondrial genome from a liverwort, *Marchantia polymorpha*: the absence of chloroplast-like tRNAs. Nucleic Acids Res 20:3773–3777

Ohtsuki T, Sato A, Watanabe Y, Watanabe K (2002) A unique serine-specific elongation factor Tu found in nematode mitochondria. Nat Struct Biol 9:669–673

Okimoto R, Wolstenholme DR (1990) A set of tRNAs that lack either the T psi C arm or the dihydrouridine arm: towards a minimal tRNA adaptor. EMBO J 9:3405–3411

Osawa S, Ohama T, Jukes TH, Watanabe K (1989) Evolution of the mitochondrial genetic code. I. Origin of AGR serine and stop codons in metazoan mitochondria. J Mol Evol 29:202–207

Pape LK, Koerner TJ, Tzagoloff A (1985) Characterization of a yeast nuclear gene (MST1) coding for the mitochondrial threonyl-tRNA1 synthetase. J Biol Chem 260:15362–15370

Pfitzinger H, Weil JH, Pillay DT, Guillemaut P (1990) Codon recognition mechanisms in plant chloroplasts. Plant Mol Biol 14:805–814

Pramateftaki PV, Kouvelis VN, Lanaridis P, Typas MA (2006) The mitochondrial genome of the wine yeast *Hanseniaspora uvarum*: a unique genome organization among yeast/fungal counterparts. FEMS Yeast Res 6:77–90

Preiser P, Williamson DH, Wilson RJ (1995) tRNA genes transcribed from the plastid-like DNA of *Plasmodium falciparum*. Nucleic Acids Res 23:4329–4336

Putz J, Dupuis B, Sissler M, Florentz C (2007) Mamit-tRNA, a database of mammalian mitochondrial tRNA primary and secondary structures. RNA 13:1184–1190

Rao BS, Maris EL, Jackman JE (2011) tRNA 5′-end repair activities of tRNAHis guanylyl-transferase (Thg1)-like proteins from Bacteria and Archaea. Nucleic Acids Res 39 (5):1833–1842

Rawlings TA, Collins TM, Bieler R (2003) Changing identities: tRNA duplication and remolding within animal mitochondrial genomes. Proc Natl Acad Sci USA 100:15700–15705

Rich A, RajBhandary UL (1976) Transfer RNA: molecular structure, sequence, and properties. Annu Rev Biochem 45:805–860

Rinehart J, Horn E, Wei D, Soll D, Schneider A (2004) Non-canonical eukaryotic glutaminyl- and glutamyl-tRNA synthetases form mitochondrial aminoacyl-tRNA in *Trypanosoma brucei*. J Biol Chem 279:1161–1166

Rinehart J, Krett B, Rubio M, Alfonzo J, Soll D (2005) *Saccharomyces cerevisiae* imports the cytosolic pathway for Gln-tRNA synthesis into the mitochondrion. Genes Dev 19:583–592

Rogalski M, Karcher D, Bock R (2008) Superwobbling facilitates translation with reduced tRNA sets. Nat Struct Mol Biol 15:192–198

Rubio MA et al (2008) Mammalian mitochondria have the innate ability to import tRNAs by a mechanism distinct from protein import. Proc Natl Acad Sci USA 105:9186–9191

Saks ME, Sampson JR, Abelson J (1998) Evolution of a transfer RNA gene through a point mutation in the anticodon. Science 279:1665–1670

Sakurai M, Watanabe Y, Watanabe K, Ohtsuki T (2006) A protein extension to shorten RNA: elongated elongation factor-Tu recognizes the D-arm of T-armless tRNAs in nematode mitochondria. Biochem J 399:249–256

Salinas T, Duchene A, Marechal-Drouard L (2008) Recent advances in tRNA mitochondrial import. Trends Biochem Sci 33:320–329

Schneider A (2001) Unique aspects of mitochondrial biogenesis in trypanosomatids. Int J Parasitol 31:1403–1415

Schneider A, Marechal-Drouard L (2000) Mitochondrial tRNA import: are there distinct mechanisms? Trends Cell Biol 10:509–513

Schultz DW, Yarus M (1994) Transfer RNA mutation and the malleability of the genetic code. J Mol Biol 235:1377–1380

Segovia R, Pett W, Trewick S, Lavrov DV. Extensive and evolutionarily persistent mitochondrial tRNA editing in velvet worms (phylum Onychophora). Mol Biol Evol. 2011 May 4. [Epub ahead of print]

Seif ER, Forget L, Martin NC, Lang BF (2003) Mitochondrial RNase P RNAs in ascomycete fungi: lineage-specific variations in RNA secondary structure. RNA 9:1073–1083

Seif E, Leigh J, Liu Y, Roewer I, Forget L, Lang BF (2005) Comparative mitochondrial genomics in zygomycetes: bacteria-like RNase P RNAs, mobile elements and a close source of the group I intron invasion in angiosperms. Nucleic Acids Res 33:734–744

Sengupta S, Yang X, Higgs PG (2007) The mechanisms of codon reassignments in mitochondrial genetic codes. J Mol Evol 64:662–688

Shi H, Moore PB (2000) The crystal structure of yeast phenylalanine tRNA at 1.93 A resolution: a classic structure revisited. RNA 6:1091–1105

Shinozaki K et al (1986) The complete nucleotide sequence of the tobacco chloroplast genome: its gene organization and expression. EMBO J 5:2043–2049

Sibler AP, Bordonne R, Dirheimer G, Martin R (1980) Primary structure of yeast mitochondrial tryptophan-tRNA capable of translating the termination U-G-A codon. C R Seances Acad Sci D 290:695–698

Sibler AP, Dirheimer G, Martin RP (1985) Yeast mitochondrial tRNAIle and tRNAMetm: nucleotide sequence and codon recognition patterns. Nucleic Acids Res 13:1341–1345

Smith DR (2009) Unparalleled GC content in the plastid DNA of Selaginella. Plant Mol Biol 71: 627–39

Söll D, RajBhandary U (1995) tRNA: structure, biosynthesis, and function. ASM Press, Washington, D.C

Steinberg S, Cedergren R (1994) Structural compensation in atypical mitochondrial tRNAs. Nat Struct Biol 1:507–510

Steinberg S, Ioudovitch A (1996) A role for the bulged nucleotide 47 in the facilitation of tertiary interactions in the tRNA structure. RNA 2:84–87

Steinberg S, Gautheret D, Cedergren R (1994) Fitting the structurally diverse animal mitochondrial tRNAs(Ser) to common three-dimensional constraints. J Mol Biol 236:982–989

Steinberg S, Leclerc F, Cedergren R (1997) Structural rules and conformational compensations in the tRNA L-form. J Mol Biol 266:269–282

Su D, Lieberman A, Lang BF, Siminovic M, Söll D, Ling J (2011) An unusual tRNAThr derived from tRNAHis reassigns in yeast mitochondria the CUN codons to threonine. Nucleic Acids Res 39(11):4866–4874

Suzuki T, Ueda T, Watanabe K (1997) The 'polysemous' codon – a codon with multiple amino acid assignment caused by dual specificity of tRNA identity. EMBO J 16:1122–1134

Tarassov I et al (2007) Import of nuclear DNA-encoded RNAs into mitochondria and mitochondrial translation. Cell Cycle 6:2473–2477

Temperley R, Richter R, Dennerlein S, Lightowlers RN, Chrzanowska-Lightowlers ZM (2010) Hungry codons promote frameshifting in human mitochondrial ribosomes. Science 327:301

Tomita K, Ueda T, Watanabe K (1996) RNA editing in the acceptor stem of squid mitochondrial tRNA(Tyr). Nucleic Acids Res 24:4987–4991

Tomita K, Ueda T, Watanabe K (1998) 7-Methylguanosine at the anticodon wobble position of squid mitochondrial tRNA(Ser)GCU: molecular basis for assignment of AGA/AGG codons as serine in invertebrate mitochondria. Biochim Biophys Acta 1399:78–82

Tomita K, Ueda T, Watanabe K (1999) The presence of pseudouridine in the anticodon alters the genetic code: a possible mechanism for assignment of the AAA lysine codon as asparagine in echinoderm mitochondria. Nucleic Acids Res 27:1683–1689

Tsuji S et al (2007) The chloroplast genome from a lycophyte (microphyllophyte), *Selaginella uncinata*, has a unique inversion, transpositions and many gene losses. J Plant Res 120:281–290

Turmel M, Otis C, Lemieux C (2002) The chloroplast and mitochondrial genome sequences of the charophyte *Chaetosphaeridium globosum*: insights into the timing of the events that restructured organelle DNAs within the green algal lineage that led to land plants. Proc Natl Acad Sci USA 99:11275–11280

Turmel M, Otis C, Lemieux C (2003) The mitochondrial genome of *Chara vulgaris*: insights into the mitochondrial DNA architecture of the last common ancestor of green algae and land plants. Plant Cell 15:1888–1903

Turmel M, Otis C, Lemieux C (2010) A deviant genetic code in the reduced mitochondrial genome of the picoplanktonic green alga *Pycnococcus provasolii*. J Mol Evol, Epub ahead of print

Vogel J, Hubschmann T, Borner T, Hess WR (1997) Splicing and intron-internal RNA editing of trnK-matK transcripts in barley plastids: support for MatK as an essential splice factor. J Mol Biol 270:179–187

Voigt O, Erpenbeck D, Worheide G (2008) A fragmented metazoan organellar genome: the two mitochondrial chromosomes of *Hydra magnipapillata*. BMC Genomics 9:350

Wakita K et al (1994) Higher-order structure of bovine mitochondrial tRNA(Phe) lacking the 'conserved' GG and T psi CG sequences as inferred by enzymatic and chemical probing. Nucleic Acids Res 22:347–353

Wang X, Lavrov DV (2008) Seventeen new complete mtDNA sequences reveal extensive mitochondrial genome evolution within the Demospongiae. PLoS One 3:e2723

Wang X, Lavrov DV (2011) Gene recruitment – a common mechanism in the evolution of transfer RNA gene families. Gene 475(1):22–29

Watanabe Y et al (1994a) Higher-order structure of bovine mitochondrial tRNA(SerUGA): chemical modification and computer modeling. Nucleic Acids Res 22:5378–5384

Watanabe Y et al (1994b) Primary and higher order structures of nematode (*Ascaris suum*) mitochondrial tRNAs lacking either the T or D stem. J Biol Chem 269:22902–22906

Weber F, Dietrich A, Weil JH, Marechal-Drouard L (1990) A potato mitochondrial isoleucine tRNA is coded for by a mitochondrial gene possessing a methionine anticodon. Nucleic Acids Res 18:5027–5030

Weixlbaumer A et al (2007) Mechanism for expanding the decoding capacity of transfer RNAs by modification of uridines. Nat Struct Mol Biol 14:498–502

Wintz H, Chen HC, Pillay DT (1988) Presence of a chloroplast-like elongator tRNAMet gene in the mitochondrial genomes of soybean and *Arabidopsis thaliana*. Curr Genet 13:255–260

Wolfe KH, Morden CW, Ems SC, Palmer JD (1992) Rapid evolution of the plastid translational apparatus in a nonphotosynthetic plant: loss or accelerated sequence evolution of tRNA and ribosomal protein genes. J Mol Evol 35:304–317

Wolstenholme DR, Macfarlane JL, Okimoto R, Clary DO, Wahleithner JA (1987) Bizarre tRNAs inferred from DNA sequences of mitochondrial genomes of nematode worms. Proc Natl Acad Sci USA 84:1324–1328

Wolstenholme DR, Okimoto R, Macfarlane JL (1994) Nucleotide correlations that suggest tertiary interactions in the TV-replacement loop-containing mitochondrial tRNAs of the nematodes, *Caenorhabditis elegans* and *Ascaris suum*. Nucleic Acids Res 22:4300–4306

Wyman SK, Jansen RK, Boore JL (2004) Automatic annotation of organellar genomes with DOGMA. Bioinformatics 20:3252–3255

Yermovsky-Kammerer AE, Hajduk SL (1999) In vitro import of a nuclearly encoded tRNA into the mitochondrion of *Trypanosoma brucei*. Mol Cell Biol 19:6253–6259

Yokobori S, Pääbo S (1995) Transfer RNA editing in land snail mitochondria. Proc Natl Acad Sci USA 92:10432–10435

Yokobori S, Pääbo S (1997) Polyadenylation creates the discriminator nucleotide of chicken mitochondrial tRNA(Tyr). J Mol Biol 265:95–99

Zagryadskaya EI, Doyon FR, Steinberg SV (2003) Importance of the reverse Hoogsteen base pair 54-58 for tRNA function. Nucleic Acids Res 31:3946–3953

Zagryadskaya EI, Kotlova N, Steinberg SV (2004) Key elements in maintenance of the tRNA L-shape. J Mol Biol 340:435–444

Zimmerly S, Guo H, Perlman PS, Lambowitz AM (1995) Group II intron mobility occurs by target DNA-primed reverse transcription. Cell 82:545–554

Index

C.E. Bullerwell (ed.), *Organelle Genetics*,
DOI 10.1007/978-3-642-22380-8, © Springer-Verlag Berlin Heidelberg 2012